STUDY GUIDE AND WORKBOOK: AN INTERACTIVE APPROACH

for Starr and Taggart's

BIOLOGY

The Unity and Diversity of Life
TENTH EDITION

RICHARD CHENEY
Christopher Newport University

MICHAEL WINDELSPECHT
Appalachian State University

JANE B. TAYLOR
Northern Virginia Community College

JOHN D. JACKSON
North Hennepin Community College

Australia • Canada • Mexico • Singapore • Spain • United Kingdom • United States

Biology Publisher: Jack Carey
Assistant Editor: Suzannah Alexander
Editorial Assistant: Jana Davis
Marketing Manager: Ann Caven
Marketing Assistant: Sandra Perin
Advertising Project Manager: Linda Yip
Project Manager, Editorial Production: Belinda Krohmer
Print Buyer: Judy Inouye

Printed in the United States of America

4 5 6 7 07 06 05

ISBN 0-534-39750-6

For more information about our products, contact us at:
Thomson Learning Academic Resource Center
1-800-423-0563

For permission to use material from this text, contact us by:
Phone: 1-800-730-2214
Fax: 1-800-730-2215
Web: www.thomsonrights.com

Permissions Editor: Bob Kauser
Production Service: G & S Typesetters, Inc.
Copy Editor: Christine Gever
Cover Designer: Kevin Schafer/Getty Images
Cover Image: From Central America, one of the tropical rain forests that may disappear in your lifetime
Compositor: G & S Typesetters, Inc.
Printing and Binding: Globus Printing Company, Inc.

Brooks/Cole—Thomson Learning
10 Davis Drive
Belmont, CA 94002
USA

Asia
Thomson Learning
5 Shenton Way, #01-01
UIC Building
Singapore 068808

Australia
Nelson Thomson Learning
102 Dodds Street
Southbank
Victoria 3006
Australia

Canada
Nelson Thomson Learning
1120 Birchmount Road
Toronto, Ontario M1K 5G4
Canada

Europe/Middle East/South Africa
Thomson Learning
High Holborn House
50/51 Bedford Row
London WC1R 4LR
United Kingdom

CONTENTS

Photo Credits:

Chapter 7
p. 83 (36, 37): David Fisher

Chapter 21
p. 245 (10): George Musil/Visuals Unlimited
p. 245 (12): Tony Brain/SPL/Photo Researchers
p. 245 (14): T.J. Beverige, University of Guelph
p. 245 (15): Tony Brain and David Parker/SPL/Photo Researchers

Chapter 22
p. 252 (20): M. Abbey/Visuals Unlimited
p. 252 (21): T.E. Adams/Visuals Unlimited
p. 252 (22): John Clegg/Ardea, London
p. 255: D.J. Patterson/Seaphot Limited: Planet Earth Pictures
p. 258 (20): Jan Hinsch/SPL/Photo Researchers
p. 258 (22): John Clegg/Ardea, London
p. 258 (23): M. Abbey, Visuals Unlimited

Chapter 23
p. 264: Jane Burton/Bruce Coleman Ltd.
p. 266: A.&E. Boomford, Ardea, London; Lee Casebere

Chapter 24
p. 279 (3): Garry T. Cole, University of Texas, Austin/BPS
p. 280 (3): Ed Reschke
p. 282 (2): John Hodgin
p. 283 (3): Jane Burton/Bruce Coleman, Ltd.
p. 283 (6): Edward S. Ross
p. 283 (7): Robert C. Simpson/Nature Stock

Chapter 25
p. 298 (36): Carolina Biological Supply Company
p. 310 (24): Herve Chaumeton/Agence Nature
p. 311 (a): Ian Took/Biofotos
p. 311 (b): John Mason/Ardea, London
p. 311 (c): Chris Huss/The Wildlife Collection
p. 311 (d): Kjell B. Sandved

Chapter 26
p. 320 (a): Bill Wood/Bruce Coleman Ltd.
p. 331 (30): Christopher Crowley
p. 331 (31): Bill Wood/Bruce Coleman Ltd.
p. 332 (33): Reinhard/ZEFA
p. 332 (34): Peter Scoones/Seaphot Ltd.: Planet Earth Pictures
p. 333 (35): Ervin Christian/ZEFA
p. 333 (36): © Norbert Wu/Peter Arnold, Inc.
p. 333 (37) Allan Power/Bruce Coleman, Ltd.
p. 334 (38) Rick M. Harbo
p. 334 (39) Herve Chaumeton/Agence Nature

Chapter 29
p. 357 (1): D.E. Akin and I.L. Rigsby, Richard B. Russel Agricultural Research Center, Agricultural Research Service, U.S. Dept. Agriculture Athens, GA
p. 357 (2): Biophoto Associates
p. 357 (3): Biophoto Associates
p. 357 (4): Kingsley R. Stern
p. 357 (5): Biophoto Associates
p. 360 (1-3): Robert & Linda Mitchell Photography
p. 361 (15-17): Carolina Biological Supply Company
p. 361 (19-22): Carolina Biological Supply Company
p. 364 (11): Jeremy Burgess/SPL/Photo Researchers
p. 364 (12): Jeremy Burgess/SPL/Photo Researchers
p. 366 (8-10)(a Inset): Chuck Brown
p. 366 (15-17): H.A. Core, W.A. Cote, and A.C. Day, *Wood Structure and Identification*, 2nd Ed., Syracuse University Press, 1979

Chapter 31
p. 384 (1-5): ©Gary Head
p. 391: Patricia Schulz

Chapter 32
p. 399: Herve Chaumeton/Agence Nature
p. 399: Larry L. Runk/Grant Heilman, Inc.; James D. Mauseth

Chapter 34
p. 431 (16-17): Manfred Kage/ Peter Arnold, Inc.
p. 435 (1-10): C. Yokochi and J. Rohen, *Photographic Anatomy of the Human Body*, 2nd Ed., Igaku-Shoin, Ltd., 1979

Chapter 44
p. 560 (2): Ed Reschke
p. 566 (9-10, 12): Photographs Lennart Nilsson, *A Child Is Born*, ©1966, 1977 Dell Publishing Company, Inc.

PREFACE

Tell me and I will forget, show me and I might remember, involve me and I will understand.
—Chinese Proverb

The proverb outlines three levels of learning, each successively more effective than the method preceding it. The writer of the proverb understood that humans learn most efficiently when they *involve* themselves in the material to be learned. This study guide is like a tutor; when properly used it increases the efficiency of your study periods. The interactive exercises involve you in the most important terms and central ideas of your text. Specific tasks ask you to recall key concepts and terms and apply them to life; they test your understanding of the facts and indicate items to reexamine or further analyze. Your performance on these tasks provides an estimate of your next test score based on specific material. Most important, though, this biology study guide and your text together help you make informed decisions about matters that affect your own well-being and that of your environment. In the years to come, human survival on planet Earth will demand administrative and managerial decisions based on an informed biological background.

HOW TO USE THIS STUDY GUIDE

Following this preface, you will find an outline that will show you how the study guide is organized and will help you use it efficiently. Each chapter begins with a title and an outline list of the 1- and 2-level headings in that chapter. The Interactive Exercises follow, wherein each chapter is divided into sections of one or more of the main (1-level) headings that are labeled 1.1, 1.2, and so on. *For easy reference to an answer or definition, each question and term in this unique study guide is accompanied by the appropriate text page(s), and appears in the form* [p.352]. The Interactive Exercises begin with a list of Selected Words (other than boldfaced terms) chosen by the authors as those that are most likely to enhance understanding. In the text chapters, the selected words appear in italics, quotation marks, or roman type. The list of Selected Words is followed by a list of Boldfaced, Page-Referenced Terms that appear in the text. These terms are essential to understanding each study guide section of a particular chapter. Space is provided beside each term for you to formulate a definition in your own words. Next is a series of different types of exercises that may include completion, short answer, true/false, fill-in-the-blanks, matching, choice, dichotomous choice, label and match, problems, labeling, sequencing, multiple choice, and completion of tables.

A Self-Quiz immediately follows the Interactive Exercises. This quiz is composed primarily of multiple-choice questions, although sometimes we present another examination device or some combination of devices. Any wrong answers in the Self-Quiz indicate portions of the text you need to reexamine. A series of Chapter Objectives/Review Questions follows each Self-Quiz. These are tasks that you should be able to accomplish if you have understood the assigned reading in the text. Some objectives require you to compose a short answer or long essay while others may require a sketch or may require you to supply the correct words.

The final part of each chapter is named Integrating and Applying Key Concepts. This section invites you to try your hand at applying major concepts to situations in which there is not necessarily a single pat answer—and so none is provided in the chapter answer section. Your text generally will provide enough clues to get you started on an answer, but this part is intended to stimulate thought and provoke group discussions.

A person's mind, once stretched by a new idea, can never return to its original dimension.
—Oliver Wendell Holmes

STRUCTURE OF THIS STUDY GUIDE

The outline below shows how each chapter in this study guide is organized.

Chapter Number ⟶ **4**

Chapter Title ⟶ **CELL STUCTURE AND FUNCTION**
Animalcules and Cells Fill'd With Juices

Chapter Outline ⟶

BASIC ASPECTS OF CELL STRUCTURE AND FUNCTION
Structural Organization of Cells
Organization of Cell Membranes
Why Aren't All Cells Big?

FOCUS ON SCIENCE: *Microscopes— Gateways to the Cell*

DEFINING FEATURES OF EUKARYOTIC CELLS
Major Cellular Components
Which Organelles Are Typical of Plants?
Which Organelles Are Typical of Animals?

THE NUCLEUS
Nuclear Envelope
Nucleolus
Chromosomes
What Happens to the Proteins Specified by DNA?

THE ENDOMEMBRANE SYSTEM
Endoplasmic Reticulum
Golgi Bodies
A Variety of Vesicles

MITOCHONDRIA

SPECIALIZED PLANT ORGANELLES
Chloroplasts and Other Plastids
Central Vacuole

SUMMARY OF TYPICAL FEATURES OF EUKARYOTIC CELLS

EVEN YOUR CELLS HAVE A SKELETON
Microtubules — The Big Ones
Microfilaments — The Thin Ones
Myosin and Other Accessory Proteins
Intermediate Filaments

HOW DO CELLS MOVE?
Chugging along with Motor Proteins
Cilia, Flagella, and False Feet

CELL SURFACE SPECIALIZATIONS
Eukaryotic Cell Walls
Matrixes between Animal Cells
Cell Junctions
Cell Communication

PROKARYOTIC CELLS

Interactive Exercises ⟶ The interactive exercises are divided into numbered sections by titles of main headings and page references. Each section begins with a list of author-selected words that appear in the text chapter in italics, quotation marks, or roman type. This is followed by a list of important boldfaced, page-referenced terms from each section of the chapter. Each section ends with interactive exercises that vary in type and require constant interaction with the important chapter information.

Self-Quiz ⟶ Usually a set of multiple-choice questions that sample important blocks of text information.

Chapter Objectives/ Review Questions ⟶ Combinations of relative objectives to be met and questions to be answered.

Integrating and Applying Key Concepts ⟶ Applications of text material to questions for which there may be more than one correct answer.

Answers to Interactive Exercises and Self-Quiz ⟶ Answers for all interactive exercises can be found at the end of this study guide by chapter and title, and the main headings with their page references, followed by answers for the Self-Quiz.

1

CONCEPTS AND METHODS IN BIOLOGY

Interactive Exercises

Note: In the answer sections of this book, a specific molecule is most often indicated by its abbreviation. For example, adenosine triphosphate is ATP.

Why Biology? [pp.2–3]

1.1. DNA, ENERGY, AND LIFE [pp.4–5]

Selected Words: *cells* [p.4], *DNA to RNA to protein* [p.4]

In addition to the boldfaced terms, the text features other important terms essential to understanding the assigned material. "Selected Words" is a list of these terms, which appear in the text in italics, in quotation marks, and occasionally in roman type.

Boldfaced, Page-Referenced Terms

The page-referenced terms are important; they are in boldface type in the text chapter. Write a definition for each term in your own words without looking at the text. Next, compare your definition with that given in the chapter or in the text glossary. If your definition seems inaccurate, allow some time to pass and repeat this procedure until you can define each term rather quickly (how fast you answer is a gauge of your learning effectiveness).

[p.3] biology ____

____

[p.4] DNA ____

____

[p.4] inheritance ____

____

[p.4] reproduction ______________________________

____

[p.4] development ______________________________

____

[p.5] energy ______________________________

____

[p.5] metabolism ______________________________

____

[p.5] receptors ______________________________

____

[p.5] stimulus ______________________________

____

[p.5] homeostasis ______________________________

____

Fill-in-the-Blanks

The fundamental unit of life is called the (1) ______________ [p.4]. Within these cells the signature, or defining, molecule is a nucleic acid known as (2) ______________ [p.4]. Encoded within the structure of the genetic material are the instructions needed for the manufacturing of (3) ______________ [p.4]. The building blocks used during the assembly of these molecules are the (4) ______________ [p.4]. When supplied with energy, a specialized group of proteins, called the (5) ______________ [p.4], act as worker molecules, which serve by rapidly building, splitting, or rearranging the wide variety of complex molecules needed for life. Another nucleic acid, called (6) ______________ [p.4], assists the enzymes in carrying out DNA's protein-building instructions. Thus, the information originates in the (7) ______________ [p.4] and flows to the (8) ______________ [p.4], resulting in the formation of a(n) (9) ______________ [p.4].

The passing of traits from the parent to the offspring by means of DNA is called (10) ______________ [p.4]. The physical mechanism by which these traits are transferred is called (11) ______________ [p.4]. For multicellular organisms such as plants and animals, which consist of specialized cells, tissues, and organs, the process of (12) ______________ [p.4] is directed by the instructions contained within the DNA.

The work of the cell in assembling, splitting, and rearranging the structure of molecules requires an input of (13) ______________ [p.5], which represents the capacity to do work. (14) ______________ [p.5] refers to the cell's capacity to obtain and convert energy from its surroundings and to use that energy to maintain itself, grow, and produce more cells. In leaves, for example, specialized cells carry out the process of (15) ______________ [p.5] by intercepting energy from the sun and converting it into chemical energy molecules called (16) ______________ [p.5]. In turn, these molecules transfer energy to the (17) ______________ [p.5] of metabolic pathways. ATP energy molecules may also be formed by

(18) ______________ ______________ [p.5]. This process can release energy that cells have tucked away in sugars and other kinds of molecules.

Organisms sense changes in their surroundings, then make controlled, compensatory (19) ______________ [p.5] to them. To accomplish this, special molecules and structures called (20) ______________ [p.5] detect (21) ______________ (plural) [p.5]. A (22) ______________ (singular) [p.5] is a specific form of energy detected by receptors. Sunlight, heat, (23) ______________ [p.5], and mechanical stress are all examples of stimuli. Cells adjust metabolic activities in response to signals from receptors. For example, after a snack, simple sugars and other molecules leave the gut and enter the blood, which is part of a(n) (24) ______________ [p.5] environment. Blood sugar levels then rise and stimulate secretion of the hormone insulin by the pancreas. Most of the cells in your body have (25) ______________ [p.5] for this hormone. Insulin stimulates cells to take up sugar molecules from the internal environment and return blood sugar concentration levels to normal. Organisms respond so exquisitely to energy changes that their internal operating conditions remain within tolerable limits. This is called a state of (26) ______________ [p.5], one of the key defining features of life.

1.2. ENERGY AND LIFE'S ORGANIZATION [pp.6–7]

Boldfaced, Page-Referenced Terms

[p.6] cell ______________________________

[p.6] multicelled organisms ______________________________

[p.6] population ______________________________

[p.6] community ______________________________

[p.6] ecosystem ______________________________

[p.6] biosphere ______________________________

[p.7] producers ______________________________

[p.7] consumers ______________________________

[p.7] decomposers ______________________________

Matching

Choose the most appropriate answer for each term.

1. ____ molecule [p.6]
2. ____ cell [p.6]
3. ____ community [p.6]
4. ____ ecosystem [p.6]
5. ____ organ system [p.6]
6. ____ organelle [p.6]
7. ____ population [p.6]
8. ____ subatomic particle [p.6]
9. ____ tissue [p.6]
10. ____ atom [p.6]
11. ____ multicelled organism [p.6]
12. ____ biosphere [p.6]
13. ____ organ [p.6]

A. The interaction of multiple tissues to perform a common task
B. An electron, neutron, or proton
C. The cellular structure where specialized reactions are conducted
D. All the regions of Earth that have the capacity to sustain life
E. The smallest unit of life capable of surviving and reproducing on its own
F. The interaction of two or more organs at the physical or chemical level to perform a common task
G. Two or more atoms bonded together
H. The interaction of all populations in a given geographic area
I. The smallest unit of an element that possesses all of the properties of that element
J. The interaction of a community and the physical environment
K. An individual composed of interdependent cells organized into tissues, organs, and organ systems
L. A group of individuals of the same species in a particular place at a particular time
M. A group of cells that work together to carry out a particular function

Sequence

Arrange the following levels of organization in nature in the correct hierarchical order. Write the letter of the most inclusive level next to 14. Write the letter of the least inclusive level next to 26.

14. ____ A. Community [p.6]
15. ____ B. Tissue [p.6]
16. ____ C. Cell [p.6]
17. ____ D. Organ [p.6]
18. ____ E. Organ system [p.6]
19. ____ F. Atom [p.6]
20. ____ G. Ecosystem [p.6]
21. ____ H. Organelle [p.6]
22. ____ I. Molecule [p.6]
23. ____ J. Population [p.6]
24. ____ K. Subatomic particle [p.6]
25. ____ L. Multicelled organism [p.6]
26. ____ M. Biosphere [p.6]

Fill-in-the-Blanks

The (27) ______________ [p.7] are the plants and other organisms that make their own food. They serve to start the flow of energy (28) ______________ [p.7] the living world. Animals, known as (29) ______________ [p.7], feed directly or indirectly on energy stored in the tissues of the producers. The (30) ______________ [p.7], consisting of bacteria and fungi, feed on the tissues or remains of other organisms and break down biological molecules to simple materials that may be cycled back to the producers. Thus, within the biosphere there is a one-way flow of (31) ______________ [p.7] through organisms and a (32) ______________ [p.7] of materials.

1.3. IF SO MUCH UNITY, WHY SO MANY SPECIES? [pp.8–9]

Selected Words: *prokaryotic* [p.8], *eukaryotic* [p.8]

Boldfaced, Page-Referenced Terms

[p.8] species ______________________________

[p.8] genus ______________________________

[p.8] archaebacteria ______________________________

[p.8] eubacteria ______________________________

[p.8] protistans ______________________________

[p.8] fungi ______________________________

[p.8] plants ______________________________

[p.8] animal ______________________________

Fill-in-the-Blanks

Different "kinds" of organisms are referred to as (1) ______________ [p.8]. The first part of the two-part name of each organism, which encompasses all the species that seem closely related by way of their recent descent from a common ancestor, is called the (2) ______________ [p.8]. The second part of the name designates a particular (3) ______________ [p.8]. For example, the red squirrel is known by the two-part name *Tamiasciurus hudsonicus.* In this name, *Tamiasciurus* represents the name of the (4) ______________ [p.8] and *hudsonicus* denotes the name of the (5) ______________ [p.8]. Biologists now recognize three general (6) ________________ [p.8] of organisms, called archae, bacteria, and eukarya. Within these three domains are six (7) ________________ [p.8] of life, which include the archaebacteria, eubacteria, protistans, (8) ________________ [p.8], plants, and animals. The adjective (9) ________________ [p.8] describes single-celled organisms that lack a nucleus, while the adjective (10) ________________ [p.8] describes single-celled and multicelled organisms whose DNA is enclosed within a nucleus.

Complete the Table

11. Complete the following table by entering the correct name of each kingdom of life described.

Kingdom	*Description*
[pp.8–9] a.	Eukaryotic, single-celled species and some multicelled forms, larger than bacteria but internally more complex
[pp.8–9] b.	Eukaryotic, multicelled, photosynthetic producers
[pp.8–9] c.	Eukaryotic, mostly multicelled decomposers and consumers; food is digested outside their cells and bodies
[pp.8–9] d.	Prokaryotic; primarily live only in extreme habitats, such as the ones that prevailed when life originated
[pp.8–9] e.	Prokaryotic, single-celled organisms that are very successful in distribution
[pp.8–9] f.	Eukaryotic, diverse, multicelled consumers; actively move during at least some stage of their life

1.4. AN EVOLUTIONARY VIEW OF DIVERSITY [pp.10–11]

Selected Words: *artificial* environment [p.10], *natural* selection [pp.10–11], selective agents [p.10]

Boldfaced, Page-Referenced Terms

[p.10] mutation ____

____

[p.10] adaptive trait ____

____

[p.10] evolution ____

____

[p.10] artificial selection ____

____

[p.11] natural selection ____

____

[p.11] antibiotics ____

____

Choice

For questions 1–10, choose from the following:

a. evolution through artificial selection b. evolution through natural selection

1. _____ Pigeon breeding [p.10]
2. _____ Antibiotics are powerful agents of this process [p.11]
3. _____ The effects of a peregrine falcon on a pigeon population [pp.10–11]
4. _____ A favoring of adaptive traits in nature [p.10]
5. _____ The selection of one form of a trait over another, taking place under contrived, manipulated conditions [p.10]
6. _____ Refers to change that is occurring within a line of descent over time [p.10]
7. _____ A difference in which individuals of a given generation survive and reproduce, the difference being an outcome of which ones have adaptive forms of traits [p.11]
8. _____ Darwin viewed this as a simple model for natural selection [p.10]
9. _____ Breeders are the "selective agents" [p.10]
10. _____ The mechanism whereby antibiotic resistance evolves [p.11]

1.5. THE NATURE OF BIOLOGICAL INQUIRY [pp.12–13]

1.6. FOCUS ON SCIENCE: *The Power of Experimental Tests* [pp.14–15]

1.7. THE LIMITS OF SCIENCE [p.15]

Selected Words: if–then process [p.12], the scientific method [p.12], *logic* [p.12], *high or low probability* [p.13], *biological therapy* [p.14], *Escherichia coli* [p.14], *quantitative* [p.15], *subjective* [p.15]

Boldfaced, Page-Referenced Terms

[p.12] hypotheses ____

____

[p.12] prediction ____

____

[p.12] test ____

____

[p.12] models ____

____

[p.12] inductive logic ____

____

[p.12] deductive logic ______________________________

____

[p.12] experiments ______________________________

____

[p.13] control group ______________________________

____

[p.13] variables ______________________________

____

[p.13] sampling error ______________________________

____

[p.13] scientific theory ______________________________

____

Sequence

Arrange the following steps of the scientific method in correct chronological sequence. Write the letter of the first step next to 1, the letter of the second step next to 2, and so on.

1. ____ A. Develop hypotheses about what the solution or answer to a problem might be [p.12]
2. ____ B. Repeat or devise new tests [p.12]
3. ____ C. Devise ways to test the accuracy of predictions drawn from the hypothesis (use of observations, models, and experiments) [p.12]
4. ____ D. Make a prediction, using hypotheses as a guide; the "if–then" process [p.12]
5. ____ E. If the tests do not provide the expected results, check to see what might have gone wrong [p.12]
6. ____ F. Objectively analyze and report the results from tests and the conclusions drawn [p.12]
7. ____ G. Identify a problem or make an observation about the natural world [p.12]

Labeling

Assume that you have to identify what object is hidden inside a sealed, opaque box. Your only tools to test the contents are a bar magnet and a triple-beam balance. Label each of the following with an "O" (for observation) or a "C" (for conclusion).

8. ____ The object has two flat surfaces.
9. ____ The object is composed of nonmagnetic metal.
10. ____ The object is not a quarter, a half-dollar, or a silver dollar.
11. ____ The object weighs x grams.
12. ____ The object is a penny.

Complete the Table

13. Complete the following table of concepts important to understanding the scientific method of problem solving. Choose from *scientific experiments, variables, prediction, inductive logic, control group, hypotheses, deductive logic,* and *scientific theory.*

Concept	*Definition*
[p.12] a.	An individual derives a general statement from specific observations
[p.13] b.	Identical to an experimental group in all respects *except* for the one variable being studied
[p.12] c.	Educated guesses about what the possible answers (or solutions) to scientific problems might be
[p.13] d.	A testable explanation of the cause or causes of a broad range of related phenomena; it remains open to tests, revision, and tentative acceptance or rejection
[p.12] e.	Simplify observation in nature or the laboratory by manipulating and controlling the conditions under which observations are made
[p.13] f.	Specific aspects of objects or events that may differ or change over time and among individuals
[p.12] g.	A statement of what one should be able to observe in nature if one looks; the "if–then" process
[p.12] h.	An individual makes inferences about specific consequences or specific predictions that must follow from a hypothesis

Dichotomous Choice

Circle one of two possible answers given between parentheses in each statement; questions 14–18 deal with the entry on spontaneous generation below, while questions 19–24 address general scientific principles.

An Italian physician, Francisco Redi, published a paper in 1688 in which he challenged the doctrine of spontaneous generation, the proposition that living things can arise from dead material. Although many examples of spontaneous generation were described in his day, Redi's work dealt particularly with disproving the notion that decaying meat could be transformed into flies. He tested his ideas in a laboratory.

14. "Two sets of jars are filled with meat or fish. One set is sealed; the other is left open so that flies can enter the jars." This description deals with a(n) (hypothesis/experiment). [p.12]
15. The sealed jar in the statement above is an example of a (variable/control). [p.13]
16. Prior to his test, Redi suggested that "worms are derived directly from the droppings of flies." This statement represents a (theory/hypothesis). [p.12]
17. "Worms (maggots) will appear only in the second set of jars" represents a (prediction/hypothesis). [p.12]
18. The statement "Mice arise from a dirty shirt and a few grains of wheat placed in a dark corner" is best called a(n) (belief/experiment). [p.15]
19. From a multitude of individual observations he made of the natural world, Charles Darwin proposed the theory of organic evolution. This was an example of (deductive logic/inductive logic). [p.12]

20. Since the time when Darwin proposed the theory of organic evolution, countless numbers of biologists have discovered evidence of various kinds that conforms to the general theory. This is an example of (deductive logic/inductive logic). [p.12]
21. The control group is identical to the experimental group except for the (hypothesis/variable) being studied. [p.13]
22. Systematic observations, model development, and conducting experiments are all methods employed to (make predictions/test predictions). [p.12]
23. Science is distinguished from faith in the supernatural by (cause and effect/experimental design). [p.12]
24. Through their failure to use large-enough samples in their experiments, scientists encounter (bias in reporting results/sampling error). [p.13]

Completion

25. Questions whose answers are ______________ in nature do not readily lend themselves to scientific analysis and experimentation. [p.15]
26. Scientists often stir up controversy when they explain a phenomenon of the world that was considered beyond natural explanation—that is, belonging to the "______________." [p.15]
27. The external world, not internal ______________, must be the testing ground for scientific beliefs. [p.15]

Self-Quiz

____ 1. Normally, the body's blood glucose levels vary between a very narrow range of 60 and 90 mg per 100 ml of blood. The body's ability to maintain this narrow range, despite irregular inputs of sugar in the diet, is an example of ______________. [p.5]
a. predictions
b. inheritance
c. metabolism
d. homeostasis

____ 2. Different species of Galápagos Island finches have different types of beaks for obtaining different kinds of food. One species removes tree bark with a sharp beak to forage for insect larvae and pupae, whereas another species has a large, powerful beak capable of crushing and eating large, heavy, coated seeds. The beaks of these birds represent a(n) ______________. [p.10]
a. adaptive trait
b. metabolic function
c. control group
d. prediction

____ 3. A boy is color-blind just as his grandfather was, even though his mother has normal vision. This situation is the result of ______________. [p.4]
a. adaptation
b. inheritance
c. metabolism
d. homeostasis

____ 4. To increase yield from a citrus crop, a farmer prunes back weak seedlings in favor of those that display strong growth characteristics. This is an example of ______________. [p.10]
a. natural selection
b. classification
c. artificial selection
d. experimental control

_____ 5. The digestion of food, the production of ATP by photosynthesis and respiration, the construction of the body's proteins, the reproduction of cells, and the contraction of muscles are all activities associated with ______________. [p.5]
a. adaptation
b. inheritance
c. metabolism
d. homeostasis

_____ 6. Which of the following involves using energy to do work? [p.5]
a. atoms bonding together to form molecules
b. the division of one cell into two cells
c. the digestion of food
d. all of these

_____ 7. The experimental group and the control group are identical except for ______________. [pp.12–13]
a. the number of variables studied
b. the variable under study
c. the two variables under study
d. the number of experiments performed on each group

_____ 8. A hypothesis should *not* be accepted as valid if ______________. [pp.12–13]
a. the sample studied is determined to be representative of the entire group
b. a variety of different tools and experimental designs yield similar observations and results
c. other investigators can obtain similar results when they conduct the experiment under similar conditions
d. several different experiments, each without a control group, systematically eliminate each of the variables except one

_____ 9. The principal point of evolution by natural selection is that ______________. [p.11]
a. it measures the difference in survival and reproduction that has occurred among individuals who differ from one another in one or more heritable traits
b. even bad mutations can improve survival and reproduction of organisms in a population
c. evolution does not occur when some forms of traits increase in frequency and others decrease or disappear with time
d. individuals lacking adaptive traits make up more of the reproductive base for each new generation

_____ 10. Which match is incorrect? [pp.8–9]
a. Kingdom Animalia—multicelled consumers; most move about
b. Kingdom Plantae—mostly multicelled producers
c. Kingdom Monera—relatively simple, multicelled organisms
d. Kingdom Fungi—mostly multicelled decomposers
e. Kingdom Protista—many complex single cells, some multicellular

_____ 11. The most inclusive of the taxonomic categories listed below is ______________. [p.8]
a. family
b. phylum
c. class
d. domain
e. genus

Chapter Objectives/Review Questions

This section lists general and detailed chapter objectives that can be used as review questions. You can make maximum use of these items by writing answers on a separate sheet of paper. Fill in answers where blanks are provided. To check for accuracy, compare your answers with information given in the chapter or glossary.

1. The fundamental unit of life is the ________. [p.4]
2. The signature molecule of cells is a nucleic acid known as ________. [p.4]
3. Describe the difference between inheritance and reproduction. [p.4]
4. Cells arise only from cells that already exist, through the process of ________. [p.4]
5. Briefly describe the role of DNA in the processes of development and inheritance. [pp.4–5]
6. Briefly explain why metabolism is a characteristic of all living organisms. [p.5]
7. ________ is an energy carrier that helps drive hundreds of metabolic activities. [p.5]
8. By the process of aerobic ________, cells can release energy stored in sugars and other kinds of molecules. [p.5]
9. ________ are certain molecules and structures that can detect stimuli. [p.5]
10. Give examples of stimuli from both the internal and the external environment that may be detected by receptors. [p.5]
11. ________ refers to the internal operating conditions of organisms remaining within tolerable limits. [p.5]
12. Arrange in order, from least inclusive to most inclusive, the levels of organization that occur in nature. Define each level as you list it. [pp.6–7]
13. Explain how the actions of producers, consumers, and decomposers create an interdependency among organisms. [p.7]
14. Describe the general pattern of energy flow through Earth's life forms and explain how Earth's resources are used again and again (cycled). [p.7]
15. Explain the use of genus and species names by considering your Latin name, *Homo sapiens.* [p.8]
16. Arrange in order, from greater to fewer organisms included, the following categories of classification: class, family, genus, kingdom, domain, order, phylum, and species. [p.8]
17. List the three general domains of life. [p.8]
18. List the six kingdoms of life; briefly describe the general characteristics of the organisms placed in each. [pp.8–9]
19. Distinguish between the terms *prokaryotic* and *eukaryotic.* [pp.8–9]
20. Explain the relationship between mutation and genetic variation. [p.10]
21. An ________ trait is any form of a trait that helps an organism survive and reproduce under a given set of environmental conditions. [p.10]
22. The term ________ means genetically based changes in a line of descent over time. [p.10]
23. Distinguish between natural and artificial selection and give an example of each. [p.10]
24. Define *natural selection* and briefly describe what is occurring when a population is said to evolve. [pp.10–11]
25. Explain what is meant by the term *diversity* and speculate about what caused the great diversity of life forms on Earth. [p.11]
26. List and explain the general steps used in scientific research. [p.12]
27. Distinguish between inductive logic and deductive logic. [p.12]
28. ________ are tests that simplify observation in nature or the laboratory by manipulating and controlling the conditions under which observations are made. [p.12]
29. Generally, members of a control group should be identical to those of the experimental group except for the one ________ being studied. [p.13]
30. Define what is meant by *scientific theory;* cite an actual example. [p.13]
31. Because of possible bias on the part of experimenters, science emphasizes presenting test results in ________ terms. [p.15]
32. Explain how the methods of science differ from answering questions by using subjective thinking and systems of belief. [p.15]

Interpreting and Applying Key Concepts

1. Humans have the ability to maintain body temperature very close to 37°C.
   a. What conditions would tend to make body temperature drop?
   b. What measures do you think your body takes to raise body temperature when it drops?
   c. What conditions would cause body temperature to rise?
   d. What measures do you think your body takes to lower body temperature when it rises?
2. Do you think that all humans on Earth today should be grouped in the same species?
3. What sorts of topics do scientists usually regard as untestable by the kinds of methods that scientists generally use?

2

CHEMICAL FOUNDATIONS FOR CELLS

Interactive Exercises

How Much Are You Worth? [pp.20–21]

2.1. REGARDING THE ATOMS [p.22]

2.2. FOCUS ON SCIENCE: *Using Radioisotopes to Track Chemicals and Save Lives* [p.23]

Selected Words: *trace* element [p.20], *atomic* number [p.22], *mass* number [p.22], *PET* [p.23], *radiation therapy* [p.23]

Boldfaced, Page-Referenced Terms

[p.20] element ______________________________

____

[p.22] atoms ______________________________

____

[p.22] protons ______________________________

____

[p.22] electrons ____

____

[p.22] neutrons ____

____

[p.22] isotopes ____

____

[p.22] radioisotope ____

____

[p.23] tracers ____

____

Matching

Choose the most appropriate answer for each.

1. ____ atoms [p.22]
2. ____ protons [p.22]
3. ____ trace element [p.20]
4. ____ PET [p.23]
5. ____ neutrons [p.22]
6. ____ electrons [p.22]
7. ____ atomic number [p.22]
8. ____ mass number [p.22]
9. ____ elements [p.20]
10. ____ isotope [p.22]
11. ____ radioisotopes [p.22]
12. ____ tracer [p.23]
13. ____ radiation therapy [p.23]

A. A compound that has a radioisotope attached that is used to determine the pathway or destination of a substance
B. Subatomic particles with a negative charge
C. Positively charged subatomic particles within the nucleus
D. Positron emission tomography; obtains images of particular body tissues
E. Atoms of a given element that differ in the number of neutrons
F. The number of protons in an atom
G. A form of isotope that contains an unstable nucleus that emits energy and particles in an attempt to stabilize its structure
H. Chemical elements representing less than 0.01 percent of body weight
I. Destroys or impairs living cancer cells
J. The number of protons and neutrons in the nucleus of one atom
K. Fundamental forms of matter that occupy space, have mass, and cannot be broken down into something else
L. Smallest units that retain the properties of a given element
M. Subatomic particles within the nucleus carrying no charge

2.3. WHAT HAPPENS WHEN ATOM BONDS WITH ATOM? [pp.24–25]

Selected Words: *lowest available energy level* [p.24], *higher energy levels* [p.24], *inert* atom [p.24], *formulas* [p.25], *chemical equations* [p.25]

Boldfaced, Page-Referenced Terms

[p.24] shell model ______________________________

____

[p.24] chemical bond ______________________________

____

[p.25] molecule ______________________________

____

[p.25] compounds ______________________________

____

[p.25] mixture ______________________________

____

Matching

Choose the most appropriate answer for each.

1. ____ mixture [p.25]
2. ____ shell model [p.24]
3. ____ lowest available energy level [p.24]
4. ____ inert atoms [p.25]
5. ____ orbitals [p.24]
6. ____ chemical bond [p.25]
7. ____ compounds [p.25]
8. ____ molecule [p.25]
9. ____ higher energy levels [p.24]

A. Regions of space around an atom's nucleus where electrons are likely to be at any one instant
B. Results when two or more atoms bond together
C. Two or more elements that may combine in various proportions
D. A union between the electron structures of atoms
E. A graphic representation of the distribution of electrons in their energy levels
F. Types of molecules composed of two or more different elements in proportions that never vary
G. Energy of electrons farther from the nucleus than the first orbital
H. Refers to those elements with no vacancies in their shells; hence they show little tendency to enter chemical reactions
I. Energy of electrons in the orbital closest to the nucleus

Complete the Table

10. Referring to Table 2.1 [p.22] and Figure 2.7 [p.25] in the text, enter the missing information for each line in the following table.

Element	*Atomic Number*	*Atomic Mass*	*Number of*		
			Protons	*Neutrons*	*Electrons*
[p.25] a.	11	23			
[p.25] b. calcium			20	20	20
[p.25] c. carbon	6			6	
[p.25] d.	1		1	0	1
[p.25] e. oxygen		16		8	8
[p.25] f. nitrogen				7	7
[p.25] g.	17	35	17		

Fill-in-the-Blanks

The expression $12H_2O + 6CO_2 \rightarrow 6O_2 + C_6H_{12}O_6 + H_2O$ is known as the chemical (11) ______________ [p.25] for photosynthesis. H_2O is the (12) ______________ [p.25] for water. The (13) ______________ [p.25] are to the left of the reaction arrow and the (14) ______________ [p.25] are to the right of the reaction arrow. In the expression one can count six carbon atoms on the left side of the arrow, so one should be able to count (15) ______________ [p.25] carbon atoms on the right side of the arrow. Since water always consists of a ratio of two hydrogen atoms to one oxygen atom, it is called a (16) ______________ [p.25], in contrast to a (17) ______________ [p.25], in which the components may vary in their proportions.

Identification

18. The example below demonstrates a shell model for a given element. The number of protons and the number of neutrons are shown within the nucleus. For each energy level, indicate the number of electrons present, using the form 2e⁻ to indicate a pair of electrons. In the space under each diagram, indicate the name of the element. [pp.24–25]

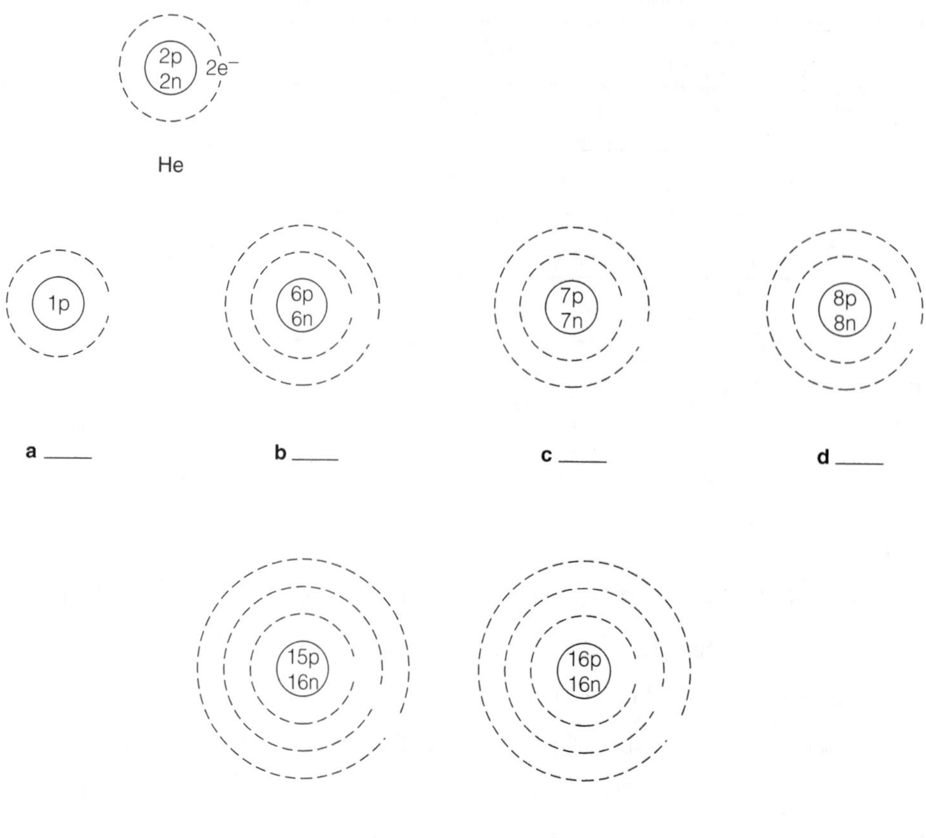

2.4. IMPORTANT BONDS IN BIOLOGICAL MOLECULES [pp.26–27]

Selected Words: *biological molecules* [p.26], *sharing* [p.26], *single* covalent bond [p.26], *double* covalent bond [p.26], *triple* covalent bond [p.26], *nonpolar* covalent bond [p.26], *polar* covalent bond [p.27], *electronegative* [p.27], no *net charge* [p.27]

Boldfaced, Page-Referenced Terms

[p.26] ion ____

____

[p.26] ionic bond ____

____

[p.26] covalent bond ____

____

[p.27] hydrogen bond ____

____

Identification

1. Following the example below, complete the diagram by adding arrows to identify the transfer of electron(s), showing how positive magnesium and negative chlorine ions form ionic bonds to create a molecule of $MgCl_2$ (magnesium chloride). [p.26]

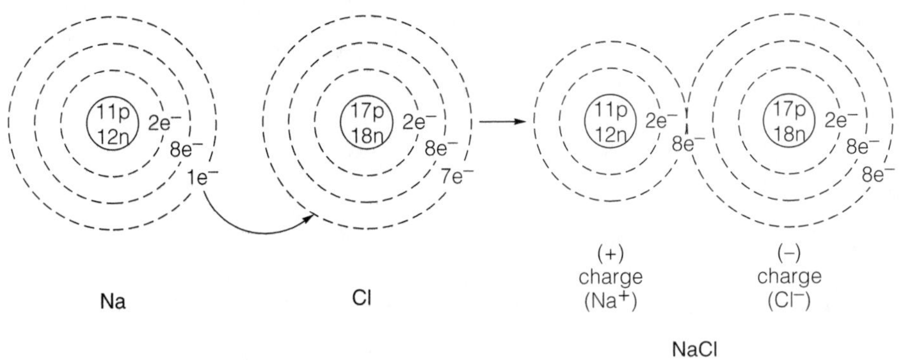

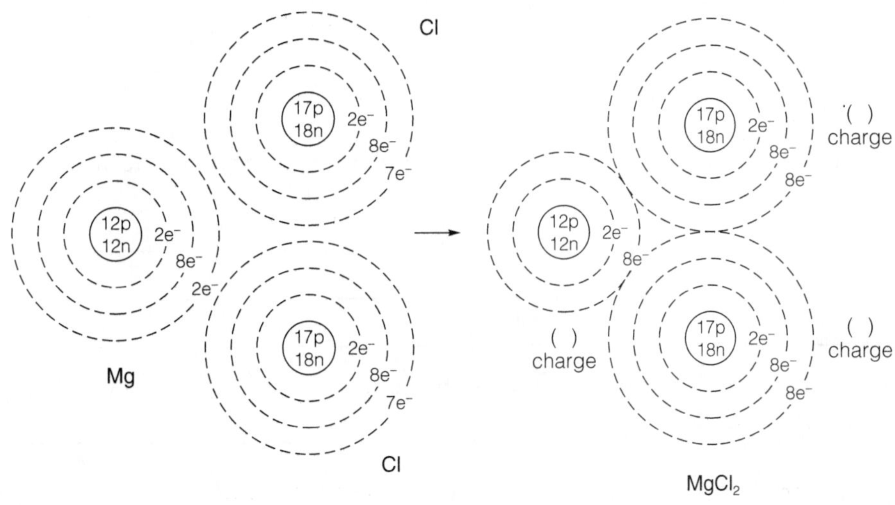

Identification

2. Referring to the example of hydrogen gas shown below, complete each diagram by placing electrons (as dots) in the outer shells to identify the nonpolar covalent bonding that forms oxygen gas and water molecules. [pp.26–27]

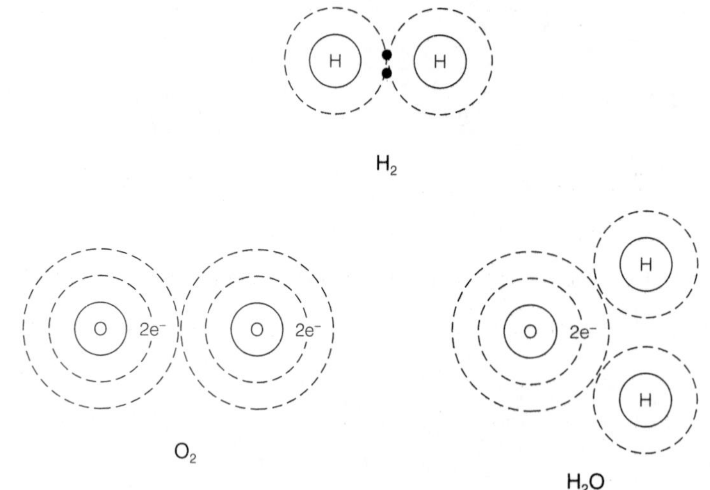

Short Answer

3. Distinguish between a nonpolar covalent bond and a polar covalent bond. Cite an example of each. [pp.26–27]

____

____

____

4. Explain the importance of hydrogen bonds in establishing the structure of DNA. [p.27]

____

____

____

2.5. PROPERTIES OF WATER [pp.28–29]

Selected Words: *dissolved* [p.29]

Boldfaced, Page-Referenced Terms

[p.28] hydrophilic substances ____

____

[p.28] hydrophobic substances ____

____

[p.29] temperature ____

____

[p.29] evaporation ____

____

[p.29] cohesion ____

____

[p.29] solute ____

____

Fill-in-the-Blanks

The (1) ______________ [p.28] of water molecules allows them to hydrogen-bond with each other. Water molecules hydrogen-bond with polar molecules, which are (2) ______________ (water-loving) substances [p.28]. Polarity causes water to repel oil and other nonpolar substances, which are (3) ______________ (water-dreading) [p.28]. The term (4) ______________ [p.29] is the measure of the molecular motion of a given substance. Liquid water changes its temperature more slowly than air because of the great amount of heat required to break the high number of (5) ______________ [pp. 28–29] bonds between water molecules; this property helps (6) ______________ [pp. 28–29] temperature in aquatic habitats and cells. (7) ______________ [p.29] occurs when large energy inputs increase molecular motion to the point where hydrogen bonds stay broken, releasing individual molecules from the water surface. Below 0°C, water molecules become locked in the latticelike bonding pattern of (8) ______________ [p.29], which is less dense than water. Collective hydrogen bonding creates a high tension on surface water molecules, resulting in (9) ______________ [p.29], the property of water that explains how long, narrow water columns rise to the tops of tall trees. Water is an excellent (10) ______________ [p.29], in which ions and polar molecules readily dissolve. Substances dissolved in water are known as (11) ______________ [p.29]. A substance is (12) ______________ in water when spheres of (13) ______________ [p.29] form around its individual ions or molecules.

2.6. ACIDS, BASES, AND BUFFERS [pp.30–31]

Selected Words: *donate* H^+ [p.30], *accept* H^+ [p.30], *acidic* solutions [p.30], *basic* solutions [p.30], alkaline [p.30], *acid stomach* [p.30], *chemical burns* [p.30], *acid rain* [p.31], *coma* [p.31], *tetany* [p.31], *acidosis* [p.31], *alkalosis* [p.31]

Boldfaced, Page-Referenced Terms

[p.30] hydrogen ions ____

____

[p.30] pH scale ____

____

[p.30] acids ____

____

[p.30] bases ____

____

[p.31] buffer system ____

____

[p.31] salts ____

____

Matching

Choose the most appropriate answer for each.

1. ____ acid stomach [p.30]
2. ____ acids [p.30]
3. ____ pH scale [p.30]
4. ____ chemical burns [p.30]
5. ____ alkaline [p.30]
6. ____ H^+ [p.30]
7. ____ bases [p.30]
8. ____ examples of basic solutions [p.30]
9. ____ acidosis [p.31]
10. ____ OH^- [p.30]
11. ____ examples of acid solutions [p.30]
12. ____ alkalosis [p.31]
13. ____ buffer system [p.31]
14. ____ acid rain [p.31]

A. CO_2 builds up in the blood, too much carbonic acid (H_2CO_3) forms, and blood pH severely decreases
B. Hydroxide ion
C. Substances that accept H^+ when dissolved in water
D. An uncorrected increase in blood pH
E. Used to represent H^+ concentration in various fluids
F. A partnership between a weak acid and the base that forms when it dissolves in water; counters slight pH shifts
G. Hydrogen ion or proton
H. Baking soda, seawater, egg white
I. A condition that can be caused by powerful acids and bases, such as ammonia, drain cleaner, and sulfuric acid in car batteries
J. Substances that donate H^+ when dissolved in water
K. Lemon juice, gastric fluid, coffee
L. An environmental condition caused by the burning of fossil fuels and the use of nitrogen fertilizers
M. A condition caused by a diet high in spicy foods
N. Another term used to describe a basic solution

Complete the Table

15. Complete the following table by consulting Figure 2.15 on page 30 of the text.

Fluid	*pH value*	*Acid/Base/Neutral*
[p.30] a. bleach		
[p.30] b. oranges		
[p.30] c. urine		
[p.30] d. blood		
[p.30] e. gastric fluid		
[p.30] f. pure water		

Self-Quiz

_____ 1. Each element has a unique __________, which refers to the number of protons present in its atoms. [p.22]
a. isotope
b. mass number
c. atomic number
d. radioisotope

_____ 2. A molecule is __________. [p.25]
a. a bonding together of two or more atoms
b. less stable than its constituent atoms separated
c. electrically charged
d. a carrier of one or more extra neutrons

_____ 3. If neon has an atomic number of 10 and an atomic mass of 20, it has __________ neutron(s) in its nucleus. [p.22]
a. one
b. two
c. five
d. ten
e. twenty

_____ 4. Substances that are nonpolar and repelled by water are __________. [p.27]
a. hydrolyzed
b. nonpolar
c. hydrophilic
d. hydrophobic

_____ 5. A hydrogen bond is __________. [p.27]
a. a sharing of a pair of electrons between a hydrogen nucleus and an oxygen nucleus
b. a sharing of a pair of electrons between a hydrogen nucleus and either an oxygen or a nitrogen nucleus
c. formed when an electronegative atom of a molecule weakly interacts with a hydrogen atom that is already participating in a separate polar covalent bond
d. none of the above

_____ 6. An ionic bond is one in which __________. [p.26]
a. electrons are shared equally
b. electrically neutral atoms have a mutual attraction
c. two charged atoms have a mutual attraction due to a transfer of electrons
d. electrons are shared unequally

_____ 7. A covalent bond is one in which __________. [p.26]
a. electrons are shared
b. electrically neutral atoms have a mutual attraction
c. two charged atoms have a mutual attraction due to a transfer of electrons
d. electrons are lost

_____ 8. A nonpolar covalent bond implies that __________. [pp.26–27]
a. one negative atom bonds with a hydrogen atom
b. it is a double bond
c. there is no difference in charge at the ends (the two poles) of the bond
d. atoms of different elements do not exert the same pull on shared electrons

_____ 9. This type of chemical bond contributes to the shape of large molecules. [p.27]
a. hydrogen
b. ionic
c. covalent
d. inert
e. single

_____ 10. A solution with a pH of 2 is __________ times as acidic as one with a pH of 5. [p.30]
a. 3
b. 10
c. 30
d. 100
e. 1,000

_____ 11. A control that minimizes unsuitable pH shifts is a(n) _____________. [p.31]

a. hydrophilic compound
b. salt
c. base
d. acid
e. buffer

_____ 12. Which of the following properties of water protects aquatic organisms during a long, cold winter? [pp. 28–29]

a. cohesion
b. solvent properties
c. temperature-stabilizing
d. none of the above

Chapter Objectives/Review Questions

1. Define an element. [p.20]
2. What are the four most abundant elements in living organisms? [p.20]
3. What is meant by the term *trace element?* Give some examples of trace elements. [p.20]
4. List and describe the three types of subatomic particles. [p.22]
5. Distinguish between the atomic mass and atomic number of an element and tell what subatomic particles contribute to the value of each. [p.22]
6. Distinguish between isotopes and radioisotopes. [p.22]
7. Describe why researchers may use a tracer in an experiment. [p.23]
8. Explain how radioisotopes are used differently in PET and radiation therapy. [p. 23]
9. What information may be obtained about an atom from a shell model? [p.24]
10. Explain the difference between shells and orbitals. Which more accurately illustrates the true structure of an atom? [pp.24–25]
11. Sketch shell models of the atoms described in the text. [text Figure 2.7, p.25]
12. For a chemical equation, such as the one for photosynthesis shown in Figure 2.8 in the text, be able to identify the reactants and products of the equation. [p. 25]
13. Explain why helium, neon, and argon are known as inert elements. [p.25]
14. Explain the difference between molecules and compounds [p.25]
15. How does a mixture differ from a compound? [p.25]
16. A(n) __________ is an atom that becomes positively or negatively charged. [p.26]
17. An association of two oppositely charged ions is a(n) __________ bond. [p.26]
18. In a(n) __________ bond, two atoms share electrons. [p.26]
19. Explain why H_2 is an example of a nonpolar covalent bond and why H_2O has two polar covalent bonds. [pp.26–27]
20. In a(n) __________ bond, a small, highly electronegative atom of a molecule weakly interacts with a hydrogen atom that is already participating in a polar covalent bond. [p.27]
21. Explain the biological importance of hydrogen bonds. [p.27]
22. Polar molecules attracted to water are __________; all nonpolar molecules are __________ and are repelled by water. [p.28]
23. What is meant by the temperature of an object at the molecular level? [pp.28–29];
24. What occurs at the molecular level during the process of evaporation? [p.29]
25. Describe the formation of ice in terms of hydrogen bonding. [p.29]
26. Completely describe the properties of water that allow it to move from the roots to the tops of the tallest trees. [p.29]
27. Explain the relationship between spheres of hydration and water's solvent properties. [p.29]
28. Describe what happens when a substance is dissolved in water. [p.29]
29. The ionization of water is the basis of the __________ scale. [p.30]
30. Explain the difference between acids and bases in terms of their relationship to hydrogen ions (H^+). [p.30]
31. Blood, with its pH of 7.34, is a(n) __________ solution, whereas gastric fluid, with a pH of 2, is a(n) __________ solution. [p.30]
32. Define *buffer system;* cite an example and describe how buffers operate. [p.31]
33. Distinguish between acidosis and alkalosis. [p.31]
34. A __________ and water are produced by a chemical reaction between an acid and a base. [p.31]

Integrating and Applying Key Concepts

1. Explain what would happen if water were a nonpolar molecule instead of a polar molecule. Would water be a good solvent for the same kinds of substances? Would the nonpolar molecule's specific heat likely be higher or lower than that of water? Would surface tension be affected? cohesive nature? ability to form hydrogen bonds? Is it likely that the nonpolar molecules could form unbroken columns of liquid? What implications would that hold for trees?
2. What would be the implications for life on Earth if water did not have temperature-stabilizing effects? What would be different within individual organisms? What would change in ecosystems such as ponds and rivers? What would change on a global (biosphere) level? Why then do scientists search for evidence of water on other planets?

3

CARBON COMPOUNDS IN CELLS

Interactive Exercises

Carbon, Carbon, in the Sky—Are You Swinging Low and High? [pp.34–35]

3.1. THE MOLECULES OF LIFE—FROM STRUCTURE TO FUNCTION [pp.36–37]

3.2. OVERVIEW OF FUNCTIONAL GROUPS [p.38]

Selected Words: "organic" substances [p.36], "inorganic" substances [p.36], *hydroxyl* groups [p.38], *carbonyl* groups [p.38], *carboxyl* groups [p.38], *phosphate* group [p.38], *sulfhydryl* group [p.38]

Boldfaced, Page-Referenced Terms

[p.34] global warming ____

____

[p.36] organic compounds ____

____

[p.36] hydrocarbons ____

____

[p.36] functional groups ____

____

[p.38] alcohols ____

____

Fill-in-the-Blanks

The change in the temperature of the lower atmosphere, a phenomenon called (1) ______________ [p.34], is associated with changes in the patterns of precipitation around the globe. There is a distinct connection between temperature and (2) ______________ [p.35]. As a result, the carbon dioxide concentration of the atmosphere (3) ______________ [p.35] in spring and rises in (4) ______________ [p.35]. The burning of (5) ______________ [p.35] by humans has also contributed to changes in atmospheric carbon. Carbon permeates the world of life, from the organization of (6) ______________ [p.35] to the chemical conditions of (7) ______________ [p.35] around the globe.

Only living (8) ______________ [p.36] have the ability to synthesize molecules such as carbohydrates and proteins. A molecule that contains carbon and at least one hydrogen atom is called a(n) (9) ______________ [p.36]. If the organic compound contains only hydrogen and carbon atoms, it is called a(n) (10) ______________ [p.36]. Many organic molecules also contain groups of atoms called (11) ______________ [p.36].

Carbon's role as the molecule of life is derived from its ability to covalently bond with (12) ______________ [p.36] other atoms. These carbon chains serve as (13) ______________ [p.36], to which other atoms are attached. This is the start of the three-dimensional shape of the organic compounds. The three different models that are used to visually present carbon compounds are (14) ______________, ______________, and ______________ [p.37]. These models give you a glimpse into the (15) ______________ [p.37] and function of organic compounds.

Matching

Choose the most appropriate answer for each.

16. ____ sulfhydryl group [p.38]
17. ____ hydroxyl group [p.38]
18. ____ carbonyl group [p.38]
19. ____ phosphate group [p.38]
20. ____ carboxyl group [p.38]
21. ____ alcohols [p.38]

A. Stabilizes the structure of proteins
B. Found within the structure of ATP and DNA
C. The functional group of the alcohols
D. Used in the building of fats and carbohydrates
E. A highly polar group found in amino acids
F. An organic molecule containing large numbers of –OH functional groups

Labeling

Study the structural formulas of the following organic compounds. Note that each carbon atom can share pairs of electrons with as many as four other atoms. Refer to Figure 3.5 in the text, page 38, to identify the circled functional groups by entering the correct name in the blanks with matching numbers. Some functional groups are represented more than once.

22. ____________ [p.36]
23. ____________ [p.36]
24. ____________ [p.36]
25. ____________ [p.36]
26. ____________ [p.36]
27. ____________ [p.36]
28. ____________ [p.36]

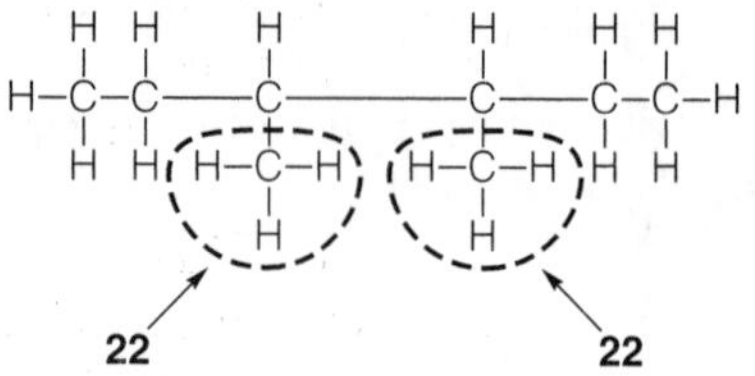

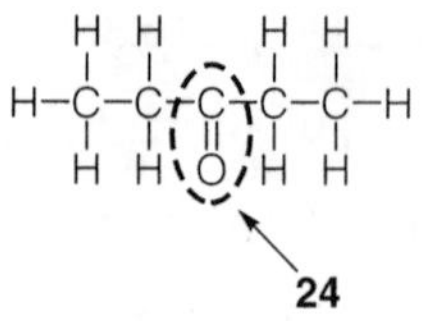

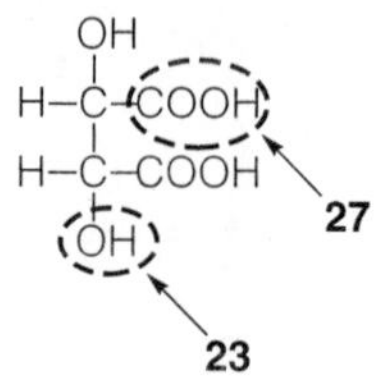

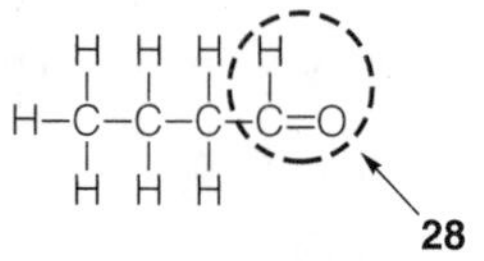

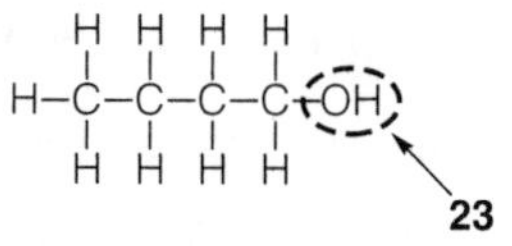

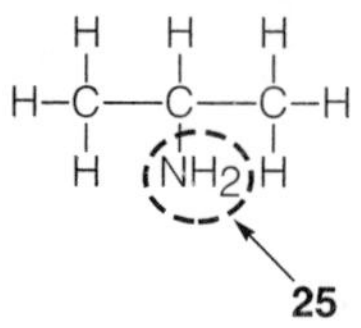

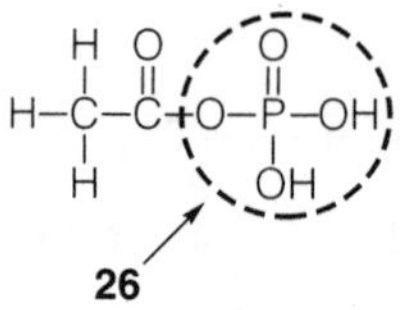

3.3. HOW DO CELLS BUILD ORGANIC COMPOUNDS? [p.39]

Selected Words: *Functional-group transfer* [p.39], *electron transfer* [p.39], *rearrangement* [p.39], *condensation* [p.39], *cleavage* [p.39]

Boldfaced, Page-Referenced Terms

[p.39] monomers ____

____

[p.39] polymers ____

____

[p.39] enzymes ____

[p.39] condensation reaction ____

[p.39] hydrolysis ____

Matching

Choose the most appropriate answer for each.

1. ____ enzymes [p.39]
2. ____ condensation reaction [p.39]
3. ____ monomers [p.39]
4. ____ hydrolysis [p.39]
5. ____ polymers [p.39]
6. ____ functional-group transfer [p.39]
7. ____ cleavage [p.39]
8. ____ rearrangement [p.39]
9. ____ electron transfer [p.39]

A. A class of proteins that make chemical reactions occur faster
B. A type of reaction that splits molecules using water
C. The individual subunits of organic molecules
D. Any reaction that splits a molecule into two smaller molecules
E. The type of chemical reaction that moves electrons between molecules
F. The movement of functional groups between molecules
G. The formation of a covalent bond by the removal of –OH and H^+ functional groups, forming water
H. Long chains of subunits, sometimes consisting of millions of individual subunits
I. A change in the internal bond structure of a molecule

10. Study the structural formulas of the two adjacent amino acids. Identify the enzyme action causing formation of a covalent bond and a water molecule (through a condensation reaction) by circling an H atom from one amino acid and an –OH group from the other amino acid. Also circle the covalent bond that formed the dipeptide [pp.39,45].

```
H   R   O                                  H   R   O
 \  |  //                                   \  |  //
  N-C-C                                      N-C-C     R   O
 /  |  \     H   R   O          ——>         /  |   \   |  //     + H2O
H   H   OH    \  |  //                     H   H    N-C-C
               N-C-C                               /  |  \
              /  |  \                             H   H   OH
             H   H   OH
```

amino acid　　amino acid　　dipeptide

3.4. CARBOHYDRATES—THE MOST ABUNDANT MOLECULES OF LIFE [pp.40–41]

Selected Words: "saccharide" [p.40], *mono*saccharide [p.40], *oligo*saccharide [p.40], *di*saccharides [p.40], "complex" carbohydrates [p.40], *poly*saccharides [p.40]

Boldfaced, Page-Referenced Terms

[p.38] carbohydrates ____

____

[p.38] monosaccharides ____

____

[p.38] oligosaccharides ____

____

[p.38] polysaccharides ____

____

Identification

1. In the following diagram, identify condensation reaction sites between the two glucose molecules by circling the components of the water removed that allow a covalent bond to form between the glucose molecules (text Figure 3.8, p.40). Note that the reverse reaction is hydrolysis and that both condensation and hydrolysis reactions require enzymes in order to proceed efficiently. [p.40]

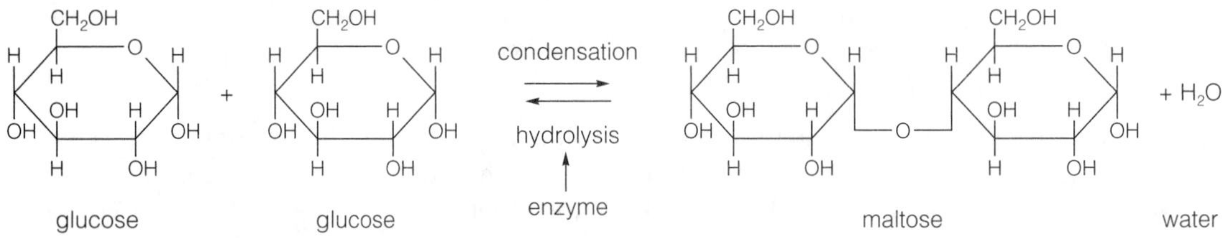

Choice

For each of the following, choose the correct class of carbohydrates with which the indicated term is associated. Answers may be used more than once.

a. oligosaccharides   b. polysaccharides   c. monosaccharides

2. ____ "complex carbohydrates" [p.40]
3. ____ chitin [p.41]
4. ____ disaccharides [p.40]
5. ____ ribose and deoxyribose [p.40]
6. ____ lactose, sucrose, and maltose [p.40]
7. ____ glucose and fructose [p.40]
8. ____ starch and glycogen [p.41]
9. ____ cellulose [pp.40–41]

Complete the Table

10. For each of the following, give the name of the carbohydrate that has the indicated function.

Carbohydrate	*Function*
[p.40] a.	Main energy source for most organisms; precursor of many organic molecules; serves as building blocks for larger carbohydrates
[p.41] b.	Storage form for photosynthetically produced sugars
[p.40] c.	Most plentiful sugar in nature; transport form of carbohydrates in plants; table sugar; formed from glucose and fructose
[p.41] d.	Animal starch; stored especially in liver and muscle tissue; formed from glucose chains
[p.41] e.	Main structural material in some external skeletons and other hard body parts of some animals and fungi
[pp.40–41] f.	Structural material of plant cell walls; formed from glucose chains
[p.40] g.	Five-carbon sugars occurring in DNA and RNA

3.5. GREASY, OILY—MUST BE LIPIDS [pp.42–43]

Selected Words: *unsaturated* tails [p.42], *saturated* tails [p.42], "vegetable oils" [p.42], "neutral" fats [p.42]

Boldfaced, Page-Referenced Terms

[p.42] lipids ______________________________

____

[p.42] fats ______________________________

____

[p.42] fatty acid ______________________________

____

[p.42] triglycerides ______________________________

____

[p.43] phospholipid ______________________________

____

[p.43] sterols ______________________________

____

[p.43] waxes ______________________________

____

Labeling

1. In the answer blanks, label the molecules shown as saturated or unsaturated. For the unsaturated molecules, circle the regions that make them unsaturated.

   a. ________________________ [p.42]

   b. ________________________ [p.42]

   c. ________________________ [p.42]

```
     O  H  H  H  H  H  H  H  H  H  H  H  H  H  H  H  H  H
     ‖  |  |  |  |  |  |  |  |  |  |  |  |  |  |  |  |  |
HO—C—C—C—C—C—C—C—C—C=C—C—C—C—C—C—C—C—C—H
        |  |  |  |  |  |  |        |  |  |  |  |  |  |  |
        H  H  H  H  H  H  H        H  H  H  H  H  H  H  H
```
a. oleic acid

```
     O  H  H  H  H  H  H  H  H  H  H  H  H  H  H  H  H  H
     ‖  |  |  |  |  |  |  |  |  |  |  |  |  |  |  |  |  |
HO—C—C—C—C—C—C—C—C—C—C—C—C—C—C—C—C—C—C—H
        |  |  |  |  |  |  |  |  |  |  |  |  |  |  |  |  |
        H  H  H  H  H  H  H  H  H  H  H  H  H  H  H  H  H
```
b. stearic acid

```
     O  H  H  H  H  H  H  H  H  H  H  H  H  H  H  H  H  H
     ‖  |  |  |  |  |  |  |  |  |  |  |  |  |  |  |  |  |
HO—C—C—C—C—C—C—C—C—C=C—C—C=C—C—C=C—C—C—H
        |  |  |  |  |  |  |        |        |        |  |
        H  H  H  H  H  H  H        H        H        H  H
```
c. linolenic acid

Identification

2. Using a condensation reaction, combine glycerol with three fatty acids to form a triglyceride. In the diagram to the left, circle the participating atoms that will identify three covalent bonds. In the diagram to the right, circle the resulting covalent bonds. [p.42]

```
   H              O                    H      O
   |              ‖                    |      ‖
H—C—OH  +  HO—C—R                H—C—O—C—R
   |                                   |
   |              O      yields        |      O
   |              ‖                    |      ‖
H—C—OH  +  HO—C—R     ——→       H—C—O—C—R  + 3H2O
   |                                   |
   |              O                    |      O
   |              ‖                    |      ‖
H—C—OH  +  HO—C—R                H—C—O—C—R
   |                                   |
   H                                   H
glycerol    three fatty            triglyceride
              acids             (a complete fat
                                    molecule)
```

Choice

For questions 3–19, choose from the following:

a. triglycerides b. phospholipids c. waxes d. sterols

3. _____ Richest source of body energy [p.42]
4. _____ The lipid found in honeycombs [p.43]
5. _____ Cholesterol belongs to this class [p.43]
6. _____ This class may have either saturated or unsaturated tails [p.42]
7. _____ Butter and lard belong to this class [p.42]
8. _____ This is the only class of lipids that lack fatty acid tails [p.43]
9. _____ The primary component of cell membranes [p.43]
10. _____ All possess a rigid backbone of four fused carbon rings [p.43]
11. _____ Found in the cuticles of plants [p.43]
12. _____ Precursors of vitamin D, steroids, and bile salts [p.43]
13. _____ This class is made from three fatty acids combined with a unit of glycerol [p.42]
14. _____ Vegetable oil belongs to this class [p.42]
15. _____ Used by vertebrates for insulation [p.42]
16. _____ Furnishes protection and lubrication for hair, skin, and feathers [p.43]
17. _____ The "neutral" fats belong to this class [p.42]
18. _____ This molecule has a phosphate functional group in place of a fatty acid chain [p.43]
19. _____ The sex hormones are formed from this class [p.43]

3.6. A STRING OF AMINO ACIDS: PROTEIN PRIMARY STRUCTURE [pp.44–45]

3.7. HOW DOES A PROTEIN'S FINAL STRUCTURE EMERGE? [pp.46–47]

3.8. WHY IS PROTEIN STRUCTURE SO IMPORTANT? [pp.48–49]

Selected Words: *primary* structure [p.44], *fibrous* proteins [p.44], *globular* proteins [p.44], *disulfide bridges* [p.45], *secondary* structure [p.46], *tertiary* structure [p.46], *quaternary* structure [p.47], "self" and "nonself" cells [p.47], *glyco*proteins [p.47], *lipo*proteins [p.47], "gene" [p.49], *sickle-cell anemia* [p.49]

Boldfaced, Page-Referenced Terms

[p.44] proteins ____

____

[p.44] amino acid ____

____

[p.44] polypeptide chain ____

____

[p.46] domain ____

____

[p.47] HLAs (human leukocyte antigens) ______________________________

____

[p.47] denaturation ______________________________

____

Labeling

1. In the model of an amino acid below, label the R group, amino group, and carboxyl group.

   a. ______________________________ [p.44]

   b. ______________________________ [p.44]

   c. ______________________________ [p.44]

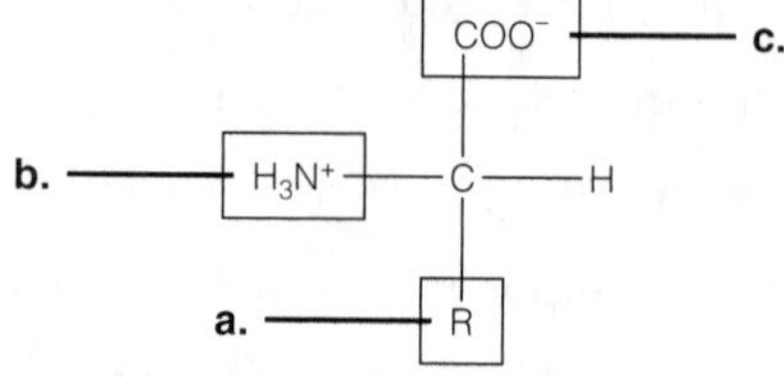

2. For the amino acid valine, label the R group, amino group, and carboxyl group.

   a. ______________________________ [p.44]

   b. ______________________________ [p.44]

   c. ______________________________ [p.44]

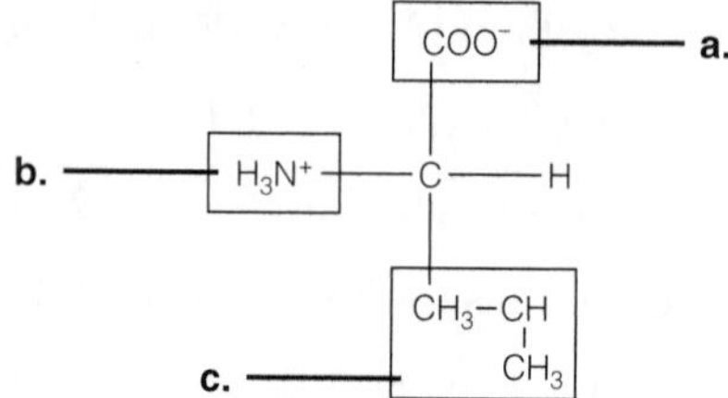

Identification

3. The diagram below illustrates four ionized amino acids (in cellular solution) forming a polypeptide chain. In the upper section, circle the atoms from each amino acid that are involved in the formation of a peptide bond. On the polypeptide chain (lower section), circle the location of the resulting peptide bonds. [p.45]

Matching

Choose the most appropriate answer for each term.

4. ____ amino acid [p.44]
5. ____ disulfide bridges [p.45]
6. ____ peptide bond [p.44]
7. ____ polypeptide chain [pp.44–45]
8. ____ primary structure [p.44]
9. ____ proteins [p.44]
10. ____ secondary structure [p.46]
11. ____ tertiary structure [p.44]
12. ____ domain [p.46]
13. ____ quaternary structure [p.47]
14. ____ lipoproteins [p.47]
15. ____ glycoproteins [p.47]
16. ____ denaturation [p.47]
17. ____ fibrous proteins [p.44]
18. ____ globular proteins [p.44]

A. A coiled or extended pattern of protein structure caused by regular intervals of H bonds
B. Three or more amino acids joined in a linear chain
C. Proteins with linear or branched oligosaccharides covalently bonded to them; found on animal cell surfaces, in cell secretion, or on blood proteins
D. Folding of a protein through interactions among R groups of a polypeptide chain
E. Form when freely circulating blood proteins encounter and combine with cholesterol, or phospholipids
F. The type of covalent bond linking one amino acid to another
G. HLAs and hemoglobin are examples of this level of protein structure
H. Breaking weak bonds in large molecules (such as protein) to change their shape so they no longer function
I. A self-organized polypeptide chain that functions as a stable unit
J. Lowest level of protein structure; has a linear, unique sequence of amino acids
K. A small organic compound having an amino group, an acid group, a hydrogen atom, and an R group
L. The most diverse of all the large biological molecules; constructed from pools of only 20 kinds of amino acids
M. A group of proteins that contribute to cell shape and organization
N. A chemical bond between polypeptide chains that uses two sulfur atoms
O. Enzymes typically belong to this group of proteins

Fill-in-the-Blanks

The (19) ______________ [p.48] structure of hemoglobin consists of four polypeptide chains called (20) ______________ [p.48]. The folded protein contains a small pocket that is chemically attractive to a (21) ______________ [p.48] group, the iron-containing portion of hemoglobin. These globins come in two forms, alpha and (22) ______________ [p.49]. A genetic mutation can cause the glutamate in the beta globin to be replaced by the amino acid (23) ______________ [p.49]. This difference can cause the genetic disorder called (24) ______________ [p.49]. In people with this disease, the abnormal hemoglobin molecules are (25) ______________ [p.49] to one another and become sickle-shaped. These cells (26) ______________ [p.49] easily, clogging capillaries and creating problems in the (27) ______________ [p.49] system.

3.9. NUCLEOTIDES AND NUCLEIC ACIDS [pp.50–51]

Selected Words: "base-pair" [p.50]

Boldfaced, Page-Referenced Terms

[p.50] nucleotides ______________________________

[p.50] ATP ______________________________

[p.50] coenzymes ______________________________

[p.50] nucleic acids ______________________________

[p.50] DNA ______________________________

[p.50] RNAs ______________________________

Labeling

1. In the diagram of a nucleotide below, label the phosphate groups, nitrogenous base, and five-carbon sugar subunits.

a. ______________________ [p.50]

b. ______________________ [p.50]

c. ______________________ [p.50]

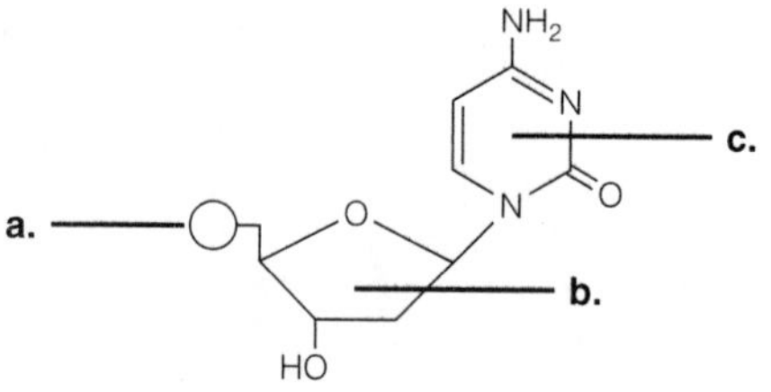

2. In the diagram of a single-stranded nucleic acid molecule, encircle as many complete nucleotides as possible. How many complete nucleotides are present? [pp.50–51]

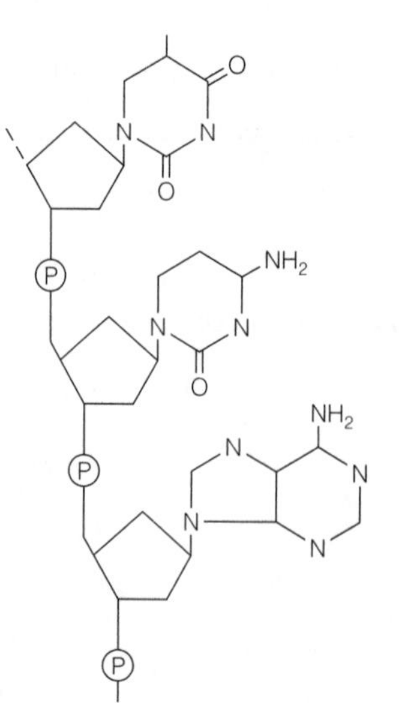

Matching

Choose the most appropriate answer for each term.

3. _____ adenosine triphosphate [p.50]
4. _____ coenzymes [p.50]
5. _____ base pairs [p.50]
6. _____ RNAs [pp.50–51]
7. _____ DNA [p.50]

A. Single nucleotide strands; function in processes by which genetic instructions are used to build proteins
B. The cellular energy carrier
C. Enzyme assistants; examples are NAD^+ and FAD
D. Double nucleotide strand; encodes genetic instructions with nucleotide sequences
E. Two nucleotides linked together by hydrogen bonds

Self-Quiz

Choice

For each of the following, choose the class of organic molecule to which the item belongs. Choices may be used more than once.

a. lipids b. nucleic acids c. proteins d. carbohydrates

1. _____ glycoproteins [p.47]
2. _____ phospholipids [p.43]
3. _____ glycogen [p.41]
4. _____ adenosine triphosphate [p.50]
5. _____ sucrose and maltose [p.40]
6. _____ triglycerides [p.42]
7. _____ DNA and RNA [pp.50–51]
8. _____ HLAs and hemoglobin [pp.47–48]
9. _____ cholesterol [p.43]
10. _____ glycogen and starch [p.41]
11. _____ waxes [p.43]

Multiple Choice

_____ 12. Amino acids are linked by __________ bonds to form the primary structure of a protein. [p.44]
a. disulfide
b. hydrogen
c. ionic
d. peptide

_____ 13. Proteins __________. [p.44]
a. are weapons against disease-causing bacteria and other invaders
b. are composed of amino acid subunits
c. may act as hormones
d. may function as enzymes
e. all of the above

_____ 14. Which of the following does not belong to the lipid class of organic molecules? [pp.42–43]
a. sterols
b. waxes
c. phospholipids
d. glycoproteins
e. triglycerides

_____ 15. DNA _____________. [p.50]
a. is one of the adenosine phosphates
b. is one of the nucleotide coenzymes
c. contains protein-building instructions
d. is composed of monosaccharides

_____ 16. Most of the chemical reactions in cells must have _____________ present before they proceed. [pp.39,44]
a. RNA
b. salt
c. enzymes
d. fats
e. chitin

_____ 17. Carbon is part of so many different substances because _____________. [p.36]
a. carbon generally forms two covalent bonds with a variety of other atoms
b. a carbon atom generally forms four covalent bonds with a variety of atoms
c. carbon ionizes easily
d. carbon is a polar compound

_____ 18. Which of the following levels of protein structure is not correctly linked to its description? [pp.44–47]
a. primary—the linear sequence of amino acids
b. secondary—coiling of a polypeptide due to the action of hydrogen bonds
c. tertiary—interactions between the domains of a protein
d. quaternary—chemical interactions between multiple polypeptide chains
e. all of the above are correct

_____ 19. _____________ are molecules used by cells as structural materials, energy transport molecules, or storage forms of energy. [p.40]
a. Lipids
b. Nucleic acids
c. Carbohydrates
d. Proteins

_____ 20. Hydrolysis could be correctly described as the _____________. [p.39]
a. heating of a compound in order to drive off its excess water and concentrate its volume
b. breaking of a long-chain compound into its subunits by adding water molecules to its structure between the subunits
c. linking of two or more molecules by the removal of one or more water molecules
d. constant removal of hydrogen atoms from the surface of a carbohydrate
e. an example of a condensation class of reactions

_____ 21. Genetic instructions are encoded in the base sequence of _____________; molecules of _____________ function in processes using genetic instructions to construct proteins. [pp.50–51]
a. DNA; DNA
b. DNA; RNA
c. RNA; DNA
d. RNA; RNA

Chapter Objectives/Review Questions

1. The molecules of life are ________ compounds. [p.38]
2. Each carbon atom can share pairs of electrons with as many as ________ other atoms. [p.38]
3. A ________ has only hydrogen atoms attached to a carbon backbone. [p.38]
4. Hydroxyl, amino, and carboxyl are examples of ________ groups. [p.38]
5. Describe the importance of enzymes to chemical reactions. [p.39]
6. Define the following: *functional-group transfer, electron transfer, rearrangement, condensation,* and *cleavage.* [p.39]
7. Define and distinguish between *condensation reactions* and *hydrolysis;* cite a general example. [p.39]
8. Give the general characteristics of a *carbohydrate;* list the two general functions of carbohydrates. [p.40]
9. Name and generally define the three classes of carbohydrates. [pp.40–41]
10. Give an example of a monosaccharide, oligosaccharide, and polysaccharide. [pp.40–41]
11. Describe the difference in structure between cellulose and starch. [pp.40–41]
12. For the complex carbohydrates, indicate which is associated with animals, which with plants, and which with fungi. [pp.40–41]
13. Give the characteristics of lipids and list their general functions. [p.42]
14. Distinguish a saturated fatty acid from an unsaturated fatty acid. [p.42]
15. Describe the structure of triglycerides; list examples and their functions. [p.42]
16. Describe how phospholipids differ in structure from triglycerides. [pp.42–43]
17. Give the biological function of phospholipids. [p.43]
18. Give the characteristics of sterols and describe how their structure differs from the other classes of lipids. [p.43]
19. What is the biological importance of sterols? [p.43]
20. How are waxes used by living organisms? [p.43]
21. Describe the structure of an amino acid. [p.44]
22. Describe the formation of a peptide bond. [pp.44–45]
23. Describe how the primary, secondary, tertiary, and quaternary structures of proteins result in complex three-dimensional structures. [pp.44–47]
24. Distinguish lipoproteins from glycoproteins. [p.47]
25. ________ refers to the loss of a molecule's three-dimensional shape through disruption of the weak bonds responsible for it. [p.47]
26. Using sickle-cell anemia as an example, explain how a single amino acid change can have a drastic effect on protein function. [pp.48–49]
27. List the three parts of every nucleotide. [p.50]
28. The nucleic acids ________ and ________, built of nucleotides, are the basis of inheritance and reproduction. [pp.50–51]
29. What is a coenzyme? [p.50]

Integrating and Applying Key Concepts

1. Humans can obtain energy from many different food sources. Do you think this ability is an advantage or a disadvantage in terms of long-term survival? Why?
2. If the ways in which atoms bond affect molecular shape, do the ways in which molecules behave toward one another influence the shape of organelles? Do the ways that organelles behave toward one another influence the structure and function of the cells?
3. Modern diets often focus on proteins as a primary energy source. What are the preferred energy molecules of the cell? What is the overall role of proteins? Do you think that the metabolic pathways have evolved toward long-term protein use for energy?

4

CELL STRUCTURE AND FUNCTION

Interactive Exercises

Animalcules and Cells Fill'd with Juices [pp.54–55]

4.1. BASIC ASPECTS OF CELL STRUCTURE AND FUNCTION [pp.56–57]

4.2. FOCUS ON SCIENCE: *Microscopes—Gateways to the Cell* [pp.58–59]

Selected Words: *cellulae* [p.54], "cell" [p.54], *compound light microscope* [p.58], *transmission electron microscope* [p.58], *scanning electron microscope* [p.58], *scanning tunneling microscope* [p.59]

Boldfaced, Page-Referenced Terms

[p.55] cell theory ____________________

[p.56] cell ____________________

[p.56] plasma membrane ____________________

[p.56] nucleus ____________________

[p.56] nucleoid ____________________

[p.56] cytoplasm ____________________

[p.56] ribosomes ____________________

[p.56] eukaryotic cells ____________________

[p.56] prokaryotic cells ____________________

[p.56] lipid bilayer ____________________

[p.57] surface-to-volume ratio ____________________

[p.58] micrograph ____________________

[p.58] wavelength ____________________

Matching

Choose the most appropriate answer for each term.

1. ____ prokaryotic cells [p.56]
2. ____ plasma membrane [p.56]
3. ____ cytoplasm [p.56]
4. ____ ribosomes [p.56]
5. ____ nucleus [p.56]
6. ____ eukaryotic cells [p.56]
7. ____ surface-to-volume ratio [p.57]
8. ____ lipid bilayer [p.56]
9. ____ nucleoid [p.56]
10. ____ phospholipid molecules [p.54]
11. ____ cell [p.56]

A. An interior region of prokaryotic cells where DNA is found
B. The structural arrangement of all cell membranes
C. The type of cell that lacks a nucleus
D. A physical relationship that constrains increases in cell size
E. The smallest unit of life that retains all the properties of life
F. Structures within the cytoplasm that are involved in building proteins
G. The most abundant components of cell membranes
H. The thin, outermost membrane of cells; maintains the cell as a distinct entity; substances and signals move across it
I. In eukaryotic cells, this is the membranous sac that contains the DNA
J. The area of a cell between the plasma membrane and the region of DNA
K. This type of cell possesses organelles and a nucleus

Short Answer

12. Explain the principles of the surface-to-volume ratio that limit the size of cells. [p.57] ____________________

____

____

____

Matching

Choose the most appropriate answer for each term.

13. ____ micrograph [p.58]
14. ____ transmission electron microscope [p.58]
15. ____ scanning tunneling microscope [p.59]
16. ____ wavelength [p.58]
17. ____ compound light microscope [p.58]
18. ____ phase-contrast and Nomarski processes [p.59]
19. ____ scanning electron microscope [p.58]

A. Glass lenses bend incoming light rays to form an enlarged image of a cell or some other specimen
B. The distance from one wave's peak to the peak of the wave behind it
C. A computer analyzes a tunnel formed in electron orbitals, produced between the tip of a needlelike probe and an atom
D. A narrow beam of electrons moves back and forth across the surface of a specimen coated with a thin metal layer
E. A photograph of an image formed with a microscope
F. Electrons pass through a thin section of cells to form an image
G. Create optical contrasts without staining the cells; enhance the usefulness of light micrographs

4.3. DEFINING FEATURES OF EUKARYOTIC CELLS [pp.58–61]

Selected Words: *compartmentalization* [p.58]

Boldfaced, Page-Referenced Terms

[p.58] organelle ____

____

Complete the Table

1. Complete the following table to identify eukaryotic organelles and their functions.

Organelle or Structure	*Main Function*
[p.60] a.	Modifying polypeptide chains into mature proteins; sorting and shipping proteins and lipids for secretion or use inside the cell
[p.60] b.	Localizing the cell's DNA
[p.60] c.	Producing many ATP molecules in highly efficient fashion
[p.60] d.	Overall cell shape and internal organization; moving the cell and its internal structures
[p.60] e.	Transporting or storing a variety of substances; digesting substances and structures in the cell; other functions
[p.60] f.	Assembling polypeptide chains
[p.60] g.	Routing and modifying newly formed polypeptide chains; lipid synthesis

Short Answer

2. Compare the illustrations of plant and animal cells shown in Figure 4.10 in the text. [p.60] List the organelles and structures that do not seem to be common to both plant and animal cells.

____

____

____

4.4. THE NUCLEUS [pp.62–63]

Selected Words: *chromatin* [p.63], *chromosomes* [p.63]

Boldfaced, Page-Referenced Terms

[p.62] nuclear envelope ____

____

[p.62] nucleolus (plural, nucleoli) ____

____

[p.63] chromatin ____

____

[p.63] chromosome ____

____

Short Answer

1. List the two major functions of the nucleus. [p.62]

____

____

____

Complete the Table

2. Complete this table about the eukaryotic nucleus by entering the name of each nuclear component described.

Nuclear Component	*Description*
[p.62] a.	Site where some forms of RNA and proteins are assembled into ribosomes
[p.63] b.	Two lipid bilayers that form a double-membrane barrier surrounding the nucleoplasm
[p.63] c.	The cell's collection of DNA, together with all proteins associated with it
[p.63] d.	Fluid interior portion of the nucleus
[p.62] e.	A single DNA molecule with its associated proteins, in either a threadlike or a condensed form

Fill-in-the-Blanks

Outside the nucleus, polypeptide chains for proteins are assembled on (3) ______________ [p.63]. Many new chains become stockpiled in the (4) ______________ [p.63] or get used at once. Many others enter the (5) ______________ [p.63] system. This system consists of different organelles, including (6) ______________ [p.63] reticulum, (7) ______________ [p.63] bodies, and vesicles. Thanks to (8) ______________ [p.63] instructions, many (9) ______________ [p.63] take on particular final forms in the endomembrane system. (10) ______________ [p.63] are also packaged and assembled in the system by (11) ______________ [p.63] and other proteins constructed according to DNA's instructions. (12) ______________ [p.63] deliver the proteins and lipids to specific sites within the cell or to the plasma membrane for export.

4.5. THE ENDOMEMBRANE SYSTEM [pp.64–65]

Selected Words: *rough* ER [p.64], *smooth* ER [p.64], *endocytic* vesicles [p.64], *exocytic* vesicles [p.64]

Boldfaced, Page-Referenced Terms

[p.64] endomembrane system ____

____

[p.64] endoplasmic reticulum, or ER ____

____

[p.64] Golgi bodies ____

____

[p.65] vesicles ____

____

[p.65] lysosomes ____

____

[p.65] peroxisomes ____

____

Matching

Study the following illustration and match each component of the endomembrane system with the description of its function. Some components may be used more than once.

1. ____ Incorporates proteins into ER membrane [p.64]
2. ____ Lipid assembly [p.64]
3. ____ DNA instructions for building polypeptide chains [p.64]
4. ____ This site modifies proteins after they have been assembled [p.64]
5. ____ Proteins and lipids take on final form [p.64]
6. ____ Sort and package lipids and proteins for transport to proper destinations following modification [p.64]
7. ____ Vesicles formed at plasma membrane transport substances into cytoplasm [p.64]
8. ____ Sacs of enzymes that break down fatty acids and amino acids, forming hydrogen peroxide [p.65]
9. ____ Special vesicles budding from Golgi bodies that become organelles of intracellular digestion [p.65]
10. ____ Transport unfinished proteins to a Golgi body [p.64]
11. ____ Transport finished Golgi products to the plasma membrane [p.64]
12. ____ Release Golgi products at the plasma membrane [p.64]
13. ____ Transport unfinished lipids to a Golgi body [p.64]

A. spaces within smooth membranes of ER
B. nucleus
C. Golgi body
D. vesicles from Golgi
E. vesicles budding from rough ER
F. endocytosis with vesicles
G. exocytosis with vesicles
H. spaces within rough ER
I. vesicles budding from smooth ER
J. lysosomes
K. peroxisomes

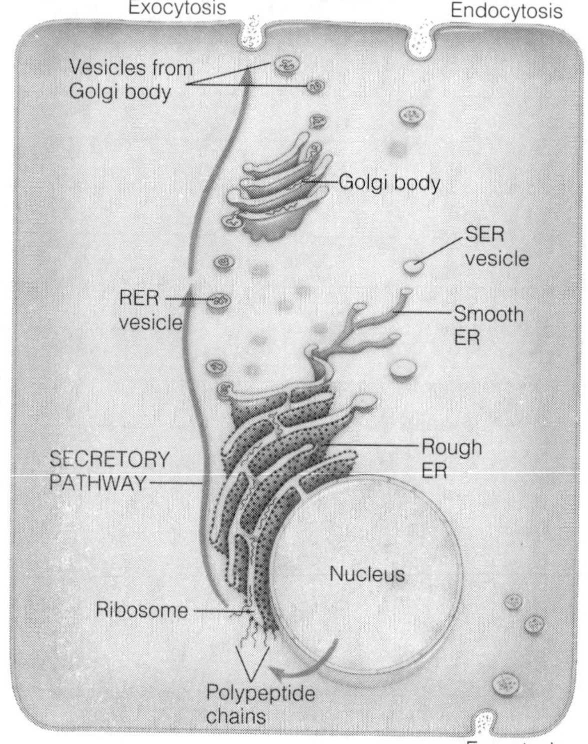

4.6. MITOCHONDRIA [p.66]

4.7. SPECIALIZED PLANT ORGANELLES [p.67]

Selected Words: *"endosymbiosis"* [p.66]

Boldfaced, Page-Referenced Terms

[p.66] mitochondrion ______________________________

____

[p.67] chloroplasts ______________________________

____

[p.67] central vacuole ______________________________

____

Choice

For questions 1–19, choose from the following:

a. mitochondria b. chloroplasts c. amyloplasts d. central vacuole e. chromoplasts

1. ____ Occur only in photosynthetic eukaryotic cells [p.67]
2. ____ ATP molecules form when organic compounds are completely broken down into carbon dioxide and water [p.66]
3. ____ Plastids that lack pigments [p.67]
4. ____ A muscle cell might have a thousand or more [p.66]
5. ____ Plastids that have an abundance of carotenoids but no chlorophylls [p.67]
6. ____ ATP-forming reactions requiring oxygen occur here [p.66]
7. ____ Causes fluid pressure to build up inside a living plant cell [p.67]
8. ____ Organelles that convert sunlight energy into the chemical energy of ATP [p.67]
9. ____ The source of the red-to-yellow colors of many flowers, autumn leaves, ripening fruits, and carrots or other roots [p.67]
10. ____ May increase so much in volume that it takes up 50 to 90 percent of the cell's interior [p.67]
11. ____ Contains internal areas known as *grana* and *stroma* [p.67]
12. ____ Plastids that resemble photosynthetic bacteria [p.67]
13. ____ Two distinct compartments are created by a double-membrane system [p.66]
14. ____ Store starch grains and are abundant in cells of stems, potato tubers, and seeds [p.67]
15. ____ May have originated by the process of endosymbiosis [p.66]
16. ____ The site of photosynthesis in plant cells [p.67]
17. ____ Fluid-filled; stores amino acids, sugars, ions, and toxic wastes [p.67]
18. ____ All eukaryotic cells have one or more [p.66]
19. ____ Contains the thylakoid membrane [p.67]

4.8. SUMMARY OF TYPICAL FEATURES OF EUKARYOTIC CELLS [pp. 68–69]

Complete the Table

1. Referring to Figure 4.18 on pages 68 and 69 of the text, summarize the function of each of the following organelles.

Organelle	*Function*
[pp.68–69] a. endoplasmic reticulum	
[p.68] b. cell wall	
[p.68] c. chloroplast	
[p.69] d. lysosome	
[pp.68–69] e. plasma membrane	
[p.69] f. Golgi body	
[p.68] g. nucleus	
[pp.68–69] h. mitochondrion	

4.9. EVEN YOUR CELLS HAVE A SKELETON [pp.70–71]

4.10. HOW DO CELLS MOVE? [pp.72–73]

4.11. CELL SURFACE SPECIALIZATIONS [pp.74–75]

Selected Words: *cytoplasmic streaming* [p.71], *motor* proteins [p.71], cross-linking proteins [p.71], *kinesins* [p.72], *dyneins* [p.72], *myosins* [p.72], "9 + 2 array" [p.73], *tight* junctions [p.73], *adhering* junctions [p.73]

Boldfaced, Page-Referenced Terms

[p.70] cytoskeleton ______________________________

____

[p.70] microtubules ______________________________

____

[p.70] microfilaments ______________________________

____

[p.70] intermediate filaments ______________________________

____

[p.71] cell cortex ______________________________

____

[p.72] motor proteins ____________________

[p.72] flagella (singular, flagellum) ____________________

[p.72] cilia (singular, cilium) ____________________

[p.72] centriole ____________________

[p.72] basal body ____________________

[p.72] pseudopods ____________________

[p.73] cell wall ____________________

[p.73] primary wall ____________________

[p.73] secondary wall ____________________

[p.74] cell junctions ____________________

[p.74] cell communication ____________________

Dichotomous Choice

Circle one of two possible answers given between parentheses in each statement.

1. (Protein/Carbohydrate) subunits form the basic components of microtubules. [p.70]
2. Microtubules and microfilaments are involved in most aspects of eukaryotic cell (movement/chemistry). [p.70]
3. Hollow microtubule cylinders consist of (tubulin/actin) protein subunits or monomers. [p.70]
4. A microtubule grows in length by adding monomers at its (minus/plus) end. [p.70]
5. Sites giving rise to microtubules are known as (MTOCs/mitochondria). [p.70]
6. Doctors have used (taxol/colchicine) to decrease the uncontrolled cell divisions underlying the growth of some forms of tumors. [pp.70–71]
7. Microfilaments consist of two polypeptide chains of (tubulin/actin) monomers helically twisted together. [p.71]
8. The fluid movement of organelles and materials within a cell is called (cytoplasmic streaming/endosymbiosis). [p.71]
9. The extensive, three-dimensional network of microfilaments and other proteins just beneath the plasma membrane is known as the *cell* (*cortex*/*MTOC*). [p.71]
10. (Intermediate filaments/Microfilaments) mechanically strengthen cells or cell parts and help maintain their shape. [p.69]

11. Kinesins, dyneins, and myosins are examples of (construction/motor) proteins. [p.72]
12. The (glucose/ATP) molecule energizes the motor proteins, allowing them to move over the surface of the microtubules and microfilaments. [p.72]
13. Sperm and many other free-living cells use (flagella/cilia) as whiplike tails for swimming. [p.73]
14. The human respiratory tract is lined with (flagella/cilia). [p.73]
15. 9 + 2 arrays of microtubules arise from a (centriole/pseudopod), one type of microtubule-organizing center. [p.73]
16. Basal bodies are derived from (centrioles/cilia). [p.73]
17. The beating action of cilia and flagella is due to the action of (intermediate filaments/microtubules). [p.73]
18. The motor protein called (dynein/actin) is responsible for the bending motion of the flagella and cilia. [p.73]
19. The irregular cellular projections of amoebas and macrophages that aid in movement are called (flagella/pseudopods). [p.73]
20. The apparent movement of a pseudopod is controlled by the action of (microtubules/microfilaments). [p.73]

Matching

Choose the most appropriate answer for each term.

21. ____ primary wall [p.74]
22. ____ secondary wall [p.74]
23. ____ plasmodesmata [p.75]
24. ____ tight junctions [p.75]
25. ____ adhering junctions [p.75]
26. ____ gap junctions [p.75]

A. Link the cells of epithelial tissues lining the body's outer surface, inner cavities, and organs
B. Numerous tiny channels that cross the adjacent primary walls of living plant cells and connect their cytoplasm
C. Joins cells that are subject to stretching
D. Link the cytoplasm of neighboring animal cells and are open channels for the rapid flow of signals and substances
E. Formed of rigid cellulose, lignin, and additional deposits; reinforces plant cell shape
F. Made from polysaccharides, these thin, pliable structures allow plant cells to respond to changing water pressure

Short Answer

27. What is the overall purpose of cell communication? [p.75] ________________

4.12. PROKARYOTIC CELLS [pp.76–77]

Selected Words: *prokaryotic* [p.76]

Fill-in-the-Blanks

The word *prokaryotic* means "before the (1) ______________" [p.76], which implies that bacteria evolved before the (2) ______________ cells. [p.76] Prokaryotic cells are typically smaller than a few (3) ______________ [p.76] in size. Most bacteria have a semirigid or rigid cell (4) ______________ [p.76] that wraps around the plasma membrane, structurally supports the cell, and imparts shape to it. Dissolved substances can move freely to and from the plasma membrane because the wall is (5) ______________ [p.76]. Often, sticky (6) ______________ [p.76] enclose the cell wall, allowing it to attach to surfaces in the cell's environment. (7) ______________ [p.76] eubacteria may also have a protective, thick, jellylike capsule around them. Prokaryotic cells have a plasma membrane that helps to (8) ______________ [p.76] the movement of substances to and from the cytoplasm. This plasma membrane has (9) ______________ [p.76] that serve as channels, transporters, and receptors for signals and substances. Prokaryotic cells also have many cytoplasmic (10) ______________ [p.76], on which polypeptide chains are assembled. The cytoplasm of a bacterium is continuous with an irregularly shaped region of (11) ______________ [p.76] named the (12) ______________ [p.76]. A circular molecule of DNA, also called the *bacterial* (13) ______________ [p.76], occupies this region. Extending from the surface of many bacterial cells are one or more threadlike motile structures known as *bacterial* (14) ______________ [p.76]. These structures help a cell move rapidly in its fluid surroundings. Other surface projections include (15) ______________ [p.76], which are the main (16) ______________ [p.76] filaments that help many kinds of bacteria attach to various surfaces, even to each other. There are two kingdoms of metabolically diverse organisms, the (17) ______________ [p.76] and the (18) ______________ [p.76]. Ancient (19) ______________ [pp.76–77] cells gave rise to all the protistans, plants, fungi, and animals ever to appear on Earth.

Self-Quiz

Labeling and Matching

First, identify each indicated part of the accompanying illustrations. Then, match and enter the letter of the proper function description in the parentheses following each label. Some letter choices must be used more than once.

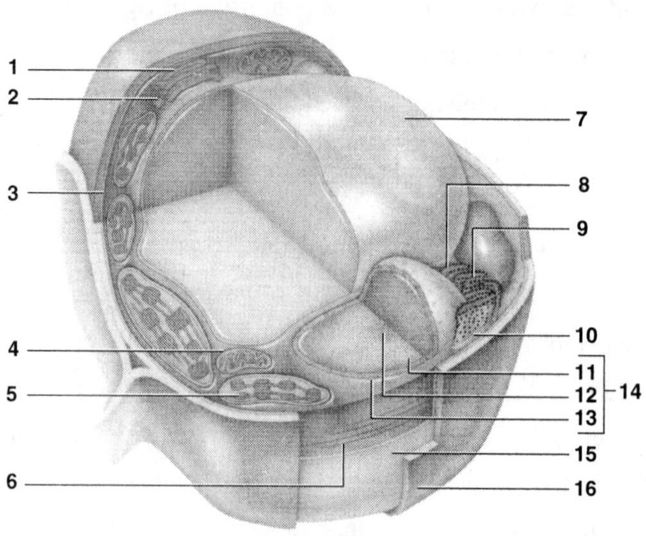

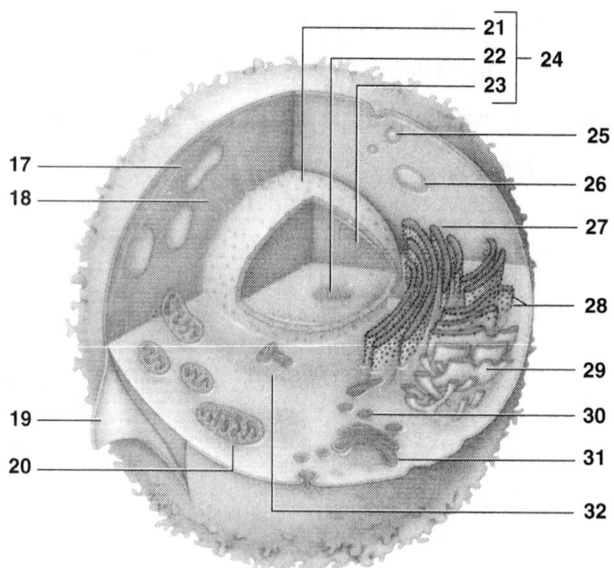

1. ______ ______ ( ) [p.64]

2. ______ ( ) [p.65]

3. ______ (cytoskeletal component) ( ) [p.71]

4. ______ ( ) [p.66]

5. ______ ( ) [p.67]

6. ______ (cytoskeletal component) ( ) [p.70]

7. ______ ______ ( ) [p.67]

8. ______ ______ ______ ( ) [p.64]

9. ______ (attached to #8) ( ) [p.64]

10. ______ ______ ______ ( ) [p.64]

11. ______ + ______ ( ) [pp.62–63]

12. ______ ( ) [p.63]

13. ______ ______ ( ) [p.62]

14. ______ ( ) [p.62]

15. ______ ______ ( ) [p.56]

16. ______ ______ ( ) [p.74]

17. ______ (cytoskeletal component) ( ) [p.71]

18. ______ (cytoskeletal component) ( ) [p.70]

19. ______ ______ ( ) [p.56]

20. ______ ( ) [p.66]

21. ______ ______ ( ) [pp.62–63]

22. ______ ( ) [p.63]

23. ______ + ______ ( ) [pp.62–63]

24. ______ ( ) [p.62]

25. ______ ( ) [p.65]

26. ______ ( ) [p.65]

27. ______ ______ ______ ( ) [p.64]

28. ______ (attached to #27) ( ) [p.64]

29. ______ ______ ______ ( ) [p.64]

30. ______ ( ) [p.65]

31. ______ ______ ( ) [p.64]

32. ______ ( ) [p.73]

A. Pore-riddled, two-membrane structure; a profusion of ribosomes is found on its outer surface
B. Porous, but provides protection and structural support for some cells
C. Present in mature, living plant cells; a fluid-filled organelle that stores amino acids, sugars, ions, and toxic wastes
D. Free of ribosomes; the main site of lipid synthesis in many cells
E. Formed as buds from Golgi membranes of animal cells and some fungal cells; an organelle of intracellular digestion
F. Barrel-shaped structures that serve as a type of microtubule-producing center
G. Tiny membranous sacs that move throughout the cytoplasm; a common type functions as an organelle of intracellular digestion
H. Take part in diverse cell movements; consist of protein subunits, called *tubulins*, which form a hollow cylinder
I. A membrane-bound compartment that houses DNA in eukaryotic cells
J. Enzymes put the finishing touches on proteins and lipids, sort them out, and package them inside vesicles for shipment to specific locations
K. Site of assembly of protein and RNA molecules into ribosomal subunits
L. Take part in diverse cell movements; consist of two chains of actin subunits, twisted together
M. Organelles that convert sunlight energy into the chemical energy of ATP, which is used to make sugars and other organic compounds
N. Thin, outermost membrane, which maintains the cell as a distinct entity; substances and signals continually move across it in highly controlled ways
O. Site of aerobic respiration; ATP-producing powerhouse of all eukaryotic cells
P. Stacks of flattened sacs with many ribosomes attached; accomplishes initial modification of protein structure after formation on ribosomes
Q. Genetic material and the fluid interior portion of the nucleus
R. Site of the synthesis of every new polypeptide chain

Multiple Choice

_____ 33. Membranes consist of _______________. [pp.56–57]
a. a lipid bilayer
b. embedded proteins
c. phospholipids
d. all of the above

_____ 34. Which of the following statements most correctly describes the relationship between cell surface area and cell volume? [p.57]
a. As a cell expands in volume, its diameter increases at a rate faster than its surface area does.
b. Volume increases with the square of the diameter, but surface area increases only with the cube.
c. If a cell were to grow four times in diameter, its volume of cytoplasm would increase 16 times and its surface area 64 times.
d. Volume increases with the cube of the diameter, but surface area increases only with the square.

_____ 35. Most cell membrane functions are carried out by _______________. [p.56]
a. carbohydrates
b. nucleic acids
c. lipids
d. proteins

_____ 36. The cellular structure in animals that is involved in the process of intracellular digestion is the _______________ [p.65]
a. nucleus
b. lysosome
c. rough endoplasmic reticulum
d. microtubules
e. mitochondria

_____ 37. The nucleolus is the site where _______________. [p.63]
a. the protein and RNA subunits of ribosomes are assembled
b. the chromatin is formed
c. chromosomes are bound to the inside of the nuclear envelope
d. chromosomes duplicate themselves prior to cell division

_____ 38. Which of the following is *not* found as a part of prokaryotic cells? [p.76]
a. ribosomes
b. DNA
c. nucleus
d. cytoplasm
e. cell wall

_____ 39. The _______________ is free of ribosomes and curves through the cytoplasm like connecting pipes; it is the main site of lipid synthesis. [p.64]
a. lysosome
b. Golgi body
c. smooth ER
d. rough ER

_____ 40. Which of the following is *not* present in all cells? [p.56]
a. cell wall
b. plasma membrane
c. ribosomes
d. DNA molecules

_____ 41. As a part of the endomembrane system, the _______________ put the finishing touches on lipids and proteins to permit sorting and packaging for specific locations. [p.64]
a. endoplasmic reticulum
b. Golgi bodies
c. peroxisomes
d. lysosomes
e. mitochondria

_____ 42. Chloroplasts _______________. [p.67]
a. are specialists in oxygen-requiring reactions
b. function as part of the cytoskeleton
c. trap sunlight energy and produce organic compounds
d. assist in carrying out cell membrane functions

_____ 43. Mitochondria convert energy stored in _______________ into forms that the cell can use, principally ATP. [p.66]
a. water
b. carbon compounds
c. sunlight
d. carbon dioxide

_____ 44. _______________ are sacs of enzymes that bud from ER; they produce potentially harmful hydrogen peroxide while breaking down fatty acids and amino acids. [p.65]
a. Lysosomes
b. Ribosomes
c. Golgi bodies
d. Peroxisomes

_____ 45. The two components of the cytoskeleton that are involved in virtually all eukaryotic cell movement are the _______________. [pp.72–73]
a. desmins and vimentins
b. centrioles and microfilaments
c. microtubules and microfilaments
d. microtubules and plasma membrane
e. intermediate filaments and nucleoplasm

_____ 46. Which of the following is not used by eukaryotic cells for locomotion? [pp.72–73]
a. cilia
b. flagella
c. pseudopods
d. gap junctions
e. all of the above are mechanisms of eukaryotic cell movement

Choice

Cells of the organisms in the six kingdoms of life share the following characteristics: plasma membrane, DNA, RNA, and ribosomes. For questions 47–60, choose the kingdom(s) possessing the characteristics listed below. Some questions will require more than one kingdom as the answer.

a. archaebacteria, eubacteria b. protistans c. fungi d. plants e. animals

47. _____ cytoskeleton [p.78]
48. _____ cell wall [p.78]
49. _____ nucleus [p.78]
50. _____ nucleolus [p.78]
51. _____ central vacuole [p.78]
52. _____ simple flagellum [p.78]
53. _____ photosynthetic pigment [p.78]
54. _____ chloroplast [p.78]
55. _____ endoplasmic reticulum [p.78]
56. _____ Golgi body [p.78]
57. _____ lysosome [p.78]
58. _____ complex flagella and cilia [p.78]
59. _____ plasma membrane [p.78]
60. _____ ribosomes [p.78]

Chapter Objectives/Review Questions

1. List the three generalizations that together constitute the cell theory. [p.55]
2. List and describe the three major regions that all cells have in common. [p.56]
3. Describe the difference between eukaryotic and prokaryotic cells in terms of the nucleus and cellular organelles. [p.56]
4. Describe the lipid bilayer arrangement for the plasma membrane. [p.56]
5. Cell size is necessarily limited because its volume increases with the ________ but surface area increases only with the ________ . [p.57]
6. Briefly describe the operating principles of light microscopes, phase-contrast microscopes, scanning tunneling microscopes, transmission electron microscopes, and scanning electron microscopes. [pp.58–59]
7. Briefly describe the cellular location and function of the organelles typical of most eukaryotic cells: nucleus, ribosomes, endoplasmic reticulum, Golgi body, various vesicles, mitochondria, and the cytoskeleton. [p.60]
8. Explain the advantage of complex compartmentalization in eukaryotic cells. [p.60]
9. Eukaryotic cells protect their ________ inside a nucleus. [p.62]
10. Describe the structure of the nuclear envelope and its function in the cell. [p.62]
11. ________ is the location where the protein and RNA subunits of ribosomes are assembled. [p.63]
12. ________ is the cell's collection of DNA molecules and associated proteins; a ________ is an individual DNA molecule and associated proteins. [p.63]
13. List the organelles that are part of the endomembrane system. [pp.64–65]
14. A protein is being manufactured that is to be exported from the cell. List, in order, the organelles of the endomembrane system that are involved in this process and their individual functions. [pp.64–65]
15. A lipid is being manufactured within a cell. List, in order, the organelles of the endomembrane system that are involved in this process and their individual functions. [pp.64–65]
16. Explain the variety of tasks that the lysosome performs for a eukaryotic cell. [p.65]
17. Define and describe the function of peroxisomes. [p.65]
18. Within ________ , energy stored in organic molecules is released by enzymes and used to form many ATP molecules in the presence of oxygen. [p.66]
19. Give the general function of the following plant organelles: chloroplasts, chromoplasts, amyloplasts, and the central vacuole. [p.67]
20. Elements of the ________ give eukaryotic cells their internal organization, overall shape, and capacity to move. [p.70]
21. List the three major structural elements of the cytoskeleton and give the general function of each. [pp.70–71]
22. ________ filaments are the most stable elements of the cytoskeleton. [p.71]
23. ________ proteins such as myosin and dynein play roles in cell movement. [p.72]
24. Explain the role that ATP has in the movement of microtubules and microfilaments. [p72]
25. Explain the three mechanisms of eukaryotic cell movement. [pp.72–73]
26. Both cilia and flagella have an internal microtubule arrangement called the "________ ________" array. [p.73]
27. A centriole remains at the base of a completed microtubule-producing center, where it is often called a ________ ________ . [p.73]
28. Explain the sliding mechanism by which flagella and cilia beat. [p.73]
29. Distinguish a primary cell wall from a secondary cell wall in leafy plants. [p.74]
30. Describe the location and function of plasmodesmata. [p.75]
31. Explain the difference in function between gap junctions, adhering junctions, and tight junctions. [p.75]
32. Why is cell communication important in complex organisms? [p.75]
33. Describe the structure of a generalized prokaryotic cell. Include the bacterial flagellum, nucleoid, pili, capsule, cell wall, plasma membrane, cytoplasm, and ribosomes. [p.76]

Integrating and Applying Key Concepts

1. Which parts of a cell constitute the minimum necessary for keeping the simplest of living cells alive?
2. How did the existence of a nucleus, compartments, and extensive internal membranes confer selective advantages on cells that developed these features?
3. Cells are frequently described as biological factories. Compare a typical eukaryotic cell to a modern automobile factory. Describe the function of the nucleus, endoplasmic reticulum, mitochondria, Golgi body, ribosomes, vesicles, lysosomes, and peroxisomes in relation to a factory.

5

A CLOSER LOOK AT CELL MEMBRANES

Interactive Exercises

One Bad Transporter and Cystic Fibrosis [pp.80–81]

5.1. MEMBRANE STRUCTURE AND FUNCTION [pp.82–83]

5.2. A GALLERY OF MEMBRANE PROTEINS [pp.84–85]

5.3. FOCUS ON SCIENCE: *Do Membrane Proteins Stay Put?* [p.86]

Selected Words: *cystic fibrosis* [p.81], *sinusitis* [p.81], *mosaic* [p.82], *fluid* [p.82], *integral* proteins [p.84], *peripheral* proteins [p.84], *nonself* and *self* [p.85], *passive transporters* [p.85], *ion-selective transporters* [p.85], *P type* ATPases [p.85], *V type* ATPases [p.85], *F type* ATPases [p.85]

Boldfaced, Page-Referenced Terms

[p.80] biofilms ____________________

____

[p.81] ABC transporters ____________________

____

[p.82] phospholipid ____________________

____

[p.82] lipid bilayer ____

____

[p.82] fluid mosaic model ____

____

[p.84] adhesion proteins ____

____

[p.84] communication proteins ____

____

[p.84] receptor proteins ____

____

[p.84] recognition proteins ____

____

[p.84] transport proteins ____

____

Matching

Choose the most appropriate answer for each term.

1. ____ fluid mosaic model [p.82]
2. ____ phospholipid [p.82]
3. ____ adhesion proteins [p.84]
4. ____ transport proteins [p.84]
5. ____ communication proteins [p.84]
6. ____ integral proteins [p. 84]
7. ____ recognition proteins [p.84]
8. ____ peripheral proteins [p.84]
9. ____ receptor proteins [p.84]
10. ____ lipid bilayer [p.82]
11. ____ passive transporters [p.85]
12. ____ ion-selective transporters [p.85]
13. ____ ATPases [p.85]

A. These let signals travel rapidly between two adjoining cells
B. These specific channels allow molecules to move through without expending energy
C. Use the energy of adenosine triphosphate to actively transport molecules across the membrane
D. A composition of phospholipids, proteins, sterols, and glycolipids
E. The general name for proteins that are physically embedded within the cell membrane
F. The primary component of the cell membrane; consists of both hydrophobic and hydrophilic regions
G. These bind extracellular substances that trigger changes in the cell's activity
H. These help cells of the same type stick together
I. This general group of proteins are positioned at the surface of the membrane
J. Contain molecular gates that move small molecules
K. The double layer of phospholipids that forms the cell membrane
L. Allow materials to pass through the cell membrane using the interior of the protein
M. Act as molecular fingerprints to identify tissues or individuals

Short-Answer

14. What is the cause of cystic fibrosis? [pp.80–81] ______

15. What is a biofilm? [p.80] ______

5.4. THINK DIFFUSION [pp.86–87]

5.5. TYPES OF CROSSING MECHANISMS [p.88]

5.6. HOW DO THE TRANSPORTERS WORK? [pp.88–89]

Selected Words: *gradient* [p.86], *dynamic equilibrium* [p.87], *passive transport* [p.88], *active transport* [p.88], "facilitated" diffusion [p.88], *exocytosis* [p.88], *endocytosis* [p.88], *net* direction [p.89]

Boldfaced, Page-Referenced Terms

[p.86] selective permeability ______

[p.86] concentration gradient ______

[p.86] diffusion ______

[p.87] electric gradient ______

[p.87] pressure gradient ______

[p.88] passive transport ______

[p.89] active transport ______

[p. 89] calcium pump ______

[p. 89] sodium–potassium pump ______

Fill-in-the-Blanks

If a membrane has selective (1) ______________ [p.86], it possesses a molecular structure that permits some substances but not others to cross it in certain ways, at certain times. If the concentration of a substance in one region differs from that in an adjoining region, it is called a (2) ______________ [p.86]. A (3) ______________ ______________ [p.86] is a difference between the number of molecules or ions of a given substance in adjoining regions. (4) ______________ [p.86] is the name for the net movement of like molecules or ions down a concentration gradient; it is a factor in the movement of substances across cell membranes and through cytoplasmic fluid. Diffusion is faster when a gradient is (5) ______________ [p.87]. A net distribution of molecules that is nearly uniform through two adjoining regions is called "dynamic (6) ______________ " [p.87]. In addition, the rates of diffusion are faster at (7) ______________ temperatures [p.87]. Molecular (8) ______________ [p.87] also affects diffusion rates. The rate and direction of diffusion may also fall under the influence of a(n) (9) ______________ [p.87] gradient, a difference between electric charges in adjoining regions. The presence of a(n) (10) ______________ [p.87] gradient may likewise affect the rate and direction of diffusion.

The (11) ______________ [p.88] transporters permit a substance to follow its concentration gradient across a membrane. This process is also sometimes called (12) ______________ [p.88] diffusion. The ATPase pumps engage in (13) ______________ transport [p.88], with the net direction of movement being (14) ______________ [p.88] the concentration gradient. Unlike passive transport, active transport requires an input of (15) ______________ [p.88] to counter the concentration gradient. In the bulk movement of substances across a membrane, the process of (16) ______________ [p.88] moves particles into the cell by forming a vesicle from the plasma membrane. In (17) ______________ [p.88], a membrane-bound vesicle inside the cell fuses with the plasma membrane, allowing particles to exit the cell.

Choice

For questions 18–29, choose from the following mechanisms of protein-mediated transport:

a. passive transport b. active transport c. applies to both active and passive transport

18. _____ The calcium pump [p.89]
19. _____ The glucose transporter [p.88]
20. _____ The transport protein must receive an energy boost, usually from ATP [p.89]
21. _____ A carrier protein has a specific site that weakly binds a substance [pp.88–89]
22. _____ Solute binding to a carrier protein leads to changes in protein shape [pp.88–89]
23. _____ The sodium–potassium pump [p.89]
24. _____ Transport proteins span the bilayer, and their interior is able to open on both sides of it [pp.88–89]
25. _____ A solute is pumped across the cell membrane *against* its concentration gradient [p.89]
26. _____ Part of the transport protein closes in behind the bound solute—and part opens up to the opposite side of the membrane [pp.88–89]

27. ______ Involves a transporter protein that is not energized [p.88]

28. ______ Net movement will be down the solute's concentration gradient [p.88]

29. ______ During a given interval, the *net* direction of movement depends on how many solute molecules make random contact with vacant binding sites in the interior of proteins [p.88]

Complete the Table

30. Complete the table below by entering the name of the correct membrane transport mechanism that will move the substance listed across a membrane. Choose from the following: *diffusion, osmosis, facilitated diffusion, passive transport, active transport, endocytosis, and exocytosis.*

Substance	*Transport Mechanism*
a. H_2O [p.87]	
b. CO_2 [p.87]	
c. Na^+ [p.89]	
d. Glucose [p.88]	
e. K^+ [p.89]	
f. O_2 [p.87]	
g. substances moved through the interior of a protein; requires energy [p.89]	
h. substances moved through the interior of a protein; no energy required [p.88]	
i. bulk movement of substances into a cell using a vesicle [p.88]	
j. bulk movement of particles out of the cell [p.88]	

5.7. WHICH WAY WILL WATER MOVE? [pp.90–91]

Selected Words: *tonicity* [p.91], *turgor* pressure [p.91]

Boldfaced, Page-Referenced Terms

[p.90] bulk flow ______________________________

____

[p.90] osmosis ______________________________

____

[p.91] hypotonic solution ______________________________

____

[p.91] hypertonic solution ____

____

[p.91] isotonic solution ____

____

[p.91] hydrostatic pressure ____

____

[p.91] osmotic pressure ____

____

Matching

Choose the most appropriate answer for each.

1. ____ bulk flow [p.90]
2. ____ osmosis [p.90]
3. ____ tonicity [p.91]
4. ____ hypotonic solution [p.91]
5. ____ hypertonic solution [p.91]
6. ____ isotonic solutions [p.91]
7. ____ hydrostatic pressure [p.91]
8. ____ osmotic pressure [p.91]
9. ____ turgor pressure [p.91]
10. ____ plasmolysis [p.91]

A. Refers to the relative solute concentrations of the fluids
B. Having the same solute concentrations
C. Mass movement of one or more substances in response to pressure, gravity, or other external force
D. The amount of force that prevents further increase in a solution's volume
E. The fluid on one side of a membrane that contains more solutes than the fluid on the other side of the membrane
F. The diffusion of water in response to a water concentration gradient between two regions separated by a selectively permeable membrane
G. Osmotically induced shrinkage of cytoplasm
H. The term for hydrostatic pressure in plants
I. The fluid on one side of a membrane that contains fewer solutes than the fluid on the other side of the membrane
J. The general term for a fluid force exerted against a cell wall and/or membrane enclosing the fluid

True/False

If the statement is true, write a "T" in the blank. If the statement is false, make it correct by changing the underlined word(s) and writing the correct word(s) in the answer blank.

____________ 11. Because membranes exhibit selective permeability, concentrations of dissolved substances can <u>increase</u> on one side of the membrane or the other. [p.90]

____________ 12. A water concentration gradient is influenced by the number of <u>solute</u> molecules present on both sides of the membrane. [p.88]

____________ 13. The relative concentrations of solutes in two fluids are referred to as <u>turgor pressure</u>. [pp.90–91]

____________ 14. An animal cell placed in a <u>hypertonic</u> solution will swell and perhaps burst. [p.91]

____________ 15. Water tends to move from <u>hypotonic</u> solutions to areas with more solutes. [p.91]

____________ 16. Physiological saline is 0.9 percent NaCl; red blood cells placed in such a solution will not gain or lose water; therefore, one could state that the fluid in red blood cells is <u>hypertonic</u>. [p.91]

_______ 17. A solution of 65 percent water, 35 percent solute is more concentrated than a solution of 70 percent water, 30 percent solute. [p.91]

_______ 18. The mass movement of one or more substances in response to pressure, gravity, or some other external force is called osmosis [p.90]

_______ 19. Plant cells placed in a hypotonic solution will swell. [p.90]

5.8. MEMBRANE TRAFFIC TO AND FROM THE CELL SURFACE [pp.92–93]

Selected Words: *receptor-mediated* endocytosis [p.92], *bulk-phase* endocytosis [p.92]

Boldfaced, Page-Referenced Terms

[p.92] exocytosis _______

[p.92] endocytosis _______

[p.92] phagocytosis _______

Matching

Choose the most appropriate answer for each.

1. ___ exocytosis [p.92]
2. ___ receptor-mediated endocytosis [p.92]
3. ___ bulk-phase endocytosis [p.92]
4. ___ phagocytosis [p.92]
5. ___ membrane cycling [p.93]

A. A cell engulfs microorganisms, large edible particles, and cellular debris
B. Membrane initially used for endocytic vesicles returns receptor proteins and lipids back to the plasma membrane
C. Vesicles form around small volumes of extracellular fluid of various content
D. A cytoplasmic vesicle moves to the cell surface; its own membrane fuses with the plasma membrane while its contents are released to the environment
E. Chemical recognition and binding of specific substances; pits of clathrin baskets sink into the cytoplasm and close on themselves

Self-Quiz

___ 1. White blood cells use _______ to devour disease agents invading your body. [p.92]
a. diffusion
b. bulk flow
c. osmosis
d. phagocytosis

___ 2. _______ proteins bind extracellular substances, such as hormones, that trigger changes in cell activities. [p.83]
a. Receptor
b. Adhesion
c. Transport
d. Recognition

_____ 3. In a lipid bilayer, the phospholipid tails point inward and form a(n) _______________ region that excludes water. [p.82]
a. acidic
b. basic
c. hydrophilic
d. hydrophobic

_____ 4. A protistan adapted to life in a freshwater pond is collected in a bottle and transferred to a saltwater bay. Which of the following is likely to happen? [p.91]
a. The cell bursts.
b. Salts flow out of the protistan cell.
c. The cell shrinks.
d. Enzymes flow out of the protistan cell.
e. Nothing; the cell would be isotonic with the external environment.

_____ 5. Which of the following is not an example of an active transport mechanism? [pp.88–89]
a. calcium pump
b. glucose transporter
c. sodium–potassium pump
d. all of the above are examples of active transport

_____ 6. Which of the following is *not* a form of passive transport? [p.92]
a. osmosis
b. passive transport
c. bulk flow
d. exocytosis

_____ 7. O_2, CO_2, H_2O, and other small, electrically neutral molecules move across the cell membrane by _______________. [p.87]
a. electric gradients
b. receptor-mediated endocytosis
c. passive transport
d. active transport

_____ 8. Ions such as H^+, Na^+, K^+, and Ca^{++} move across cell membranes by _______________. [p.89]
a. receptor-mediated endocytosis
b. pressure gradients
c. passive transport
d. active transport

_____ 9. Receptors, pits, and clathrin baskets participate in _______________. [p.92]
a. passive transport
b. receptor-mediated endocytosis
c. bulk-phase endocytosis
d. active transport

_____ 10. The fluid mosaic model is used to describe _______________. [pp.82–83]
a. the process by which particles are exported from the cell
b. the structure of the cell membrane
c. the movement of water across a membrane
d. the action of transport proteins

_____ 11. A cell is placed in a beaker containing a solution of 40 percent NaCl and 60 percent water. After a few minutes you notice that the cytoplasm of the cell is shrinking in size. The cell is _______________ in relation to the contents of the beaker. [pp.90–91]
a. isotonic
b. hypertonic
c. hypotonic
d. saturated

_____ 12. Which of the following types of membrane proteins would be used as a molecular identification tag? [pp.84–85]
a. recognition proteins
b. transport proteins
c. adhesion proteins
d. receptor proteins

Chapter Objectives/Review Questions

1. ________ molecules are the most abundant component of cell membranes. [p.82]
2. Describe what is meant by the term *fluid mosaic model.* [p.82]
3. Explain why the structure of a phospholipid results in the formation of a lipid bilayer. [p.82]
4. Explain the difference between peripheral and integral proteins. [p.84]
5. State the general functions of transport proteins, receptor proteins, recognition proteins, and adhesion proteins. [p.84]
6. Explain the concept of selective permeability as it applies to cell membrane function. [p.86]
7. The difference in the number of molecules or ions of a given substance in two adjoining regions is known as ________ ________. [p.86]
8. The net movement of like molecules or ions down a concentration gradient is called ________. [p.86]
9. List the factors that can have an influence on the rate of diffusion. [pp.86–87]
10. Generally distinguish passive transport from active transport. [p.88]
11. ________ involves fusion between the plasma membrane and a membrane-bound vesicle that formed inside the cytoplasm; ________ involves an inward sinking of a small patch of plasma membrane, which then seals back on itself to form a cytoplasmic vesicle. [p.88]
12. Describe the mechanism and result of solute binding to a transport protein. [p.88]
13. ________ transport is the name for a flow of solutes through the interior of transport proteins, down their concentration gradients. [p.86]
14. In passive transport, ________ movement will be down the concentration gradient (from higher to lower) until the concentrations are the ________ on both sides of the membrane. [p.88]
15. Explain why the glucose transporter is an example of passive transport. [p.88]
16. Describe how ATP is used in active transport and how it improves the passage of solutes. [p.89]
17. The calcium and sodium–potassium pumps are examples of ________ transport, wherein net solute movement is against the concentration gradient. [p.89]
18. Define and cite an example of *bulk flow.* [p.90]
19. The diffusion of water molecules in response to water concentration gradients between two regions separated by a selectively permeable membrane is known as ________. [p.90]
20. Define *tonicity.* [pp.90–91]
21. Water tends to move from a(n) ________ solution (less solutes) to a(n) ________ solution (more solutes). [p.91]
22. Water shows no net osmotic movement in a(n) ________ solution. [p.91]
23. A fluid force directed against a wall or membrane is ________ pressure. [p.89]
24. What is turgor pressure? [p.91]
25. Define *osmotic pressure.* [p.91]
26. Describe mechanisms involved in receptor-mediated endocytosis. [p.92]
27. How does bulk-phase endocytosis differ from phagocytosis? [p.92]
28. Explain the role of endocytosis and exocytosis in membrane recycling; cite an example. [p.93]

Integrating and Applying Key Concepts

1. If there were no such thing as active transport, how would the lives of organisms be affected?

6

GROUND RULES OF METABOLISM

Interactive Exercises

Growing Old with Molecular Mayhem [pp.96–97]

6.1. ENERGY AND THE UNDERLYING ORGANIZATION OF LIFE [pp.98–99]

Selected Words: *antioxidants* [p.96], superoxide dismutase [p.96], catalase [p.96], hydrogen peroxide [p.96], *thermal* energy [p.98], *chemical* work [p.98], *mechanical* work [p.98], *electrochemical* work [p.98]

Boldfaced, Page-Referenced Terms

[p.96] free radical ____

____

[p.97] metabolism ____

____

[p.98] potential energy ____

____

[p.98] kinetic energy ____

____

[p.98] heat ____

____

[p.98] chemical energy ____

____

[p.98] first law of thermodynamics ____

____

[p.99] entropy ____

____

[p.99] second law of thermodynamics ____

____

Fill-in-the-Blanks

Living cells must get (1) ______________ [p.97] from their environment so that they can perform various kinds of tasks (referred to as (2) ______________ [p.97] by scientists). Cells need to build specific kinds of molecules and store them until they need to be used, at which time the cells may need to rearrange the molecules or break them apart; these activities are a kind of chemical (2) known as (3) ______________ [p.97]. (4) ______________ [p.97], the main energy carrier in cells, couples energy-releasing reactions with energy-requiring reactions. When you rise from sitting in a chair, some of the potential energy stored in food molecules is transformed into (5) ______________ ______________ [p.98]: the energy of motion involved in the doing of work. One hundred percent of the potential energy cannot be transformed into (5) because some will be released as (6) ______________ [p.98] during the conversion.

Choice

In the blank preceding each item, indicate whether the first law of thermodynamics (I) or the second law of thermodynamics (II) is more applicable.

7. ____ Apple trees absorbing energy from the sun and storing it in the chemical bonds of starch and sugar [p.98]
8. ____ A hydroelectric plant at a waterfall, producing electricity [p.98]
9. ____ Egyptian pyramids crumbling slowly to dust over long periods of time [p.99]
10. ____ The glow of an incandescent bulb following the flow of electrons through a wire [p.98]
11. ____ Earth's sun is continuously losing energy to its surroundings [p.99]
12. ____ The movement of a gasoline-powered automobile [logic; p.98]
13. ____ Humans running the 100-meter dash following usual food intake [p.98]
14. ____ The death and decay of an organism [p.99]

Choice

For questions 15–19, choose from these possibilities:

a. chemical energy b. entropy c. heat d. kilocalorie e. metabolism

15. _____ A measure of the amount of disorder in a system [p.99]

16. _____ Includes all of the activities by which a cell acquires energy and materials and uses them to build, break apart, store, and release substances in controlled processes that are typical for that cell [p.97]

17. _____ The measure of energy required to heat 1,000 grams of water from 14.5°C to 15.5°C at standard pressure [p.97]

18. _____ The potential energy stored in the attractive forces (bonds) that cause atoms to group together into molecules [p.98]

19. _____ Results from collisions among molecules and their surroundings; a kind of kinetic energy also called *thermal energy* [p.98]

True/False

If the statement is true, write a "T" in the blank. If the statement is false, make it correct by changing the underlined word(s) and writing the correct word(s) in the answer blank.

20. ______________ The first law of thermodynamics states that entropy is constantly increasing in the universe. [p.98]

21. ______________ At rest your body steadily gives off heat equal to that from a 100-watt light bulb. [p.98]

22. ______________ When you eat a potato, some of the stored chemical energy of the food is converted into kinetic energy that moves your muscles. [p.98]

23. ______________ Energy is the capacity to accomplish work. [p.98]

24. ______________ The amount of low-quality energy in the universe is decreasing. [p.99]

25. ______________ No energy conversion can ever be 100 percent efficient. [p.99]

Fill-in-the-Blanks

During metabolism, an electron can be added to O_2 to form a (26) ______________ ______________ [p.96]. (26) are highly reactive and can (27) ______________ [p.96] the structure and function of molecules. The enzyme (28) ______________ ______________ [p.96] can combine (26) with hydrogen ions to form O_2 and (29) ______________ [p.96]. A second enzyme, (30) ______________ [p.96], splits (29) into O_2 and water. These two enzymes, along with vitamins C and E, are called (31) ______________ [p.96]. As we age, lower amounts of (31) are produced, and (32) ______________ ______________ [p.97] seen on the skin are evidence of (26) attack.

6.2. ENERGY INPUTS, OUTPUTS, AND CELLULAR WORK [pp.100–101]

6.3. CELLS JUGGLE SUBSTANCES AS WELL AS ENERGY [pp.102–103]

6.4. ELECTRON TRANSFER CHAINS IN THE MAIN METABOLIC PATHWAYS [p.104]

Selected Words: energy inputs [p.98], *endergonic* [p.100], *exergonic* [p.100], oxidation [p.101], reduction [p.101], *biosynthetic pathway* [p.102], *degradative pathway* [p.102], enzymes [p.102], reversible [p.102]

Boldfaced, Page-Referenced Terms

[p.100] ATP, adenosine triphosphate ______________________________

[p.101] phosphorylation ______________________________

[p.101] ADP, adenosine diphosphate ______________________________

[p.101] ATP/ADP cycle ______________________________

[p.102] reactants ______________________________

[p.102] intermediates ______________________________

[p.102] products ______________________________

[p.102] energy carriers ______________________________

[p.102] cofactors ______________________________

[p.102] transport proteins ______________________________

[p.102] metabolic pathways ______________________________

[p.103] chemical equilibrium ______________________________

[p.103] law of conservation of mass ______________________________

[p.104] electron transport system ______________________________

Labeling

Classify each of the following reactions as *endergonic* or *exergonic.* [p.100]

_______________ 1. Burning wood at a campfire

_______________ 2. The products of a chemical reaction have more energy than the reactants

_______________ 3. Glucose + oxygen → carbon dioxide + water plus energy

_______________ 4. The reactants of a chemical reaction have more energy than the products

_______________ 5. The reaction releases energy

Fill-in-the-Blanks

Photosynthetic cells possess the chemical mechanisms to convert light energy into the chemical energy of a molecule known as (6) _______________ [p.101], a type of nucleotide. It is important to recognize that all cells lack the ability to extract energy *directly* from food molecules such as glucose. Cells harness the energy by causing the release of (7) _______________ [p.101] from these complex molecules. A(n) (6) molecule consists of (8) _______________ [p.100] (a nitrogen-containing compound), the five-carbon sugar (9) _______________ [p.100], and a string of three (10) _______________ [p.100] groups. These components are held together by (11) _______________ [p.100] bonds, some of which are rather unstable. Many different enzymes carry out reactions to release the outer (12) _______________ [p.100] group. This releases (13) _______________ [p.101] that can be used by cells to carry out many types of biological tasks. ATP molecules serve as the cell's renewable (14) _______________ [p.101] carrier. In the ATP/ADP cycle, an energy input drives the binding of ADP to a phosphate group or to unbound phosphate (P_i), forming (15) _______________ [p.101]. Attaching a phosphate group to a molecule is called (16) _______________ [p.101]; this energizes that molecule and prepares it to enter a specific reaction in a metabolic pathway.

Labeling

Identify each of the following molecules and label its parts.

17. _______________ [p.101]

18. _______________ [p.101]

19. _______________ [p.101]

20. _______________ The name of this molecule is _______________ _______________. [p.101]

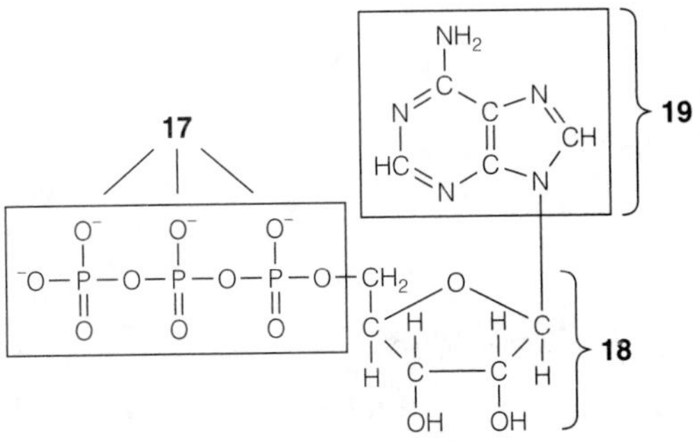

17
19
18
NH_2
N
C
CH
HC
O^-
O
P
CH_2
H
OH

Fill-in-the-Blanks

Oxidation–reduction refers to (21) ______________ [p.101] transfers. In terms of oxidation–reduction reactions, a molecule in the sequence that donates electrons is said to be (22) ______________ [p.101], while molecules accepting electrons are said to be (23) ______________ [p.101]. Electron transfer chains "intercept" excited electrons and make use of the (24) ______________ [p.101] they release. If we think of the electron transport system as a staircase, excited electrons at the top of the staircase have the (25) [choose one] □ most □ least energy [p.104]. As the electrons are transferred from one electron carrier to another, some (26) ______________ [p.104] can be harnessed to do biological (27) ______________ [p.104]. One type of biological (27) occurs when energy released during electron transfers (down the steps) is used to move ions in ways that set up ion concentration and electric gradients across a membrane. These gradients are essential for the formation of (28) ______________ [p.104].

Matching

Match the lettered statements to the numbered items on the sketch. [All from p.104]

29. _____
30. _____
31. _____
32. _____
33. _____

A. Represent the cytochrome molecules in an electron transport system
B. Electrons at their highest energy level
C. Released energy harnessed and used to produce ATP
D. Electrons at their lowest level
E. The separation of hydrogen atoms into protons and electrons

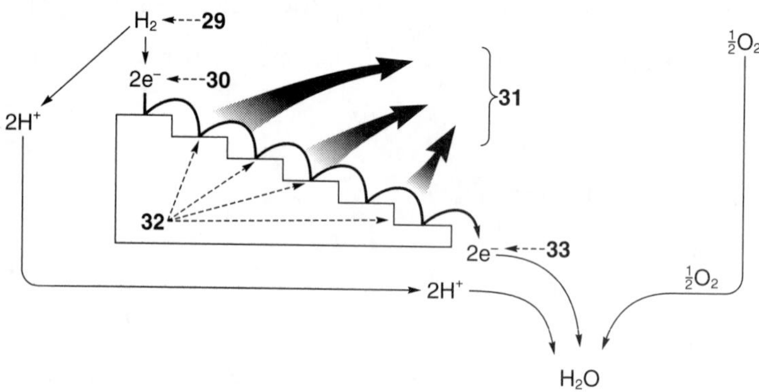

Fill-in-the-Blanks

The substances present at the end of a chemical reaction, the (34) ______________ [p.102], may have less or more energy than did the starting substances, the (35) ______________ [p.102]. Most reactions are (36) ______________ [p.102], in that they can proceed in both forward and reverse directions. Such reactions tend to approach chemical (37) ______________ [p.103], a state in which the reactions are proceeding at about the same (38) ______________ [p.103] in both directions.

Matching

Study the sequence of reactions below. Identify the components of the reactions by selecting items from the following list and entering the correct letter in the appropriate blank.

39. ____ [p.102]

40. ____ [p.102]

41. ____ [p.102]

42. ____ [p.102]

43. ____ [p.102]

44. ____ [p.102]

A. cofactor
B. intermediates
C. reactants
D. product
E. reversible reaction
F. enzymes

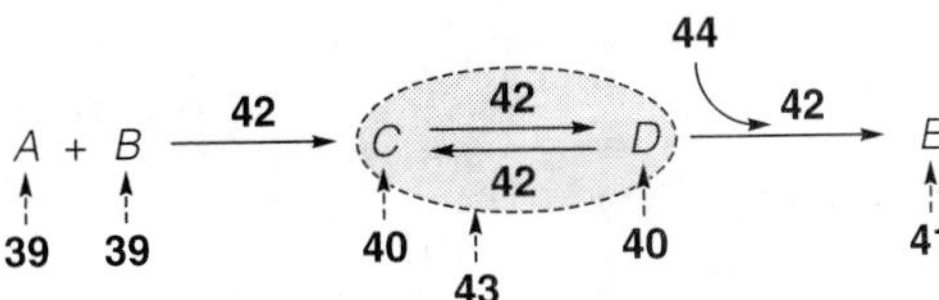

Labeling

Study the diagram at the right; choose the most appropriate answer for each: A or B.

45. ____ A degradative reaction [p.102]

46. ____ An endergonic reaction [p.100]

47. ____ A biosynthetic reaction [p.102]

48. ____ An exergonic reaction [p.100]

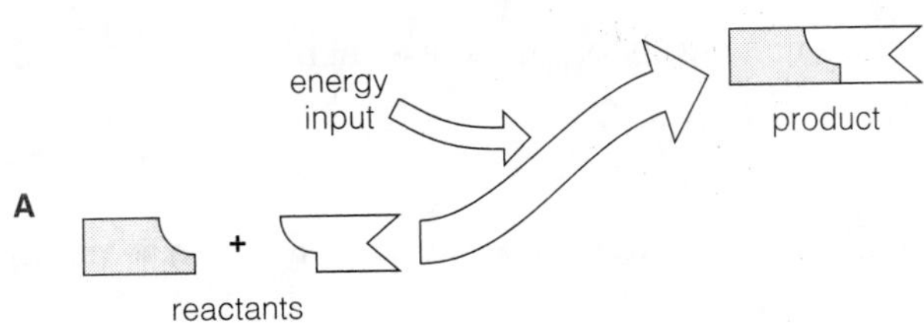

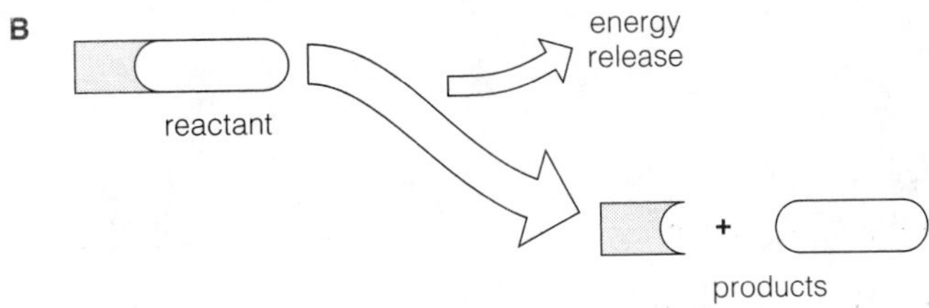

Matching

Match the most appropriate letter to its number.

49. ____ intermediates [p.102]
50. ____ degradative pathways [p.102]
51. ____ chemical equilibrium [p.103]
52. ____ cofactors [p.102]
53. ____ transport proteins [p.102]
54. ____ metabolic pathway [p.102]
55. ____ reactants (substrates) [p.102]
56. ____ energy carriers [p.102]
57. ____ biosynthetic pathways [p.102]
58. ____ enzymes [p.102]

A. An orderly series of reactions catalyzed by enzymes
B. Small organic molecules are assembled into larger organic molecules
C. Mainly ATP; donate(s) energy to reactions
D. Small molecules and metal ions that assist enzymes or serve as carriers
E. Substances able to enter into a reaction
F. Compounds formed between the beginning and the end of a metabolic pathway
G. Organic compounds are broken down in stepwise reactions
H. Proteins (usually) that catalyze reactions
I. Rate of forward reaction = rate of reverse reaction
J. Membrane-bound substances that adjust concentration gradients in ways that influence the direction of metabolic reactions

Fill-in-the-Blanks

In cells, the release of energy from glucose proceeds in controlled steps of a degradative pathway, so that (59) ______________ [p.103] molecules form along the route from glucose to carbon dioxide and water. Each step of the pathway is controlled and helped by a(n) (60) ______________ [p.102], which is usually a protein that speeds up a specific reaction. At each step in the pathway, only some of the bond energy is released. In the internal membrane systems of chloroplasts and mitochondria, the liberated electrons released from the breaking of chemical bonds are sent through (61) ______________ ______________ [p.104] systems; these organized systems consist of enzymes and (62) ______________ [p.104], bound in a cell membrane, that transfer electrons in a highly organized sequence.

Identification

Name the types of pathways in the following diagram of chemical reaction sequences. [p.102]

63. ______________ pathway
64. ______________ pathway
65. ______________ pathway

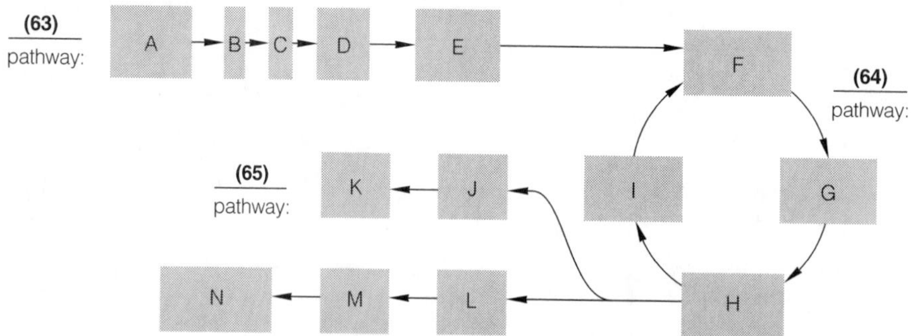

6.5. ENZYMES HELP WITH ENERGY HILLS [p.105]

6.6. HOW DO ENZYMES LOWER ENERGY HILLS? [pp.106–107]

6.7. ENZYMES DON'T WORK IN A VACUUM [pp.108–109]

6.8. FOCUS ON HEALTH: *BEER, ENZYMES, AND YOUR LIVER* [p. 109]

6.9. CONNECTIONS: *LIGHT UP THE NIGHT—AND THE LAB—WITH ENZYMES* [pp. 110–111]

Selected Words: *prosthetic* group [p.106], enzyme–substrate complex [p.106], *allosteric* control [p.108], *alcoholic hepatitis* [p.109], *alcoholic cirrhosis* [p.109], luciferase [p.110], *fluorescent* light [p.110]

Boldfaced, Page-Referenced Terms

[p.105] enzymes ____

____

[p.105] activation energy ____

____

[p.106] active sites ____

____

[p.106] substrates ____

____

[p.106] transition state ____

____

[p.106] binding energy ____

____

[p.107] induced-fit model ____

____

[p.108] feedback inhibition ____

____

[p.110] bioluminescence ____

____

Fill-in-the-Blanks

At each step in a metabolic pathway, a specific (1) ________________ [p.105] lowers the activation energy for the formation of an intermediate compound. (2) ________________ [p.105] are highly selective proteins that act as catalysts, which means that they greatly enhance the rate at which specific reactions approach (3) ________________ [p.105]. The specific substance upon which a particular enzyme acts is called its (4) ________________ [p.106]; this substance fits into the enzyme's crevice, which is called its (5) ________________ ________________ [p.106]. The (6) ________________ - ________________ [p.107] model describes how a substrate contacts the site without a perfect fit. Enzymes increase reaction rates by lowering the required (7) ________________ ________________ [p.105]. Sometimes (7) is lowered because the reactant molecule(s) are oriented by the enzyme into positions that put their mutually attractive chemical groups on precise (8) ________________ [p.106] courses much more frequently than if the enzyme weren't there. In other situations, (9) ________________ [p.106] molecules are squeezed out of the active site, lowering the activation energy. Electron and hydrogen transfers are usually mediated by (10) ________________ [p.107], which are also important to enzyme function. Enzymes change the (11) ________________ [p.105], not the outcome, of a chemical reaction.

Identification

Below is an "energy hill" reaction diagram. Identify the different amounts of energy in nos. 12, 13, and 16. Identify the two reaction pathways in nos. 14 and 15. [p.105]

12. ________________ ________________ ________________ ________________

13. ________________ ________________ ________________ ________________

14. ________________

15. ________________

16. ________________ ________________ ________________ ________________ ________________

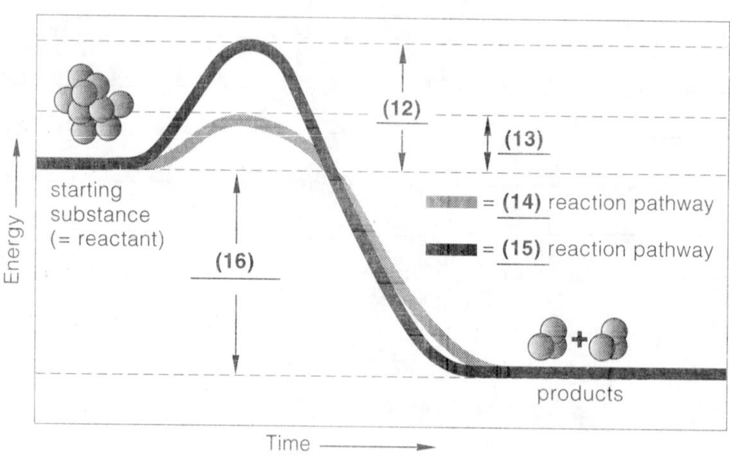

Fill-in-the-Blanks

In the graph (below left), maximum enzyme activity occurs at (17) ______________ [p.109] °C, and minimum activity occurs at (18) ______________ [p.109] °C. In the graph (below right), enzyme (19) ______________ [p.109] functions best in basic solutions, enzyme (20) ______________ [p.109] functions best in neutral solutions, and enzyme (21) ______________ [p.109] functions best in acidic solutions.

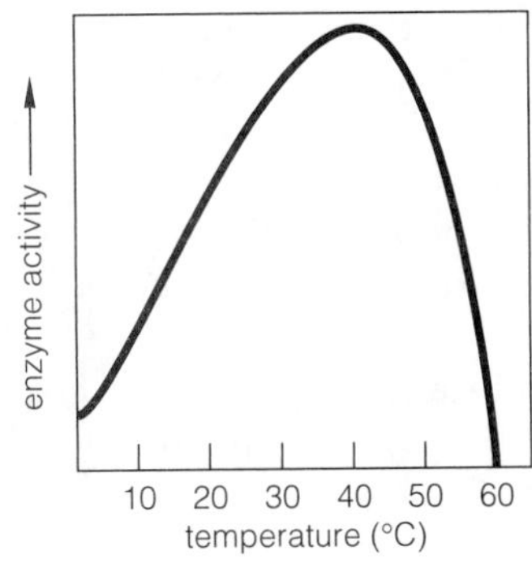

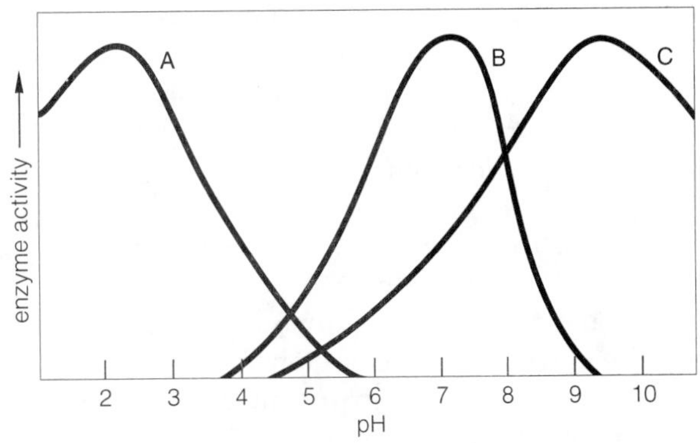

(22) ______________ [p.108] and (23) ______________ [p.108] are two factors that influence the rates of enzyme activity. Extremely high fevers can destroy the three-dimensional shape of an enzyme, which may adversely affect (24) ______________ [p.108] and cause death. The increased heat energy disrupts weak (25) ______________ [p.108] holding the enzyme in its three-dimensional shape. Most enzymes function best at about pH (26) ______________ [p.109].

One example of a control governing enzyme activity is the (27) ______________ [p.108] pathway in bacteria. The bacterium synthesizes tryptophan and other (28) ______________ ______________ [p.108] used to construct its proteins. When no more tryptophan is needed, the rate of protein synthesis slows and the cellular concentration of tryptophan (the end product) (29) [choose one] ☐ rises ☐ falls [p.108], as a control mechanism called (30) ______________ ______________ [p.108] begins to operate. In this case, excess tryptophan molecules shut down their own production when the unused molecules of tryptophan act to inhibit a key (31) ______________ [p.108] in the biochemical pathway. The inhibited enzyme is governed by (32) ______________ [p.108] control. Such enzymes have an active site and a(n) (33) ______________ ______________ [p.108] site where specific substances may bind and alter enzyme activity. Another type of biochemical control in humans and other multicelled organisms operates through signaling agents called (34) ______________ [p.108]. Several enzymes produced by the (35) ______________ [p.109] are responsible for (36) ______________ [p.109] ethyl alcohol. Excessive amounts of alcohol can (37) ______________ [p.109] liver tissue, greatly reducing the liver's ability to function. (38) ______________ ______________ [p.109] and (39) ______________ ______________ [p.109] are two liver problems related to alcohol consumption.

The enzyme (40) ______________ [p.110] can convert chemical energy to light energy in a process called (41) ______________ [p.110]. When genes for this process are inserted into bacteria, the resulting glow is a visual sign of their (42) ______________ [p.111]. This can be used to test the effectiveness of antibiotics because only (43) ______________ [p.111] bacteria are able to exhibit this light.

Self-Quiz

____ 1. An important principle of the second law of thermodynamics states that ______________. [p. 99]
   a. energy can be transformed into matter, and because of this, we can get something for nothing
   b. energy can only be destroyed during nuclear reactions, such as those that occur inside the sun
   c. if energy is gained by one region of the universe, another region must also gain energy in order to maintain the balance of nature
   d. matter tends to become increasingly more disorganized

____ 2. Essentially, the first law of thermodynamics states that ______________. [p.98]
   a. one form of energy cannot be converted into another
   b. entropy is increasing in the universe
   c. energy cannot be created or destroyed
   d. energy cannot be converted into matter or matter into energy

____ 3. An enzyme is best described as ______________. [p.105]
   a. an acid
   b. a protein
   c. a catalyst
   d. a fat
   e. both b and c

____ 4. Which is *not* true of enzyme behavior? [p.105]
   a. Enzyme shape may change during catalysis.
   b. The active site of an enzyme orients its substrate molecules, thereby promoting interaction of their reactive parts.
   c. All enzymes have an active site where substrates are temporarily bound.
   d. An individual enzyme can catalyze a wide variety of different reactions.

____ 5. When NAD^+ combines with hydrogen and electrons, the NAD^+ is ______________. [p.104]
   a. reduced
   b. oxidized
   c. phosphorylated
   d. denatured

____ 6. A substance that gains electrons is ______________. [p.101]
   a. oxidized
   b. a catalyst
   c. reduced
   d. a substrate

____ 7. In ______________ pathways, carbohydrates, lipids, and proteins are broken down in stepwise reactions that lead to products of lower energy. [p.102]
   a. intermediate
   b. biosynthetic
   c. induced
   d. degradative

____ 8. As to major function, NAD^+, FAD, and $NADP^+$ are classified as ______________. [p.104]
   a. enzymes
   b. phosphate carriers
   c. cofactors that function as coenzymes
   d. end products of metabolic pathways

____ 9. The outer phosphate bond in ATP ______________. [pp.100–101]
   a. absorbs a large amount of free energy when the phosphate group is attached during hydrolysis
   b. is formed when ATP is hydrolyzed to ADP and one phosphate group
   c. is usually found in each glucose molecule; that is why glucose is chosen as the starting point for glycolysis
   d. releases a large amount of usable energy when the phosphate group is split off during hydrolysis

____ 10. An allosteric enzyme __________.
[p.108]
a. has an active site where substrate molecules bind and another site that binds with intermediate or end-product molecules
b. is an important energy-carrying nucleotide
c. carries out either oxidation reactions or reduction reactions but not both
d. raises the activation energy of the chemical reaction it catalyzes

Chapter Objectives/Review Questions

1. __________ is the controlled capacity to acquire and use energy for stockpiling, breaking apart, building, and eliminating substances in ways that contribute to survival and reproduction. [p.102]
2. Explain what is meant by a *system* as related to the laws of thermodynamics. [p.99]
3. Define *energy;* be able to state the first and second laws of thermodynamics. [p.99]
4. __________ is a measure of the degree of randomness or disorder of systems. [p.99]
5. Explain how the world of life maintains a high degree of organization. [p.99]
6. Explain the functioning of the ATP/ADP cycle. [pp.100–101]
7. Adding a phosphate to a molecule is called __________. [p.101]
8. __________ molecules are the cell's main, renewable energy carriers between sites of metabolic reactions in cells. [p.101]
9. __________ reactions refer to the electron transfers occurring in an electron transport system. [p.101]
10. What is meant by a *reversible reaction*? [p.102]
11. A __________ pathway is an orderly sequence of reactions with specific enzymes acting at each step; the pathways are always linear or circular. [p.102]
12. Give the function of each of the following participants in metabolic pathways: substrates, intermediates, enzymes, cofactors, energy carriers, and end products. [p.102]
13. Describe the condition known as *chemical equilibrium.* [p.103]
14. Explain the differences between NAD^+ and NADH; $NADP^+$ and NADPH; FAD and $FADH_2$. [p.104]
15. What are enzymes? Explain their importance. [p.105]
16. The location on the enzyme where specific reactions are catalyzed is the active __________. [p.106]
17. According to the __________-__________ model, each substrate has a surface region that almost but not quite matches chemical groups in an active site. [p.107]
18. When the "energy hill" is made smaller by enzymes so that particular reactions may proceed, it may be said that the enzyme has lowered the __________ energy. [p.105]
19. Explain what happens to enzymes in the presence of extreme temperatures and pH. [pp.108–109]
20. An enzyme control called __________ inhibition operates in the tryptophan pathway when excess tryptophan molecules inhibit a key allosteric enzyme in the pathway. [p.108]
21. __________ are small molecules or metal ions that assist enzymes or carry atoms or electrons from one reaction site to another. [p.107]
22. __________ is toxic to the liver and can destroy liver tissue. [p.109]
23. Fireflies, various beetles, and some other organisms display __________ when luciferases excite the electrons of luciferins. [p.111]

Integrating and Applying Key Concepts

A piece of dry ice left sitting on a table at room temperature vaporizes. As the dry ice vaporizes into CO_2 gas, does its entropy increase or decrease? Explain your answer.

7

HOW CELLS ACQUIRE ENERGY

Interactive Exercises

Sunlight and Survival [pp.114–115]

7.1. PHOTOSYNTHESIS—AN OVERVIEW [pp.116–117]

Selected Words: Engelmann [p.114], *Spirogyra* [p.114], aerobic respiration [p.114], *light-dependent* reactions [p.116], *light-independent* reactions [p.116], $NADP^+$ [p.116], NADPH [p.116], glucose [p.116]

Boldfaced, Page-Referenced Terms

[p.114] autotrophs ____

[p.114] photoautotrophs ____

[p.114] photosynthesis ____

[p.114] heterotrophs ____

[p.116] chloroplasts ____ [p.116] stroma ____

____

[p.116] thylakoid membrane ____

____

Fill-in-the-Blanks

(1) ____________ [p.114] obtain carbon and energy from the physical environment; their carbon source is (2) ____________ ____________ [p.114]. (3) ____________ [p.114] autotrophs obtain energy from sunlight. T. (4) ____________ [p.114] demonstrated that aerobic bacteria moved to the regions of a strand of *Spirogyra* where red and (5) ____________ [p.114] light was falling on the strand; O_2 was produced here. (6) ____________ [p.114] feed on autotrophs, each other, and organic wastes; representatives include (7) ____________ [p.114], fungi, many protistans, and most bacteria. Although energy stored in organic compounds such as glucose may be released by several pathways, the pathway known as (8) ____________ ____________ [p.114] releases the most energy.

9. In the space below, supply the missing information to complete the summary equation for photosynthesis: [all from p.115]

$$12 \_\_\_\_\_\_ + \_\_\_\_\_\_ CO_2 \rightarrow \_\_\_\_\_\_ O_2 + C_6H_{12}O_6 + 6 \_\_\_\_\_\_$$

10. Supply the appropriate information to state the equation for photosynthesis (above) in words:

(a) ____________ [p.115] molecules of water plus six molecules of (b) ____________ ____________ [p.115] (in the presence of pigments, enzymes, and sunlight) yield six molecules of (c) ____________ [p.115] plus one molecule of (d) ____________ [p.115] plus (e) ____________ [p.115] molecules of water.

The two major sets of reactions of photosynthesis are the (11) ____________-____________ [p.116] reactions and the (12) ____________-____________ [p.116] reactions. (13) ____________ ____________ [p.116] and (14) ____________ [p.116] are the reactants of photosynthesis, and the end product is usually given as (15) ____________ [p.116]. The internal membranes and channels of the chloroplast form the (16) ____________ [p.116] membrane and are organized into stacks, called (17) ____________ [p.116]. Spaces inside the thylakoid disks and channels form a continuous compartment where (18) ____________ [p.116] ions accumulate to be used to produce ATP. The semifluid interior area surrounding the thylakoid is known as the (19) ____________ [p.116] and is the area where the products of photosynthesis are produced.

Although glucose is commonly listed as the end product of photosynthesis, little glucose is actually found in the cells, since the glucose phosphate produced is quickly converted into (20) ____________ [p.116], (21) ____________ [p.116], and (22) ____________ [p.116].

Choice

For questions 23–31, choose the area of the chloroplast that correctly relates to the listed structures and events.

a. thylakoid membrane   b. stroma

23. ____ Light-independent reactions [p.116]
24. ____ Sugars are assembled [p.116]
25. ____ Light-dependent reactions [p.116]
26. ____ ATP production [p.116]
27. ____ Carbon dioxide provides the carbon [p.116]
28. ____ Sunlight energy is absorbed [p.116]
29. ____ Water molecules are split [p.116]
30. ____ NADPH delivers the hydrogen received from water [p.116]
31. ____ Oxygen is formed [p.115]

Labeling

Identify the following structures/chemicals.

32. ________________ [p.117]

33. ________________ [p.117]

34. ________________ [p.117]

35. ________________ [p.117]

36. ________________ [p.116]

37. ________________ ________________ ________________[p.117]

38. ________________ [p.117]

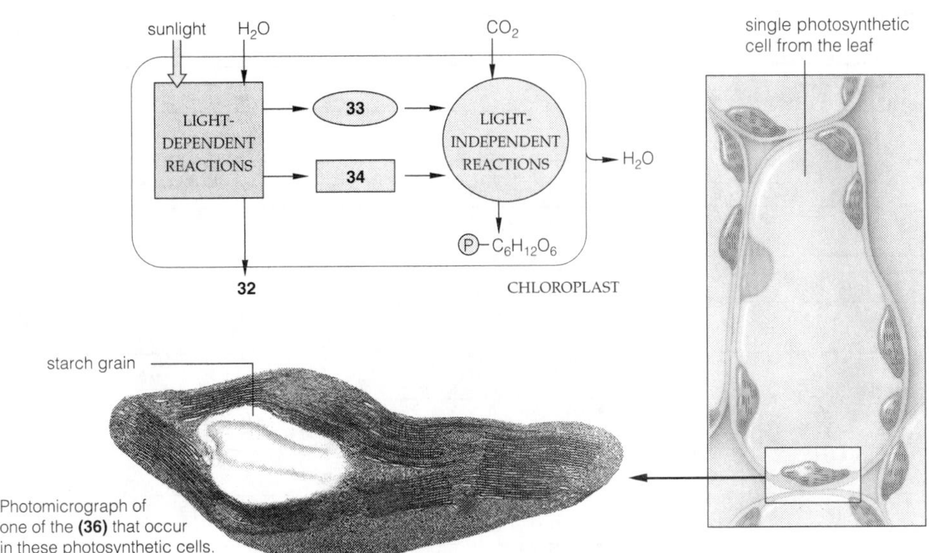

Photomicrograph of one of the **(36)** that occur in these photosynthetic cells.

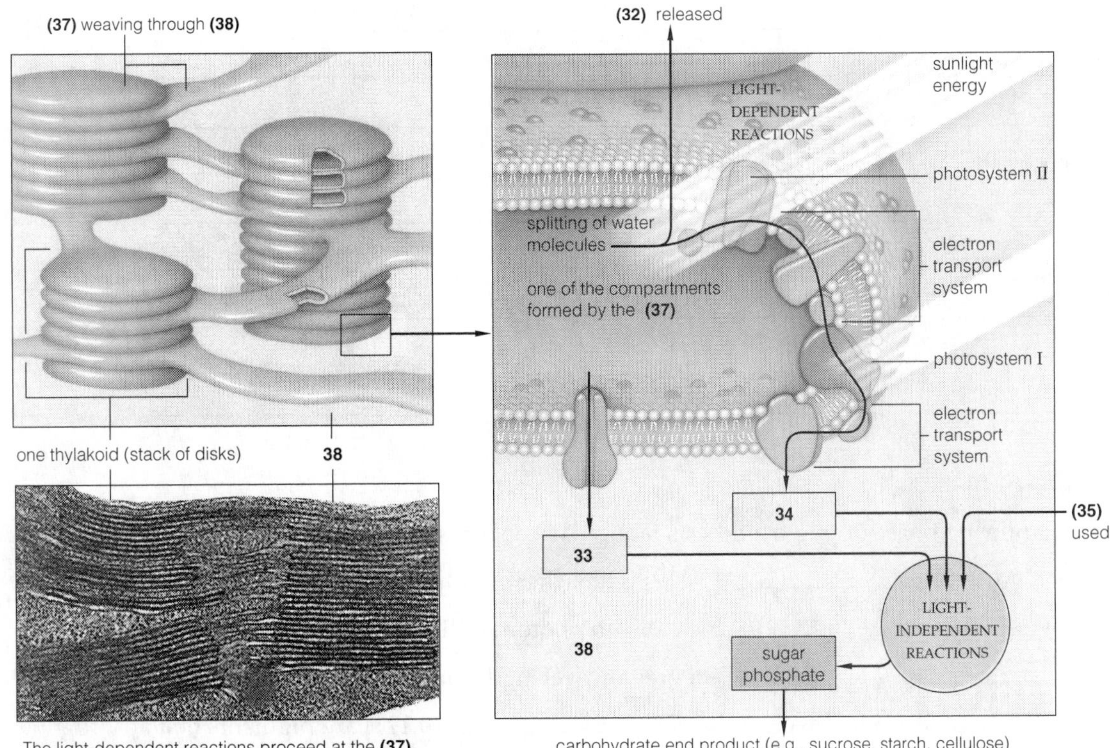

The light-dependent reactions proceed at the **(37)**.
The light-independent reactions proceed in the **(38)**.

carbohydrate end product (e.g., sucrose, starch, cellulose)

Overview of the sites where the key steps of both stages of reactions proceed. The sections to follow will fill in the details.

7.2. SUNLIGHT AS AN ENERGY SOURCE [pp.118–119]

7.3. THE RAINBOW CATCHERS [pp.120–121]

Selected Words: radiant energy [p.118], visible light [p.118], ultraviolet radiation [p.118], ozone (O_3) [p.118], *accessory* pigments [p.119], melanin [p.119], chlorophyll *a* [p.121], chlorophyll *b* [p.121], beta-carotene [p.121], plasma membrane [p.121], bacteriorhodopsin [p.121]

Boldfaced, Page-Referenced Terms

[p.118] wavelength ______________________________

____

[p.118] electromagnetic spectrum ______________________________

____

[p.119] photons ______________________________

____

[p.119] pigments ______________________________

____

[p.119] absorption spectrum ______________________________

____

[p.120] fluorescence ______________________________

____

[p.121] chlorophylls ______________________________

____

[p.121] carotenoids ______________________________

____

[p.121] anthocyanins ______________________________

____

[p.121] phycobilins ______________________________

____

[p.121] photosystems ______________________________

____

Fill-in-the-Blanks

The light-capturing phase of photosynthesis takes place on a system of (1) ______________ [p.121] membranes. A(n) (2) ______________ [p.119] is a packet of light energy. Thylakoid membranes contain (3) ______________ [p.119], which absorb photons of light. The principal pigments are the (4) ______________ [p.121], which reflect green wavelengths but absorb (5) ______________ [p.121] and (6) ______________ [p.121] wavelengths. (7) ______________ [p.121] are pigments that absorb violet and blue wavelengths but reflect yellow, orange, and red.

A cluster of 200 to 300 of these pigment proteins is a(n) (8) ______________ [p.121].

Matching

Choose the most appropriate answer.

9. ____ chlorophylls [p.121]
10. ____ accessory pigments [p.119]
11. ____ carotenoids [p.121]
12. ____ violet–blue–green–yellow–red [p.118]
13. ____ photons [p.119]
14. ____ chloroplast [p.116]
15. ____ phycobilins [p.121]
16. ____ thylakoid [recall p.116]
17. ____ pigments [p.119]

A. Packets of energy that have an undulating motion through space
B. The two stages of photosynthesis occur here
C. Molecules that can absorb light
D. Absorb violet and blue wavelengths but transmit red, orange, and yellow
E. Visible light portion of the electromagnetic spectrum
F. Pigments that transfer energy to chlorophyll *a*
G. Absorb violet-to-blue and red wavelengths; the reason leaves appear green
H. Red and blue pigments
I. The site of the first stage of photosynthesis

7.4. THE LIGHT-DEPENDENT REACTIONS [pp.122–123]

7.5. CASE STUDY: *A Controlled Release of Energy* [p.124]

Selected Words: electron acceptor molecule [p.122], NADPH [p.122], *type I* photosystem [p.122], P700 [p.122], *cyclic* pathway of ATP formation [p.122], *type II* photosystem [p.122], *noncyclic* pathway of ATP and NADPH formation [p.122], P680 [p.122], ATP synthases [p.123], chemiosmotic model of ATP formation [p.124]

Boldfaced, Page-Referenced Terms

[p.122] light-dependent reactions ____________________

____

[p.122] reaction center ____________________

____

[p.122] electron transfer chain ____________________

____

[p.122] photolysis ____________________

____

Complete the Table

1. Complete the following table on elements of the cyclic pathway of the light-dependent reactions.

Component	*Function*
a. Type I photosystem [p.122]	
b. Electrons [pp.122–123]	
c. P700 [p.122]	
d. Electron acceptor [p.122]	
e. Electron transfer chain [p.120]	
f. ADP [p.124]	

Labeling

The following diagram illustrates noncyclic photophosphorylation. Identify each numbered part of the illustration.

2. ______________ ______________ ______________ [p.123]
3. ______________ ______________ ______________ [p.123]
4. ______________ ______________ [p.123]
5. ______________ ______________ [p.123]
6. ______________ [p.122]
7. ______________ [p.123]
8. ______________ [p.124]

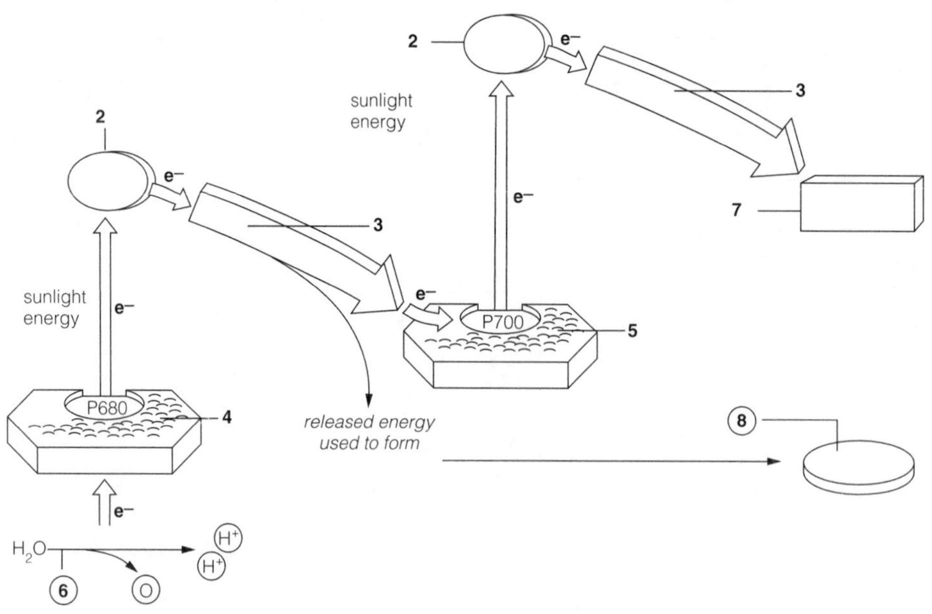

Fill-in-the-Blanks

ATP forms in both the cyclic and noncyclic pathways. When (9) ______________ [p.124] flow through the membrane-bound transport systems, the systems also pick up hydrogen ions (H^+) outside the membrane and shunt them into the (10) ______________ [p.124] compartment. This sets up H^+-concentration and electric (11) ______________ [p.124] across the membrane. Hydrogen ions that were split away from (12) ______________ [p.124] molecules increase the gradients. The ions respond by flowing out through the interior of (13) ______________ ______________ [p.124] proteins that span the membrane. Energy associated with the flow drives the binding of unbound phosphate (P_i) to ADP, the result being (14) ______________ [p.124]. The above description is known as the (15) ______________ [p.124] theory of ATP formation.

The noncyclic pathway also produces (16) ______________ [p.122] by using (17) ______________ [p.122] from water and H^+ ions from the thylakoid compartment to reduce $NADP^+$.

Complete the Table

With a checkmark (√), indicate for each phase of the light-dependent reactions all items from the left-hand column that are applicable.

Light-Dependent Reactions:	*Cyclic Pathway*	*Noncyclic Pathway*	*Photolysis Alone*
Uses H_2O as a reactant [p.123]	(18)	(29)	(40)
Photosystem I involved (P700) [pp.122–123]	(19)	(30)	(41)
Photosystem II involved (P680) [pp.122–123]	(20)	(31)	(42)
ATP produced [pp.123–124]	(21)	(32)	(43)
NADPH produced [pp.122–123]	(22)	(33)	(44)
Uses CO_2 as a reactant [logic; not mentioned on pp.122–124]	(23)	(34)	(45)
Causes H^+ to be shunted into the thylakoid compartments from the stroma [p.124]	(24)	(35)	(46)
Produces O_2 as a product [p.123]	(25)	(36)	(47)
Produces H^+ by breaking apart H_2O [p.123]	(26)	(37)	(48)
Uses ADP and P_i as reactants [pp.122–124]	(27)	(38)	(49)
Uses $NADP^+$ as a reactant [p.122]	(28)	(39)	(50)

7.6. THE LIGHT-INDEPENDENT REACTIONS [p.125]

7.7. FIXING CARBON—SO NEAR, YET SO FAR [pp.126–127]

7.8. CONNECTIONS: *Autotrophs, Humans, and the Biosphere* [p.128]

Selected Words: "synthesis" [p.125], *photorespiration* [p.126], glycolate [p.126], mesophyll cells [p.126], *bundle-sheath* cells [p.126], succulents [p.127], archaebacteria [p.128]

Boldfaced, Page-Referenced Terms

[p.125] light-independent reactions ____________________

____

[p.125] Calvin–Benson cycle ____________________

____

[p.125] RuBP (ribulose bisphosphate) ____________________

____

[p.125] Rubisco (RuBP carboxylase) ____________________

____

[p.125] PGA (phosphoglycerate) ____________________

____

[p.125] carbon fixation ____________________

____

[p.125] PGAL (phosphoglyceraldehyde) ____________________

____

[p.126] stomata (singular, stoma) ____________________

____

[p.126] C3 plants ____________________

____

[p.126] C4 plants ____________________

____

[p.127] CAM plants ____________________

____

[p.128] chemoautotrophs ____________________

____

[p.128] hydrothermal vent ____________________

____

Labeling and Matching

Identify each part of the following illustration. Complete the exercise by matching and entering the letter of the proper function description in the parentheses following each label. [All from p.125]

1.__________ __________ ( )

2.__________ __________ ( )

3.__________ ( )

4.__________ __________ ( )

5.__________ ( )

6.__________ ( )

7.__________ __________ ( )

8.__________-__________ __________ ( )

9.__________ __________ ( )

A. A three-carbon sugar, the first sugar produced; goes on to form sugar phosphate and RuBP

B. Typically used at once to form carbohydrate end products of photosynthesis

C. A five-carbon compound produced from PGALs; attaches to incoming CO_2

D. A compound that diffuses into leaves; attached to RuBP by enzymes in photosynthetic cells

E. Includes all the chemical reactions that "fix" carbon into an organic compound

F. Three-carbon compounds formed from the splitting of the six-carbon intermediate compound

G. A molecule that was reduced in the noncyclic pathway; furnishes hydrogen atoms to construct sugar molecules

H. A product of the light-dependent reactions; necessary in the light-independent reactions to energize molecules in metabolic pathways

I. Includes all the chemistry that fixes CO_2; converts PGA to PGAL and PGAL to RuBP and sugar phosphates

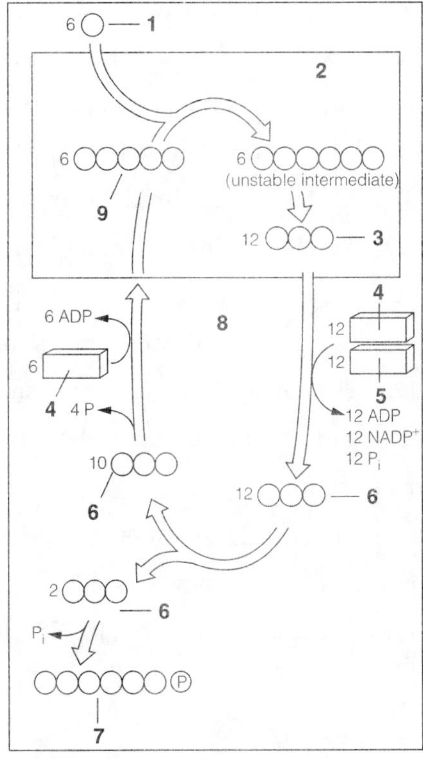

Fill-in-the-Blanks

The light-independent reactions can proceed without sunlight as long as (10) ______________ [p.125] and (11) ______________ [p.125] are available. The reactions begin when an enzyme links (12) ______________ ______________ [p.125] to (13) ______________ ______________ [p.125], a five-carbon compound. The resulting six-carbon compound is highly unstable and breaks apart at once into two molecules of a three-carbon compound, (14) ______________ [p.125]. This entire reaction sequence is called carbon (15) ______________ [p.125]. Each ATP gives a phosphate group to each (16) ______________ [p.125]. This intermediate compound takes on H^+ and electrons from NADPH to form (17) ______________ [p.125]. It takes (18) ______________ [p.125] carbon dioxide molecules to produce 12 PGAL. Most of the PGAL becomes rearranged into new (19) ______________ [p.125] molecules—which can be used to fix more (20) ______________ ______________ [p.125]. Two (21) ______________ [p.125] are joined together to form a (22) ______________ ______________ [p.125], primed for further reactions. The Calvin–Benson cycle yields enough RuBP molecules to replace the ones used in carbon (23) ______________ [p.125]. ADP, $NADP^+$, and phosphate leftovers are sent back to the (24) ______________-______________ [p.125] reaction sites, where they are again converted to (25) ______________ [p.125] and (26) ______________ [p.125]. (27) ______________ ______________ [p.125] formed in the cycle serves as a building block for the plant's main carbohydrates. When RuBP attaches to oxygen instead of carbon dioxide, (28) ______________ [p.126] results; this is typical of (29) ______________ [p.126] plants in hot, dry conditions. If less PGA is available, leaves produce a reduced amount of (30) ______________ [p.126]. C4 plants can still construct carbohydrates when the ratio of carbon dioxide to (31) ______________ [p.126] is unfavorable because of the attachment of carbon dioxide to form (32) ______________ [p.126] in certain leaf cells. CAM plants generally live in hot climates; they often have the fleshy, water-storing tissues and thick, water-restricting surface structure characteristic of (33) ______________ [p.127] plants. CAM plants open their stomata and fix CO_2 at (34) ______________ [p.127], thereby minimizing water loss.

Nearly half of the (35) ______________ ______________ [p.128] produced by humans is taken up by autotrophs that live in the (36) ______________ [p.128]. Most of these are so small that they are only visible through a (37) ______________ [p.128]. The vast majority are organisms that obtain energy from sunlight and are known as (38) ______________ -autotrophs [p.128]. Without them, global levels of CO_2 would increase more rapidly, possibly contributing to (39) ______________ ______________ [p.128]. On the seafloor there are organisms that obtain energy from oxidation of inorganic substances such as ammonium compounds and iron or sulfur compounds; they are known as (40) ______________ -autotrophs [p.128]. Such organisms use this energy to build (41) ______________ [p.128] compounds.

Complete the Table

With a checkmark (√), indicate for each phase of the light-independent reactions all items from the left-hand column that are applicable.

Light-Independent Reactions	*CO_2 Fixation Alone*	*Conversion of PGA to PGAL*	*Regeneration of RuBP*	*Formation of Glucose and Other Organic Compounds*
Requires RuBP as a reactant [p.123]	(42)	(53)	(64)	(75)
Requires ATP as a reactant [p.123]	(43)	(54)	(65)	(76)
Produces ADP as a product [p.123]	(44)	(55)	(66)	(77)
Requires NADPH as a reactant [p.123]	(45)	(56)	(67)	(78)
Produces $NADP^+$ as a reactant [p.123]	(46)	(57)	(68)	(79)
Produces PGA [p.125]	(47)	(58)	(69)	(80)
Produces PGAL [p.125]	(48)	(59)	(70)	(81)
Requires PGAL as a reactant [p.123]	(49)	(60)	(71)	(82)
Produces P_i as a product [p.123]	(50)	(61)	(72)	(83)
Produces H_2O as a product [p.123]	(51)	(62)	(73)	(84)
Requires CO_2 as a reactant [p.123]	(52)	(63)	(74)	(85)

Self-Quiz

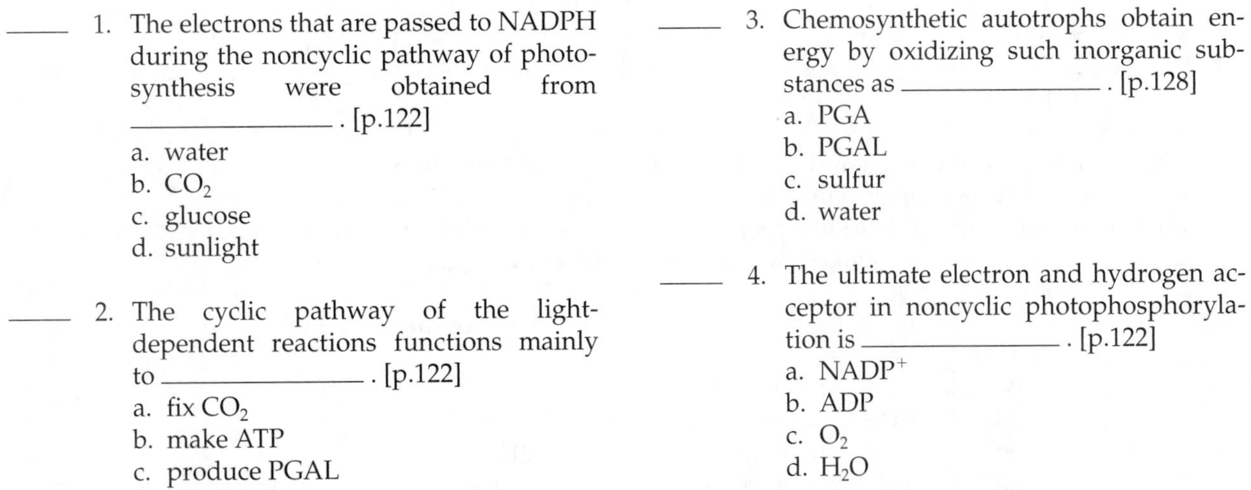

_____ 1. The electrons that are passed to NADPH during the noncyclic pathway of photosynthesis were obtained from _____________. [p.122]
a. water
b. CO_2
c. glucose
d. sunlight

_____ 2. The cyclic pathway of the light-dependent reactions functions mainly to _____________. [p.122]
a. fix CO_2
b. make ATP
c. produce PGAL
d. regenerate ribulose bisphosphate

_____ 3. Chemosynthetic autotrophs obtain energy by oxidizing such inorganic substances as _____________. [p.128]
a. PGA
b. PGAL
c. sulfur
d. water

_____ 4. The ultimate electron and hydrogen acceptor in noncyclic photophosphorylation is _____________. [p.122]
a. $NADP^+$
b. ADP
c. O_2
d. H_2O

____ 5. C4 plants have an advantage in hot, dry conditions because ______________ . [p.126]
a. their leaves are covered with thicker wax layers than those of C3 plants
b. their stomates open wider than those of C3 plants, thus cooling their surfaces
c. they have a two-step CO_2 fixation that reduces photorespiration
d. they are also capable of carrying on photorespiration

____ 6. Chlorophyll is ______________ . [p.121]
a. on the outer chloroplast membrane
b. inside the mitochondria
c. in the stroma
d. in the thylakoid membrane system

____ 7. Which of the following is applicable to C3 plants? [p.126]
a. At the end of carbon fixation, the intermediate compound is PGA.
b. At the end of carbon fixation, the intermediate compound is oxaloacetate.
c. They are more sensitive to cold temperatures than are C4 plants.
d. Corn, crabgrass, and sugarcane are examples of C3 plants.

____ 8. Plant cells produce O_2 during photosynthesis by ______________ . [p.122]
a. splitting CO_2
b. splitting water
c. degradation of the stroma
d. breaking up sugar molecules

____ 9. Plants need ______________ and ______________ to carry on photosynthesis. [p.116]
a. oxygen; water
b. oxygen; CO_2
c. CO_2; H_2O
d. sugar; water

____ 10. The two products of the light-dependent reactions that are required for the light-independent chemistry are ______________ and ______________ . [p.125]
a. CO_2; H_2O
b. O_2; NADPH
c. O_2; ATP
d. ATP; NADPH

Chapter Objectives/Review Questions

1. Distinguish between organisms known as *autotrophs* and those known as *heterotrophs*. [p.114]
2. State what T. Engelmann's 1882 experiment with *Spirogyra* revealed. [p.114]
3. Study the general equation for photosynthesis until you can remember the reactants and products. Reproduce the equation from memory on another piece of paper. [p.115]
4. List the two major stages of photosynthesis as well as where in the cell they occur and what reactions occur there. [p.116]
5. Describe the details of a familiar site of photosynthesis, the green leaf. Begin with the layers of a leaf cross section and complete your description with the minute structural sites within the chloroplast where the major sets of photosynthetic reactions occur. [pp.116–117]
6. The flattened channels and disklike compartments inside the chloroplast are organized into stacks, the __________, which are surrounded by a semifluid interior, the __________; this is the __________ membrane system. [p.116]
7. The energy-poor molecules that act as raw materials in the photosynthetic equation are __________ and __________. [p.116]
8. __________ are packets of light energy. [p.119]
9. __________ absorbs violet-to-blue as well as red wavelengths. [p.121]
10. The main pigments of photosynthesis are __________. [p.121]
11. What pigments are responsible for the red and blue colors of red algae and cyanobacteria, respectively? [p.121]
12. Name the wavelengths absorbed and transmitted by the carotenoids. [p.121]

13. Describe how the pigments found on thylakoid membranes are organized into photosystems and how they relate to photon light energy. [pp.121–122]
14. Contrast the components and functioning of the cyclic and noncyclic pathways of the light-dependent reactions. [pp.122–123]
15. Explain what the water split during photolysis contributes to the noncyclic pathway of the light-dependent reactions. [pp.122–123]
16. Two energy-carrying molecules produced in the noncyclic pathways are ________ and ________; explain why these molecules are necessary for the light-independent reactions. [p.122]
17. ________ ________ is the metabolic pathway most efficient in releasing the energy stored in organic compounds. [p.123]
18. Following evolution of the noncyclic pathway, ________ accumulated in the atmosphere and made ________ respiration possible. [p.123]
19. Explain how the chemiosmotic theory is related to thylakoid compartments and the production of ATP. [p.124]
20. Explain why the light-independent reactions are called by that name. [p.125]
21. Describe the process of carbon dioxide fixation by naming the reactants necessary to initiate the process and the stable products that result. [p.125]
22. Describe the Calvin–Benson cycle in terms of its reactants and products. [p.125]
23. What is the fate of all the phosphorylated glucose produced by photosynthetic reactions in photoautotrophs? [pp.116,125]
24. When carbon fixation occurs, sunlight energy, which excited electrons during the light-dependent reactions, becomes stored as ________ energy in an organic compound. [p.125]
25. Describe the mechanism by which C4 plants thrive under hot, dry conditions; distinguish this CO_2-capturing mechanism from that of C3 plants. [pp.126–127]
26. Describe the carbon-fixing adaptation of the CAM plants living in arid environments. [p.127]
27. Bacteria that are able to obtain energy from ammonium ions, iron, or sulfur compounds are known as ________. [p.128]

Integrating and Applying Key Concepts

Suppose that humans acquired all the enzymes needed to carry out photosynthesis. Speculate about the attendant changes in human anatomy, physiology, and behavior that would be necessary for those enzymes actually to carry out photosynthetic reactions.

8

HOW CELLS RELEASE STORED ENERGY

Interactive Exercises

The Killers Are Coming! The Killers Are Coming! [pp.132–133]

8.1. HOW DO CELLS MAKE ATP? [pp.134–135]

Selected Words: *anaerobic* [p.134], mitochondrion [p. 134], coenzymes [p.135]

Boldfaced, Page-Referenced Terms

[p.134] aerobic respiration ____

____

[p.134] glycolysis ____

____

[p.134] pyruvate ____

____

[p.135] the Krebs cycle ____

____

[p.135] NAD^+ (nicotinamide adenine dinucleotide) ____

____

[p.135] FAD (flavin adenine dinucleotide) ____

____

[p.135] electron transfer phosphorylation ____

____

Short Answer

1. Although various organisms use different energy sources, what is the usual form of chemical energy that will drive metabolic reactions? [p.133] ____

____

2. Describe the function of oxygen in the main degradative pathway, aerobic respiration. [p.133] ________

____

____

____

3. List the most common anaerobic pathways and describe the conditions under which they function. [pp.133–134] ____

____

____

____

Fill-in-the-Blanks

Virtually all forms of life depend on a molecule called (4) ______________ [p.134] as their primary energy carrier. Plants produce adenosine triphosphate during (5) ______________ [p.134], but plants and all other organisms can also produce ATP through chemical pathways that degrade (take apart) food molecules. The main degradative pathway requires free oxygen and is called (6) ______________ ______________ [p.134].

There are three stages of aerobic respiration. In the first stage, (7) ______________ [p.134], glucose is partially degraded to (8) ______________ [p.134]. By the end of the second stage, which includes the (9) ______________ [p.135] cycle, glucose has been completely degraded to carbon dioxide and (10) ______________ [p.135]. Neither of the first two stages produces much (11) ______________ [p.135]. During both stages, protons and (12) ______________ [p.135] are stripped from intermediate compounds and delivered to a(n) (13) ______________ ______________ [p.135] chain. This system is used in the third stage of reactions, electron transfer (14) ______________ [p.135]; passage of electrons along the transfer chain drives the enzymatic "machinery" that phosphorylates ADP to produce a high yield of (15) ______________ [p.135]. (16) ______________ [p.135] accepts "spent" electrons from the transfer chain and keeps the pathway clear for repeated ATP production.

Other degradative pathways are (17) ______________ [p.134], in that something other than oxygen serves as the final electron acceptor in energy-releasing reactions. (18) ______________ [pp.133–134] and anaerobic (19) ______________ ______________ [pp.133–134] are the most common anaerobic pathways.

Completion

20. Complete the following equation, which summarizes the degradative pathway known as *aerobic respiration:* [p.134]

$$\_\_\_\_\_\_ + \_\_\_\_\_\_ O_2 \rightarrow 6 \_\_\_\_\_\_ + 6 \_\_\_\_\_\_$$

21. Supply the appropriate information to state the equation for aerobic respiration [see no. 20] in words: One molecule of glucose plus six molecules of ______________ [p.134] (in the presence of appropriate enzymes) yield ______________ [p.134] molecules of carbon dioxide plus ______________ [p.134] molecules of water.

Matching

Choose the most appropriate answer for each.

22. ____ Krebs cycle [p.135]
23. ____ oxygen [p.134]
24. ____ mitochondrion [p.132]
25. ____ electron transfer phosphorylation [p.135]
26. ____ enzymes [p.134]
27. ____ ATP [p.135]
28. ____ glycolysis [p.134]
29. ____ aerobic respiration [p.134]
30. ____ cytoplasm [p.134]
31. ____ fermentation pathways and anaerobic electron transport [p.134]

A. Starting point for three energy-releasing pathways
B. Main energy-releasing pathway for ATP formation
C. Site of glycolysis
D. Third and final stage of aerobic respiration; high ATP yield
E. Oxygen is not the final electron acceptor
F. Catalyze each reaction step in the energy releasing pathways
G. Second stage of aerobic respiration; pyruvate is broken down into CO_2 and H_2O
H. Site of the second and third stages of the aerobic pathway
I. The final electron acceptor in aerobic pathways
J. The energy form that drives metabolic reactions

8.2. GLYCOLYSIS: FIRST STAGE OF ENERGY-RELEASING PATHWAYS [pp.136–137]

Selected Words: pyruvate [p.136], *energy-requiring* steps [p.136], *energy-releasing* steps [p.136], NAD^+ [p.136], NADH [p.136], *net* energy yield [p.136]

Boldfaced, Page-Referenced Terms

[p.136] substrate-level phosphorylation ______________________________

Fill-in-the-Blanks

(1) ______________ [recall Ch.7] organisms can synthesize and stockpile energy-rich carbohydrates and other food molecules from inorganic raw materials. (2) ______________ [p.136] is partially broken down by the glycolytic pathway; at the end of this process some of its stored energy remains in two (3) ______________ [p.136] molecules. Some of the energy of glucose is released during the breakdown reactions and used in forming the energy carrier (4) ______________ [p.136] and the reduced coenzyme (5) ______________ [p.136]. These reactions take place in the cytoplasm. Glycolysis begins with two phosphate groups being transferred to (6) ______________ [pp.136–137] from two (7) ______________ [pp.136–137] molecules. The addition of two phosphate groups to (6) energizes it and causes it to become unstable and split apart, forming two molecules of (8) ______________ [pp.136–137]. Each (8) gains one

(9) _______________ [pp.136–137] group from the cytoplasm, then (10) _______________ [pp.136–137] atoms and electrons from each PGAL are transferred to NAD^+, changing this coenzyme to NADH. At the same time, two (11) _______________ [pp.136–137] molecules form by substrate-level phosphorylation; the cell's energy investment is paid off. One (12) _______________ [p.137] molecule is released from each 2-PGA as a waste product. The resulting intermediates are rather unstable; each gives up a(n) (13) _______________ [pp.136–137] group to ADP. Once again, two (14) _______________ [pp.136–137] molecules have formed by (15) _______________ - _______________ [pp.136–137] phosphorylation. For each (16) _______________ [p.136] molecule entering glycolysis, the net energy yield is two ATP molecules that the cell can use anytime to do work. The end products of glycolysis are two molecules of (17) _______________ [pp.136–137], each with a (18) _______________ (number) -carbon [pp.136–137] backbone.

Sequence

Arrange the following events of the glycolysis pathway in correct chronological sequence. Write the letter of the first step next to 19, the letter of the second step next to 20, and so on. [All from p.137]

19. ____
20. ____
21. ____
22. ____
23. ____
24. ____
25. ____
26. ____

A. The first two ATPs form by substrate-level phosphorylation; the cell's energy debt is paid off.
B. Diphosphorylated glucose (fructose 1,6-bisphosphate) molecules split to form 2 PGALs; this is the first energy-releasing step.
C. Two three-carbon pyruvate molecules form as the end products of glycolysis.
D. Glucose is present in the cytoplasm.
E. Two more ATPs form by substrate-level phosphorylation; the cell gains ATP; net yield of ATP from glycolysis is 2 ATPs.
F. The cell invests two ATPs; one phosphate group is attached to each end of the glucose molecule (fructose 1,6-bisphosphate).
G. Two PGALs gain two phosphate groups from the cytoplasm.
H. Hydrogen atoms and electrons from each PGAL are transferred to NAD^+, reducing this carrier to NADH.

8.3. SECOND STAGE OF THE AEROBIC PATHWAY [pp. 138–139]

8.4. THIRD STAGE OF THE AEROBIC PATHWAY [pp. 140–141]

Selected Words: coenzymes [p.138], coenzyme A [p.138], ATP synthases [pp.138,140], NADH [p.138], FAD [p.138], $FADH_2$ [p.138], electron transfer *phosphorylation* [p.140], chemiosmotic model [p.140], kilocalorie [p.141]

Boldfaced, Page-Referenced Terms

[p.138] mitochondrion _______________

____

[p.138] acetyl-CoA _______________

____

[p.138] oxaloacetate _______________

____

Fill-in-the-Blanks

If sufficient oxygen is present, the end product of glycolysis enters a preparatory step, (1) ______________ - ______________ [p.138] formation. This step converts pyruvate into acetyl CoA, the molecule that enters the (2) ______________ [p.138] cycle, which is followed by (3) ______________ ______________ [p.140] phosphorylation. During these three processes, a total of (4) ______________ (number) [p.140] (5) ______________ [p.140] (energy-carrier molecules) are generated. In the preparatory conversions, prior to and within the Krebs cycle, the food molecule fragments are further broken down into (6) ______________ ______________ [p.140]. During these reactions, hydrogen atoms (with their (7) ______________ [p.141]) are stripped from the fragments and transferred to the coenzymes (8) ______________ [pp.140–141] and (9) ______________ [pp.140–141].

Labeling

In exercises 10–14, identify the structure or location; in exercises 15–18, identify the chemical substance involved. In exercise 19, name the metabolic pathway. [All from p.138]

10. __________ __________ of mitochondrion
11. __________ __________ of mitochondrion
12. __________ __________ of mitochondrion
13. __________ __________ of mitochondrion
14. __________
15. __________
16. __________
17. __________
18. __________
19. __________ __________ __________

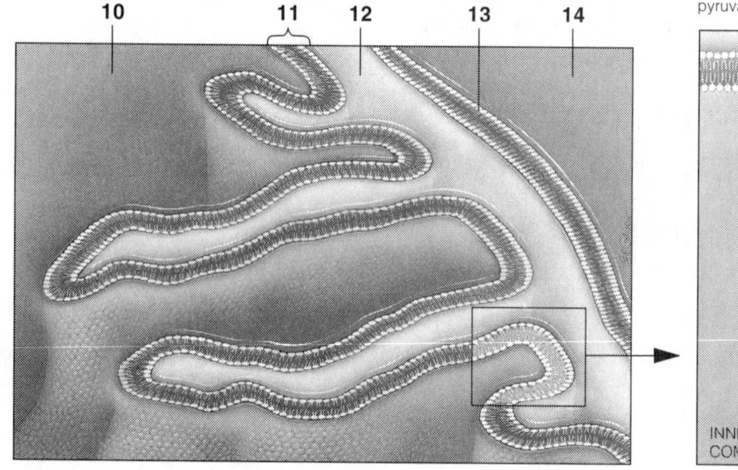

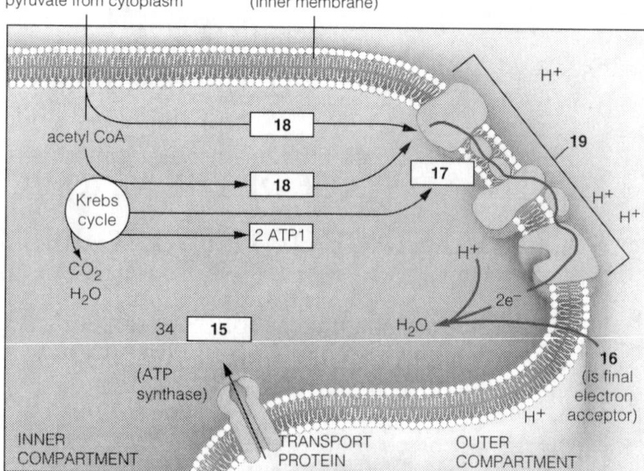

Fill-in-the-Blanks

NADH delivers its electrons to the highest possible point of entry into a transfer chain; from each NADH, enough H^+ is pumped to produce (20) ______________ (number) [p.141] ATP molecules. $FADH_2$ delivers its electrons at a lower point of entry into the transfer chain; fewer H^+ are pumped, and (21) ______________ (number) [p.141] ATPs are produced. The electrons are then sent down highly organized (22) ______________

_______________ [p.140] chains located in the inner membrane of the mitochondrion; hydrogen ions are pumped into the outer mitochondrial compartment. According to the (23) _______________ [p.140] model, the hydrogen ions accumulate and then follow a gradient to flow through channel proteins, called ATP (24) _______________ [p.140], that lead into the inner compartment. The energy of the hydrogen ion flow across the membrane is used to phosphorylate ADP in order to produce (25) _______________ [p.140]. Electrons leaving the electron transfer chain combine with hydrogen ions and (26) _______________ [p.140] to form water. In eukaryotic cells, these reactions occur only in (27) _______________ [p.140]. From glycolysis (in the cytoplasm) to the final reactions occurring in the mitochondria, the aerobic pathway commonly produces a total of (28) _______________ (number) [pp.140–141] or (29) _______________ (number) [pp.140–141] ATP(s) for every glucose molecule degraded.

Choice

For questions 30–50, choose from the following. Some correct answers may require more than one letter.

a. preparatory steps to Krebs cycle b. Krebs cycle c. electron transfer phosphorylation

30. _____ Chemiosmosis occurs to form ATP molecules [p.140]
31. _____ Three carbon atoms in pyruvate leave as three CO_2 molecules [p.139]
32. _____ Chemical reactions occur at transfer chains [pp.140–141]
33. _____ Coenzyme A picks up a two-carbon acetyl group [p.139]
34. _____ Makes two turns for each glucose molecule entering glycolysis [p.139]
35. _____ Two NADH molecules form for each glucose entering glycolysis [p.139]
36. _____ Oxaloacetate forms from intermediate molecules [pp.138–139]
37. _____ Named for a scientist who worked out its chemical details [p.138]
38. _____ Occurs within the mitochondrion [p.141]
39. _____ Two $FADH_2$ and six NADH form from one glucose molecule entering glycolysis [p.139]
40. _____ Hydrogens collect in the mitochondrion's outer compartment [p.140]
41. _____ Hydrogens and electrons are transferred to NAD^+ and FAD [p.139]
42. _____ Two ATP molecules form by substrate-level phosphorylation [p.139]
43. _____ Free oxygen withdraws electrons from the system and then combines with H^+ to form water molecules [pp.138,140]
44. _____ No ATP is produced [p.139]
45. _____ Thirty-two or thirty-four ATPs are produced [pp.140–141]
46. _____ Delivery point of NADH and $FADH_2$ [p.140]
47. _____ Two pyruvates enter for each glucose molecule entering glycolysis [p.139]
48. _____ The carbons in the acetyl group leave as CO_2 [p.139]
49. _____ One carbon in pyruvate leaves as CO_2 [p.139]
50. _____ An electron transfer chain and channel proteins are involved [p.140]

8.5. ANAERORIC ROUTES OF ATP FORMATION [pp.142–143]

Selected Words: pyruvate [p.142], *Lactobacillus* [p.142], *Saccharomyces cerevisiae* [p.142], *Saccharomyces ellipsoideus* [p.142].

Boldfaced, Page-Referenced Terms

[p.142] lactate fermentation ____

____

[p.142] alcoholic fermentation ____

____

[p.143] anaerobic electron transfer ____

____

Fill-in-the-Blanks

If (1) ______________ [p.142] is not present in sufficient amounts, the end product of glycolysis enters (2) ______________ [p.142] pathways; in some bacteria and muscle cells, pyruvate is converted into such products as (3) ______________ [p.142], while in yeast cells it is converted into (4) ______________ [p.142] and (5) ______________ ______________ [p.142].

(6) ______________ [p.142] pathways do not use oxygen as the final (7) ______________ [p.142] acceptor that ultimately drives the ATP-forming machinery. Anaerobic routes must be used by many bacteria and protistans that live in an oxygen-free environment. (8) ______________ [p.142] precedes any of the fermentation pathways. During (8), a glucose molecule is split into two (9) ______________ [p.142] molecules, two energy-rich (10) ______________ [p.142] intermediate molecules form, and the net energy yield from one glucose molecule is two ATPs.

In one kind of fermentation pathway, (11) ______________ [p.142] itself accepts hydrogen and electrons from NADH. (11) is then converted to a three-carbon compound, (12) ______________ [p.142], during this process in a few species of bacteria and some animal cells. Human muscle cells can carry on (12) fermentation in times of oxygen depletion; this provides a low yield of ATP.

In yeast cells, each pyruvate molecule from glycolysis forms an intermediate called (13) ______________ [p.142], as a gas, (14) ______________ ______________ [p.142], is detached from pyruvate with the help of an enzyme. This intermediate accepts hydrogen and electrons from NADH and is then converted to (15) ______________ [p.142], the end product of alcoholic fermentation.

In both types of fermentation pathways, the net energy yield of two ATPs is formed during (16) ______________ [p.142]. The reactions of the fermentation chemistry regenerate the (17) ______________ [p.142] needed for glycolysis to occur.

Anaerobic electron transport is an energy-releasing pathway occurring among the (18) ______________ [p.143]. For example, sulfate-reducing bacteria living in soil or water produce (19) ______________ [p.143] by stripping electrons from a variety of compounds and sending them through membrane transport systems. The inorganic compound (20) ______________ [p.143] ($SO_4^=$) serves as the final electron acceptor and is converted into foul-smelling hydrogen sulfide gas (H_2S).

Complete the Table

Include a checkmark (√) in each box that correctly links an occurrence (left-hand column) with a process (or processes).

	Glycolysis	*Lactate Fermentation*	*Alcoholic Fermentation*	*Anaerobic Electron Transport*
6-C → 3C + 3C [p.142]	(21)	(33)	(45)	(57)
$NADH \rightarrow NAD^+$ [p.142]	(22)	(34)	(46)	(58)
$NAD^+ \rightarrow NADH$ [p.142]	(23)	(35)	(47)	(59)
$SO_4^= \rightarrow H_2S$ [p.143]	(24)	(36)	(48)	(60)
CO_2 is a waste product [p.142]	(25)	(37)	(49)	(61)
3-C → 3-C [p.142]	(26)	(38)	(50)	(62)
3-C → 2-C [p.142]	(27)	(39)	(51)	(63)
ATP is used as a reactant [p.142]	(28)	(40)	(52)	(64)
ATP is produced [pp.142–143]	(29)	(41)	(53)	(65)
pyruvate → ethanol + CO_2 [p.142]	(30)	(42)	(54)	(66)
Occurs in animal cells [p.142]	(31)	(43)	(55)	(67)
Occurs in yeast cells [p.142]	(32)	(44)	(56)	(68)

8.6. ALTERNATIVE ENERGY SOURCES IN THE HUMAN BODY [pp.144–145]

8.7. CONNECTIONS: *Perspective on the Molecular Unity of Life* [p.146]

True/False

If the statement is true, write a "T" in the blank. If the statement is false, explain why in the answer blanks below the statement.

_____ 1. Glucose is the only carbon-containing molecule that can be fed into the glycolytic pathway. [p.144] ____________________

____

____

____

_____ 2. Simple sugars, fatty acids, and glycerol that remain after a cell's biosynthetic and storage needs have been met are generally sent to the cell's respiratory pathways for energy extraction. [p.144] ____________________

____

____

____

_____ 3. Carbon dioxide and water, the products of aerobic respiration, generally get into the blood and are carried to gills or lungs, kidneys, and skin, where they are expelled from the animal's body. [p.144; common knowledge] ____________________

____

____

____

_____ 4. Energy is recycled along with materials. [p.146] ____________________

____

____

____

_____ 5. The first forms of life on Earth were most probably photosynthetic eukaryotes. [p.146] __________

____

____

____

Labeling

Identify the process or substance indicated in the illustration below. [All from p.145]

6. ______________ ______________
7. ______________
8. ______________
9. ______________ ______________
10. ______________ ______________
11. ______________ ______________
12. ______________

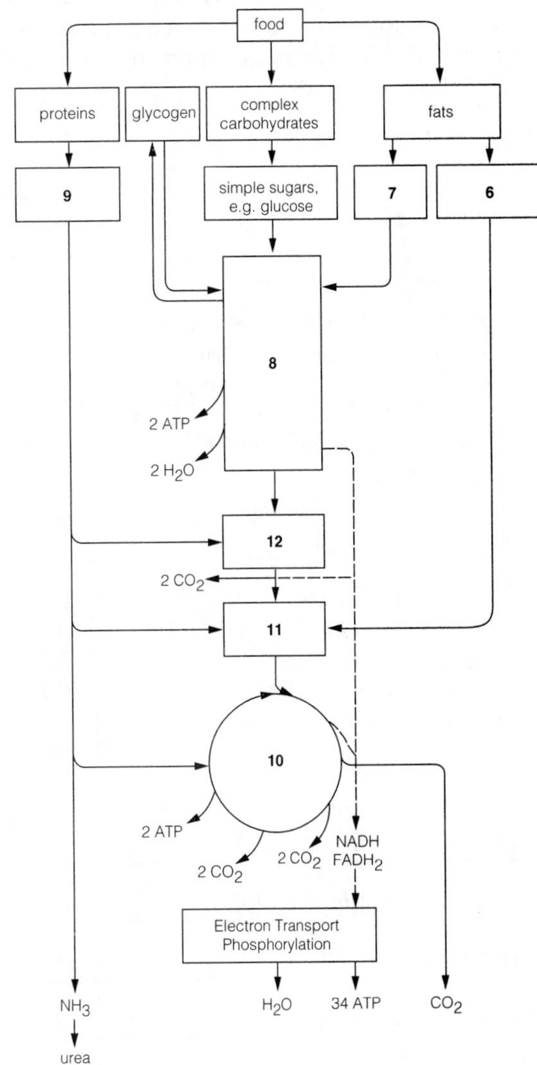

Complete the Table

Across the top of each column are the principal phases of degradative pathways into which food molecules (in various stages of breakdown) enter or in which specific events occur. Put a checkmark (✓) in each box that indicates the phase into which a specific food molecule is fed or in which a specific event occurs. For example, if simple sugars can enter the glycolytic pathway, put a checkmark in the top left-hand box; if not, let the box remain blank.

	Glycolysis (includes pyruvate)	*Acetyl-CoA Formation*	*Krebs Cycle*	*Electron Transfer Phosphorylation*	*Fermentation*	
					Alcoholic	*Lactate*
Complex Carbohydrates ⟶ Simple Sugars ⟶ which enter [p.144]	(13)	(28)	(43)	(58)	(73)	(88)
Fats → Fatty acids, which enter	(14)	(29)	(44)	(59)	(74)	(89)
Fats → Glycerol, which enters [pp.144–145]	(15)	(30)	(45)	(60)	(75)	(90)
Proteins → Amino acids, which enter [pp.144–145]	(16)	(31)	(46)	(61)	(76)	(91)
Intermediate energy carriers (NADH, $FADH_2$) are produced [pp.137,139,145]	(17) (18)	(32) (33)	(47) (48)	(62) (63)	(77) (78)	(92) (93)
ATPs produced directly as a result of this process alone [p.137]	(19)	(34)	(49)	(64)	(79)	(94)
NAD^+ produced [pp.140–143]	(20)	(35)	(50)	(65)	(80)	(95)
FAD produced [pp.138,140]	(21)	(36)	(51)	(66)	(81)	(96)
ADP produced [p.137]	(22)	(37)	(52)	(67)	(82)	(97)
Unbound phosphate (P_i) required [pp.137–139]	(23)	(38)	(53)	(68)	(83)	(98)
CO_2 produced (waste product) [pp.139,141,143,145]	(24)	(39)	(54)	(69)	(84)	(99)
H_2O produced (waste product) [pp.137,140]	(25)	(40)	(55)	(70)	(85)	(100)
One-half O_2 reacts with H^+ [p.140]	(26)	(41)	(56)	(71)	(86)	(101)
NADH required to drive this process [pp.140–143]	(27)	(42)	(57)	(72)	(87)	(102)

Choice

For questions 103–117, refer to the text and Figure 8.12; choose from the following: [All pp.144–145]

a. glucose b. glucose-6-phosphate c. glycogen d. fatty acids
e. triglycerides f. PGAL g. acetyl-CoA h. amino acids i. glycerol j. proteins

103. _____ Fats that are broken down between meals or during exercise as alternatives to glucose

104. _____ Used between meals when free glucose supply dwindles; enters glycolysis after conversion

105. _____ Its breakdown yields much more ATP than glucose

106. _____ Absorbed in large amounts immediately following a meal

107. _____ Represents only 1 percent or so of the total stored energy in the body

108. _____ Following removal of amino groups, the carbon backbones may be converted to fats or carbohydrates or they may enter the Krebs cycle

109. _____ On the average, represents 78 percent of the body's stored food

110. _____ Between meals liver cells can convert it back to free glucose and release it

111. _____ Can be stored in cells but not transported across plasma membranes

112. _____ Amino groups undergo conversions that produce urea, a nitrogen-containing waste product excreted in urine

113. _____ Converted to PGAL in the liver; a key intermediate of glycolysis

114. _____ Accumulate inside the fat cells of adipose tissues, at strategic points under the skin

115. _____ A storage polysaccharide produced from glucose-6-phosphate following food intake that exceeds cellular energy demand (and increases ATP production to inhibit glycolysis)

116. _____ Building blocks of the compounds that represent 21 percent of the body's stored food

117. _____ A product resulting from enzymes cleaving circulating fatty acids; enters the Krebs cycle

---

Self-Quiz

_____ 1. Glycolysis would quickly halt if the process ran out of _____________, which serves as the hydrogen and electron acceptor. [p.136]
a. $NADP^+$
b. ADP
c. NAD^+
d. H_2O

_____ 2. The ultimate electron acceptor in aerobic respiration is _____________. [p.140]
a. NADH
b. carbon dioxide (CO_2)
c. oxygen (½ O_2)
d. ATP

_____ 3. When glucose is used as an energy source, the largest amount of ATP is generated by the _____________ portion of the entire respiratory process. [p.140]
a. glycolytic pathway
b. acetyl-CoA formation
c. Krebs cycle
d. electron transfer phosphorylation

_____ 4. The process by which about 10 percent of the energy stored in a sugar molecule is released as it is converted into two small organic-acid molecules is _____________. [p.136]
a. photolysis
b. glycolysis
c. fermentation
d. the dark reactions

_____ 5. During which of the following phases of respiration is ATP produced directly by substrate-level phosphorylation? [pp.137,139]
a. glycolysis
b. Krebs cycle
c. both a and b
d. neither a nor b

_____ 6. What is the name of the process by which reduced NADH transfers electrons along a chain of acceptors to oxygen so as to form water and in which the energy released along the way is used to generate ATP? [p.140]
a. glycolysis
b. acetyl-CoA formation
c. the Krebs cycle
d. electron transfer phosphorylation

_____ 7. Pyruvic acid can be regarded as the end product of _______________. [p.137]
a. glycolysis
b. acetyl-CoA formation
c. fermentation
d. the Krebs cycle

_____ 8. Which of the following is *not* ordinarily capable of being reduced at any time? [p.140]
a. NAD^+
b. FAD
c. oxygen, O_2
d. water

_____ 9. ATP production by chemiosmosis involves _______________. [p.140]
a. H^+ concentration and electric gradients across a membrane
b. ATP synthases
c. both a and b
d. neither a nor b

_____ 10. During the fermentation pathways, a net yield of two ATPs is produced from _______________; the NAD^+ necessary for _______________ is regenerated during the fermentation reactions. [p.142]
a. the Krebs cycle; glycolysis
b. glycolysis; electron transport phosphorylation
c. the Krebs cycle; electron transport phosphorylation
d. glycolysis; glycolysis

Matching

Match the following components of respiration to the list of words below. Some components may have more than one answer.

11. _____ lactic acid, lactate [p.142]
12. _____ $NAD^+ \rightarrow NADH$ [pp.137,139]
13. _____ carbon dioxide is a product [p.139]
14. _____ $NADH \rightarrow NAD^+$ [p.140]
15. _____ pyruvate used as a reactant [p.139]
16. _____ ATP produced by substrate-level phosphorylation [pp.137,139]
17. _____ glucose [p.137]
18. _____ acetyl-CoA is either a reactant or a product [p.139]
19. _____ oxygen [p.140]
20. _____ water is a product [p.140]

A. Glycolysis
B. Preparatory conversions prior to the Krebs cycle
C. Fermentation
D. Krebs cycle
E. Electron transport phosphorylation

Chapter Objectives/Review Questions

1. No matter what the source of energy may be, organisms must convert it to __________, a form of chemical energy that can drive metabolic reactions. [p.134]
2. The main energy-releasing pathway is __________ respiration. [p.134]
3. State the overall equation for the aerobic respiratory route. [p.134]
4. In the first of the three stages of aerobic respiration, one __________ is partially degraded to two pyruvate molecules. [pp.135–136]
5. By the end of the second stage of aerobic respiration, which includes the __________ cycle, __________ has been completely degraded to carbon dioxide and water. [pp.135,139]
6. Do the first two stages of aerobic respiration yield a high or low quantity of ATP? [p.135]
7. The third stage of aerobic respiration is called *electron transport* __________, which yields many ATP molecules. [pp.135,141]
8. Explain, in general terms, the role of oxygen in aerobic respiration. [p.135]
9. Glycolysis occurs in the __________ of the cell. [p.136]
10. Explain the purpose served by the cell's investment of two ATP molecules in the chemistry of glycolysis. [p.136]
11. Four ATP molecules are produced by __________-__________ phosphorylation for every two used during glycolysis. (Consult Fig. 8.4 in the text.) [pp.136–137]
12. Glycolysis produces __________ [number] NADH, __________ [number] ATP (net), and __________ [number] pyruvate molecules for each glucose molecule entering the reactions. [p.137]
13. Consult Figure 8.6 in the main text. Describe the events that occur during acetyl-Coenzyme A formation and explain how the process of acetyl-CoA formation relates glycolysis to the Krebs cycle. [p.139]
14. What happens to the CO_2 produced during acetyl-CoA formation and the Krebs cycle? [p.139]
15. Calculate the number of ATP molecules produced during the Krebs cycle for each glucose molecule that enters glycolysis. [p.139]
16. Explain how chemiosmosis theory operates in the mitochondrion to account for the production of ATP molecules. [p.140]
17. Consult Figure 8.8 in the text and predict what will happen to the NADH produced during acetyl-CoA formation and the Krebs cycle. [p.141]
18. Briefly describe the process of electron transfer phosphorylation by stating what reactants are needed and what the products are. State how many ATP molecules are produced through operation of the transport system. [p.141]
19. Account for the total *net yield* of 36 ATP molecules produced through aerobic respiration; that is, state how many ATPs are produced in glycolysis, the Krebs cycle, and the electron transfer chain. [p.141]
20. List some environments where there is very little oxygen present, hence anaerobic organisms might be found. [p.142]
21. In fermentation chemistry __________ molecules from glycolysis are accepted to construct either lactate or ethyl alcohol; thus, a low yield of __________ molecules continues in the absence of oxygen. [pp.142–143]
22. List the main anaerobic energy-releasing pathways and examples of organisms that use them. [pp.142–143]
23. Describe what happens to pyruvate in anaerobic organisms. Then explain the necessity for pyruvate to be converted into a fermentation product. [pp.142–143]
24. State which factors determine whether the pyruvate (pyruvic acid) produced at the end of glycolysis will enter into the alcoholic fermentation pathway, the lactate fermentation pathway, or the acetyl-CoA formation pathway. [pp.142–143]
25. List some sources of energy (other than glucose) that can be fed into the respiratory pathways. [pp.144–145]
26. Predict what your body would do to synthesize its needed carbohydrates and fats if you switched to a diet of 100 percent protein. [pp.144–145]

27. After reading "Perspective on Life" in the main text, outline the supposed evolutionary sequence of energy-extraction processes. [p.146]
28. Scrutinize the sketch in "*Connections*" in the main text closely; then reproduce the carbon cycle from memory. [p.146]

Integrating and Applying Key Concepts

How is the "oxygen debt" experienced by runners and sprinters related to aerobic respiration and fermentation in humans?

9

CELL DIVISION AND MITOSIS

Interactive Exercises

From Cell to Silver Salmon [pp.150–151]

9.1. DIVIDING CELLS: THE BRIDGE BETWEEN GENERATIONS [pp.152–153]

9.2. THE CELL CYCLE [pp.154–155]

Selected Words: *cell divisions* [p.150], *nuclear* division [p.152], *asexual* reproduction [p.152], *2n* [p.153], *haploid* chromosome number [p.153], *n* [p.153], "bipolar" spindle [p.155]

Boldfaced, Page-Referenced Terms

[p.150] reproduction ______________________________

____

[p.152] mitosis ______________________________

____

[p.152] meiosis ______________________________

____

[p.152] somatic cells ______________________________

____

[p.152] germ cells ______________________________

____

[p.152] chromosome ______________________________

____

[p.152] sister chromatids ______________________________

____

[p.152] histones ______________________________

____

[p.152] nucleosome ______________________________

____

[p.152] centromere ______________________________

____

[p.152] chromosome number ______________________________

____

[p.153] diploid ______________________________

____

[p.154] cell cycle ______________________________

____

[p.154] interphase ______________________________

____

[p.154] prophase ______________________________

____

[p.154] metaphase ______________________________

____

[p.154] anaphase ______________________________

____

[p.154] telophase ______________________________

____

[pp.154–155] spindle apparatus ______________________________

____

Matching

Choose the most appropriate answer for each term.

1. ____ centromere [p.152]
2. ____ haploid or *n* cell [p.153]
3. ____ diploid or 2*n* cell [p.153]
4. ____ chromosome [p.152]
5. ____ nuclear division [p.152]
6. ____ germ cells [p.152]
7. ____ somatic cells [p.152]
8. ____ sister chromatids [p.152]
9. ____ mitosis and meiosis [p.152]
10. ____ chromosome number [p.152]
11. ____ nucleosomes [p.152]
12. ____ histones [p.152]

A. Cell lineage set aside for forming gametes and sexual reproduction
B. Necessary for the reproduction of eukaryotic cells
C. Any cell having two of each type of chromosome characteristic of the species
D. The number of each type of chromosome present in a cell (*n* or 2*n*)
E. Each DNA molecule with attached proteins
F. The chromosomal proteins
G. A histone with a spool of DNA around it
H. Mechanisms by which cells sort out and package parent DNA molecules into new cell nuclei
I. A small chromosome region with attachment sites for microtubules
J. The two attached DNA molecules of a duplicated chromosome
K. Body cells that reproduce by mitosis and cytoplasmic division
L. Possess only one of each type of chromosome characteristic of the species

Labeling

For each number in the accompanying figure, identify the appropriate component of the cell cycle.

13. ________________ [p.154]
14. ________________ [p.154]
15. ________________ [p.154]
16. ________________ [p.154]
17. ________________ [p.154]
18. ________________ [p.154]
19. ________________ [p.154]
20. ________________ [p.154]
21. ________________ [p.154]
22. ________________ [p.154]
23. ________________ [p.154]

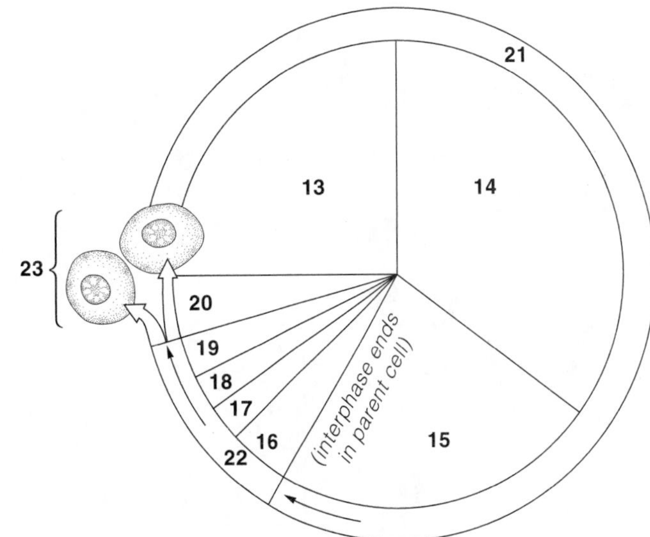

Matching

Link each time span identified below with the most appropriate number in the preceding Labeling section.

24. ____ Period after duplication of DNA during which the cell prepares for division [p.154]
25. ____ The complete period of nuclear division, which is followed by cytoplasmic division (a separate event) [p.154]
26. ____ DNA duplication occurs now; a time for "synthesis" of DNA and proteins [p.154]
27. ____ Period of cell growth before DNA duplication; a "gap" of interphase [p.154]
28. ____ Usually the longest part of a cell cycle [p.154]
29. ____ Period of cytoplasmic division [p.154]
30. ____ Period that includes G1, S, G2 [p.154]
31. ____ Period of nuclear division [p.154]

9.3. MITOSIS [pp.156–157]

Selected Words: "mitosis" [p.156], *mitos* [p.156], "prometaphase" [p.156], *meta-* [p.156]

Boldfaced, Page-Referenced Terms

[p.156] centrioles ____

____

Labeling and Matching

Identify each of the mitotic stages shown by entering the correct stage in the blank beneath the sketch below. Select from the following: late prophase, transition to metaphase (prometaphase), cell at interphase, metaphase, early prophase, telophase, interphase–daughter cells, and anaphase. Complete the exercise by matching and entering the letter of the correct phase description in the parentheses following each label.

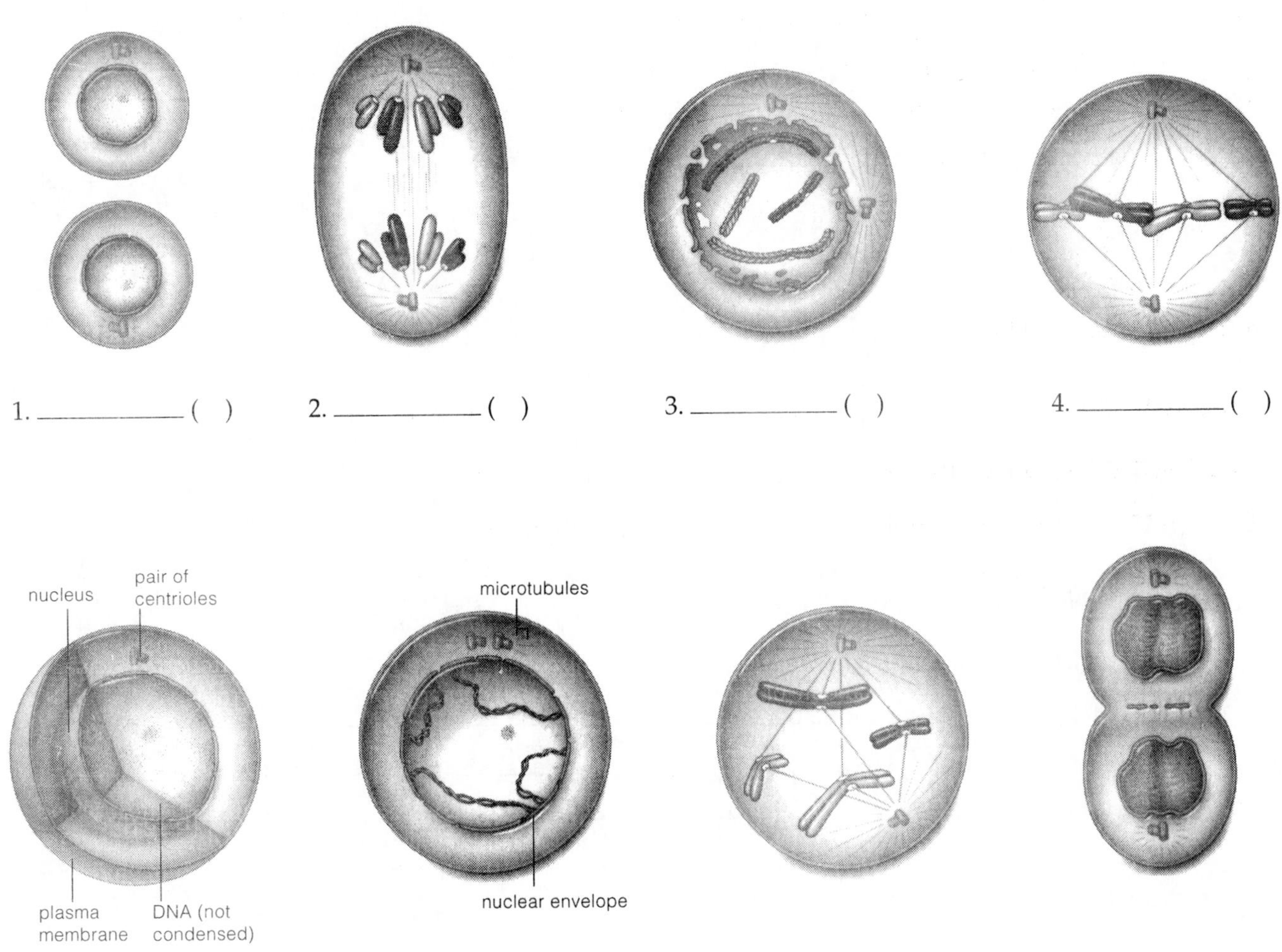

A. Attachments between two sister chromatids of each chromosome break; the two are now separate chromosomes that move to opposite spindle poles.
B. Microtubules penetrate the nuclear region and collectively form the spindle apparatus; microtubules become attached to the two sister chromatids of each chromosome.
C. The DNA and its associated proteins have started to condense.
D. All the chromosomes are now fully condensed and lined up at the equator of the spindle.
E. DNA is duplicated and the cell prepares for nuclear division.
F. Two daughter cells have formed, each diploid with two of each type of chromosome, just like the parent cell's nucleus.
G. Chromosomes continue to condense. New microtubules are assembled, and they move one of two centriole pairs toward the opposite end of the cell. The nuclear envelope begins to break up.
H. Patches of new membrane fuse to form a new nuclear envelope around the decondensing chromosomes.

Chronological Order

Using the illustration from the previous page, place the phases of mitosis in their correct chronological order. First, in the parentheses, place the number corresponding to the mitosis diagram. Then place the correct name of the stage in the answer blank following. The first step is done as an example. [pp.156–157]

9. (5) interphase
10. ( ) ____________
11. ( ) ____________
12. ( ) ____________
13. ( ) ____________
14. ( ) ____________
15. ( ) ____________
16. ( ) ____________

9.4. DIVISION OF THE CYTOPLASM [pp.158–159]

9.5. FOCUS ON SCIENCE: *Henrietta's Immortal Cells* [p.160]

Selected Words: *Discoverer XVII* satellite [p.160]

Boldfaced, Page-Referenced Terms

[p.158] cytoplasmic division ____________

____

[p.158] cell plate formation ____________

____

[p.159] cleavage furrow ____________

____

[p.160] HeLa cells ____________

____

Choice

For questions 1–10 on cytoplasmic division, choose from the following:

a. plant cells b. animal cells

1. ____ Formation of a cell plate [p.158]
2. ____ Contractions continue and in time will divide the cell in two [pp.158–159]
3. ____ Cellulose deposits form a crosswall between the two daughter cells [p.158]
4. ____ Possess rigid walls that cannot be pinched in two [p.158]

5. ____ A cleavage furrow [p.159]

6. ____ Deposits form a cementing middle lamella [p.158]

7. ____ A band of microfilaments beneath the cell's plasma membrane generates the force for cytoplasmic division [p.159]

8. ____ Two daughter nuclei are cut off in separate cells, each with its own cytoplasm and plasma membrane [p.159]

9. ____ At the location of a disklike structure, deposits of cellulose accumulate [p.158]

10. ____ The structure grows at its margins until it fuses with the parent cell's plasma membrane [p.158]

Short Answer

11. Describe the importance of HeLa cells. [p.160] ____

____

____

____

____

Self-Quiz

____ 1. The replication of DNA occurs ______________. [p.154]
   a between the growth phases of interphase
   b. immediately before prophase of mitosis
   c. during prophase of mitosis
   d. during prophase of meiosis

____ 2. In the cell life cycle of a particular cell, ______________. [p.154]
   a. mitosis occurs immediately prior to G1
   b. G2 precedes S
   c. G1 precedes S
   d. mitosis and S precede G1

____ 3. Diploid refers to ______________. [p.153]
   a. having two chromosomes of each type in somatic cells
   b. twice the parental chromosome number
   c. half the parental chromosome number
   d. having one chromosome of each type in somatic cells

____ 4. Somatic cells are ______________ cells; germ cells are ______________ cells. [p.152]
   a. meiotic; body
   b. body; body
   c. meiotic; meiotic
   d. body; meiotic

____ 5. If a parent cell has 16 chromosomes and undergoes mitosis, the resulting cells will have ______________ chromosomes. [p.152]
   a. 64
   b. 32
   c. 16
   d. 8
   e. 4

____ 6. Which of the following terms identifies a constricted region on the chromosome to which microtubules bind during cell division? [p.152]
   a. sister chromatids
   b. nucleosome
   c. centromere
   d. diploid
   e. cell membrane

_____ 7. The correct order of the stages of mitosis is _____________. [pp.156–157]
a. prophase, metaphase, telophase, anaphase
b. telophase, anaphase, metaphase, prophase
c. telophase, prophase, metaphase, anaphase
d. anaphase, prophase, telophase, metaphase
e. prophase, metaphase, anaphase, telophase

_____ 8. The stage of mitosis that is characterized by the alignment of the chromosomes along a central line in the cell is called _____________. [pp.156–157]
a. prophase
b. metaphase
c. prometaphase
d. anaphase
e. telophase

_____ 9. During _____________, sister chromatids of each chromosome are separated from each other, and those former partners, now chromosomes, move to opposite poles. [pp.156–157]
a. prophase
b. metaphase
c. anaphase
d. telophase

_____ 10. Each histone–DNA spool is a single structural unit called a _____________. [p.152]
a. somatic cell
b. motor protein
c. centromere
d. nucleosome

_____ 11. In the process of cytokinesis, cleavage furrows are associated with _____________ cell division, and cell plate formation is associated with _____________ cell division. [pp.158–159]
a. animal; animal
b. plant; animal
c. plant; plant
d. animal; plant

Chapter Objectives/Review Questions

1. Define the word *reproduction*. [p.150]
2. The terms *mitosis* and *meiosis* refer to the division of the cell's _________. [p.152]
3. Distinguish between somatic cells and germ cells as to their location and function. [p.152]
4. The eukaryotic chromosome is composed of _________ and _________. [p.152]
5. The two attached threads of a duplicated chromosome are known as sister _________. [p.152]
6. The _________ is a small region with attachment sites for the microtubules that move the chromosome during nuclear division. [p.152]
7. Any cell having two of each type of chromosome is a _________ cell; a _________ cell is one set aside for the formation of gametes. [p.153]
8. List and describe, in order, the various activities occurring in the eukaryotic cell life cycle. [p.154]
9. Interphase of the cell cycle consists of G1, _________, and G2. [p.154]
10. S is the time in the cell cycle when _________ replication occurs. [p.154]
11. Describe the structure and function of the spindle apparatus. [pp.154–155]
12. Describe the number and movements of centrioles in the cell division of some cells. [p.156]
13. The _________ is a time of transition when the nuclear envelope breaks up into tiny, flattened vesicles prior to metaphase. [p.156]
14. Give a detailed description of the cellular events occurring in the prophase, metaphase, anaphase, and telophase of mitosis. [pp.156–157]
15. Compare and contrast cytokinesis as it occurs in plant and animal cell division; use the following concepts: cleavage furrow, microfilaments at the cell's midsection, and cell plate formation. [pp.158–159]
16. Explain how cells from Henrietta Lacks continue to benefit humans everywhere more than 40 years after her death. [p.160]

Integrating and Applying Key Concepts

1. Runaway cell division is characteristic of cancer. Imagine the various points of the mitotic process that might be sabotaged in cancerous cells in order to halt their multiplication. Then try to imagine how one might discriminate between cancerous and normal cells in order to guide those methods of sabotage most effective in combating cancer.
2. What differentiates nerve cell division from that of normal somatic cells? With regard to the cell cycle, how might scientists one day help patients with neurological diseases?

10

MEIOSIS

Interactive Exercises

Octopus Sex and Other Stories [pp.162–163]

10.1. COMPARING SEXUAL WITH ASEXUAL REPRODUCTION [p.164]

10.2. HOW MEIOSIS HALVES THE CHROMOSOME NUMBER [pp.164–165]

Selected Words: *unfertilized* egg cells [p.163], *sexual* reproduction [p.163], *clones* [p.164], *pairs of genes* [p.164], *hom-* [p.164], *two consecutive divisions* [p.165], *homologue to homologue* [p.165]

Boldfaced, Page-Referenced Terms

[p.163] germ cells ____

____

[p.163] gametes ____

____

[p.164] asexual reproduction ____

____

[p.164] genes ____

____

[p.164] sexual reproduction ____

____

[p.164] allele ____

____

[p.164] meiosis ____

____

[p.164] chromosome number ____

____

[p.164] diploid number ($2n$) ____

____

[p.164] homologous chromosomes ____

____

[p.165] haploid number (n) ____

____

[p.165] sister chromatids ____

____

Choice

For questions 1–10, choose from the following:

a. asexual reproduction   b. sexual reproduction

1. ____ Offspring inherit new combinations of alleles [p.164]
2. ____ One parent alone produces offspring [p.164]
3. ____ Each offspring inherits the same number and kinds of genes as its parent [p.164]
4. ____ Commonly involves two parents [p.164]
5. ____ The production of "clones" [p.164]
6. ____ Involves meiosis, formation of gametes, and fertilization [p.164]
7. ____ Offspring are genetically identical copies of the parent [p.164]
8. ____ Produces the variation in traits that is the foundation of evolutionary change [p.164]
9. ____ Change can only occur by mutations [p.164]
10. ____ Offspring inherit new combinations of alleles, which lead to variations in the details of their traits [p.164]

Dichotomous Choice

Circle one of two possible answers given between parentheses in each statement.

11. (Meiosis/Mitosis) divides chromosomes into separate parcels not once but twice prior to cell division. [p.164]
12. Sperm and eggs are sex cells known as (germ/gamete) cells. [p.164]
13. (Haploid/Diploid) cells possess pairs of homologous chromosomes. [p.164]
14. (Haploid/Diploid) germ cells produce haploid gametes. [pp.164–165]
15. (Meiosis/Mitosis) produces cells that have one member of each pair of homologous chromosomes possessed by the species. [p.165]
16. Identical alleles are found on (homologous chromosomes/sister chromatids). [pp.164–165]
17. Two attached DNA molecules are known as (sister chromatids/homologous chromosomes). [p.165]
18. Two attached sister chromatids represent (one/two) chromosome(s). [p.165]
19. One pair of duplicated chromosomes would be composed of (two/four) chromatids. [p.165]
20. With meiosis, chromosomes proceed through (one/two) divisions to yield four haploid nuclei. [p.165]
21. During meiosis I, each duplicated (chromosome/chromatid) lines up with its partner, homologue to homologue, and then the partners are moved apart from one another. [p.165]
22. Cytoplasmic division following meiosis II results in four (diploid/haploid) cells. [p.165]
23. During (meiosis I/meiosis II), the two sister chromatids of each chromosome are separated from each other; each sister chromatid is now a chromosome in its own right. [p.165]
24. If human body cell nuclei contain 23 pairs of homologous chromosomes, each resulting gamete will contain (23/46) chromosomes. [p.165]

Chronological Order

Starting with a single germ cell, place the following items in chronological order. Use "1" to indicate the first step in the sequence, "2" for the second, and so on. [p.165]

25. _____ Two haploid daughter cells are formed, each having one of each type of chromosome.
26. _____ The cytoplasm divides, producing four haploid cells.
27. _____ Following interphase, each duplicated chromosome is aligned with its partner.
28. _____ The two sister chromatids of each chromosome separate from each other.
29. _____ The DNA is duplicated during the S stage of interphase.

10.3. A VISUAL TOUR OF THE STAGES OF MEIOSIS [pp.166–167]

10.4. A CLOSER LOOK AT KEY EVENTS OF MEIOSIS I [pp.168–169]

Selected Words: "sister chromatids" [p.167], *non*sister chromatids [p.168], *maternal* chromosomes [p.169], *paternal* chromosomes [p.169]

Boldfaced, Page-Referenced Terms

[p.168] crossing over ____

____

Labeling and Matching

Identify each of the meiotic stages shown on the opposite page by entering the correct stage of either meiosis I or meiosis II in the blank beneath the sketch. Choose from the following: prophase I, metaphase I, anaphase I, telophase I, prophase II, metaphase II, anaphase II, and telophase II. Complete the exercise by matching and entering the letter of the correct stage description in the parentheses following each label. [pp.166–167]

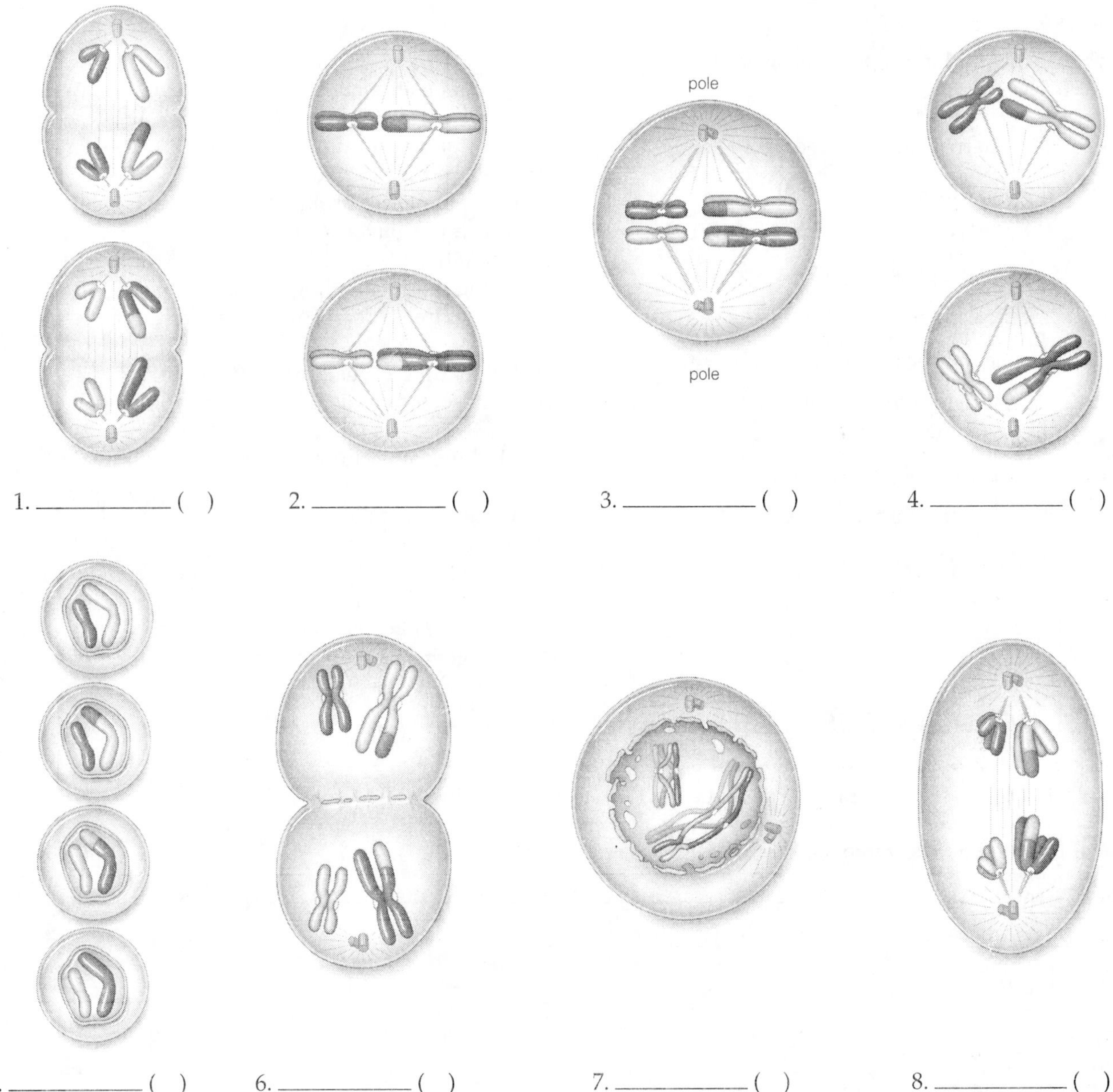

1. ____________ ( )
2. ____________ ( )
3. ____________ ( )
4. ____________ ( )
5. ____________ ( )
6. ____________ ( )
7. ____________ ( )
8. ____________ ( )

A. The spindle is now fully formed; all chromosomes are positioned midway between the poles of one cell.
B. In each of two daughter cells, microtubules attach to the kinetochores of chromosomes, and motor proteins drive the movement of chromosomes toward the spindle's equator.
C. Four daughter nuclei form; when the cytoplasm divides, each new cell has a haploid chromosome number, all in the unduplicated state; the cells may develop into gametes in animals or spores in plants.
D. In one cell, each duplicated chromosome is pulled away from its homologous partner; the partners are moved to opposite spindle poles.
E. Duplicated chromosomes condense; each chromosome pairs with its homologous partner; crossing over and genetic recombination (swapping of gene segments) occur; each chromosome becomes attached to some microtubules of a newly forming spindle.
F. Motor proteins and spindle microtubule interactions have moved all the duplicated chromosomes so that they are positioned at the spindle equator, midway between the poles.
G. Two haploid cells form, each having one of each type of chromosome that was present in the parent cell; the chromosomes are still in the duplicated state.
H. Attachment between the two chromatids of each chromosome breaks; former "sister chromatids" are now chromosomes in their own right and are moved to opposite poles by motor proteins.

Matching

Choose the most appropriate answer for each.

9. ____ 2^{23} [p.169]
10. ____ paternal chromosomes [p.169]
11. ____ early prophase I [p.168]
12. ____ nonsister chromatids [pp.168–169]
13. ____ late prophase I [p.168]
14. ____ crossing over [pp.168–169]
15. ____ metaphase I [p.169]
16. ____ each chromosome zippers to its homologue [p.168]
17. ____ function of meiosis [p.168]
18. ____ maternal chromosomes [p.169]

A. Random positioning of maternal and paternal chromosomes at the spindle equator
B. Chromosomes continue to condense and become thicker, rodlike forms
C. Reduction of the chromosome number by half for forthcoming gametes
D. Break at the same places along their length and then exchange corresponding segments
E. Twenty-three chromosomes inherited from your mother
F. Combinations of maternal and paternal chromosomes possible in gametes from one germ cell
G. Each duplicated chromosome is in thin, threadlike form
H. An intimate parallel array of homologues that favors crossing over
I. Twenty-three chromosomes inherited from your father
J. Breaks up old combinations of alleles and puts new ones together in pairs of homologous chromosomes

10.5. FROM GAMETES TO OFFSPRING [pp.170–171]

Selected Words: *gamete*-producing bodies [p.170], *spore*-producing bodies [p.170]

Boldfaced, Page-Referenced Terms

[p.170] spores ____________________

[p.170] sperm ____________________

[p.170] oocyte ____________________

[p.170] egg ____________________

[p.170] fertilization ____________________

Choice

For questions 1–14, choose from the following:

a. animal life cycle b. plant life cycle c. both animal and plant life cycles

1. ____ Meiosis results in the production of haploid spores. [p.170]
2. ____ A zygote divides by mitosis. [p.170]
3. ____ Meiosis results in the production of haploid gametes. [p.170]
4. ____ Haploid gametes fuse in fertilization to form a diploid zygote. [p.170]
5. ____ A zygote divides by mitosis to form a diploid spore-producing body (sporophyte). [p.170]
6. ____ During meiosis, one egg and three polar bodies form. [p.170]
7. ____ A spore divides by mitosis to produce a haploid gamete-producing body (gametophyte). [p.170]
8. ____ Gametes form by oogenesis and spermatogenesis. [p.170]
9. ____ A haploid gamete-producing body (gametophyte) divides by mitosis to produce haploid gametes. [p.170]
10. ____ A secondary oocyte gets nearly all the cytoplasm. [p.170]
11. ____ A haploid spore divides by mitosis to produce a gamete-producing body (gametophyte). [p.170]
12. ____ A diploid body forms from mitosis of a zygote. [p.170]
13. ____ A gamete-producing body and a spore-producing body develop during the life cycle. [p.170]
14. ____ One daughter cell of the secondary oocyte develops into a second polar body. [p.170]

Sequence

Arrange the following entities in correct order of development, entering a "1" by the stage that appears first and a "5" by the stage that completes the process of spermatogenesis. Complete the exercise by indicating, in the parentheses following each blank, if each cell is n or $2n$.

15. ____ ( ) primary spermatocyte [pp.170–171]
16. ____ ( ) sperm [pp.170–171]
17. ____ ( ) spermatid [pp.170–171]
18. ____ ( ) spermatogonium [pp.170–171]
19. ____ ( ) secondary spermatocyte [pp.170–171]

Matching

Choose the most appropriate answer to match with each oogenesis concept.

20. ____ primary oocyte [pp.170–171]
21. ____ oogonium [pp.170–171]
22. ____ secondary oocyte [pp.170–171]
23. ____ ovum and three polar bodies [pp.170–171]
24. ____ first polar body [pp.170–171]

A. The cell in which synapsis, crossing over, and recombination occur
B. A cell that is equivalent to a diploid germ cell
C. A haploid cell formed after division of the primary oocyte that does not form an ovum at second division
D. Haploid cells, only one of which functions as an egg
E. A haploid cell formed after division of the primary oocyte, the division of which forms a functional ovum

Short Answer

25. List the three mechanisms discussed in the chapter that contribute to the huge number of new gene combinations that may result from fertilization. [pp.170–171] ______

______

______

______

10.6. MEIOSIS AND MITOSIS COMPARED [pp.172–173]

Selected Words: *germ cell* [p.172], *crossing over* [p.172], *somatic cell* [p.172]

Complete the Table

1. Complete the following table by entering the word *mitosis* or the word *meiosis* in the blank adjacent to the statement describing one of these processes.

Description	*Mitosis/Meiosis*
[pp.172–173]: a. Involves one division cycle	
[p.172] b. Functions in growth and tissue repair	
[p.173] c. Daughter cells are haploid	
[p.172] d. Occurs only in germ cells	
[pp.172–173] e. Involves two division cycles	
[p.173] f. Daughter cells have one chromosome from each homologous pair	
[p.172] g. Introduces variation in traits among the offspring	
[p.173] h. Daughter cells have the diploid chromosome number	
[p.173] i. Completed when four daughter cells are formed	
[p.172] j. Produces clones	
[p.172] k. Involves the process of crossing over	

Matching

The cell model used in this exercise has two pairs of homologous chromosomes, one long pair and one short pair. Match the descriptions to the numbers of chromosomes shown in the following sketches.

2. ____ one cell at the beginning of meiosis II [p.172]
3. ____ a daughter cell at the end of meiosis II [p.173]
4. ____ metaphase I of meiosis [p.172]
5. ____ metaphase of mitosis [p.173]
6. ____ G1 in a daughter cell after mitosis [p.173]
7. ____ prophase of mitosis [p.173]

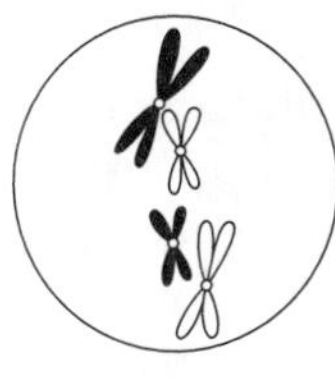

A

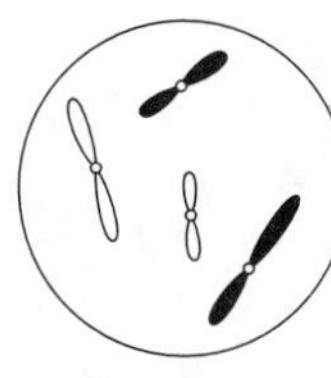

B

C

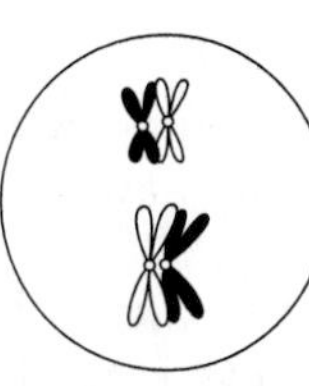

D

E

F

Short Answer

The following questions refer to the sketches above; enter your answer in the blank following each question.

8. How many chromosomes are present in cell E? ________
9. How many chromatids are present in cell E? ________
10. How many chromatids are present in cell C? ________
11. How many chromatids are present in cell D? ________
12. How many chromosomes are present in cell F? ________

Self-Quiz

_____ 1. Which of the following does not occur in prophase I of meiosis? [p.168]
a. a cytoplasmic division
b. a cluster of four chromatids
c. homologues pairing tightly
d. crossing over

_____ 2. Crossing over is one of the most important events in meiosis because _______________. [pp.168–169]
a. it establishes new genetic combinations not present in the parents
b. homologous chromosomes must be separated into different daughter cells
c. the number of chromosomes allotted to each daughter cell must be halved
d. it sorts the chromatids into gametes for fertilization

_____ 3. Crossing over _______________. [pp.168–169]
a. generally results in pairing of homologues and binary fission
b. is accompanied by gene-copying events
c. involves breakages and exchanges between sister chromatids
d. is a molecular interaction between two of the nonsister chromatids of a pair of homologous chromosomes

_____ 4. If an organism has a chromosome number of 5, how many different combinations of homologous chromosomes are possible at metaphase I? [p.169]
a. 2^5
b. 5^2
c. 5
d. 10

_____ 5. Which of the following does not increase genetic variation? [pp.172–173]
a. crossing over
b. random fertilization
c. prophase of mitosis
d. random homologue alignments at metaphase I

_____ 6. Which of the following is the most correct sequence of events in animal life cycles? [p.170]
a. meiosis—fertilization—gametes—diploid organism
b. diploid organism—meiosis—gametes—fertilization
c. fertilization—gametes—diploid organism—meiosis
d. diploid organism—fertilization—meiosis—gametes

_____ 7. In sexually reproducing organisms, the zygote is _______________. [p.172]
a. an exact genetic copy of the female parent
b. an exact genetic copy of the male parent
c. completely unlike either parent genetically
d. a genetic mixture of male and female parents

_____ 8. Which of the following is the most correct sequence of events in plant life cycles? [p.170]
a. fertilization—zygote—sporophyte—meiosis—spores—gametophytes—gametes
b. fertilization—sporophyte—zygote—meiosis—spores—gametophytes—gametes
c. fertilization—zygote—sporophyte—meiosis—gametes—gametophyte—spores
d. fertilization—zygote—gametophyte—meiosis—gametes—sporophyte—spores

_____ 9. While observing a given cell under a microscope, you notice that pairs of condensed homologous chromosomes appear to be aligning in the center of the cell. The most likely stage of meiosis for this cell is _______________.
a. anaphase II
b. metaphase II
c. metaphase I
d. prophase II
e. telophase I

_____ 10. The end result of meiosis in human females is _____________. [p.171]
a. two diploid clones of the parent cell
b. four haploid ovum
c. one haploid ovum and three diploid polar bodies
d. one haploid ovum and three haploid polar bodies

Chapter Objectives/Review Questions

1. Distinguish between germ cells and gametes. [p.163]
2. "One parent alone produces offspring, and each offspring inherits the same number and kinds of genes as its parent" describes _________ reproduction. [p.164]
3. _________ reproduction involves meiosis, gamete formation, and fertilization. [p.164]
4. _________ divides chromosomes into separate parcels not once but twice prior to cell division. [p.164]
5. Explain the terms *allele* and *gene.* [p.164]
6. Describe the relationships between homologous chromosomes and sister chromatids. [pp.164–165]
7. Explain what is meant by the terms *diploid chromosome number* and *haploid chromosome number.* [pp.164–165]
8. If the diploid chromosome number for a particular plant species is 8, the haploid gamete number is _________. [p.165]
9. During meiosis _________, each duplicated chromosome lines up with its partner, homologue to homologue. [p.165]
10. During meiosis II, the two sister _________ of each _________ are separated from each other. [p.165]
11. Sketch cell models of the various stages of meiosis I and meiosis II. Using colored pencils or crayons to distinguish between two pairs of homologous chromosomes (a long pair and a short pair) is helpful. [pp.166–167]
12. _________ _________ breaks up old combinations of alleles and puts new ones together during prophase I of meiosis. [p.168]
13. The _________ attachment and subsequent positioning of each pair of maternal and paternal chromosomes at metaphase I lead to different _________ of maternal and paternal traits in each generation of offspring. [p.169]
14. Meiosis in the animal life cycle results in haploid _________; meiosis in the plant life cycle results in haploid _________. [p.170]
15. Describe the differences in gamete production between male and female animals. [pp.170–171]
16. Explain how crossing over, the distribution of random mixes of homologous chromosomes into gametes, and fertilization all contribute to variation in the traits of offspring. [p.171]
17. Mitotic cell division produces only _________; meiotic cell division, in conjunction with subsequent fertilization, promotes _________ in traits among offspring. [p.172]

Integrating and Applying Key Concepts

In 1997, the first mammal was cloned. It was a sheep named Dolly, which resulted from a cloning experiment in Scotland. Within a few years pigs were successfully cloned. Scientists now believe that they have the ability to clone higher primates, including humans. From your knowledge of mitosis and meiosis, what would be the result of variation in the human population if cloning became widespread? What would be the advantages of being able to clone humans? The disadvantages?

11

OBSERVABLE PATTERNS OF INHERITANCE

Interactive Exercises

A Smorgasbord of Ears and Other Traits [pp.176–177]

11.1. MENDEL'S INSIGHT INTO PATTERNS OF INHERITANCE [pp.178–179]

Selected Words: *attached* earlobes [p.176], *straight* nose [p.176], *flat* feet [p.176], *observable* evidence [pp.176,178], "blended" [p.178], "units" of information [p.178], *identical* alleles [p.179], *nonidentical* alleles [p.179], *homozygous* condition [p.179], *heterozygous* condition [p.179], *dominant* [p.179], *recessive* [p.179], pair of homologous chromosomes [p.179], gene locus [p.179], pair of alleles [p.179], pairs of genes [p.179]

Boldfaced, Page-Referenced Terms

[p.179] genes ____

____

[p.179] alleles ____

____

[p.179] true-breeding lineage ____

____

[p.179] hybrids ____________________

____

[p.179] homozygous dominant ____________________

____

[p.179] homozygous recessive ____________________

____

[p.179] heterozygous ____________________

____

[p.179] genotype ____________________

____

[p.179] phenotype ____________________

____

[p.179] P, F_1, F_2 ____________________

____

Matching

Choose the most appropriate answer for each.

1. ____ genotype [p.179]
2. ____ alleles [p.179]
3. ____ heterozygous [p.179]
4. ____ dominant allele [p.179]
5. ____ phenotype [p.179]
6. ____ genes [p.179]
7. ____ true-breeding lineage [p.179]
8. ____ homozygous recessive [p.179]
9. ____ recessive allele [p.179]
10. ____ homozygous [p.179]
11. ____ *P*, F_1, F_2 [p.179]
12. ____ hybrids [p.179]
13. ____ diploid organism [p.179]
14. ____ gene locus [p.179]
15. ____ homozygous dominant [p.179]
16. ____ homologous chromosomes [p.179]

A. Parental, first-generation, and second-generation offspring
B. All the different molecular forms of the same gene
C. Particular location of a gene on a chromosome
D. Describes an individual having a pair of nonidentical alleles
E. An individual with a pair of recessive alleles, such as *aa*
F. Allele whose effect is masked by the effect of the dominant allele paired with it
G. Offspring of a genetic cross that inherit a pair of nonidentical alleles for a trait
H. Refers to an individual's observable traits
I. Refers to the particular genes an individual carries
J. When the effect of an allele on a trait masks that of any recessive allele paired with it
K. When both alleles of a pair are identical
L. An individual with a pair of dominant alleles, such as *AA*
M. Units of information about specific traits; passed from parents to offspring
N. Has a pair of genes for each trait, one on each of two homologous chromosomes
O. When offspring of genetic crosses inherit a pair of identical alleles for a trait, generation after generation
P. A pair of similar chromosomes, one obtained from the father and the other from the mother

11.2. MENDEL'S THEORY OF SEGREGATION [pp.180–181]

11.3. INDEPENDENT ASSORTMENT [pp.182–183]

Selected Words: "monohybrids" [p.180]

Boldfaced, Page-Referenced Terms

[p.180] monohybrid crosses ______________________________

____

[p.181] probability ______________________________

____

[p.181] Punnett-square method ______________________________

____

[p.181] testcrosses ______________________________

____

[p.181] segregation ______________________________

____

[p.182] dihybrid crosses ______________________________

____

[p.183] independent assortment ______________________________

____

Matching

Choose the most appropriate answer for each.

1. ____ Independent assortment [p.183]
2. ____ Probability [p.181]
3. ____ Punnett-square method [p.181]
4. ____ Testcross [p.181]
5. ____ Monohybrid cross [p.180]
6. ____ Dihybrid cross [p.182]
7. ____ Segregation [p.181]

A. Used to determine the genotype of an individual when it displays the dominant phenotype
B. The mathematical chance that a given event will occur
C. The separation of traits during a genetic cross
D. Genetic crosses that examine the inheritance of a single trait
E. The process by which each pair of homologous chromosomes is sorted out into gametes
F. Genetic crosses that examine the inheritance of two traits
G. A graphic means of representing the distribution of gametes and possible zygotes in a genetic cross

Problems

8. In garden pea plants, tall (*T*) is dominant over dwarf (*t*). In the cross $Tt \times tt$, the *Tt* parent would produce a gamete carrying *T* (tall) and a gamete carrying *t* (dwarf) through segregation; the *tt* parent could only produce gametes carrying the *t* (dwarf) gene. Use the Punnett-square method (refer to Figures 11.6 and 11.7 in the text) to determine the genotype and phenotype probabilities of offspring from the cross $Tt \times tt$. [pp.178–179]

a. phenotype ____

____

b. genotype ____

____

Although the Punnett-square (checkerboard) method is a common method for solving single-factor genetics problems, there is a quicker way. Only six different outcomes are possible from single-factor crosses. Studying the following relationships allows one to obtain the result of any such cross by inspection:

a. $AA \times AA$ = all AA
(Each of the four blocks of the Punnett square would be AA.)
b. $aa \times aa$ = all aa
c. $AA \times aa$ = all Aa
d. $AA \times Aa$ = 1/2 AA; 1/2 Aa
or $Aa \times AA$
(Two blocks of the Punnett square are AA, and two blocks are Aa.)
e. $aa \times Aa$ = 1/2 aa; 1/2 Aa
or $Aa \times aa$
f. $Aa \times Aa$ = 1/4 AA; 1/2 Aa; 1/4 aa
(One block in the Punnett square is AA, two blocks are Aa, and one block is aa.)

Complete the Table

9. Using the gene symbols (tall and dwarf pea plants) from question 8, determine the genotypic and phenotypic ratios of the crosses below. For assistance, apply the six Mendelian ratios listed above. [pp.180–181]

Cross	*Phenotype Ratio*	*Genotype Ratio*
a. $Tt \times tt$		
b. $TT \times Tt$		
c. $tt \times tt$		
d. $Tt \times Tt$		
e. $tt \times Tt$		
f. $TT \times tt$		
g. $TT \times TT$		
h. $Tt \times TT$		

When working genetics problems dealing with two gene pairs, you can visualize the independent assortment of gene pairs located on nonhomologous chromosomes into gametes by using a fork-line device. Assume that in humans, pigmented eyes (B) (an eye color other than blue) are dominant over blue (b) and that right-handedness (R) is dominant over left-handedness (r). To learn to solve a problem, cross the parents $BbRr \times BbRr$. A 16-block Punnett square is required, with gametes from each parent arrayed on two sides of the square (refer to Figure 11.9 in the text). The gametes receive genes through independent assortment using a fork-line method, as follows. [pp.182–183]

10. Array the gametes at the right on two sides of the Punnett square; combine these haploid gametes to form diploid zygotes within the squares. In the blank spaces below, enter the probability ratios derived within the Punnett square for the phenotypes listed. [pp.182–183]

B b R r × B b R r

BR, Br, bR, br   BR, Br, bR, br

a. ____ pigmented eyes, right-handed

b. ____ pigmented eyes, left-handed

c. ____ blue-eyed, right-handed

d. ____ blue-eyed, left-handed

11. Albinos cannot form the pigments that normally produce skin, hair, and eye color, so albinos exhibit white hair and pink eyes and skin (because the blood shows through). To be an albino, one must be homozygous recessive (*aa*) for the pair of genes that code for the key enzyme in pigment production. Suppose a woman of normal pigmentation (*A* ____) with an albino mother marries an albino man. State the kinds of pigmentation possible for this couple's children and specify the ratio of each kind of child the couple is likely to have. Show the genotype(s) and state the phenotype(s): [pp.180–181]

____

____

____

____

12. In horses, black coat color is influenced by the dominant allele (*B*), while chestnut coat color is influenced by the recessive allele (*b*). Trotting gait is due to a dominant gene (*T*), pacing gait to the recessive allele (*t*). A homozygous black trotter is crossed to a chestnut pacer. [pp.182–183]

a. What will be the appearance and gait of the F_1 and F_2 generations? ____________________

____

b. Which phenotype will be most common? ____________________

c. Which genotype will be most common? ____________________

d. Which of the potential offspring will be certain to breed true? ____________________

11.4. DOMINANCE RELATIONS [p.184]

11.5. MULTIPLE EFFECTS OF SINGLE GENES [p.185]

11.6. INTERACTIONS BETWEEN GENE PAIRS [pp.186–187]

Selected Words: *ABO blood typing* [p.184], *transfusions* [p.184], *pleio-* [p.185], *-tropic* [p.185], *Marfan syndrome* [p.185], *albinism* [p.186]

Boldfaced, Page-Referenced Terms

[p.184] incomplete dominance ____________________

____

[p.184] codominance ____________________

____

[p.184] multiple allele system ______________________________

____

[p.185] pleiotropy ______________________________

____

[p.186] epistasis ______________________________

____

Complete the Table

1. Complete the following table by supplying the type of inheritance illustrated by each example. Choose from the following gene interactions: pleiotropy, multiple allele system, incomplete dominance, codominance, and epistasis.

Type of Inheritance	*Example*
[p.184] a.	A gene with three or more alleles, such as the ABO blood-typing alleles
[p.184] b.	Pink-flowered snapdragons produced from red- and white-flowered parents
[p.184] c.	AB type blood from a gene system of three alleles—A, B, and O
[p.185] d.	The multiple phenotypic effects of the gene causing Marfan syndrome
[p.186] e.	Black, brown, or yellow fur of Labrador retrievers and comb shape in poultry

Problems

2. Genes that are not always dominant or recessive may blend to produce a phenotype of a different appearance. This is termed *incomplete dominance.* In four o'clock plants, red flower color is determined by gene R and white flower color by R', while the heterozygous condition, RR', is pink. Complete the table below by determining the phenotypes and genotypes of the offspring of the following crosses: [p.184]

Cross	*Phenotype*	*Genotype*
a. $RR \times R'R' =$		
b. $R'R' \times R'R' =$		
c. $R'R \times RR =$		
d. $RR \times RR =$		

Sickle-cell anemia is a genetic disease in which children who are homozygous for a defective gene produce defective hemoglobin (Hb^S/Hb^S). The genotypes of normal persons are Hb^A/Hb^A. If the level of blood oxygen drops below a certain level in a person with the Hb^S/Hb^S genotype, the hemoglobin chains stiffen and cause the red blood cells to form sickle, or crescent, shapes. These cells clog and rupture capillaries, resulting in oxygen-deficient tissues where metabolic wastes collect. Several body functions are badly damaged. Severe anemia and other symptoms develop, and death nearly always occurs before adulthood. The sickle-cell gene is considered *pleiotropic.* Persons who are heterozygous (Hb^A/Hb^S) are said to possess *sickle-cell trait.* They are able to produce enough normal hemoglobin molecules to appear normal, but their red blood cells will sickle if they encounter oxygen tension (such as at high altitudes).

3. A man whose sister died of sickle-cell anemia married a woman whose blood is found to be normal. What advice would you give this couple about the inheritance of this disease as they plan their family? [p.185] ______

______

______

______

4. If a man and a woman, each with sickle-cell trait, were planning to marry, what information could you provide them regarding the genotypes and phenotypes of their future children? [p.185] ______

______

______

______

In one example of a *multiple allele system* with *codominance,* the three genes I^Ai, I^Bi, and i produce proteins found on the surfaces of red blood cells that determine the four blood types in the ABO system, namely, A, B, AB, and O. Genes I^A and I^B are both dominant over i but not over each other; they are codominant. Recognize that blood types A and B may be heterozygous or homozygous (I^AI^A, I^Ai or I^BI^B, I^Bi), whereas blood type O is homozygous (ii). Indicate the genotypes and phenotypes of the offspring and their probabilities from the parental combinations in exercises 5–9. [p.184]

5. $I^Ai \times I^AI^B =$ ______

6. $I^Bi \times I^Ai =$ ______

7. $I^AI^A \times ii =$ ______

8. $ii \times ii =$ ______

9. $I^AI^B \times I^AI^B =$ ______

In one type of gene interaction, two alleles of a gene mask the expression of alleles of another gene, so that some expected phenotypes never appear. *Epistasis* is the term for such interactions. Work the following problems on scratch paper to understand epistatic interactions. In sweet peas, genes *C* and *P* are necessary for colored flowers. In the absence of either (_ _ *pp* or *cc*_ _) or both (*ccpp*), the flowers are white. What will be the color of the offspring of the following crosses, and in what proportions will they appear? [p.186]

10. $CcPp \times ccpp =$ ______

11. $CcPP \times Ccpp =$ ______

12. $Ccpp \times ccPp =$ ______

In the inheritance of the coat (fur) color of Labrador retrievers, allele *B* specifies black, which is dominant over brown (chocolate), *b*. Allele *E* permits full deposition of color pigment, but two recessive alleles, *ee*, reduce deposition, and a yellow coat results.

13. Predict the phenotypes of the coat color and their proportions resulting from the following cross: [p.186]

$BbEe \times Bbee =$ ______

______

In poultry, an epistatic interaction occurs in which two genes produce a phenotype that neither gene can produce alone. The two interacting genes (*R* and *P*) produce comb shape in chickens. The possible genotypes and phenotypes are as follows:

Genotypes	*Phenotypes*
*R_ P_*	walnut comb
R_pp	rose comb
*rrP_*	pea comb
rrpp	single comb

Hint: Where a blank appears in the preceding genotypes, both the dominant and recessive symbol in that blank yield the same phenotype.

14. What are the genotype and phenotype ratios of the offspring of a heterozygous walnut-combed male and a single-combed female? [p.187] ____

____

____

____

15. Cross a homozygous rose-combed rooster with a homozygous single-combed hen, and list the genotype and phenotype ratios of their offspring. [p.187] ____

____

____

____

11.7. HOW CAN WE EXPLAIN LESS PREDICTABLE VARIATIONS? [pp.188–189]

11.8. ENVIRONMENTAL EFFECTS ON PHENOTYPE [pp.190–191]

Selected Words: *camptodactyly* [p.188], "bell-curves" [p.189]

Boldfaced, Page-Referenced Terms

[p.188] continuous variation ____

____

Choice

For questions 1–5, choose from the following primary contributing factors:

a. environment b. a number of genes affect a trait

1. ____ Height of human beings [p.189]
2. ____ Continuous variation in a trait [p.188]
3. ____ Flower color in *Hydrangea macrophylla* [p.191]
4. ____ The range of eye colors in the human population [p.188]
5. ____ Heat-sensitive version of one of the enzymes required for melanin production in Himalayan rabbits [p.190]

Self-Quiz

____ 1. The best statement of Mendel's principle of independent assortment is that ________. [pp.182–183]
   a. one allele is always dominant to another
   b. hereditary units from the male and female parents are blended in the offspring
   c. the two hereditary units that influence a certain trait separate during gamete formation
   d. each hereditary unit is inherited separately from other hereditary units

____ 2. All the different molecular forms of the same gene are called ________. [p.179]
   a. hybrids
   b. alleles
   c. autosome
   d. locus

____ 3. If two heterozygous individuals are crossed in a monohybrid cross involving complete dominance, the expected phenotypic ratio is ________. [pp.180–181]
   a. 3:1
   b. 1:1:1:1
   c. 1:2:1
   d. 1:1
   e. 9:3:3:1

____ 4. In the F_2 generation of a cross between a red-flowered snapdragon (homozygous) and a white-flowered snapdragon, the expected phenotypic ratio of the offspring is ________. [p.184]
   a. 3/4 red, 1/4 white
   b. 100 percent red
   c. 1/4 red, 1/2 pink, 1/4 white
   d. 100 percent pink

____ 5. In a testcross, F_1 hybrids are crossed to an individual known to be ________ for the trait. [p.181]
   a. heterozygous
   b. homozygous dominant
   c. homozygous
   d. homozygous recessive

____ 6. The tendency for dogs to bark while trailing is determined by a dominant gene, *S*, whereas silent trailing is due to the recessive gene, *s*. In addition, erect ears, *D*, is dominant over drooping ears, *d*. What combination of offspring would be expected from a cross between two erect-eared barkers who are heterozygous for both genes? [pp.182–183]
   a. 1/4 erect barkers, 1/4 drooping barkers, 1/4 erect silent, 1/4 drooping silent
   b. 9/16 erect barkers, 3/16 drooping barkers, 3/16 erect silent, 1/16 drooping silent
   c. 1/2 erect barkers, 1/2 drooping barkers
   d. 9/16 drooping barkers, 3/16 erect barkers, 3/16 drooping silent, 1/16 erect silent

____ 7. If a mother has type O blood, which of the following blood types could not be present in her children? [p.184]
   a. type A
   b. type B
   c. type O
   d. type AB
   e. all of the above are possible

____ 8. A single gene that affects several seemingly unrelated aspects of an individual's phenotype is said to be ________. [p.185]
   a. pleiotropic
   b. epistatic
   c. allelic
   d. continuous

____ 9. Suppose two individuals, each heterozygous for the same characteristic, are crossed. The characteristic involves complete dominance. The expected genotype ratio of their progeny is ________. [pp.180–181]
   a. 1:2:1
   b. 1:1
   c. 100 percent of one genotype
   d. 3:1

_____ 10. If the two homozygous classes in the F_1 generation of the cross in exercise 9 are allowed to mate, the observed genotype ratio of the offspring will be _____________. [pp.180–181]

a. 1:1
b. 1:2:1
c. 100 percent of one genotype
d. 3:1

_____ 11. Applying the types of inheritance studied in this chapter of the text, the skin color trait in humans exhibits _____________. [pp.188–189]

a. pleiotropy
b. epistasis
c. environmental effects
d. continuous variation

Chapter Objectives/Review Questions

1. What was the prevailing method of explaining the inheritance of traits before Mendel's work with pea plants? [p.178]
2. Explain why Mendel had the ideal background for studying inheritance in plants. [pp.178–179]
3. _________ are units of information about specific traits; they are passed from parents to offspring. [p.179]
4. What is the general term applied to the location of a gene on a chromosome? [p.179]
5. Define *allele;* how many types of alleles are present in the genotypes *Cc? cc? CC?* [p.179]
6. Explain the meaning of a *true-breeding lineage.* [p.179]
7. When two alleles of a pair are identical, it is a _________ condition; if the two alleles are different, it is a _________ condition. [p.179]
8. Define the terms *dominant* and *recessive.* [p.179]
9. Distinguish between the genotype and the phenotype of an individual. [p.179]
10. Offspring of _________ crosses are heterozygous for the one trait being studied. [p.180]
11. Mendel's theory of _________ states that during meiosis, the two genes of each pair separate from each other and end up in different gametes. [p.180]
12. Explain why probability is useful to genetics. [p.181]
13. Use the Punnett-square method of solving genetics problems. [p.181]
14. Explain the purpose of a testcross. [p.181]
15. Solve dihybrid genetic crosses. [pp.182–183]
16. Mendel's theory of _________ _________ states that gene pairs on homologous chromosomes tend to be sorted into one gamete or another independent of how gene pairs on other chromosomes are sorted out. [pp.182–183]
17. Distinguish among complete dominance, incomplete dominance, and codominance. [p.184]
18. Why are human blood groups used as an example of a multiple allele system? [p.184]
19. Explain why Marfan's syndrome is given as an example of pleiotropy. [p.185]
20. Gene interaction involving two alleles of a gene that mask expression of another gene's alleles is called _________. [p.186]
21. List possible explanations for less predictable trait variations that are observed. [pp.188–189]
22. Explain what is meant by *continuous variation.* [pp.188–189]
23. Himalayan rabbits and garden hydrangeas are good examples of environmental effects on _________ expression. [pp.190–191]

Integrating and Applying Key Concepts

1. Solve the following genetics problem: In garden peas, one pair of alleles controls the height of the plant and a second pair of alleles controls flower color. The allele for tall (*D*) is dominant to the allele for dwarf (*d*), and the allele for purple (*P*) is dominant to the allele for white (*p*). A tall plant with purple flowers crossed with a tall plant with white flowers produces offspring that are 3/8 tall purple, 3/8 tall white, 1/8 dwarf purple, and 1/8 dwarf white. What are the genotypes of the parents?

2. As a breeder of exotic plants, you advertise that you have produced a true-breeding yellow flowered line. Assuming that yellow flower color (*Y*) is dominant over the less-desirable color (w), how do you test to ensure that your line is not heterozygous for the trait?
3. Height and eye color in humans are frequently given as examples of continuous variation. What are some other human traits that may display continuous variation? What about in other species?

12

HUMAN GENETICS

Interactive Exercises

The Philadelphia Story [pp.194–195]

12.1. CHROMOSOMES AND INHERITANCE [p.196]

12.2. FOCUS ON SCIENCE: *Karyotyping Made Easy* [p.197]

Selected Words: *Philadelphia chromosome* [p.194], leukemia [p.194], *spectral* karyotyping [p.194], Walther Flemming [p.195], August Weismann [p.195], *wild-type* allele [p.196], *mutant* [p.196], in vitro [p.197], *Colchicum* [p.197]

Boldfaced, Page-Referenced Terms

[p.194] karyotype ____

____

[p.196] genes ____

____

[p.196] homologous chromosomes ______________________________

____

[p.196] alleles ______________________________

____

[p.196] crossing over ______________________________

____

[p.196] genetic recombination ______________________________

____

[p.196] independent assortment ______________________________

____

[p.196] X chromosome ______________________________

____

[p.196] Y chromosome ______________________________

____

[p.196] sex chromosomes ______________________________

____

[p.196] autosomes ______________________________

____

Fill-in-the-Blanks

The first abnormal chromosome to be associated with cancer was named the (1) ______________ [p.194] chromosome. (2) ______________ [p.194] arise when stem cells in bone marrow overproduce the white blood cells necessary for the body's housekeeping and defense. A preparation of metaphase chromosomes based on their defining features is a (3) ______________ [p.194]. A Philadelphia chromosome is not easy to identify in standard karyotypes but should be easier using the newer (4) ______________ [p.194] karyotyping, which artificially colors chromosomes in an unambiguous way.

Heritable traits arise from units of molecular information (located on chromosomes) that are known as (5) ______________ [p.196]. Paired chromosomes that have the same length, shape, and gene sequence and interact during meiosis are known as (6) ______________ [p.196] chromosomes. (7) ______________ [p.196] are different molecular forms of the same gene that have arisen through mutation. The most common form of a gene is called a (8) ______________-type [p.196] allele. An event known as (9) ______________ ______________ [p.196] functions to exchange gene segments between the members of a homologous pair and thus provide genetic (10) ______________ [p.196]. Human X and Y chromosomes that determine gender are examples of (11) ______________ [p.196] chromosomes. Chromosomes that are identical in both sexes of a species are called (12) ______________ [p.196].

The (13) ______________ [p.197] of an individual (or species) is a preparation of sorted metaphase chromosomes. To prepare a karyotype, technicians culture cells (14) ______________ ______________ [p.197] (in glass). During culture, a chemical compound named (15) ______________ [p.197] is added to the culture medium. The purpose of this chemical is to block spindle formation and arrest nuclear divisions at the (16) ______________ [p.197] stage of mitosis, when the chromosomes are the most condensed and easiest to identify in cells. Cells are then moved to the bottoms of culture tubes by a spinning force called (17) ______________ [p.197]. Following saline treatment to separate the chromosomes, cells are fixed, stained, and photographed for study. The photograph is cut apart, one chromosome at a time, and the individual cutouts are arranged according to size, shape, and length of arms. Then all the pairs of homologous chromosomes are horizontally aligned by their (18) ______________ [p.197].

12.3. SEX DETERMINATION IN HUMANS [pp.198–199]

12.4. WHAT MENDEL DIDN'T KNOW: CROSSOVERS AND RECOMBINATION [pp.200–201]

Selected Words: *nonsexual* traits [p.198], *SRY* gene [p.198], *Drosophila melanogaster* [p.200], *X-linked genes* [p.200], *Y-linked genes* [p.200], Thomas Hunt Morgan [p.200], *Zea mays* [p.200], *color blindness* [p.201], *hemophilia* [p. 201]

Boldfaced, Page-Referenced Terms

[p.200] linkage group ____

____

[p.200] reciprocal crosses ____

____

Matching

Choose the most appropriate answer for each term.

1. _____ X-linked genes [p.200]
2. _____ linkage group [p.200]
3. _____ *SRY* [p.198]
4. _____ Y-linked genes [p.200]
5. _____ reciprocal crosses [p.200]

A. Male-determining gene found on the Y chromosome; expression leads to testes formation
B. In the first cross, one parent displays the trait of interest, whereas in the second cross, the other parent displays it
C. The block of genes located on each type of chromosome; these genes tend to travel together in inheritance
D. Found only on the Y chromosome
E. Found only on the X chromosome

Complete the Table

6. Complete the Punnett-square table below, which brings Y-bearing and X-bearing sperm together randomly in fertilization. [p.198]

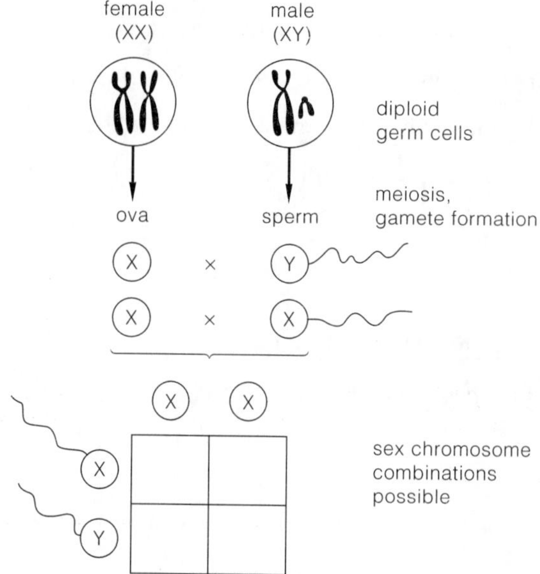

Dichotomous Choice

Answer the following questions related to the Punnett-square table just completed.

7. Male humans transmit their Y chromosome only to their (sons/daughters). [p.198]
8. Male humans receive their X chromosome only from their (mothers/fathers). [p.198]
9. Human mothers and fathers each provide an X chromosome for their (sons/daughters). [p.198]

Problems

10. In Morgan's experiments with *Drosophila,* white eyes are determined by a recessive X-linked gene, while the wild-type or normal brick-red eyes are due to its dominant allele. Use symbols of the following types: X^wY = a white-eyed male; X^+X^+ = a homozygous normal red female. [p.200]

    a. What offspring can be expected from a cross of a white-eyed male and a homozygous normal female? [p.200] ____________________

    ____

    ____

    ____

    b. In addition, show the genotypes and phenotypes of the F_2 offspring. [p.200] ____________________

    ____

    ____

    ____

11. Which of the following represents a chromosome that has undergone crossover and recombination? It is assumed that the organism involved is heterozygous with the following genotype: [p.201] ______________

____

____

____

A| |a
B| |b

|A a| B| |a
|B b| A| |B

a. b. c. d.

12. If genes A and B are twice as far apart on a chromosome as genes C and D, how often would you expect that crossing over occurs between genes A and B as compared to between genes C and D? [pp.200–201]

____

____

____

13. Explain how a child can have the genotype *Ab/Ab* when one parent has a chromosome with genes *A* and *b* linked (*Ab/aB*) and the other does not (*AB/ab*). [pp.200–201] ______________

____

____

____

12.5. HUMAN GENETIC ANALYSIS [pp.202–203]

Selected Words: *polydactyly* [p.202], *Huntington disease* [p.202], *genetic* disease [p.203]

Boldfaced, Page-Referenced Terms

[p.202] pedigree ______________

____

[p.203] genetic abnormality ______________

____

[p.203] genetic disorder ______________

____

[p.203] syndrome ______________

____

[p.203] disease ______________

____

Matching

The following standardized symbols are used in constructing pedigree charts. Choose the appropriate description for each. [p.202]

1. ____ ■●
2. ____ ●
3. ____ ◆
4. ____ ■
5. ____ **I, II, III, IV. . .**
6. ____ ■-●
7. ____ ● ■ ■ ● (joined by a line above)

A. Individual showing the trait being tracked
B. Male
C. Successive generations
D. Offspring with birth order left to right
E. Female
F. Sex unspecified; numerals present indicate number of children
G. Marriage/mating

Short Answer

8. Distinguish among these terms: *genetic abnormality, genetic disorder, syndrome,* and *genetic disease.* [p.203]

____

____

____

Complete the Table

9. Complete the following table by indicating whether the genetic disorder listed is caused by inheritance that is autosomal recessive, autosomal dominant, or X-linked recessive or that involves changes in chromosome number or in chromosome structure. [all from p.203]

Genetic Disorder	*Inheritance Pattern*
a. Galactosemia	
b. Achondroplasia	
c. Hemophilia	
d. Anhidrotic dysplasia	
e. Huntington disorder	
f. Turner and Down syndrome	
g. Achoo syndrome	
h. Cri-du-chat syndrome	
i. XYY condition	
j. Color blindness	
k. Fragile X syndrome	
l. Progeria	
m. Klinefelter syndrome	
n. Androgen insensitive syndrome	
o. Phenylketonuria	

12.6. EXAMPLES OF INHERITANCE PATTERNS [pp.204–205]

12.7. FOCUS ON HEALTH: *Progeria—Too Young to Be Old* [p.206]

Selected Words: *galactosemia* [p.204], *Huntington disease* [p.204], *achondroplasia* [p.204], *color blindness* [p.205], *hemophilia A* [p.205], fragile X syndrome [p.205], expansion mutations [p.205], *Hutchinson–Gilford progeria syndrome* [p.206]

Choice

For questions 1–17, choose from the following patterns of inheritance; some items may require more than one letter.

a. autosomal recessive b. autosomal dominant c. X-linked recessive

1. _____ The trait is expressed in heterozygous females. [p.204]
2. _____ Heterozygotes can remain undetected. [pp.204–205]
3. _____ The trait appears in each generation. [p.204]
4. _____ The recessive phenotype shows up far more often in males than in females. [p.205]
5. _____ Both parents may be heterozygous normal. [p.204]
6. _____ If one parent is heterozygous and the other homozygous recessive, there is a 50 percent chance any child of theirs will be heterozygous. [p.204]
7. _____ The allele is usually expressed, even in heterozygotes. [p.204]
8. _____ The trait is expressed in heterozygous females. [p.204]
9. _____ Heterozygous normal parents can expect that one-fourth of their children will be affected by the disorder. [p.204]
10. _____ A son cannot inherit the recessive allele from his father, but his daughter can. [p.205]
11. _____ Females can mask this gene; males cannot. [p.205]
12. _____ The trait is expressed in heterozygotes of either sex. [p.204]
13. _____ Heterozygous women transmit the allele to half their offspring, regardless of sex. [p.204]
14. _____ Individuals displaying this type of disorder will always be homozygous for the trait. [p.204]
15. _____ The trait is expressed in both the homozygote and the heterozygote. [p.204]
16. _____ Heterozygous females will transmit the recessive gene to half their sons and half their daughters. [p.205]
17. _____ If both parents are heterozygous, there is a 50 percent chance that each child will be heterozygous. [p.204]

Problems

18. The autosomal allele that causes galactosemia (*g*) is recessive to the allele for normal lactose metabolism (*G*). A normal woman whose father had galactosemia marries a man with galactosemia who had normal parents. They have three children, two normal and one with galactosemia. List the genotypes for each person involved. [p.204] _____

19. Huntington disease is a rare form of autosomal dominant inheritance, *H;* the normal gene is *h*. The disease causes progressive degeneration of the nervous system, with onset exhibited near middle age. An apparently normal man in his early twenties learns that his father has recently been diagnosed as having Huntington disorder. What are the chances that the son will develop this disorder? [p.204] _____

20. A color-blind man and a woman with normal vision whose father was color blind have a son. Color blindness, in this case, is caused by an X-linked recessive gene. If only the male offspring are considered, what is the probability that their son is color blind? [p.205] ______

21. Hemophilia A is caused by an X-linked recessive gene. A woman who is seemingly normal but whose father was a hemophiliac marries a normal man. What proportion of their sons will have hemophilia? What proportion of their daughters will have hemophilia? What proportion of their daughters will be carriers? [p.205] ______

22. The accompanying pedigree shows the pattern of inheritance of color blindness in a family (people with the trait are indicated by black circles). What is the chance that the third-generation female indicated by the arrow (below) will have a color-blind son if she marries a normal male? a color-blind male? [p.205]

______

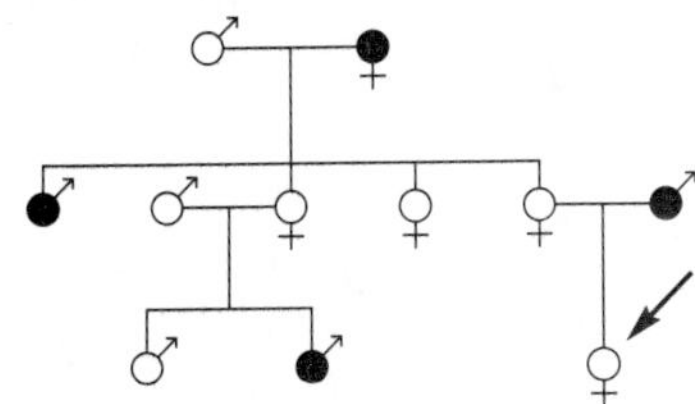

12.8. CHANGES IN CHROMOSOME STRUCTURE [pp.206–207]

12.9. CHANGES IN CHROMOSOME NUMBER [pp.208–209]

12.10. CASE STUDIES: CHANGES IN THE NUMBER OF SEX CHROMOSOMES [pp. 210–211]

Selected Words: *non*homologous chromosome [p.206], *cri-du-chat* [p.207], *miscarriages* [p.208], *tetra*ploid [p.208], trisomic [p.208], monosomic [p.208], *Down syndrome* [p.208], prenatal diagnosis [p.209], *Turner syndrome* [p.210], *XXX syndrome* [p.210], *Klinefelter syndrome* [pp.210–211], *XYY condition* [p.211]

Boldfaced, Page-Referenced Terms

[p.206] duplications ______

______

[p.206] inversion ______

______

[p.206] translocation ______

______

[p.207] deletion ______

______

[p.208] aneuploidy ______

______

[p.208] polyploidy ______

______

[p.208] nondisjunction ______________________________

____

[p. 208] mosaicism ______________________________

____

[p.209] double-blind studies ______________________________

____

Labeling and Matching

On rare occasions, chromosome structure becomes abnormally rearranged. Such changes may have profound effects on the phenotype of an organism. Label the following diagrams of abnormal chromosome structure as a deletion, a duplication, an inversion, or a translocation. Complete the exercise by matching and entering the letter of the proper description in the parentheses following each label.

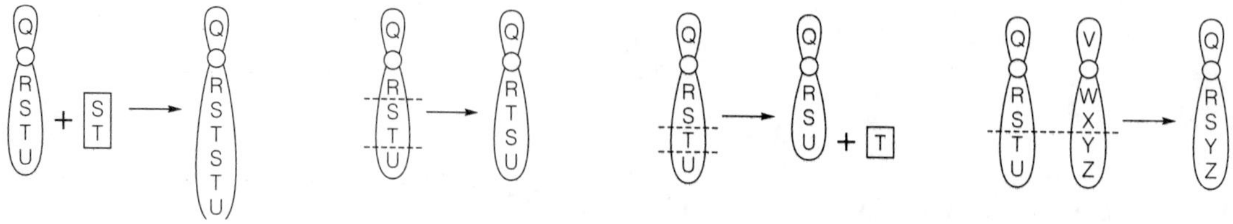

Q R S T U + S T → Q R S T S T U   Q R S T U → Q R T S U   Q R S T U → Q R S U + T   Q R S T U   V W X Y Z → Q R S Y Z

1. ______________ ( ) [p.206]
2. ______________ ( ) [p.206]
3. ______________ ( ) [p.207]
4. ______________ ( ) [p.206]

A. The loss of a chromosome segment; an example is the cri-du-chat disorder
B. A gene sequence in excess of its normal amount in a chromosome; an example is the fragile X syndrome
C. A chromosome segment that separated from the chromosome and then was inserted at the same place but in reverse; this alters the position and order of the chromosome's genes; possibly promoted human evolution
D. The transfer of part of one chromosome to a nonhomologous chromosome; an example is when chromosome 14 ends up with a segment of chromosome 8; the Philadelphia chromosome is also an example

Complete the Table

5. Complete the following table to summarize the major categories and mechanisms of chromosome number change in organisms. [all p.208]

Category of Change	*Description*
a. Aneuploidy	
b. Polyploidy	
c. Nondisjunction	

Short Answer

6. If a nondisjunction occurs at anaphase I of the first meiotic division, what will be the proportion of abnormal gametes (for the chromosomes involved in the nondisjunction)? [p.208] ______________________

____

____

____

7. If a nondisjunction occurs at anaphase II of the second meiotic division, what will be the proportion of abnormal gametes (for the chromosomes involved in the nondisjunction)? [p.208] ______________________

____

____

____

8. Contrast the effects of polyploidy in plants and humans. [p.208] ______________________

____

____

____

9. Define the following terms: *tetraploid, trisomic,* and *monosomic.* [p.208] ______________________

____

____

____

Choice

For questions 10–19, choose from the following:

a. Down syndrome   b. Turner syndrome   c. Klinefelter syndrome   d. XYY condition

10. ____ XXY male [p.211]
11. ____ Ovaries nonfunctional and secondary sexual traits fail to develop at puberty [p.210]
12. ____ Testes smaller than normal, sparse body hair, and some breast enlargement [pp.210–211]
13. ____ Could only be caused by a nondisjunction in males [p.211]
14. ____ As a group, they tend to be cheerful and affectionate; about 40 percent develop heart defects [pp.208–209]
15. ____ XO female; often abort early; distorted female phenotype [p.210]
16. ____ Males that tend to be taller than average; some mildly retarded but most are phenotypically normal [p.211]
17. ____ Injections of testosterone reverse feminized traits but not the fertility [pp.210–211]
18. ____ Trisomy 21; muscles and muscle reflexes weaker than normal [pp.208–209]
19. ____ At one time these males were thought to be genetically predisposed to become criminals [p.211]

12.11. FOCUS ON BIOETHICS: *Prospects in Human Genetics* [pp.212–213]

Selected Words: *phenylketonuria* [p.212], *cleft lip* [p.212], *genetic counseling* [p.212], *prenatal diagnosis* [p.212], *embryo* [p.212], *fetus* [p.212], *amniocentesis* [p.212], *chorionic villi sampling* [pp.212–213], *fetoscopy* [p.213], *preimplantation diagnosis* [p.213], *pre*-pregnancy stage [p.213], *"test-tube babies"* [p.213], *Fanconi anemia* [p. 213]

Boldfaced, Page-Referenced Terms

[p.212] abortion ____

____

[p.213] in vitro fertilization ____

____

Complete the Table

1. Complete the following table, which summarizes methods of dealing with the problems of human genetics. Choose from phenotypic treatments, abortion, genetic screening, preimplantation diagnosis, genetic counseling, and prenatal diagnosis.

Method	*Description*
[pp.212–213] a.	Detects embryo or fetal conditions before birth; may use amniocentesis, CVS, in vitro fertilization, and/or fetoscopy; often used in a pregnancy at risk in a mother 45 years or older
[p.213] b.	A controversial option if prenatal diagnosis reveals a severe heritable problem; "pro-life" and "pro-choice" factions
[p.212] c.	If a severe heritable problem exists, it includes diagnosis of parental genotypes, detailed pedigrees, and genetic testing for known metabolic disorders; geneticists may be contacted for predictions
[p.213] d.	Relies on in vitro fertilization; a fertilized egg mitotically divides into a ball of eight cells that provides a cell to be analyzed for genetic defects
[p.213] e.	Suppressing or minimizing symptoms of genetic disorders by surgical intervention or by controlling diet or environment; PKU and cleft lip are examples of dietary control and surgery, respectively
[p. 212] f.	Large-scale programs to detect affected persons or carriers in a population; early detection may allow introduction of preventive measures before symptoms develop; PKU screening for newborns is an example

Self-Quiz

____ 1. All the genes located on a given chromosome compose a ________. [p.200]
a. karyotype
b. bridging cross
c. wild-type allele
d. linkage group

____ 2. Chromosomes other than those involved in sex determination are known as ________. [p.196]
a. nucleosomes
b. heterosomes
c. alleles
d. autosomes

____ 3. The farther apart two genes are on a chromosome, ________. [p.201]
a. the less likely that crossing over and recombination will occur between them
b. the greater will be the frequency of crossing over and recombination between them
c. the more likely they are to be in two different linkage groups
d. the more likely they are to be segregated into different gametes when meiosis occurs

____ 4. Karyotype analysis is ________. [p.197]
a. a means of detecting and reducing mutagenic agents
b. a surgical technique that separates chromosomes which have failed to segregate properly during meiosis II
c. used in prenatal diagnosis to detect chromosomal mutations and metabolic disorders in embryos
d. a process that substitutes normal alleles for defective ones

____ 5. Which of the following did Morgan and his research group not do? [pp.198–199]
a. They isolated and kept under culture fruit flies with the sex-linked recessive white-eyed trait.
b. They developed the technique of amniocentesis.
c. They discovered X-linked genes.
d. Their work reinforced the concept that each gene is located on a specific chromosome.

____ 6. Red–green color blindness is a sex-linked recessive trait in humans. A color-blind woman and a man with normal vision have a son. What are the chances that the son is color-blind? If the parents ever have a daughter, what is the chance that a daughter will be color-blind? (Consider only the female offspring.) [p.205]
a. 100 percent, 0 percent
b. 50 percent, 0 percent
c. 100 percent, 100 percent
d. 50 percent, 100 percent
e. none of the above

____ 7. Suppose that a hemophilic male (X-linked recessive allele) and a female carrier for the hemophilic trait have a nonhemophilic daughter with Turner syndrome. Nondisjunction could have occurred in ________. [pp.205,210]
a. either parent
b. neither parent
c. the father only
d. the mother only
e. the nonhemophilic daughter

____ 8. Nondisjunction involving the X chromosome occurs during oogenesis and produces two kinds of eggs, XX and O (no X chromosome). If normal Y sperm fertilize the two types, which genotypes are possible? [p.210]
a. XX and XY
b. XXY and YO
c. XYY and XO
d. XYY and YO
e. YY and XO

_____ 9. Of all phenotypically normal males in prison, the type once thought to be genetically predisposed to becoming criminals was the group with _____________ . [p.211]
   a. XXY disorder
   b. XYY disorder
   c. Turner syndrome
   d. Down syndrome
   e. Klinefelter syndrome

_____ 10. Amniocentesis is _____________ . [pp.212–213]
   a. a surgical means of repairing deformities
   b. a form of chemotherapy that modifies or inhibits gene expression or the function of gene products
   c. used in prenatal diagnosis; a small sample of amniotic fluid is drawn to detect chromosomal mutations and metabolic disorders in embryos
   d. a form of gene-replacement therapy
   e. a diagnostic procedure; cells for analysis are withdrawn from the chorion

Chapter Objectives/Review Questions

1. A _________ is a preparation of metaphase chromosomes based on their defining features. [p.194]
2. The units of information about heritable traits are known as _________ . [p.196]
3. Diploid (2*n*) cells have pairs of _________ chromosomes. [p.196]
4. _________ are different molecular forms of the same gene that arise through mutation; a _________-type allele is the most common form of a gene. [p.196]
5. State the circumstances required for crossing over and describe the results. [p.196]
6. Name and describe the sex chromosomes in human males and females. [p.196]
7. Human X and Y chromosomes fall in the general category of _________ chromosomes; all other chromosomes in an individual's cells are the same in both sexes and are called _________ . [p.196]
8. Define *karyotype;* briefly describe its preparation and value. [p.197]
9. Explain meiotic segregation of sex chromosomes to gametes and the subsequent random fertilization that determines sex in many organisms. [p.198]
10. A newly identified region of the Y chromosome called _________ appears to be the master gene for male sex determination. [p.198]
11. All the genes on a specific chromosome are called a _________ group. [p.200]
12. Define the terms *X-linked* and *Y-linked genes.* [p.200]
13. In whose laboratory was sex linkage in fruit flies discovered? When? [p.200]
14. State the relationship between crossover frequency and the location of genes on a chromosome. [p.200]
15. The probability that a crossover will disrupt gene linkage is proportional to the _________ that separates them. [p.201]
16. The farther apart two genes are on a chromosome, the _________ will be the frequency of crossing over and therefore of genetic recombination between them. [p.201]
17. A _________ chart or diagram is used to study genetic connections between individuals. [p.202]
18. A genetic _________ is a rare version of a trait, whereas an inherited genetic _________ causes mild to severe medical problems. [p.203]
19. A _________ is a recognized set of symptoms that characterize a given disorder. [p.203]
20. Describe what is meant by a genetic disease. [p.203]
21. Carefully characterize patterns of autosomal recessive inheritance, autosomal dominant inheritance, and X-linked recessive inheritance. [pp.204–205]
22. Describe the Hutchinson–Gilford progeria syndrome and its probable mode of inheritance. [p.206]
23. A(n) _________ is the loss of a chromosome segment; a(n) _________ is a gene sequence separated from a chromosome, then inserted at the same place but in reverse; a(n) _________ is a repeat of several gene sequences on the same chromosome; a(n) _________ is the transfer of part of one chromosome to a nonhomologous chromosome. [pp.206–207]

24. When gametes or cells of an affected individual end up with one more or one less than the parental number of chromosomes, the condition is known as ________; relate this concept to monosomy and trisomy. [p.208]
25. Having three or more complete sets of chromosomes is called ________. [p.208]
26. ________ is the failure of the chromosomes to separate in either meiosis or mitosis. [p.208]
27. Trisomy 21 is known as ________ syndrome; Turner syndrome has the chromosome constitution ________; XXY chromosome constitution is ________ syndrome; taller-than-average males with sometimes slightly lowered IQs have the ________ condition. [pp.208–211]
28. What are the general characteristics of a person with Down syndrome? [pp.208–209]
29. Do all individuals with Down syndrome develop the same symptoms? [p.209]
30. What is the relationship between maternal age and frequency of Down syndrome births? [p.209]
31. Explain what is meant by double-blind studies. [p.211]
32. Define *phenotypic treatment* and describe one example. [p.212]
33. List some benefits to society of genetic screening and genetic counseling. [p.212]
34. Explain the procedures and purpose of three types of prenatal diagnosis: amniocentesis, chorionic villi analysis, and fetoscopy; compare the risks. [pp.212–213]
35. Discuss some of the ethical considerations that might be associated with a decision to induce abortion. [p.213]
36. A procedure known as *preimplantation diagnosis* relies on ________ - ________ fertilization. [p.213]

Integrating and Applying Key Concepts

1. The parents of a young boy bring him to their doctor. They explain that the boy does not seem to be going through the same vocal developmental stages as his older brother. The doctor orders a common cytogenetics test to be done, and it reveals that the young boy's cells contain two X chromosomes and one Y chromosome. Describe the test that the doctor ordered and explain how and when such a genetic result, XXY, most logically occurred.
2. Solve the following genetics problem. Show rationale, genotypes, and phenotypes. A husband sues his wife for divorce, arguing that she has been unfaithful. His wife gave birth to a girl with a fissure in the iris of her eye, an X-linked recessive trait. Both parents have normal eye structure. Can the genetic facts be used to argue for the husband's suit? Explain your answer.

13

DNA STRUCTURE AND FUNCTION

Interactive Exercises

Cardboard Atoms and Bent-Wire Bonds [pp.216–217]

13.1. DISCOVERY OF DNA FUNCTION [pp.218–219]

Selected Words: J. F. Miescher [p.216], L. Pauling [p.216], J. Watson and F. Crick [p.216], F. Griffith [p.218], *Streptococcus pneumoniae* [p.218], O. Avery [p.218], A. Hershey and M. Chase [pp.218–219], *Escherichia coli* [pp.218–219], ^{35}S and ^{32}P [p.219]

Boldfaced, Page-Referenced Terms

[p.216] deoxyribonucleic acid, DNA ________________________________

____

[p.218] bacteriophages ________________________________

____

Complete the Table

1. Complete the following table, which traces the discovery of DNA function.

Investigators	*Year*	*Contribution*
[p.216] a. Miescher	1868	
[p.218] b. Griffith	1928	
[p.218] c. Avery (also MacLeod and McCarty)	1944	
[pp.218–219] d. Hershey and Chase	1952	

Fill-in-the-Blanks

A bacteriophage is a kind of (2) ________________ [p.218] that can infect (3) ________________ [p.218] cells. Enzymes from the (2) take over enough of the host cell's metabolic processes to make substances that are necessary to construct new (4) ________________ [p.218]. Some of these substances are (5) ________________ [p.219] that can be labeled with a radioisotope of sulfur, ^{35}S. The genetic material of the bacteriophage used in the experiments by Hershey and Chase was labeled with the radioisotope of phosphorus known as (6) ________________ [p.219]. When (7) ________________ [p.219] particles were allowed to infect *Escherichia coli* cells, the (8) ________________ [p.219] radioisotope remained outside the bacterial cells; the (9) ________________ [p.219] was part of the (10) ________________ ________________ [p.219] injected *into* the bacterial cells. Through many experiments, researchers accumulated strong evidence that (11) ________________ [p.219], not (12) ________________ [p.219], serves as the molecule of inheritance in all living cells.

13.2. DNA STRUCTURE [pp.220–221]

13.3. FOCUS ON BIOETHICS: *Rosalind's Story* [p.222]

Selected Words: E. Chargaff [p.220], R. Franklin and M. Wilkins [p.220], DNA double helix [p.221], X-ray crystallography [p.222]

Boldfaced, Page-Referenced Terms

[p.220] nucleotide ____

____

[p.220] adenine, A ____

____

[p.220] guanine, G ____

____

[p.220] thymine, T ____

____

[p.220] cytosine, C ____

____

[p.220] X-ray diffraction images ____

____

Short Answer

1. List the three parts of a nucleotide. [p.220] ____

____

____

____

Labeling

Four nucleotides are illustrated below. [All are from text p.218.] In the blank, label each nitrogen-containing base as guanine, thymine, cytosine, or adenine. In the parentheses following each blank, indicate whether that nucleotide base is a purine (pu) or a pyrimidine (py). (See next section for definitions.)

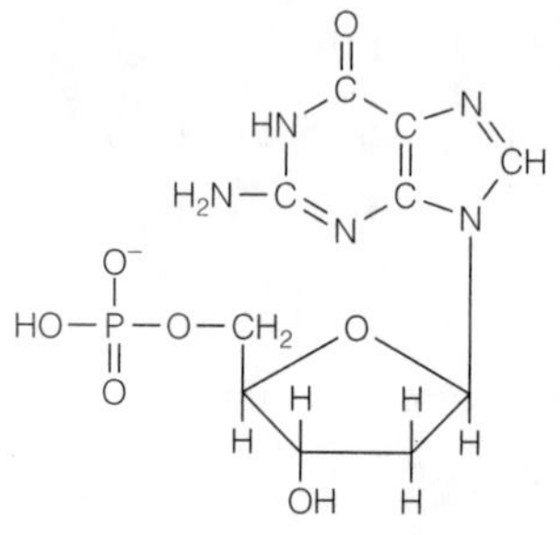

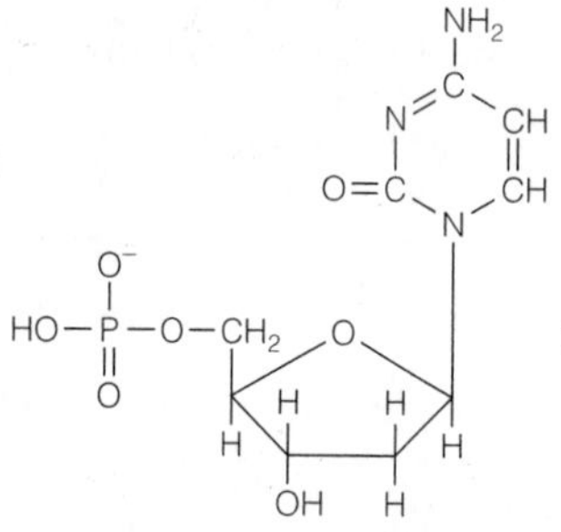

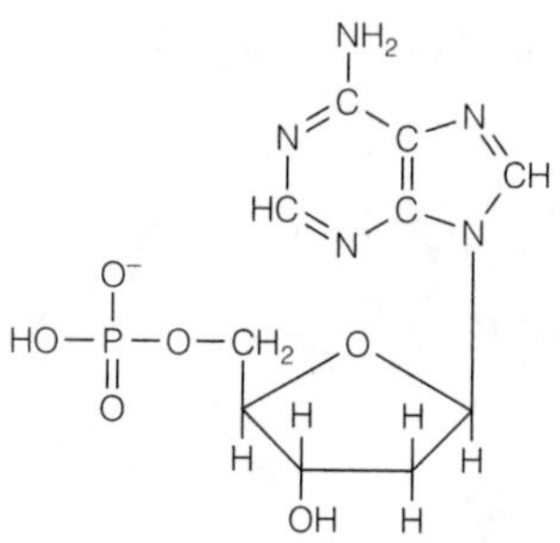

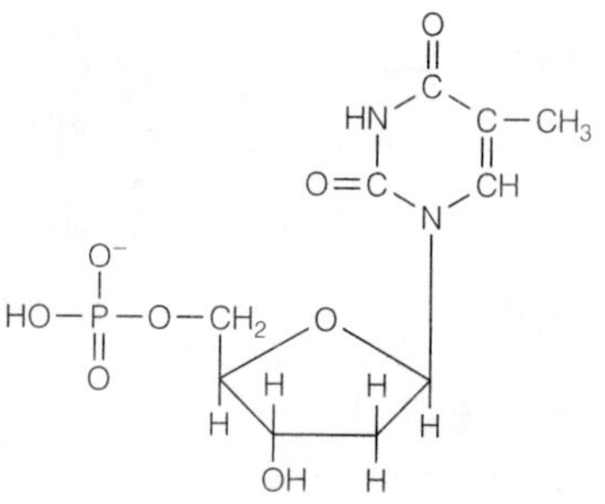

2. ________________ ( ) 3. ________________ ( ) 4. ________________ ( ) 5. ________________ ( )

Labeling and Matching

Identify each indicated part of the following DNA illustration. Choose from these answers: phosphate group, double-ring nitrogen base, single-ring nitrogen base, nucleotide, and deoxyribose. Complete the exercise by matching and entering the letter of the proper structure description in the parentheses following each label.

The following memory devices may be helpful: use *pyrCUT* to remember that the single-ring nucleotides (pyrimidines) are cytosine, uracil (in RNA), and thymine. Use *purAG* to remember that the double-ring nucleotides (purines) are adenine and guanine. To help recall the number of hydrogen bonds between the DNA bases, remember that AT = 2 and CG = 3.

6. ____________ (   ) [p.221]
7. ____________ ____________ (   ) [p.221]
8. ____________-ring nitrogen base (   ) [p.221]
9. ____________-ring nitrogen base (   ) [p.221]
10. ____________-ring nitrogen base (   ) [p.221]
11. ____________-ring nitrogen base (   ) [p.221]
12. A complete ____________ (   ) [p.221]

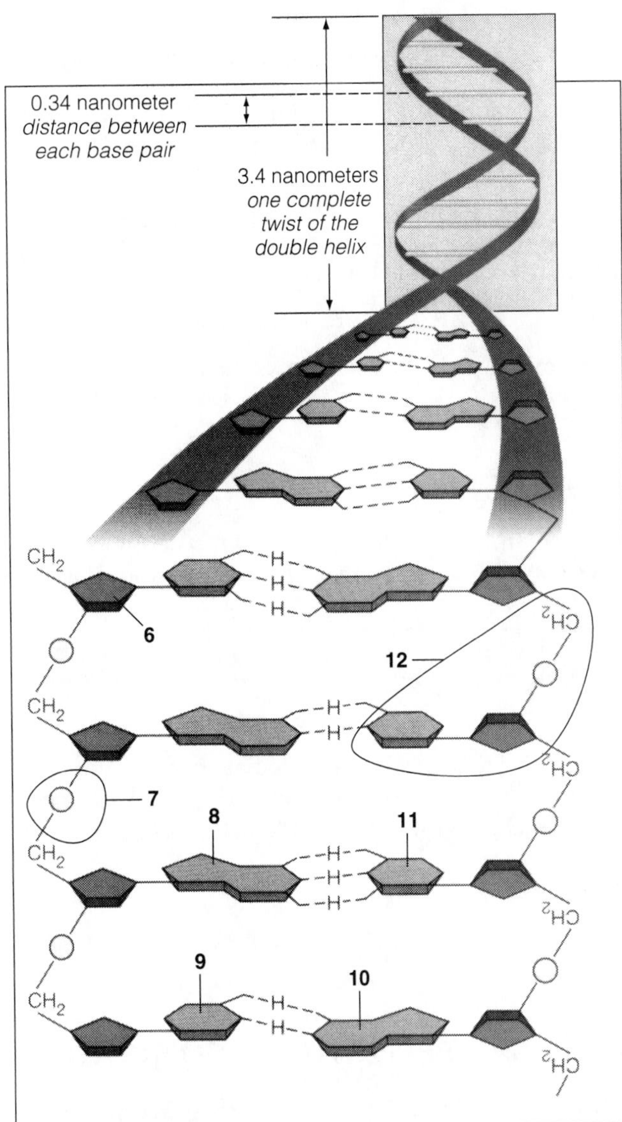

A. The single-ring nitrogen base is thymine, because it has two hydrogen bonds
B. A five-carbon sugar joined to two phosphate groups in the upright portion of the DNA ladder
C. The double-ring nitrogen base is guanine, because it has three hydrogen bonds
D. The single-ring nitrogen base is cytosine, because it has three hydrogen bonds
E. The double-ring nitrogen base is adenine, because it has two hydrogen bonds
F. Composed of three smaller molecules: a phosphate group, five-carbon deoxyribose sugar, and a nitrogenous base (in this case, a single-ring nitrogen base)
G. A chemical group that joins two sugars in the upright portion of the DNA ladder

True/False

If a statement below is true, write "T" in the blank. If false, explain why by changing one or more of the underlined words.

_______________ 13. DNA is composed of <u>four</u> different types of nucleotides. [p.220]

_______________ 14. In the DNA of every species, the amount of adenine present always equals the amount of <u>thymine</u>, and the amount of cytosine always equals the amount of <u>guanine</u> (A = T and C = G). [p.220]

_______________ 15. In a nucleotide, the phosphate group is attached to the <u>nitrogen-containing base</u>. [p.220]

_______________ 16. Watson and Crick built their model of DNA in the early <u>1950s</u>. [recall; p.216]

_______________ 17. Guanine pairs with <u>cytosine</u> and adenine pairs with <u>thymine</u> by forming hydrogen bonds between them. [p.221]

Fill-in-the-Blanks

The pattern for base (18) _______________ [p.221] between the two nucleotide strands in DNA is (19) _______________ [p.221] for all species (A–T; G–C). The base (20) _______________ [p.221] (determining which base follows the next in a nucleotide strand) is (21) _______________ [p.221] from species to species.

Short Answer

22. Explain why understanding the structure of DNA helps scientists understand how living organisms can have so much in common at the molecular level and yet be so diverse at the whole organism level. [p.221]

____

____

____

23. The term *antiparallel* means that parallel strands of a material run in opposite directions. Study Figure 13.7 in the text, read the descriptive messages, and identify the features of the DNA molecule that make biochemists describe its structure as antiparallel. [p.221] _______________

____

____

____

13.4. DNA REPLICATION AND REPAIR [pp.222–223]

13.5. FOCUS ON SCIENCE: *Cloning Mammals—A Question of Reprogramming DNA* [p.224]

Selected Words: *semiconservative* replication [p.223], R. Okazaki [p.223], *continuous* assembly [p.223], *discontinuous* assembly [p.223], complementary strand [p.223], clone [p.224], differentiated [p.224], Ian Wilmut [p.224], uncommitted [p.224]

Boldfaced, Page-Referenced Terms

[p.222] DNA replication ____

____

[p.223] DNA polymerases ____

____

[p.223] DNA ligases ____

____

[p.223] DNA repair ____

____

[p.224] cloning ____

____

Labeling

1. The term *semiconservative replication* refers to the fact that each new DNA molecule resulting from the replication process is "half old, half new." In the following illustration, complete the replication required in the middle of the molecule by adding the letters representing the missing nucleotide bases. Recall that ATP energy and the appropriate enzymes are actually required in order to complete this process. [p.222]

T– ____	____–A
G– ____	____–C
A– ____	____–T
C– ____	____–G
C– ____	____–G
C– ____	____–G
old new	new old

True/False

If a statement below is true, write "T" in the blank. If false, explain why by changing one or more of the underlined words.

____________ 2. The hydrogen bonding of adenine to <u>guanine</u> is an example of complementary base pairing. [p.222]

____________ 3. The replication of DNA is considered a <u>conservative</u> process because each new molecule is really half new and half old. [pp.222–223]

____________ 4. Each parent single strand remains intact during replication, and a new companion strand is assembled <u>on each of those parent strands</u>. [p.222]

____________ 5. DNA <u>ligases</u> govern the assembly of nucleotides on a parent strand. [p.223]

____________ 6. DNA polymerases, DNA ligases, and other enzymes engage in DNA <u>replication</u>. [p.223]

Short Answer

7. Describe the process used to clone an animal. [p.224] ________________________________

____

____

____

Self-Quiz

_____ 1. Each DNA strand has a backbone that consists of alternating _______________. [p.220]
a. purines and pyrimidines
b. nitrogen-containing bases
c. hydrogen bonds
d. sugar and phosphate molecules

_____ 2. In DNA, complementary base-pairing occurs between _______________. [p.220]
a. cytosine and uracil
b. adenine and guanine
c. adenine and uracil
d. adenine and thymine

_____ 3. Adenine and guanine are _______________. [p.220]
a. double-ringed purines
b. single-ringed purines
c. double-ringed pyrimidines
d. single-ringed pyrimidines

_____ 4. Franklin used the technique known as _______________ to determine many of the physical characteristics of DNA. [p.222]
a. transformation
b. cloning
c. density-gradient centrifugation
d. X-ray diffraction

_____ 5. The significance of Griffith's experiment in which he used two strains of pneumonia-causing bacteria is that _______________. [p.218]
a. the conserving nature of DNA replication was finally demonstrated
b. it demonstrated that harmless cells had become permanently transformed into pathogens through a change in the bacterial hereditary material
c. it established that pure DNA extracted from disease-causing bacteria and injected into harmless strains transformed them into "pathogenic strains"
d. it demonstrated that radioactively labeled bacteriophages transfer their DNA but not their protein coats to their host bacteria

_____ 6. The significance of the experiments in which ^{32}P and ^{35}S were used is that _______________. [p.219]
a. the semiconservative nature of DNA replication was finally demonstrated
b. they demonstrated that harmless cells had become permanently transformed through a change in the bacterial hereditary system
c. they established that pure DNA extracted from disease-causing bacteria transform harmless strains into "killer strains"
d. they demonstrated that radioactively labeled bacteriophages transfer their DNA but not their protein coats to their host bacteria

_____ 7. Franklin's research contribution was essential in _______________. [pp.220–222]
a. establishing the principle of base pairing
b. establishing most of the principal structural features of DNA
c. both a and b
d. neither a nor b

_____ 8. Chargaff's requirement that A = T and G = C suggested that _______________. [p.220]
a. cytosine molecules pair up with guanine molecules and thymine molecules pair up with adenine molecules
b. the two strands in DNA run in opposite directions (are antiparallel)
c. the number of adenine molecules in DNA relative to the number of guanine molecules differs from one species to the next
d. the replication process must necessarily be semiconservative

_____ 9. A single strand of DNA with the base-pairing sequence C–G–A–T–T–G is compatible only with the sequence _______________. [p.220]
a. C–G–A–T–T–G
b. G–C–T–A–A–G
c. T–A–G–C–C–T
d. G–C–T–A–A–C

_____ 10. Rosalind Franklin's data indicated that the DNA molecule had to be long and thin with a width (diameter) of 2 nanometers along its length. Watson and Crick declared that _____________ ensured that the width of the DNA molecule must be uniform. [p.221]

a. the antiparallel nature of DNA
b. semiconservative replication processes
c. hydrogen bonding of the sugar–phosphate backbones
d. complementary base-pairing processes that match purine with pyrimidine

---

Chapter Objectives/Review Questions

1. Before 1952, __________ molecules were suspected of housing the genetic code. [p.216]
2. The two scientists who assembled the clues to DNA structure and produced the first model were __________ and __________. [p.216]
3. Summarize the research carried out by Miescher, Griffith, Avery and colleagues, and Hershey and Chase; state the specific advances made by each in the understanding of genetics. [pp.216,218–219]
4. Viruses called __________ were used in early research efforts to discover the genetic material. [p.218]
5. Summarize the specific research which demonstrated that DNA, not protein, governed inheritance. [pp.218–219]
6. Draw the basic shape of a deoxyribose molecule and show how a phosphate group is joined to it when forming a nucleotide. [p.221]
7. Show how a nucleotide base would be joined to the sugar–phosphate combination drawn in objective 6. [p.221]
8. DNA is composed of double-ring nucleotides known as __________ and single-ring nucleotides known as __________; the two purines are __________ and __________, whereas the two pyrimidines are __________ and __________. [p.220]
9. Assume that the two parent strands of DNA have been separated and that the base sequence on one parent strand is A–T–T–C–G–C; the base sequence that will complement that parent strand is __________. [p.221]
10. List the pieces of information about DNA structure that Rosalind Franklin discovered through her X-ray diffraction research. [pp.220–222]
11. Explain what is meant by the pairing of nitrogen-containing bases (base pairing), and explain the mechanism that causes bases of one DNA strand to join with bases of the other strand. [pp.220–221]
12. Generally describe how double-stranded DNA replicates from stockpiles of nucleotides. [p.222]
13. Explain what is meant by "each parent strand is conserved in each new DNA molecule." [pp.222–223]
14. During DNA replication, enzymes called DNA __________ assemble new DNA strands. [p.223]
15. Distinguish between continuous strand assembly and discontinuous strand assembly. [p.223]
16. Describe the process of making a genetically identical copy of yourself. [p.224]

Integrating and Applying Key Concepts

Review the stages of mitosis and meiosis as well as the process of fertilization. Relate what was learned in the chapter about DNA replication and the relationship of DNA to a chromosome. As you pass through fertilization and the stages of both types of cell division, use a diploid number of $2n = 4$. Show the proper number of DNA threads in each cell at each of the stages of mitosis and meiosis. Include the cells that represent the end products of mitosis and meiosis.

14

FROM DNA TO PROTEINS

Interactive Exercises

Beyond Byssus [pp.226–227]

14.1. HOW IS DNA TRANSCRIBED INTO RNA? [pp.228–229]

Selected Words: byssus [p.226], keratin [p.226], *protein-building* instructions [p.228], pre-mRNA [p.228], alternate splicing [p.229], cap [p.229], tail [p.229]

Boldfaced, Page-Referenced Terms

[p.227] base sequence ______________________________

____

[p.227] transcription ______________________________

____

[p.227] translation ______________________________

____

[p.227] ribonucleic acid, RNA ______________________________

____

[p.228] messenger RNA, mRNA ______________________________

____

[p.228] ribosomal RNA, rRNA ______________________________

____

[p.228] transfer RNA, tRNA ____

____

[p.228] uracil ____

____

[p.228] RNA polymerase ____

____

[p.228] promoter ____

____

[p.229] introns ____

____

[p.229] exons ____

____

Fill-in-the-Blanks

The pattern of which base follows the next in a strand of DNA is referred to as the base (1) ______________ [p.227]. A region of DNA that calls for the assembly of specific amino acids into a polypeptide chain is a(n) (2) ______________ [p.227]. The two steps from genes to proteins are called (3) ______________ [p.227] and (4) ______________ [p.227]. In (5) ______________ [p.227], single-stranded molecules of RNA are assembled on DNA templates in the nucleus. In (6) ______________ [p.227], the RNA molecules are shipped from the nucleus into the cytoplasm, where they are used as templates for assembling (7) ______________ [p.227] chains. Following translation, one or more chains become (8) ______________ [p.227] into the three-dimensional shape of protein molecules. Proteins have (9) ______________ [p.227] and (10) ______________ [p.227] roles in cells, including control of DNA.

Complete the Table

11. Three types of RNA are transcribed from DNA in the nucleus (two are from genes that code only for RNA). Complete the following table, which summarizes information about these molecules. [p.228]

RNA Molecule	*Abbreviation*	*Description/Function*
a. Ribosomal RNA		
b. Messenger RNA		
c. Transfer RNA		

Short Answer

12. List three ways in which a molecule of RNA differs structurally from a molecule of DNA. [p.228] ______

______

______

______

13. Cite two similarities between DNA replication and transcription. [p.228] ______

______

______

______

14. What are the three key ways in which transcription differs from DNA replication? [p.228] ______

______

______

______

Sequence

Arrange the steps of transcription in correct chronological sequence. Write the letter of the first step next to 15, the letter of the second step next to 16, and so on.

15. ___ A. The RNA strand grows along exposed bases until RNA polymerase meets a DNA base sequence that signals "stop." [p.228]

16. ___ B. RNA polymerase binds with the DNA promoter region to open up a local region of the DNA double helix. [pp.228–229]

17. ___ C. An RNA polymerase enzyme locates the DNA bases of the promoter region of one DNA strand by recognizing DNA-associated proteins near a promoter. [pp.228–229]

18. ___ D. RNA is released from the DNA template as a free, single-stranded transcript. [pp.228–229]

19. ___ E. RNA polymerase moves stepwise along exposed nucleotides of one DNA strand; as it moves, the DNA double helix keeps unwinding. [p.229]

Completion

20. Suppose the following line represents the DNA strand that will act as a template for the production of mRNA through the process of transcription. Fill in the blanks below the DNA strand with the sequence of complementary bases that will represent the message carried from DNA by mRNA to the ribosome in the cytoplasm [see Figure 14.4 on pp.228–229 of the text].

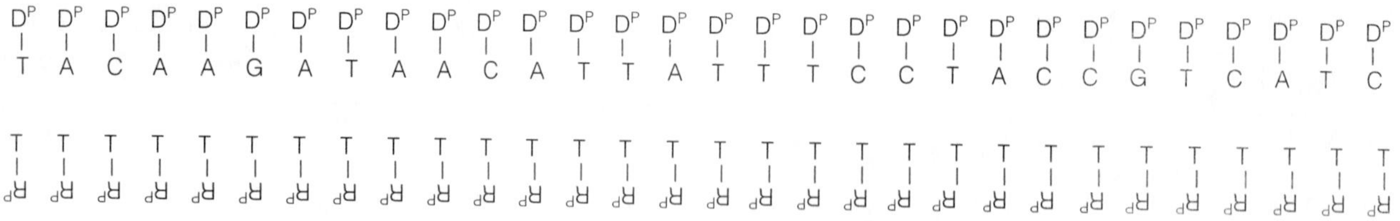

```
DP DP DP DP DP DP DP DP DP DP DP DP DP DP DP DP DP DP DP DP DP DP DP DP DP DP DP DP DP DP
|  |  |  |  |  |  |  |  |  |  |  |  |  |  |  |  |  |  |  |  |  |  |  |  |  |  |  |  |  |
T  A  C  A  A  G  A  T  A  A  C  A  T  T  A  T  T  T  C  C  T  A  C  C  G  T  C  A  T  C

T  T  T  T  T  T  T  T  T  T  T  T  T  T  T  T  T  T  T  T  T  T  T  T  T  T  T  T  T  T
|  |  |  |  |  |  |  |  |  |  |  |  |  |  |  |  |  |  |  |  |  |  |  |  |  |  |  |  |  |
RP RP RP RP RP RP RP RP RP RP RP RP RP RP RP RP RP RP RP RP RP RP RP RP RP RP RP RP RP RP
```

Labeling and Matching

Newly transcribed mRNA contains more genetic information than is necessary to code for a chain of amino acids. Before the mRNA leaves the nucleus for its ribosome destination, an editing process occurs as certain portions of nonessential information are snipped out. Identify each indicated part of the illustration below; use abbreviations for the nucleic acids. Complete the exercise by matching and entering the letter of the description in the parentheses following each label.

21. ______ ( ) [p.229]
22. ______ ( ) [p.229]
23. ______ ( ) [p.229]
24. ______ ( ) [p.229]
25. ______ ( ) [p.229]
26. ______ ( ) [p.229]

A. The actual coding portions of mRNA
B. Noncoding portions of the newly transcribed mRNA
C. Presence of cap and tail, introns snipped out, and exons spliced together
D. Acquisition of a tail by the modified mRNA transcript
E. The region of the DNA template strand to be copied
F. Reception of a nucleotide cap by the 5′ end of mRNA (the first synthesized)

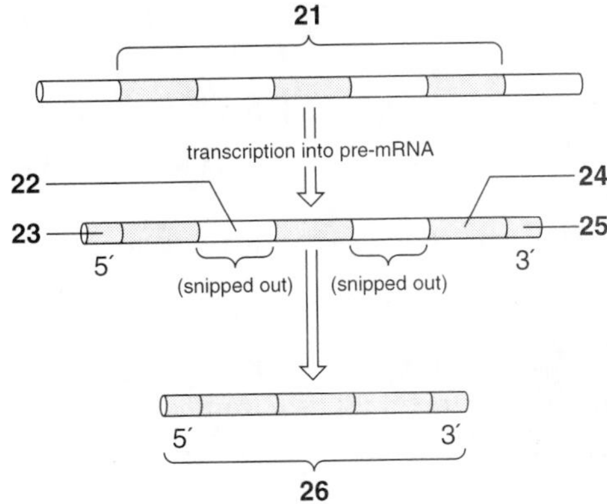

14.2. DECIPHERING THE mRNA TRANSCRIPTS [pp.230–231]

14.3. HOW IS mRNA TRANSLATED? [pp.232–233]

Selected Words: bases *three at a time* [p.230], tRNA "hook" [p.230], "wobble effect" [p.231], *initiation* [p.232], START codon [p.232], *elongation* [p.232], *termination* [p.232], STOP codon [p.232], release factors [p.232], polysome [p.233], polypeptide chains [p.233]

Boldfaced, Page-Referenced Terms

[p.230] codons ____

____

[p.230] genetic code ____

____

[p.230] anticodon ____

____

Matching

Choose the most appropriate answer for each term.

1. ____ codon [p.230]
2. ____ three bases at a time [p.230]
3. ____ sixty-one [p.230]
4. ____ the START codon [pp.230,232]
5. ____ molecular "hook" [p.230]
6. ____ ribosome [p.231]
7. ____ anticodon [p.230]
8. ____ the STOP codons [pp.230,232]

A. Composed of two subunits, the small subunit with P and A amino acid binding sites as well as a binding site for mRNA
B. How the nucleotide bases in mRNA are read
C. On tRNA, an attachment site for an amino acid
D. UAA, UAG, UGA
E. A sequence of three nucleotide bases that can pair with a specific mRNA codon
F. Name for each base triplet in mRNA
G. The number of codons that actually specify amino acids
H. AUG

Complete the Table

9. Complete the following table, which distinguishes the stages of translation. [p.232]

Translation Stage	*Description*
a.	Special initiator tRNA loads onto small ribosomal subunit and recognizes AUG; small subunit binds with mRNA, and large ribosomal subunit joins small one.
b.	Amino acids are strung together in sequence dictated by mRNA codons as the mRNA strand passes through the two ribosomal subunits; two tRNAs interact at P and A sites.
c.	mRNA STOP codon signals the end of the polypeptide chain; release factors detach the ribosome and polypeptide chain from the mRNA.

Completion

10. Given the following DNA sequence, deduce the composition of the mRNA transcript: [p.230]

TAC AAG ATA ACA TTA TTT CCT ACC GTC ATC

___ ___ ___ ___ ___ ___ ___ ___ ___ ___

(mRNA transcript)

11. Deduce the composition of the tRNA anticodons that would pair with the specific mRNA codons from question 10, as these tRNAs deliver the amino acids (identified below) to the P and A binding sites of the small ribosomal subunit. [p.230]

___ ___ ___ ___ ___ ___ ___ ___ ___ ___

(tRNA anticodons)

12. From the mRNA transcript in question 10, use Figure 14.7 in the text to deduce the composition of the amino acids of the polypeptide sequence. [p.230]

___ ___ ___ ___ ___ ___ ___ ___ ___

(amino acids)

Fill-in-the-Blanks

The order of (13) ______________ ______________ [p.230] in a protein is specified by a sequence of nucleotide bases. The genetic code is read in units of (14) ______________ [p.230] nucleotides; each unit of three codes for (15) ______________ [p.230] amino acid(s). In the table that showed which triplet specified a particular amino acid, the triplet code was incorporated in (16) ______________ [p.230] molecules. Each of these triplets is referred to as a(n) (17) ______________ [p.230]. (18) ______________ [p.230] alone carries the instructions for assembling a particular sequence of amino acids from the DNA to the ribosomes in the cytoplasm, where (19) ______________ [p.231] of the polypeptide occurs. (20) ______________ [pp.230–231] RNA acts as a shuttle molecule, as each type brings its particular (21) ______________ ______________ [pp.230–231] to the ribosome where it is to be incorporated into the growing (22) ______________ [p.231]. A(n) (23) ______________ [p.230] is a triplet on mRNA that forms hydrogen bonds with a(n) (24) ______________ [p.230], which is a triplet on tRNA. During the stage of translation called (25) ______________ [p.232], a particular tRNA that can start transcription and an mRNA transcript are both loaded onto a ribosome. In the (26) ______________ [p.232] stage of translation, a polypeptide chain is assembled as the mRNA passes between two ribosomal subunits, like a thread being moved through the eye of a needle. During the last stage of translation, (27) ______________ [p.232], a STOP codon in the mRNA moves onto the platform, and no tRNA has a corresponding anticodon. Now proteins called (28) ______________ [p.232] factors bind to the ribosome. They trigger (29) ______________ [p.232] activity that detaches the mRNA and the chain from the ribosome. Often several ribosomes translate the same mRNA simultaneously. The resulting structure is referred to as a (30) ______________ [p.233]. After production, some proteins enter the (31) ______________ ______________ [p.233] for further maturation.

14.4. DO MUTATIONS AFFECT PROTEIN SYNTHESIS? [pp.234–235]

14.5. SUMMARY [p.236]

Selected Words: *frameshift* mutation [p.234], Barbara McClintock [pp.234–235], mutation *frequency* [p.235], mutagens [p.235], repair enzymes [p.235], spontaneous mutations [p.235]

Boldfaced, Page-Referenced Terms

[p.234] gene mutations ____

____

[p.234] base-pair substitution ____

____

[p.234] insertions ____

____

[p.234] deletions ____

____

[pp.234–235] transposons ____

____

[p.235] mutation rate ____

____

[p.235] ionizing radiation ____

____

[p.235] nonionizing radiation ____

____

[p.235] alkylating agents ____

____

[p.235] carcinogens ____

____

Fill-in-the-Blanks

In addition to changes in chromosomes (crossing over, recombination, deletion, addition, translocation, and inversion), changes can also occur in the structure of DNA; these modifications are referred to as *gene mutations.* Complete the following exercise on types of spontaneous gene mutations.

Ultraviolet radiation, gamma rays, X-rays, and certain natural and synthetic chemicals are examples of environmental agents called (1) ______________ [p.235], which may enter cells and damage strands of DNA. If A becomes paired with C instead of T during DNA replication, this spontaneous mutation is a base-pair (2) ______________ [p.234]. Sickle-cell anemia is a genetic disease whose cause has been traced to a single DNA base pair; the result is that one (3) ______________ ______________ [p.234] is substituted for another during protein synthesis. When an extra base becomes inserted into a gene region, it shifts the "three-bases-at-a-time" reading frame by (4) ______________ [p.234] base; hence the name (5) ______________ [p.234] mutation. Some DNA regions "jump" to new DNA locations and often inactivate the genes in their new environment; such (6) ______________ [pp.234–235] elements may give rise to observable changes in the phenotype of an organism.

Each gene has a characteristic (7) ______________ [p.235] rate, which is the probability that it will change spontaneously during a specified interval. Not all mutations are spontaneous. Many result after exposure to (8) ______________ [p.235], or mutation-causing agents in the environment. Gamma rays and X-rays are mutagens. They can ionize water and other molecules around the DNA, so that (9) ______________ [p.235] radicals form that may attack the structure of DNA. The nonionizing radiation that DNA absorbs most strongly is (10) ______________ [p.235] radiation. It may induce alterations in the base pairing properties of (11) ______________ [p.235] and (12) ______________ [p.235] neighbors on the same DNA strand. Substances called (13) ______________ [p.235] agents transfer methyl or ethyl groups to reactive sites on the bases or phosphate groups of DNA. At an alkylated site, DNA becomes more susceptible to base-pair disruptions that invite (14) ______________. [p.235]. Many (15) ______________ [p.235], or cancer-causing agents, operate by alkylating the DNA. A (16) ______________ [p.235] that is specified by a heritable mutation may have harmful, neutral, or beneficial effects on an individual's ability to function in the prevailing environment. Thus, the outcomes of gene mutation can have powerful (17) ______________ [p.235] consequences.

Labeling and Matching

A summary of the flow of genetic information in protein synthesis is useful as an overview. Identify the indicated parts of the illustration below by filling in the blanks with the names of the appropriate structures or functions. Choose from the following: DNA, mRNA, rRNA, tRNA, amino acids, anticodons, intron, exon, polypeptide, ribosomal subunits, transcription, translation. Complete the exercise by matching and entering the letter of the description in the parentheses following each label. [all from p.236]

18. ________________ ( )
19. ________________ ( ) (process)
20. ________________ ( )
21. ________________ ( )
22. ________________ ( )
23. ________________ ( )
24. ________________ ( )
25. ________________ ________________ ( )
26. ________________ ________________ ( )
27. ________________ ( )
28. ________________ ( ) (process)
29. ________________ ( )

A. Coding portion of pre-mRNA that will translate into proteins
B. Carries a form of the genetic code from DNA in the nucleus to the cytoplasm
C. Transports amino acids to the ribosome and mRNA
D. The building blocks of polypeptides
E. Noncoding portions of pre-mRNA
F. Combines with proteins to form the ribosomal subunits
G. Join during the initiation step of protein synthesis
H. Holds the genetic code for protein production
I. Amino acids are joined together
J. RNA synthesized on a DNA template
K. A sequence of three bases that can pair with a specific mRNA codon
L. May serve as a functional protein (enzyme) or a structural protein

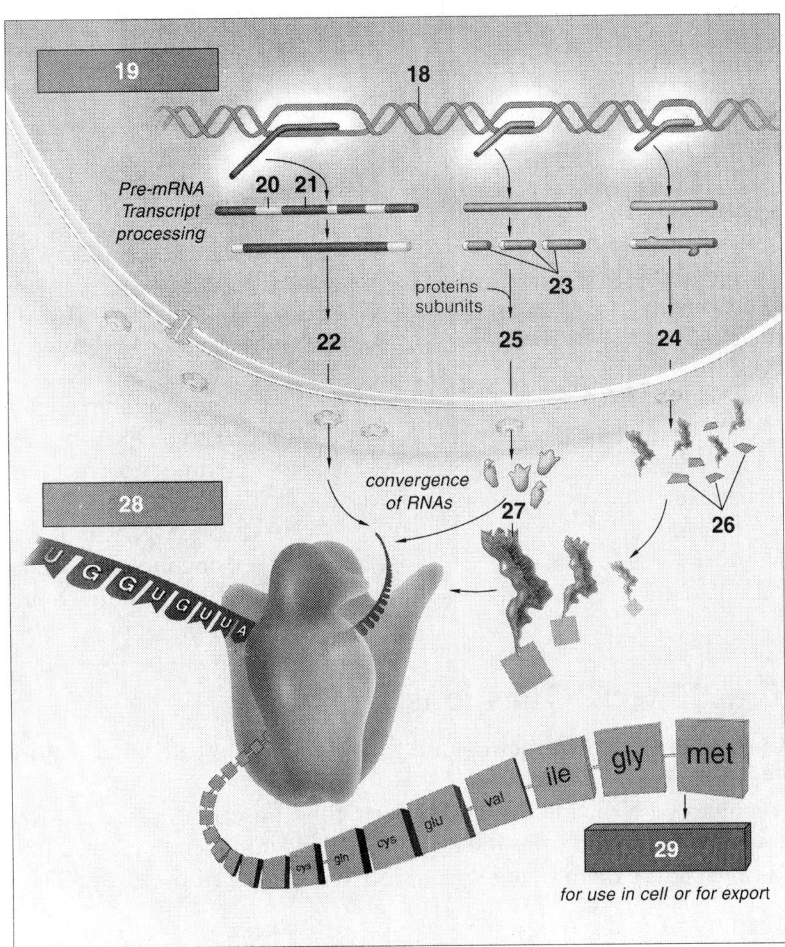

Self-Quiz

_____ 1. Transcription _______________. [p.228]
a. occurs on the surface of a ribosome
b. is the final process in the assembly of a protein DNA template
c. occurs during the synthesis of any type of RNA by use of a DNA template
d. is catalyzed by DNA polymerase

_____ 2. _______________ carries the actual instructions for a protein's sequence to the ribosome. [p.228]
a. DNA
b. mRNA
c. rRNA
d. tRNA

_____ 3. _______________ carry(ies) amino acids to ribosomes, where amino acids are linked into the primary structure of a polypeptide. [p.228]
a. mRNA
b. tRNA
c. Introns
d. rRNA

_____ 4. Transfer RNA differs from other types of RNA because it _______________. [p.231]
a. transfers genetic instructions from cell nucleus to cytoplasm
b. specifies the amino acid sequence of a particular protein
c. carries an amino acid at one end
d. contains codons

_____ 5. _______________ dominates the process of transcription. [p.228]
a. RNA polymerase
b. DNA polymerase
c. Phenylketonuria
d. Transfer RNA

_____ 6. _______________ and _______________ are found in RNA but not in DNA. [p.228]
a. Deoxyribose; uracil
b. Uracil; ribose
c. Deoxyribose; thymine
d. Thymine; ribose

_____ 7. Each "word" in the mRNA language consists of _______________ letters. [p.230]
a. three
b. four
c. five
d. more than five

_____ 8. If each kind of nucleotide were able to code for only one kind of amino acid, how many different types of amino acids could be selected? [p.230]
a. four
b. sixteen
c. twenty
d. sixty-four

_____ 9. The genetic code is composed of _______________ codons. [p.230]
a. three
b. twenty
c. sixteen
d. sixty-four

_____ 10. The cause of sickle-cell anemia has been traced to _______________. [p.234]
a. a mosquito-transmitted virus
b. two DNA mutations that result in two incorrect amino acids in a hemoglobin chain
c. three DNA mutations that result in three incorrect amino acids in a hemoglobin chain
d. one DNA mutation that results in one incorrect amino acid in a hemoglobin chain

Chapter Objectives/Review Questions

1. State how RNA differs from DNA in structure and function and indicate what features RNA has in common with DNA. [p.228]
2. __________ RNA combines with certain proteins to form the ribosome; __________ RNA carries genetic information for protein construction from the nucleus to the cytoplasm; __________ RNA picks up specific amino acids and moves them to the area of mRNA and the ribosome. [p.228]

3. Describe the process of transcription and indicate three ways in which it differs from replication. [p.228]
4. Transcription starts at a(n) ________, a specific sequence of bases on one of the two DNA strands that signals the start of a gene. [p.228]
5. What RNA code would be formed from the following DNA code: TAC–CTC–GTT–CCC–GAA? [p.229]
6. Describe how the three types of RNA participate in the process of translation. [pp.228–229]
7. The first end of the mRNA to be synthesized is the ________ end; at the opposite end, the most mature transcripts acquire a(n) ________ tail. [p.228]
8. Distinguish introns from exons. [p.229]
9. Each base triplet in mRNA is called a(n) ________ [p.230]
10. State the relationship between the DNA genetic code and the order of amino acids in a protein chain. [pp.228–233]
11. Scrutinize Figure 14.7 in the text and decide whether the genetic code in this instance applies to DNA, mRNA, or tRNA. [p.230]
12. Explain how the DNA message TAC–CTC–GTT–CCC–GAA would be used to code for a segment of protein and state what its amino acid sequence would be. [p.230]
13. Describe the "wobble effect." [p.231]
14. Describe events occurring in the following stages of translation: initiation, elongation, and termination. [p.232]
15. What is the fate of the new polypeptides produced by protein synthesis? [p.233]
16. Cite an example of a change in one DNA base pair that has profound effects on the human phenotype. [p.234]
17. Briefly describe the spontaneous DNA mutations known as *base-pair substitution*, *frameshift mutation*, and *transposable element*. [pp.234–235]
18. List some of the environmental agents, or mutagens, that can cause mutations. [p.235]
19. Using a diagram, summarize the steps involved in the transformation of genetic messages into proteins. [Use Figure 14.14 on p.236 of the text.]

Integrating and Applying Key Concepts

Genes code for specific polypeptide sequences. Not every substance in living cells is a polypeptide. Explain how genes might be involved in the production of a storage starch (such as glycogen) that is constructed from simple sugars.

15

CONTROLS OVER GENES

Interactive Exercises

When DNA Can't Be Fixed [pp.238–239]

15.1. TYPES OF CONTROL MECHANISMS [p.240]

15.2. BACTERIAL CONTROL OF TRANSCRIPTION [pp.240–241]

Selected Words: *malignant melanoma* [p.238], *basal cell carcinoma* [p.238], *squamous cell carcinoma* [p.238], repressor [p.240], *Escherichia coli (E. coli)* [p.240], CAP [p.241], cAMP [p.241], lactase [p.241], *lactose intolerance* [p.241]

Boldfaced, Page-Referenced Terms

[p.238] neoplasms ____________________

____

[p.238] cancers ____________________

____

[p.239] gene controls ____________________

____

[p.240] regulatory proteins ____________________

____

[p.240] negative control ____________________

____

[p.240] positive control ____________________

[p.240] promoter ____________________

[p.240] enhancer ____________________

[p.240] methylation ____________________

[p.240] acetylation ____________________

[p.240] operator ____________________

[p.240] operon ____________________

[p.241] activator proteins ____________________

Complete the Table

1. All the diploid cells in an organism possess the same genes, and every cell uses most of the same genes; yet specialized cells must activate only certain genes. Some agents of gene control have been discovered. Transcriptional controls are the most common. Complete the following table to summarize the agents of gene control.

Agents of Gene Control	*Method of Gene Control*
[p.240] a. Repressor protein	
[p.241] b.	Encourage binding of RNA polymerases to DNA; this is positive control of transcription
[p.240] c. Promoter	
[p.240] d.	Short DNA base sequences adjacent to the promoter; a binding site for control agents that turn off transcription of mRNA

2. When do gene controls come into play? [p.240] ____________________

Fill-in-the-Blanks

A promoter and operator provide (3) ______________ control [p.240]. The (4) ______________ ______________ [p.241] is transcribed as the mRNA that codes for a repressor protein. The affinity of the (5) ______________ [p.240] for RNA polymerase dictates the rate at which a particular operon will be transcribed. Repressor protein allows (6) ______________ ______________ [p.240] over the lactose operon. Repressor binds with operators and makes the promoter unavailable when lactose concentrations are (7) ______________ [p.240]. This blocks (8) ______________ ______________ [p.240] from transcribing the genes that code for enzymes needed to process lactose. This (9) [choose one] ☐ blocks, ☐ promotes [p.243] production of lactose-processing enzymes. When lactose is present, it is converted to allolactose, which binds with the (10) ______________ [p.240]. Thus, (10) cannot bind to (11) ______________ [p.240], and RNA polymerase has access to the lactose-processing genes. This gene control works well because lactose-degrading enzymes are not produced unless they are (12) ______________ [p.240].

Labeling and Matching

Escherichia coli, a bacterial cell living in mammalian digestive tracts, uses a negative type of gene control over lactose metabolism. Use the numbered blanks to identify each part of the illustration below. Use abbreviations for nucleic acids. Choose from the following:

lactose lactose enzyme genes regulatory gene, repressor–operator complex promoter mRNA lactose operon repressor protein RNA polymerase operator

Complete the exercise by matching and entering the letter of the proper function description in the parentheses following each label. [p.241]

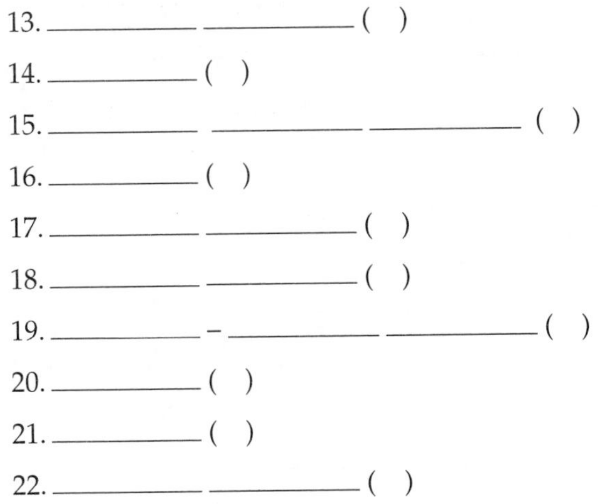

13. ______ ______ ( )
14. ______ ( )
15. ______ ______ ______ ( )
16. ______ ( )
17. ______ ______ ( )
18. ______ ______ ( )
19. ______ – ______ ______ ( )
20. ______ ( )
21. ______ ( )
22. ______ ______ ( )

A. Includes promoter, operator, and the genes coding for lactose-metabolizing enzymes
B. Short DNA base sequences on both sides of the promoter
C. The nutrient molecule in the lactose operon
D. Major enzyme that catalyzes transcription
E. Capable of preventing RNA polymerases from binding with DNA
F. Genes that code for lactose-metabolizing enzymes
G. Binds to operator and prevents RNA polymerase from binding to DNA and initiating transcription
H. Specific base sequence that signals the beginning of a gene; binding site for RNA polymerase
I. Gene that contains coding for production of repressor protein
J. Carries genetic instructions to ribosomes for production of lactose enzymes

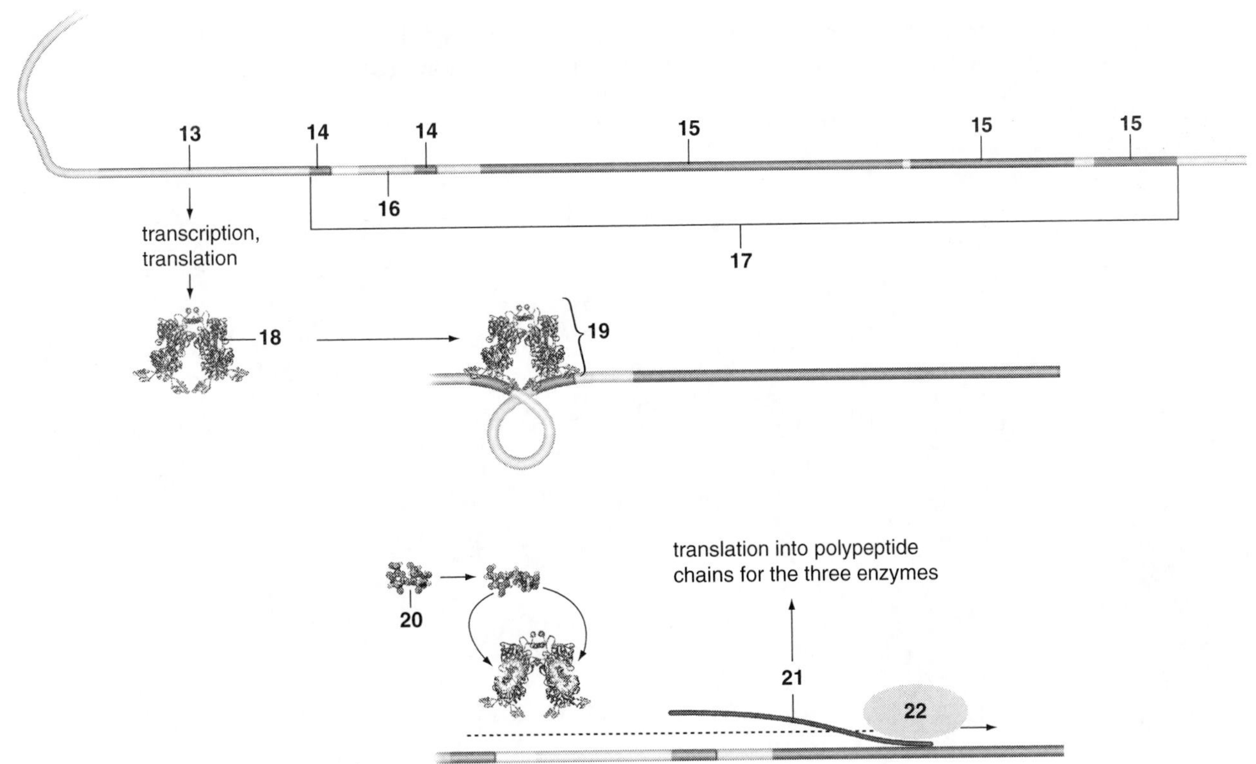

15.3. GENE CONTROLS IN EUKARYOTIC CELLS [pp.242–243]

15.4. TYPES OF CONTROL MECHANISMS [pp.244–245]

Selected Words: *gene amplification* [p.242], *DNA rearrangements* [p.242], *chemical modification* [p.242], *alternative splicing* [p.243], "masked messengers" [p.243], *antennapedia* [p.244], "homeobox" [p.244], *"calico" cat* [p.244], *"spotting gene"* [p.245], *anhidrotic ectodermal dysplasia* [p.245], XIST [p.245]

Boldfaced, Page-Referenced Terms

[p.242] cell differentiation ____

____

[p.244] homeotic genes ____

____

[p.244] X chromosome inactivation ____

____

[p.244] Barr body ____

____

[p.245] mosaic tissue effect ____

____

[p.245] dosage compensation ____

____

Short Answer

1. Although a complex organism such as a human being arises from a single cell, the zygote, differentiation occurs in development. Define *differentiation* and relate it to a definition of *selective gene expression.* [p.244]

____

____

____

Labeling and Matching

Identify each numbered part of the illustration showing eukaryotic gene control. Choose from translational control, pretranscriptional controls, transcriptional controls, transcript processing controls, and posttranslational controls. Complete the exercise by correctly matching and entering the letter of the corresponding gene control description in the parentheses following each label.

2. ________ ________ ( ) [p.242]
3. ________ ________ ( ) [pp.242–243]
4. ________ ________ ________ ( ) [p.243]
5. ________ ________ ( ) [p.243]
6. ________ ________ ( ) [p.243]

A. Govern the rates at which mRNA transcripts that reach the cytoplasm will be translated into polypeptide chains at ribosomes
B. Govern modification of the initial mRNA transcripts in the nucleus
C. Govern how the polypeptide chains become modified into functional proteins
D. Gene amplification can produce many copies of specific genes needed to produce a lot of a specific protein
E. Influence when and to what degree a particular gene will be transcribed (if at all)

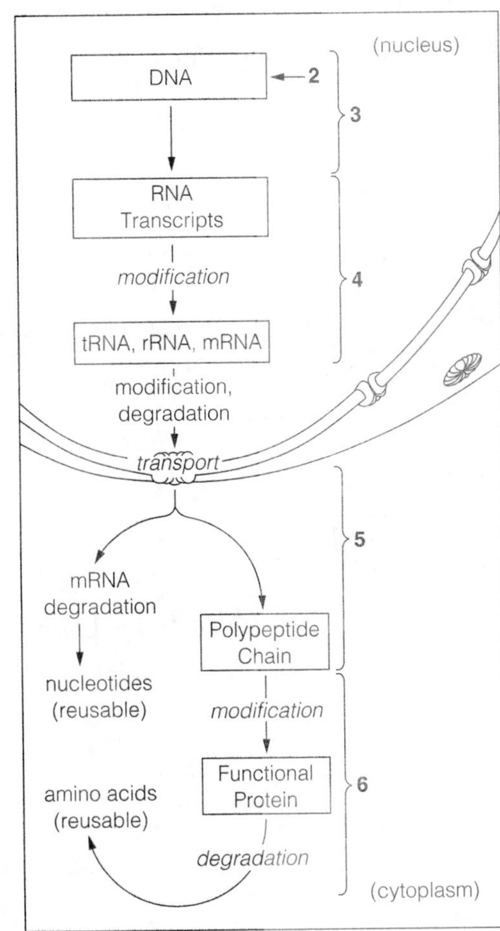

Fill-in-the-Blanks

All the cells in our bodies contain copies of the same (7) ______________ [p.242]. These genetically identical cells become structurally and functionally distinct from one another through a process called (8) ______________ ______________ [p.242], which arises through (9) ______________ [p.242] gene expression in different cells. Cells depend on (10) ______________ [p.242], which govern transcription, translation, and enzyme activity.

Controls that operate during transcription and transcript processing use (11) ______________ [recall; p.241] proteins, especially (12) ______________ [recall; p.241] that are turned on and off by the attachment and detachment of cAMP. (13) ______________ [p.242] is also controlled by the way eukaryotic DNA is packed with proteins in chromosomes. A (14) ______________ [p.244] body is a condensed X chromosome; the process by which an XX-containing nucleus randomly inactivates an X chromosome is sometimes called Lyonization. X chromosome inactivation produces adult human females who are (15) ______________ [p.245] for X-linked traits. This effect is shown in human females affected by (16) ______________ ______________ ______________ [p.245]; the disorder provides evidence that the process of (17) ______________ [p.245] gene expression causes cells to differentiate.

True/False

If the statement is true, write a "T" in the blank. If false, explain why.

_____ 18. Homeotic genes in yeasts are very different from homeotic genes in humans. [p.244] ______________

____

_____ 19. When the same primary mRNA transcript is spliced in alternative ways, the result may be different mRNAs, each coding for a slightly different protein. [pp.242–243] ______________

____

15.5. EXAMPLES OF SIGNALING MECHANISMS [pp.246–247]

15.6. FOCUS ON SCIENCE: *Lost Controls and Cancer* [pp.248–249]

Selected Words: target cell [p.246], amplification [p.246], somatotropin [p.246], prolactin [p.246], *p53* [p.248], *bcl-2* [p.249]

Boldfaced, Page-Referenced Terms

[p.246] hormones ____

____

[p.246] ecdysone ____

____

[p.246] polytene chromosome ____

____

[p.247] phytochrome ____

____

[p.248] checkpoint genes ____

____

[p.248] kinases ____

____

[p.248] growth factors ____

____

[p.248] oncogene ____

____

[p. 248] metastasis ____

____

[p.249] apoptosis ____

____

[p.249] ICE-like proteases ____

____

Fill-in-the-Blanks

Several kinds of (1) ______________ [p.246] influence gene activity. In both animals and plants, molecules called (2) ______________ [p.246] stimulate or inhibit gene activity in their target cells. Any cell with (3) ______________ [p.246] for a specific hormone is a target. Sometimes a hormone molecule must first bind with a(n) (4) ______________ ______________ [p.246] and then a(n) (5) ______________ [p.246] (generally, a base sequence on the same DNA molecule that contains the promoter to which RNA polymerase will attach and begin transcription) before transcription can be either turned on or turned off. Many insect larvae produce a hormone known as (6) ______________ [p.246], which binds to a receptor on salivary gland cells and triggers very rapid transcription of the genes that contain the instructions for assembling the protein components of saliva. Thus insect larvae can usually digest leafy vegetation as fast as they consume it by producing large amounts of salivary proteins. The genes responsible for salivary protein production have been copied many times by (7) ______________ [p.246] and lie parallel to each other, causing that portion of the (8) ______________ [p.246] chromosome to appear puffy during transcription.

In humans and other mammals, mammary gland cells do not begin producing milk until the hormone (9) ______________ [p.246] attaches to receptors on the surfaces of the gland cells and activates the genes responsible for producing milk proteins. Other cells in mammalian bodies have these genes but lack the (10) ______________ [p.246] to which the hormone attaches. Red wavelengths of sunlight activate (11) ______________ [p.247], a blue–green pigment whose activity influences transcription of certain genes at certain times of the day and season as enzymes and other proteins play key roles in germination, growth, and the formation of flowers, fruits, and seeds.

Most cells respond to signals that tell them it's time to die, but (12) ______________ [p.249] cells do not.

Many (12) cells form a tissue mass called a (13) ________________ [recall; p.238], which may be either benign or malignant. Malignant cells can break loose from their original growth area, travel through the body, and invade other tissues—a process called (14) ________________ [p.248].

True/False

If the statement is true, write a "T" in the blank. If false, make it true by changing the underlined word.

________________ 15. <u>Polytene</u> chromosomes of many insect larvae are unusual in that their DNA has been repeatedly replicated; thus, these chromosomes have multiple copies of the same genes. When these genes are undergoing transcription, they "puff." [p.246]

________________ 16. When cells become cancerous, cell populations <u>decrease</u> to very <u>low</u> densities and <u>stop</u> dividing. [p.248]

________________ 17. <u>All</u> abnormal growths and massings of new tissue in any region of the body are called <u>tumors</u>. [recall; p.238]

________________ 18. Malignant tumors have cells that <u>migrate and divide</u> in other organs. [recall; p.238]

________________ 19. Oncogenes are genes that <u>combat</u> cancerous transformations. [p.248]

________________ 20. Proto-oncogenes <u>rarely</u> trigger cancer. [p.248]

________________ 21. The <u>normal</u> expression of proto-oncogenes is vital, even though their normal expression may be lethal. [p.248]

Self-Quiz

____ 1. ________________ refers to the processes by which cells with identical genotypes become structurally and functionally distinct from one another according to the genetically controlled developmental program of the species. [p.242]
a. Metamorphosis
b. Metastasis
c. Cleavage
d. Differentiation

____ 2. ________________ binds to operator whenever lactose concentrations are low. [p.240]
a. Operon
b. Repressor
c. Promoter
d. Operator

____ 3. Any gene or group of genes together with its promoter and operator sequence is a(n) ________________. [p.240]
a. repressor
b. operator
c. promoter
d. operon

____ 4. The operon model explains the regulation of ________________ in prokaryotes. [p.240]
a. replication
b. transcription
c. induction
d. Lyonization

____ 5. In multicelled eukaryotes, cell differentiation occurs as a result of ________________. [p.242]
a. growth
b. selective gene expression
c. repressor molecules
d. the death of certain cells

____ 6. One type of gene control discovered in female mammals is ________________. [p.244]
a. a conflict in maternal and paternal alleles
b. slow embryo development
c. X chromosome inactivation
d. operon

_____ 7. Due to X inactivation of either the paternal or maternal X chromosome, human females with anhidrotic ectodermal dysplasia _____________ . [p.245]

a. completely lack sweat glands
b. develop benign growths
c. have mosaic patches of skin that lack sweat glands
d. develop malignant growths

_____ 8. Which of the following characteristics seems to be most uniquely correlated with metastasis? [p.248]

a. loss of nuclear-cytoplasm controls governing cell growth and division
b. changes in recognition proteins on membrane surfaces
c. "puffing" in the chromosomes
d. the massive production of benign tumors

_____ 9. Genes with the potential to induce cancerous formations are known as _____________ . [p.248]

a. proto-oncogenes
b. oncogenes
c. carcinogens
d. malignant genes

_____ 10. _____________ controls govern the rates at which mRNA transcripts that reach the cytoplasm will be translated into polypeptide chains at the ribosomes. [p.243]

a. Transport
b. Transcript processing
c. Translational
d. Transcriptional

Chapter Objectives/Review Questions

1. Define *tumor* and distinguish between benign and malignant tumors. [p.238, pp.248–249]
2. _________ is a process in which a cancer cell leaves its proper place and invades other tissues to form new growths. [p.248]
3. A _________ is a specific base sequence that signals the beginning of a gene in a DNA strand. [p.240]
4. Some control agents bind to _________ , which are short regulatory genes on either side of a promoter. [p.240]
5. A gene-control arrangement in which the same promoter–operator sequence services more than one gene is called a(n) _________ . [p.240]
6. Gene expression is controlled through regulatory _________ and _________ . [p.240]
7. In negative control, a(n) _________ protein prevents the enzymes of transcription from binding to DNA; in positive control a(n) _________ protein enhances the binding of RNA polymerases to DNA. [pp.240–241]
8. Describe the sequence of events that occurs on the chromosome of *E. coli* after you drink a glass of milk. [pp.240–241]
9. The cells of *E. coli* manage to produce enzymes to degrade lactose when those molecules are _________ and to stop production of lactose-degrading enzymes when lactose is _________. [p.242]
10. Define *selective gene expression* and explain how this concept relates to cell differentiation in multicelled eukaryotes. [p.242]
11. Cell _________ occurs in multicelled eukaryotes as a result of _________ gene expression. [p.242]
12. List and define the levels of gene control in eukaryotes. [pp.242–243]
13. Genes that control the basic body plans of organisms are called _________ _________ . [p.244]
14. The condensed X chromosome seen on the edge of each nucleus of female mammals is known as the _________ body. [p.244]
15. Explain how X chromosome inactivation provides evidence for selective gene expression; use the example of anhidrotic ectodermal dysplasia. [p.245]
16. Explain how hormones act as a major agent of gene control. [p.246]
17. Describe the gene control mechanism afforded many insect larvae by polytene chromosomes. [p.246]
18. Describe the relationship of proto-oncogenes, environmental irritants, and oncogenes. [pp.248–249]
19. Tobacco smoke, X-rays, gamma rays, and ultraviolet radiation are examples of _________ . [p.248]

Integrating and Applying Key Concepts

Suppose you have been restricting yourself to a completely vegetarian diet for the past six months. Quite unexpectedly, you find yourself in a social situation that requires you to eat a half-pound sirloin steak. Would you expect to digest the steak as easily as you digest soybean burgers? Explain your yes or no answer in terms of transcriptional controls or feedback inhibition.

16

RECOMBINANT DNA AND GENETIC ENGINEERING

Interactive Exercises

Mom, Dad, and Clogged Arteries [pp.252–253]

16.1. A TOOLKIT FOR MAKING RECOMBINANT DNA [pp.254–255]

Selected Words: *HDLs* [p.252], *LDLs* [p.252], *familial cholesterolemia* [p.252], *Haemophilus influenzae* [p.254], *staggered* cuts [p.254], *Taq I* [p.254], *Eco*RI [p.254], *Not*I [p.254], *sticky* ends [p.254], "cloning factory" [p.255], introns [p.255], exons [p. 255]

Boldfaced, Page-Referenced Terms

[p.252] gene therapy ____

____

[p.253] recombinant DNA technology ________________________________

____

[p.253] genetic engineering ______________________________________

____

[p.254] restriction enzyme ____

____

[p.254] genome ____

____

[p.254] DNA ligase ____

____

[p.254] plasmid ____

____

[p.255] DNA clone ____

____

[p.255] cloning vector ____

____

[p.255] cDNA ____

____

[p.255] reverse transcriptase ____

____

Short Answer

1. Describe and distinguish between the bacterial chromosome and plasmids present in a bacterial cell. [pp.254–255] ____

____

____

____

True/False

If the statement is true, write "T" in the blank. If the statement is false, make it correct by changing the underlined words and writing the correct word(s) in the answer blank.

______________ 2. Plasmids are <u>organelles on the surfaces of which amino acids are assembled into polypeptides.</u> [p.254]

______________ 3. <u>Gene mutations</u> and <u>recombination</u> are common in nature. [p.253]

Fill-in-the-Blanks

Genetic experiments have been occurring in nature for billions of years as a result of gene (4) ______________ [p.253], crossing over and recombination, and other events. Humans now are causing genetic change by using (5) ______________ ______________ [p.253] technology, in which researchers cut and splice together gene regions from different (6) ______________ [p.253] then greatly (7) ______________ [p.253] the number of copies of the genes that interest them. The genes, and in some cases their (8) ______________ [p.253] products, are produced in quantities that are large enough for (9) ______________ [p.253] and for practical applications. (10) ______________ ______________ [p.253] involves isolating, modifying, and inserting particular genes back into the same organism or a different one.

True/False

A genetic engineer used restriction enzymes to prepare fragments of DNA from two different species that were then mixed. Four of these fragments are illustrated below. Fragments (a) and (c) are from one species, (b) and (d) from the other species. Answer exercises 11–15. If false, change the underlined portion of the statement to make it true.

a.	b.	c.	d.
AGCT	TCGA	CGTA	AGCT

_____ 11. <u>Some</u> of the fragments represent sticky ends. [p.254]

_____ 12. The same restriction enzyme was used to cut fragments <u>(b), (c), and (d)</u>. [p.254]

_____ 13. <u>Different</u> restriction enzymes were used to cut fragments (a) and (d). [p.254]

_____ 14. Fragment <u>(a)</u> will base-pair with fragment (d) but not with fragment (b). [p.254]

_____ 15. The same restriction enzyme was used to make <u>all</u> the cuts in the DNA of the two species shown. [p.254]

Matching

Match the steps in the formation of a DNA library with the parts of the illustration below. [all from p.257]

16. _____
17. _____
18. _____
19. _____
20. _____
21. _____

A. Joining of chromosomal and plasmid DNA using DNA ligase
B. Restriction enzyme cuts chromosomal DNA at specific recognition sites
C. Cut plasmid DNA
D. Recombinant plasmids containing cloned library
E. Fragments of chromosomal DNA
F. Same restriction enzyme is used to cut plasmids

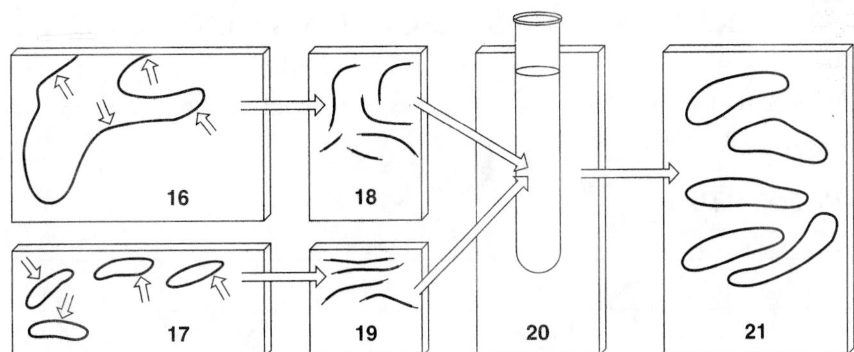

Fill-in-the-Blanks

Synthesizing the appropriate (22) ______________ [p.255] does not automatically follow the successful taking in of a modified gene by a host cell; genes must be suitably (23) ______________ [p.255] first, which involves getting rid of the (24) ______________ [p.255] from the mRNA transcripts, then using the enzyme reverse (25) ______________ [p.255] to produce cDNA. The process goes like this:

a.

b.

c.

d.

a. An mRNA transcript of a desired gene is used as a(n) (26) ______________ [p.255] for assembling a DNA strand. An enzyme (25) does the assembling.

b. An mRNA- (27) ______________ [p.255] hybrid molecule results.

c. (28) ______________ [p.255] action removes the mRNA and assembles a second strand of (29) ______________ [p.255] on the first strand.

d. The result is double-stranded (30) ______________ [p.255], "copied" from an mRNA (31) _____ [p.255].

16.2. PCR—A FASTER WAY TO AMPLIFY DNA [p.256]

16.3. FOCUS ON BIOETHICS: *DNA Fingerprints* [p.257]

16.4. HOW IS DNA SEQUENCED? [p.258]

16.5. FROM HAYSTACKS TO NEEDLES—ISOLATING GENES OF INTEREST [p.259]

Selected Words: DNA polymerase [p.256], restriction fragment length polymorphisms (RFLPs) [p.257], *genomic* library [p.259], *cDNA* library [p.259]

Boldfaced, Page-Referenced Terms

[p.256] PCR, polymerase chain reaction ____

____

[p.256] primers ____

____

[p.257] DNA fingerprint ____

____

[p.257] tandem repeats ____

____

[p.257] gel electrophoresis ______________________________

____

[p.258] automated DNA sequencing ______________________________

____

[p.258] gene library ______________________________

____

[p.259] probe ______________________________

____

[p.259] nucleic acid hybridization ______________________________

____

Complete the Table

1. Complete the following table, which summarizes some of the basic tools and procedures used in recombinant DNA technology.

Tool/Procedure	*Definition and Role in Recombinant DNA Technology*
[p.258] a. Automated DNA sequencing	
[p.256] b. Primers	
[p.259] c. cDNA library	
[p.259] d. Gel electrophoresis	
[p.256] e. PCR	
[p.258] f. DNA sequencing	
[p.259] g. Probe	
[p.257] h. DNA fingerprint	

Matching

Match the most appropriate letter with each numbered partner.

2. ____ DNA polymerase [p.256]
3. ____ A*, T*, C*, G* [p.258]
4. ____ A, T, C, G [p.258]
5. ____ RFLPs [p.257]
6. ____ tandem repeats [p.257]
7. ____ nucleic acid hybridization [p.259]

A. Identical short lengths of DNA arranged in sequence that differ from one person to the next
B. Recognizes primers as START tags and assembles complementary sequences using the template as a guide
C. DNA fragments of different lengths from a long strand of DNA cleaved by restriction enzymes
D. Nitrogen bases labeled with a molecule that fluoresces a specific color when passing through a laser beam
E. Unlabeled nitrogen bases
F. Pairing of nitrogen bases that occurs between DNA (or RNA) from different sources

16.6. USING THE GENETIC SCRIPTS [p.260]

16.7. DESIGNER PLANTS [pp.260–261]

16.8. GENE TRANSFERS IN ANIMALS [pp.262–263]

16.9. SAFETY ISSUES [p.263]

16.10. CONNECTIONS: *Bioethics in a Brave New World* [pp.264–265]

Selected Words: *Diabetics* [p.260], insulin [p.260], human somatotropin [pp.260,262], hemoglobin [p.260], interferon [p.260], *Southern corn leaf blight* [p.260], *Agrobacterium tumefaciens* [p.260], Ti plasmid [pp.260–261], CFTR protein [p.262], TPA [p.262], human collagen [p.262], *Dolly* [p.262], bovine spongiform encephalopathy [p.262], human serum albumin [p.262], Human Genome Initiative [p.262], EST [p.262], TIGR Assembler [p.262], comparative genomics [p.263], "fail-safe" genes [p.263], *hok* gene [p.263], "ice-minus" bacteria [p.263], severe combined immune deficiency (*SCID-X1*) [p.264], *eugenic engineering* [p.264], "knockout cells" [p.265], clone [p.265]

Boldfaced, Page-Referenced Terms

[p.260] seed banks ____

____

[p.263] genomics ____

____

[p.264] DNA microarrays ____

____

[p.264] xenotransplantation ____

____

Complete the Table

1. Complete the following table by providing examples of genetically engineered beneficial organisms for uses in:

[p.260] a. medicine	
[p.260] b. industry	
[p.260] c. agriculture	
[p.260] d. environmental remediation	

Short Answer

2. Explain how knowing about the genetic makeup of Earth's organisms can help us reconstruct the evolutionary history of life. [pp.260,263]__________

3. Explain how knowing the composition of genes can help scientists derive counterattacks against rapidly mutating pathogens. [pp.260,263]__________

4. Explain why it is important for farmers to plant many varieties of food and fiber plants instead of just a few varieties. [p.260]__________

5. Describe how genetic engineers transfer specific genes into plant cells. [pp.260–261]__________

6. Researchers are trying to insert the gene for human serum albumin into the chromosomes of dairy cattle. Why might that achievement be useful? [p.262]__________

7. Explain the goal of the Human Genome Initiative. [pp.262–263]__________

In exercises 8–13, summarize the results/promise of the given experimentation dealing with genetic modifications of plants and animals.

8. The bacterium *Agrobacterium tumefaciens:* [pp.260–261]________________________

9. Cotton plants: [p.261]________________________

10. Cattle may soon be producing human collagen: [p.262]________________________

11. Introduction of the rat and human somatotropin gene into fertilized mouse eggs: [p.262]____________

12. "Ice-minus" bacteria and strawberry plants [p.263]________________________

13. Dolly, the cloned lamb: [p.262]________________________

Fill-in-the-Blanks

(14) ____________ ____________ [p.263] can be used to look at evolutionary relationships of organisms. The gene harmful to strawberries is called the "ice-forming" gene, and the bacteria are known as (15) ____________ - ____________ [p.263] bacteria; genetic engineers were able to remove the harmful gene and test the modified bacterium on strawberry plants with no adverse effects. Inserting one or more genes into the (16) ____________ ____________ [recall; p.252] of an organism for the purpose of correcting genetic defects is known as (17) ____________ ____________ [recall; p.252]. Attempting to modify a human trait by inserting genes into sperm or eggs is called (18) ____________ ____________ [p.264].

Self-Quiz

____ 1. Small, circular molecules of DNA in bacteria are called __________. [p.254]
a. plasmids
b. desmids
c. pili
d. F particles
e. transferins

____ 2. Enzymes used to cut genes in recombinant DNA research are __________. [p.254]
a. ligases
b. restriction enzymes
c. transcriptases
d. DNA polymerases
e. replicases

____ 3. The total DNA in a haploid set of chromosomes of a species is its __________. [p.254]
a. plasmid
b. enzyme potential
c. genome
d. DNA library
e. none of the above

____ 4. Any DNA molecule that is copied from mRNA is known as __________. [p.255]
a. cloned DNA
b. cDNA
c. DNA ligase
d. hybrid DNA

____ 5. The most commonly used method of DNA amplification is __________. [p.256]
a. polymerase chain reaction
b. gene expression
c. genome mapping
d. RFLPs

____ 6. Tandem repeats are valuable because __________. [p.257]
a. they reduce the risks of genetic engineering
b. they provide an easy way to sequence the human genome
c. they allow fragmenting of DNA without enzymes
d. they provide DNA fragment sizes unique to each person

____ 7. __________ __________ __________ can rapidly reveal the base sequence of cloned DNA or PCR-amplified DNA fragments. [p.258]
a. Polymerase chain reaction
b. Nucleic acid hybridization
c. Recombinant DNA technology
d. Automated DNA sequencing

____ 8. A cDNA library is __________. [p.259]
a. a collection of DNA fragments derived from mRNA and free of introns
b. cDNA plus the required restriction enzymes
c. mRNA–cDNA
d. composed of mature mRNA transcripts

____ 9. The study of the genomes of humans and other organisms is called __________. [p.263]
a. cloning
b. genetic engineering
c. gene therapy
d. genomics

____ 10. A knockout cell is __________. [p.263]
a. a fertilized egg that can lead to an exceptionally attractive person
b. a cell in which a particular gene sequence has been excised
c. any cell that has been removed from an organism
d. any cell that has been genetically engineered

Chapter Objectives/Review Questions

1. List the means by which natural genetic recombination occurs. [p.253]
2. Define *recombinant DNA technology*. [p.253]
3. Some bacteria produce ________ enzymes that cut apart DNA molecules injected into the cell by viruses; such DNA fragments, or "________ ends," often have staggered cuts capable of base-pairing with other DNA molecules cut by the same ________ enzymes. [p.254]
4. Base-pairing between chromosomal fragments and cut plasmids is made permanent by DNA ________. [p.254]
5. ________ are small, circular, self-replicating molecules of DNA or RNA within a bacterial cell. [p.254]
6. Define *cDNA*. [p.255]
7. A special viral enzyme, ________ ________, presides over the process by which mRNA is transcribed into DNA. [p.255]
8. Why do researchers prefer to work with cDNA when working with human genes? [p.255]
9. Multiple, identical copies of DNA fragments produced by restriction enzymes are known as ________ DNA. [p.257]
10. List and define the two major methods of DNA amplification. [pp.255–256]
11. Polymerase chain reaction is the most commonly used method of DNA ________. [p.256]
12. Explain how gel electrophoresis is used in DNA fingerprinting. [pp.257–258]
13. How is a cDNA probe used to identify a desired gene carried by a modified host cell? [p.259]
14. Be able to explain what a DNA library is; review the steps used in creating such a library. [p.259]
15. List some practical genetic uses of RFLPs. [p.260]
16. How is *Agrobacterium tumefaciens* used in gene transfer in plants? [pp.260–261]
17. Tell about the Human Genome Initiative and its implications. [pp.262–263]
18. How could gene therapy reduce the number of infant deaths and hospital admissions? [p.264]
19. Define *gene therapy* and *eugenic engineering*. [pp.252,264]
20. Describe how pig organs can be altered so as not to be rejected by a human recipient's immune system. [p.265]

Integrating and Applying Key Concepts

How could scientists guarantee that *Escherichia coli,* the human intestinal bacterium, will not be transformed into a severely pathogenic form and released into the environment if researchers use the bacterium in recombinant DNA experiments?

17

MICROEVOLUTION

Interactive Exercises

Designer Dogs [pp.270–271]

17.1. EARLY BELIEFS, CONFOUNDING DISCOVERIES [pp.272–273]

Selected Words: *microevolution* [p.271], "species" [p.272], *sequences* of fossils [p.273], "useless" body parts [p.273]

Boldfaced, Page-Referenced Terms

[p.271] evolution ____

____

[p.272] biogeography ____

____

[pp.272–273] comparative morphology ______________________________

____

[p.273] fossils ______________________________

____

Matching

Choose the most appropriate answer for each term.

1. ____ comparative anatomy [pp.272–273]
2. ____ biogeography [p.272]
3. ____ fossils [p.273]
4. ____ school of Hippocrates [p.272]
5. ____ evolution [p.271]
6. ____ Chain of Being [p.272]
7. ____ Buffon [p.273]
8. ____ species [p.272]
9. ____ Aristotle [p.272]
10. ____ sequences of fossils [p.273]
11. ____ microevolution [p.271]

A. The increasing complexity of organisms as indicated in fossil beds
B. The person who came to view nature as a continuum of organization, from lifeless matter through complex forms of plant and animal life
C. Each kind of being that represents a link in Aristotle's Chain of Being
D. Genetic change in a line of descent over time
E. Extended from the lowest forms of life to humans and on to spiritual beings
F. Suggested that perhaps species originated in more than one place and perhaps had been modified over time
G. Studies of similarities and differences in body plans
H. Studies of the world distribution of plants and animals
I. First to perceive the link between cause and effect in nature
J. Used as evidence of life in ancient times
K. Changes in the allele frequencies of a population over time

17.2. A FLURRY OF NEW THEORIES [pp.274–275]

17.3. DARWIN'S THEORY TAKES FORM [pp.276–277]

Selected Words: *acquired* characteristics [p.274], *fluida* [p.274], H.M.S. *Beagle* [pp.274–275], *Principles of Geology* [p.275], "select" [p.277], *artificial* selection [p.277], *natural* selection [p.277], *On the Origin of Species* [p.277]

Boldfaced, Page-Referenced Terms

[p.274] catastrophism ______________________________

____

[p.275] theory of uniformity ______________________________

____

Choice

For questions 1–12, choose from the following:

a. Georges Cuvier b. Jean Lamarck c. Charles Darwin

1. ____ Stretching directed "fluida" to the necks of giraffes, which lengthened permanently [p.274]
2. ____ Acknowledged there were abrupt changes in the fossil record that corresponded to discontinuities between certain layers of sedimentary beds; thought this was evidence of change in populations of ancient organisms [p.274]
3. ____ The force for change in organisms is the drive for perfection [p.274]
4. ____ There was only one time of creation that populated the world with all species [p.274]
5. ____ Catastrophism [p.274]
6. ____ Suggested that artificial selection was a way of explaining natural selection [p.277]
7. ____ Permanently stretched giraffe necks were bestowed on offspring [p.274]
8. ____ When a global catastrophe destroyed many organisms, a few survivors repopulated the world [p.274]
9. ____ Theories were influenced by the work of Lyell and Malthus [pp.275–276]
10. ____ Theory of inheritance of acquired characteristics [p.274]
11. ____ There are no new species, only those whose fossil records had not been identified yet [p.274]
12. ____ During a lifetime, environmental pressures and internal "desires" bring about permanent changes [p.274]
13. ____ The force for change is a drive for perfection, up the Chain of Being [p.274]
14. ____ Populations evolve, not individuals [p.277]

Complete the Table

15. Several key players and events in the life of Charles Darwin led him to his conclusions about natural selection and evolution. Summarize these influences by completing the following table.

Event/Person	*Importance to Synthesis of Evolutionary Theory*
[p.274] a.	Botanist at Cambridge University who perceived Darwin's real interests and arranged for Darwin to become a ship's naturalist
[p.274] b.	British ship that carried Darwin (as a naturalist) on a five-year voyage around the world
[p.275] c.	Wrote *Principles of Geology;* advanced the theory of uniformity; suggested that Earth was much older than 6,000 years
[p.276] d.	Wrote an influential essay (read by Darwin) on human populations asserting that people tend to produce children faster than food supplies, living space, and other resources can be sustained
[p.276] e.	Volcanic islands 900 kilometers from the South American coast where Darwin correlated differences in various species of finches with their environmental challenges
[p.277] f.	English naturalist contemporary with Darwin; independently developed Darwin's theory of evolution before Darwin published
[p.274] g.	Where Darwin earned a degree in theology but also developed his love for natural history
[p.277] h.	The type of selective force Darwin used to explain his concept of natural selection
[p.277] i.	The key point in Darwin's theory of evolution; involves reproductive capacity, heritable variations, and adaptive traits

Sequence

Read ideas A–G through first, then put them in an order that logically develops the Darwin–Wallace theory of evolution in correct sequence. Number 14 is the first, most fundamental idea. Idea number 15 sets the stage for and underlies the remaining ideas.

16. ______ The first idea in the series [pp.276–277]
17. ______ The second idea in the series [pp.276–277]
18. ______ The third idea in the series [pp.276–277]
19. ______ The fourth idea in the series [pp.276–277]
20. ______ The fifth idea in the series [pp.276–277]
21. ______ The sixth idea in the series [pp.276–277]
22. ______ The concluding idea in the series [pp.276–277]

A. Nature "selects" those individuals with traits that allow them to obtain the resources they need. They live longer and produce more offspring than others in the population who cannot get the resources they need to live and reproduce.
B. As population size increases, available resources dwindle.
C. A population is evolving when the forms of its heritable traits are changing over successive generations.
D. Animal populations tend to reproduce faster than food supplies, living space, and other resources can sustain the populations.
E. The struggle for existence intensifies.
F. There is genetic variation in all sexually reproducing populations; variations in traits might affect individuals' abilities to get resources and therefore to survive and reproduce in particular environments.
G. Over time, the more successful phenotypes will dominate the population that exists in that particular environment.

17.4. INDIVIDUALS DON'T EVOLVE — POPULATIONS DO [pp.278–279]

17.5. FOCUS ON SCIENCE: *When Is a Population* Not *Evolving?* [pp.280–281]

Selected Words: *morphological* traits [p.278], *morpho-* [p.278], *physiological* traits [p.278], *behavioral* traits [p.278], *qualitatively different* [p.278], *quantitatively different* [p.278], *natural selection, gene flow,* and *genetic drift* [p.279], *allele frequencies* [p.279]

Boldfaced, Page-Referenced Terms

[p.278] population ______________________________

____

[p.278] polymorphism ______________________________

____

[p.278] gene pool ______________________________

____

[p.278] alleles ______________________________

____

[p.278] allele frequencies ______________________________

____

[pp.278–279] genetic equilibrium ______________________________

____

[p.279] microevolution ______________________________

____

[p.279] mutation rate ______________________________

____

[p.279] lethal mutation ______________________________

____

[p.279] neutral mutation ______________________________

____

[p.280] Hardy–Weinberg rule ______________________________

____

Fill-in-the-Blanks

Unlike earlier scientists, Charles Darwin recognized that (1) ______________ [p.278] do not evolve, (2) ______________ [p.278] do. A population is a group of individuals of the same (3) ______________ [p.278] that are occupying a given area. Within a population all members share the same body plan, or (4) ______________ [p.278] traits. (5) ______________ [p.278] traits relate to how the body functions in its environment, while (6) ______________ [p.278] traits indicate how members of a population respond to basic stimuli.

As is the case with (7) ______________ [p.278] reproducing species, most traits (8) ______________ [p.278] among individuals. A trait may come in two or more distinct forms. These traits are said to be (9) ______________ [p.278] different, and the differences in these traits are called (10) ______________ [p.278]. If the individuals of a population show continuous, small, incremental differences in a trait, they are said to be (11) ______________ [p.278] different.

The sum of all of the genes in an entire population is called the (12) ______________ ______________ [p.278]. Each gene may be represented by two or more different molecular forms, called (13) ______________ [p.278]. The abundance of each allele in the gene pool is called the (14) ______________ [p.278]. If the frequencies of these alleles remain stable over several generations, the population is said to be in (15) ______________ [p.278].

A (16) ______________ [p.279] is a heritable change in the DNA that usually gives rise to an altered gene product. The probability that a gene will mutate in or between DNA replications is called the (17) ______________ ______________ [p.279]. If the mutation has a severe effect on the organism, leading to death, it is called a (18) ______________ [p.279] mutation. A (19) ______________ mutation does not provide an advantage or disadvantage to the organism. However, mutations are so (20) ______________ [p.279] that they usually have little or no immediate effect on the population's (21) ______________ [p.279] frequencies.

Matching

Select the one most appropriate answer to match the sources of genetic variation. [p.282]

22. ____ fertilization
23. ____ changes in chromosome structure or number
24. ____ crossing over at meiosis
25. ____ gene mutation
26. ____ independent assortment at meiosis

A. Leads to mixes of paternal and maternal chromosomes in gametes
B. Produces new alleles
C. Leads to the loss, duplication, or alteration of alleles
D. Brings together combinations of alleles from two parents
E. Leads to new combinations of alleles in chromosomes

Short Answer

27. List the five conditions (in any order) that must be met before genetic equilibrium (or nonevolution) will occur. [pp.279–280] ______________________________

____

____

____

28. For the following situation, assume that the conditions listed in question 27 do exist; therefore, there should be no change in gene frequency, generation after generation. Consider a population of hamsters in which dominant gene *B* produces black coat color and recessive gene *b* produces gray coat color (two alleles are responsible for color). The dominant gene has a frequency of 80 percent (or .80). It would follow that the frequency of the recessive gene is 20 percent (or .20). From this, the assumption is made that 80 percent of all sperm and eggs have gene *B*. Also, 20 percent of all sperm and eggs carry gene *b*. [p.280]

a. Calculate the probabilities of all possible matings using the Punnett square. ______

______

______

______

b. Summarize the genotype and phenotype frequencies of the F_1 generation. ______

______

______

______

c. Further assume that the individuals of the F_1 generation produce another generation and that the assumptions of the Hardy–Weinberg rule still hold. What are the frequencies of the sperm produced? [p.280] ______

______

______

______

Genotypes	*Phenotypes*
__ *BB*	
__ *Bb*	__ % black
__ *bb*	__ % gray

		Sperm 0.80 *B*	Sperm 0.20 *b*
Eggs	0.80 *B*	*BB*	*Bb*
	0.20 *b*	*Bb*	*bb*

Parents (F_1)	*B sperm*	*b sperm*
__ *BB*	__	__
__ *Bb*	__	__
__ *bb*	__	__
Totals =	__	__

The egg frequencies may be similarly calculated. Note that the gamete frequencies of the F_2 are the same as the gamete frequencies of the last generation. Phenotype percentage also remains the same. Thus, the gene frequencies did not change between the F_1 and F_2 generations. Again, given the assumptions of the Hardy–Weinberg equilibrium, gene frequencies do not change generation after generation.

29. In a population, 81 percent of the organisms are homozygous dominant, and 1 percent are homozygous recessive. Find the following: [p.280]

    a. the percentage of heterozygotes ______

    b. the frequency of the dominant allele ______

    c. the frequency of the recessive allele ______

30. In a population of 200 individuals, determine the following for a particular locus if $p = 0.80$: [p.280]

    a. the number of homozygous dominant individuals ______

    b. the number of homozygous recessive individuals ______

    c. the number of heterozygous individuals ______

31. If the percentage of gene *D* is 70 percent in a gene pool, find the percentage of gene *d*. [p.280] ______

______

32. If the frequency of gene R in a population is 0.60, what percentage of the individuals are heterozygous Rr? [p.280] ____

____

Fill-in-the-Blanks

The (33) ______________–______________ [p.284] rule allows researchers to establish a theoretical reference point (baseline) against which changes in allele frequency can be measured. The formula in this rule states that (34) ______________ [p.280] represents the frequency of allele A in the population, while (35) ______________ [p.280] represents the frequency of allele a. The basic premise of the Hardy–Weinberg rule is that the (36) ______________ ______________ [p.280] will not change through successive generations if there is no (37) ______________ [p.280], if the population is infinitely (38) ______________ [p.280], if (39) ______________ [p.280] is random, and if all individuals survive and (40) ______________ [p.280] equally. As long as these assumptions are met, the allele frequencies should (41) ______________ ______________ [p.280] through the generations.

17.6. NATURAL SELECTION REVISITED [p.281]

17.7. DIRECTIONAL CHANGE IN THE RANGE OF VARIATION [pp.282–283]

17.8. SELECTION AGAINST OR IN FAVOR OF EXTREME PHENOTYPES [pp.284–285]

Selected Words: *mark-release-recapture method* [p.282], *pest resurgence* [p.283], *biological controls* [p.283], *intermediate-sized galls* [p.284]

Boldfaced, Page-Referenced Terms

[p.281] natural selection ____

____

[p.282] directional selection ____

____

[p.283] antibiotics ____

____

[p.284] stabilizing selection ____

____

[p.285] disruptive selection ____

____

Labeling

1. For each of the three curves below, first identify the curve as an example of stabilizing selection, directional selection, or disruptive selection; then give the general characteristics of each form of selection. [pp.282–285]

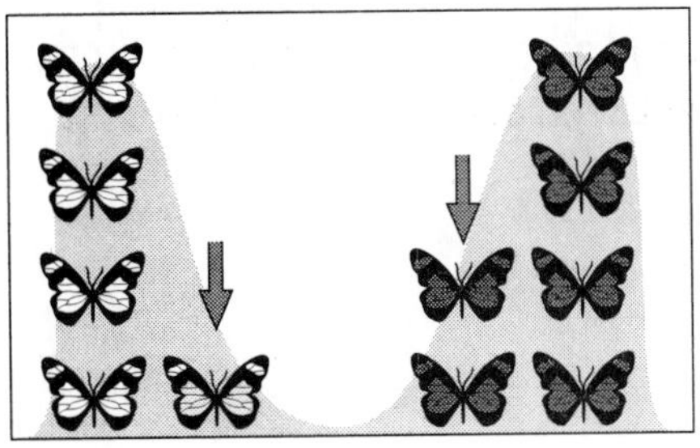

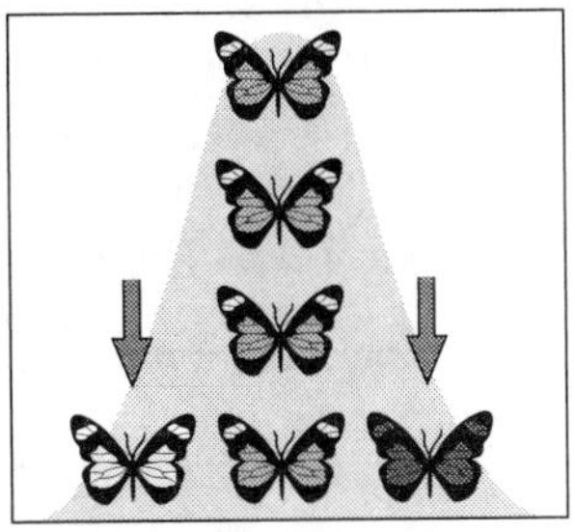

a. ______________________

____

b. ______________________

____

c. ______________________

____

Choice

For each of the following, choose from one of the following forms of selection: [pp.282–285]

a. stabilizing selection   b. directional selection   c. disruptive selection

2. _____ Intermediate forms of a trait in a population are favored
3. _____ The range of variation in a phenotype tends to shift in a consistent direction
4. _____ The most frequent wing color of peppered moths shifted from a light form to a dark form as tree trunks became soot-darkened because coal was used for fuel during the industrial revolution
5. _____ Intermediate forms of a phenotype are selected against, while forms at the end of the range of variation are favored
6. _____ Antibiotic resistance is an example
7. _____ Examples are species of gall-making flies
8. _____ The development of insecticide resistance in a pest population is an example
9. _____ An example is the selection of bill size in finches of West Africa

Short Answer

10. What are the similarities between antibiotic resistance and insecticide resistance with regard to selection? ______________________

____

____

____

11. How is the mark-release-recapture method used to study populations? ______________________

____

____

____

17.9. MAINTAINING VARIABILITY IN A POPULATION [pp.286–287]

17.10. GENE FLOW [p.287]

17.11. GENETIC DRIFT [pp.288–289]

Selected Words: *dimorphos* [p.286], *balancing* [p.286], *polymorphos* [p.286], *sickle-cell anemia* [pp.286–287], *malaria* [p.287], *nonmutated* molecules [p.287], *altered* molecules [p.287], *emigration* [p.287], *immigration* [p.287], *Ellis–van Creveld syndrome* [p.289], *feline infectious peritonitis* [p.289]

Boldfaced, Page-Referenced Terms

[p.286] sexual selection ____

____

[p.286] balanced polymorphism ____

____

[p.287] gene flow ____

____

[p.288] genetic drift ____

____

[p.288] sampling error ____

____

[p.288] fixation ____

____

[p.288] bottleneck ____

____

[p.289] founder effect ____

____

[p.289] inbreeding ____

____

Fill-in-the-Blanks

Individuals of most (1) ______________ [p.286] reproducing species have a distinctively male or female (2) ______________ [p.286]. Differences in appearance between males and females of a species are known as (3) ______________ [p.286]. (4) ______________ [p.286] selection is based on any trait that gives the individual a competitive edge in mating and producing offspring. Among many species of birds and mammals, females act as agents of (5) ______________ [p.286] when they choose their mates.

A form of selection called (6) ______________ [p.286] selection refers to any form of selection that maintains two or more alleles in a population. This type of genetic variation in a population is called (7) ______________ [p.286]. An example of a balanced polymorphism in human populations is (8) ______________ ______________ [p.286], a genetic disorder that produces an abnormal form of hemoglobin. The normal allele for hemoglobin is (9) ______________ [p.286], while the mutant allele is

Hb^s. Individuals who are (10) ______________ [p.286] for the mutant allele die early in life. However, individuals who are heterozygous are more likely to survive (11) ______________ [p.286], a disease that is transmitted by mosquitoes in tropical areas.

Choice

For questions 12–22, choose from the following microevolutionary forces that can change gene frequency:

a. genetic drift   b. gene flow

12. _____ A random change in allele frequencies over the generations, brought about by chance alone [p.288]
13. _____ Emigration [p.287]
14. _____ Prior to the turn of the twentieth century, hunters killed all but 20 of a large population of northern elephant seals [p.289]
15. _____ When allele frequencies change as the result of individuals leaving a population or new individuals entering it [p.287]
16. _____ In the absence of other forces, random change in allele frequencies leads to the homozygous condition and the loss of genetic diversity over generations [p.288]
17. _____ An example is the colonization of the Galápagos Islands by finches long ago [p.289]
18. _____ Sometimes results in endangered species [p.288]
19. _____ Founder effects and bottlenecks are two extreme cases [pp.288–289]
20. _____ Immigration [p.287]
21. _____ Only 30,000 elephant seals survived hunting in the 1890s [p.289]
22. _____ Blue jays make hundreds of round trips carrying acorns from oak trees as much as a mile away to deposit on soil in their home territories for winter storage; this introduces new alleles to oak forests [p.287]

Short Answer

23. Compare the two extreme cases of genetic drift: the founder effect and bottlenecks. Explain the relative amount of genetic diversity in each case. [pp.288–289] ______________________________

____

____

____

Fill-in-the-Blanks

Random fluctuations in allele frequencies over time due to chance occurrence alone are called (24) ______________ ______________; [p.288] it is more pronounced in (25) ______________ [p.289] populations than in large ones. (26) ______________ [p.288] flow associated with immigration and/or emigration also changes allele frequencies. A severe reduction in population size is called a (27) ______________ [p.288], which results in an alteration of the (28) ______________ ______________ [p.288] of the population. The nonrandom mating of closely related individuals is called (29) ______________ [p. 289], which can lead to a (30) ______________ [p.289] condition in the population over time.

Self-Quiz

____ 1. An acceptable definition of evolution is ____________. [p.271]
a. changes in organisms that are extinct
b. changes in organisms since the flood
c. changes in organisms over time
d. changes in organisms in only one place

____ 2. The two scientists most closely associated with the concept of evolution are ____________. [pp.276–277]
a. Lyell and Malthus
b. Henslow and Cuvier
c. Henslow and Malthus
d. Darwin and Wallace

____ 3. Studies of the distribution of organisms on the Earth is known as ____________. [p.272]
a. comparative morphology
b. biogeography
c. catastrophism
d. artificial selection

____ 4. Which of the scientists below is correctly matched to his theory of evolution? [pp.272–277]
a. Aristotle—catastrophism
b. Cuvier—Chain of Beings
c. Lamarck—Theory of Acquired Characteristics
d. Lyell—Natural Selection

____ 5. Lyell's book, *Principles of Geology*, suggested which of the following? [p.275]
a. tail bones in humans have no place in a perfectly designed body
b. the fossil record could be explained as the result of a single great flood
c. gradual changes currently underway on the Earth have been at work for millions of years
d. all species were created at one time

____ 6. Which of the following is not correct concerning Lamarck's theory of evolutionary processes? [p.274]
a. environmental pressures cause changes in organisms
b. individuals pass acquired characteristics on to their offspring
c. nerves directed an unknown "fluida" to parts of the body in need of change
d. populations evolve

____ 7. Which of the following was responsible for the concept that as a population grows, it outstrips its resources, forcing individuals to compete for limited food and water? [p.276]
a. Lyell
b. Wallace
c. Malthus
d. Henslow

____ 8. Differences in the combinations of alleles carried by the individuals of a population would be its ____________. [p.278]
a. gene pool
b. genetic variation
c. gene flow
d. allele frequency

____ 9. The sum of all of the genes in the entire population is the ____________. [p.278]
a. gene pool
b. genetic variation
c. gene flow
d. allele frequency

____ 10. The relative abundance of each type of allele in a population is the ____________. [p.278]
a. gene pool
b. genetic variation
c. gene flow
d. allele frequency

_____ 11. According to the Hardy–Weinberg rule, the allele frequencies of a population will not change over successive generations if which of the following are true? [p.280]
a. the population is infinitely large
b. there is random mating
c. there is no mutation
d. the population is isolated
e. all of the above

_____ 12. The unit used in studying evolution is the _______________ . [p.278]
a. population
b. individual
c. fossil
d. missing link

_____ 13. Changes in allele frequencies brought about by mutation, genetic drift, gene flow, and natural selection are called _______________ processes. [p.279]
a. genetic equilibrium
b. microevolution
c. founder effects
d. independent assortment

_____ 14. An insect population that becomes increasingly resistant to a class of insecticides is an example of _______________ selection. [pp.282–285]
a. directional
b. sexual
c. disruptive
d. stabilizing

_____ 15. If a trait in one sex of a species is favored by the opposite sex of the species, it is called _______________ . [p.286]
a. sexual selection
b. genetic equilibrium
c. catastrophism
d. theory of use and disuse

Chapter Objectives/Review Questions

1. Explain how biogeography and comparative morphology challenged ancient views of evolution. [p.272]
2. Explain the importance of fossils to the development of early evolutionary theories. [p.273]
3. What is catastrophism? [p.274]
4. Lamarck's theory of acquired characteristics was incorrect as to the fact that _________ evolved, but correct as to the fact that the _________ was involved in change. [p.274]
5. Explain how both Lyell and Malthus contributed to the development of the theory of natural selection. [pp.275–276]
6. What was the significance of Alfred Wallace? [p.277]
7. A _________ is a group of individuals occupying a given area and belonging to the same species. [p.278]
8. Distinguish among morphological, physiological, and behavioral traits. [p.278]
9. All the genes of an entire population belong to a _________ _________ . [p.278]
10. Each kind of gene usually exists in one or more molecular forms, called _________ . [p.278]
11. Review the five categories through which genetic variation occurs among individuals. [p.278]
12. The abundance of each kind of allele in the whole population is referred to as *allele* _________ . [p.278]
13. A point at which allele frequencies for a trait remain stable through the generations is called genetic _________ . [pp.278–279]
14. Changes in allele frequencies brought about by mutation, genetic drift, gene flow, and natural selection are called _________ . [p.279]
15. Describe the differences between neutral and lethal mutations. [p.279]
16. List the five conditions that must be met for the Hardy–Weinberg rule to apply. [p.280]
17. Calculate allele and other genotype frequencies when provided with the homozygous recessive genotype frequency. (For example, assume q2 = .36.) [p.280]
18. Define and provide an example of directional selection, stabilizing selection, and disruptive selection. [pp.282–285]

19. Define balanced polymorphism. [p.286]
20. ________ selection is based on any trait that gives an individual a competitive edge in mating and producing offspring. [p.286]
21. Allele frequencies change as individuals leave or enter a population; this is gene ________ . [p.287]
22. Random fluctuation in allele frequencies over time, due to chance, is called ________ ________. [p.288]
23. Distinguish the founder effect from a bottleneck. [pp.288–289]
24. Relate bottlenecks to endangered species. [p.293]

Integrating and Applying Key Concepts

1. Can you imagine any way in which directional selection may have occurred or may be occurring in humans? Which factors do you suppose are the driving forces that sustain the trend? Do you think the trend could be reversed? If so, by what factor(s)?
2. Endangered animals are frequently confined to zoos for protection. How does this contribute to a bottleneck effect? How could gene flow be increased and what would be the benefit of doing so?

18

SPECIATION

Interactive Exercises

The Case of the Road-Killed Snails [pp.292–293]

18.1. ON THE ROAD TO SPECIATION [pp.294–295]

Selected Words: *reproduction* [p.294], *prezygotic* mechanisms [p.294], *behavioral isolation* [p.295], *temporal isolation* [p.295], "sibling species" [p.295], *mechanical isolation* [p.295], *ecological isolation* [p.295], *gametic mortality* [p.295], *postzygotic* isolating mechanisms [p.295]

Boldfaced, Page-Referenced Terms

[p.293] speciation ____

____

[p.294] species ____

____

[p.294] biological species concept ____

____

[p.294] gene flow ____

____

[p.294] genetic divergence ____

____

[p.294] reproductive isolating mechanisms ____

____

Matching

Choose the most appropriate answer for each question.

1. ____ biological species concept [p.294]
2. ____ gene flow [p.294]
3. ____ speciation [p.293]
4. ____ sibling species [p.295]
5. ____ reproductive isolating mechanisms [p.294]

A. When the descendant species differs significantly in its alleles from the ancestral species
B. The movement of alleles into and out of a population
C. States that species are groups of interbreeding natural populations that are reproductively isolated from other groups
D. Two species that are similar morphologically
E. An inheritable form of body form, function, or behavior that prevents interbreeding between two genetically divergent populations

Matching

Select the most appropriate answer to match the isolating mechanisms; complete the exercise by writing "pre" in the parentheses if the mechanism is prezygotic and "post" if the mechanism is postzygotic. [p.295]

6. ____ ecological (   )
7. ____ temporal (   )
8. ____ early death of hybrids (   )
9. ____ mechanical (   )
10. ____ sterility (   )
11. ____ gametic mortality (   )
12. ____ behavioral (   )
13. ____ low survival of hybrids (   )

A. Potential mates occupy overlapping ranges but reproduce at different times.
B. The first-generation hybrid forms but lack the ability to reproduce.
C. Potential mates occupy different local habitats within the same area.
D. Potential mates meet but cannot figure out what to do about it.
E. Sperm is transferred, but the egg is not fertilized (gametes die or gametes are incompatible).
F. The hybrid is weak.
G. Potential mates attempt engagement, but sperm cannot be successfully transferred.
H. The egg is fertilized, but the zygote or embryo dies.

Choice

For questions 14–19, choose from the following isolating mechanisms: [p.295]

a. temporal   b. behavioral   c. mechanical   d. gametic mortality   e. ecological   f. postzygotic

14. ____ Sterile mules
15. ____ Two sage species, each of which has its flower petals arranged as a "landing platform" for a different pollinator
16. ____ Two species of cicada; one matures, emerges, and reproduces every thirteen years, the other every seventeen years
17. ____ Populations of the manzanita shrubs *Arctostaphylos patula* and *A. viscida* demonstrate differing tolerances to water stress at varying distances from water sources; speciation may be underway
18. ____ Pollen grains from one flowering plant species molecularly mismatched with gametes of a different flowering plant species
19. ____ Before copulation, male and female birds engage in complex courtship rituals recognized only by birds of their own species

Identification

20. The following illustration depicts the divergence of one species as time passes. Each horizontal line represents a population of that species, and each vertical line represents a point in time. Answer the questions below the illustration. [p.294]

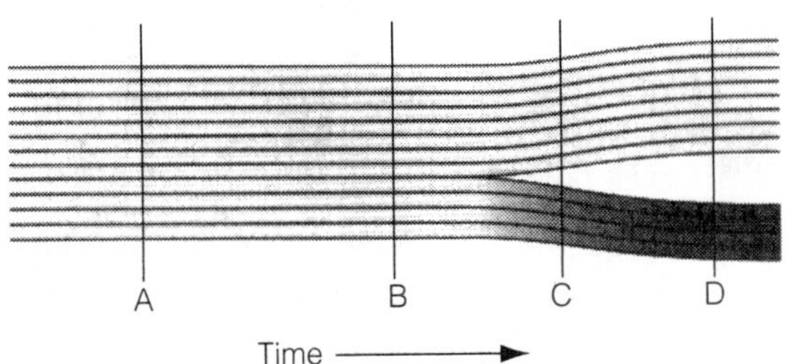

a. How many species are represented at time A? ____________

b. How many species exist at time D? ____________

c. Assuming that the observed genetic divergence leads to speciation, what letter represents the last time that all of the populations may be considered the same species? ____________

d. What letter represents the time that complete divergence is reached? ____________

e. Between what two times (letters) is divergence clearly underway but not complete? ____________

18.2. SPECIATION IN GEOGRAPHICALLY ISOLATED POPULATIONS [pp.296–297]

18.3. MODELS FOR OTHER SPECIATION ROUTES [pp.298–299]

Selected Words: *physical separation* [p.296], *allo-* [p.296], *-patric* [p.296], "Pacific" enzymes [p.296], "Atlantic" enzymes [p.296], *sym-* [p.298], *ecological* separation [p.298], *breed* [p.298], *para-* [p.299], *subspecies* [p.299], "secondary contact" [p.299]

Boldfaced, Page-Referenced Terms

[p.296] allopatric speciation ____________

[p.297] archipelago ____________

[p.298] sympatric speciation ____________

[p.298] polyploidy ____________

[p.298] dosage compensation ____________

[p.305] parapatric speciation ____________

[p.305] hybrid zone ____________

Choice

For questions 1–10, choose from the following answers:

a. sympatric speciation [p.298] b. parapatric speciation [p.299] c. allopatric speciation [p.297]

1. _____ Neighboring populations become separate species while maintaining contact along a common border
2. _____ The changing of the path of a river or formation of new mountain ranges
3. _____ An example are the cichlid species inhabiting lakes in West Africa
4. _____ One example is oriole populations in the United States
5. _____ Occurs by the physical separation of populations
6. _____ Polyploidy in plants is an example of this type of speciation
7. _____ One species forms within the home range of another species, in the absence of a physical barrier
8. _____ This type of speciation is common among populations on archipelagos
9. _____ The formation of new species due to the rise of the Isthmus of Panama
10. _____ An example is the Galápagos finches

Short Answer

11. A new four-lane highway is being built between two towns. The road will pass directly through a grassland habitat that contains a wide variety of plant and animal species. Assuming that the new road is heavily traveled and that the road will bisect the populations, answer the following questions: [pp.296–297]
    a. What type of speciation does this represent?
    b. For each of the following, indicate whether the new road will disrupt gene flow between the populations:
    _____ earthworms
    _____ hawks
    _____ plants that disperse their seeds using wind
    _____ plants that disperse their seeds with the assistance of small mammals

18.4. PATTERNS OF SPECIATION [pp.300–301]

Selected Words: *klados* [p.300], *genesis* [p.300], *ana-* [p.300], *branch* [p.300], *branch point* [p.300], *physical* access [p.301], *evolutionary* access [p.301], *ecological* access [p.301]

Boldfaced, Page-Referenced Terms

[p.300] cladogenesis ____

____

[p.300] anagenesis ____

____

[p.300] evolutionary trees ____

____

[p.300] gradual model of speciation ______

______

[p.300] punctuation model of speciation ______

______

[p.300] adaptive radiation ______

______

[pp.300–301] adaptive zone ______

______

[p.301] key innovation ______

______

[p.301] extinction ______

______

[p.301] mass extinction ______

______

Choice

For questions 1–5, choose one of the diagrams below. [p.300]

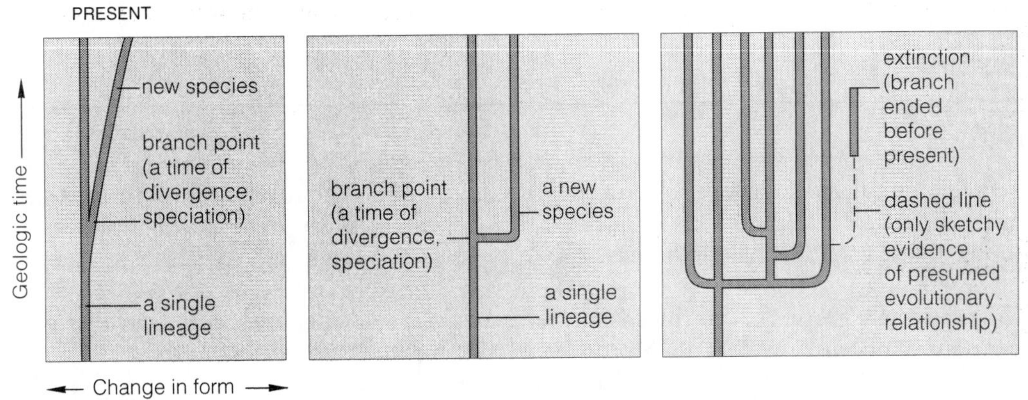

1. ____ Indicates that traits changed rapidly around the time of speciation
2. ____ Indicates that an adaptive radiation occurred
3. ____ Indicates that an extinction occurred
4. ____ Suggests that speciation occurred through gradual changes in traits over geologic time
5. ____ Indicates that evidence of this presumed evolutionary relationship is only sketchy

Matching

Select the most appropriate answer for each.

6. ____ adaptive radiation [p.300]
7. ____ adaptive zones [pp.300–301]
8. ____ anagenesis [p.300]
9. ____ cladogenesis [p.300]
10. ____ gradual model of speciation [p.300]
11. ____ key innovation [p.301]
12. ____ mass extinction [p.301]
13. ____ physical, evolutionary, and ecological access [p.301]
14. ____ punctuation model of speciation [p.300]
15. ____ evolutionary trees [p.300]

A. Modification of some structure or function that permits a lineage to exploit the environment in new or more efficient ways
B. Conditions allowing a lineage to radiate into an adaptive zone
C. Speciation in which most morphological changes are compressed into a brief period of hundreds or thousands of years when populations first start to diverge; evolutionary tree branches make abrupt, 90-degree turns
D. Ways of life such as "burrowing in the seafloor"
E. Speciation with slight angles on the evolutionary tree, which conveys many small changes in form over long time spans
F. An abrupt rise in the rates of species disappearance above background level
G. Speciation occurring within a single, unbranched line of descent
H. A method of summarizing information about the continuities of relationships among species
I. A branching speciation pattern; isolated populations diverge
J. A burst of microevolutionary activity within a lineage; results in the formation of new species in a wide range of habitats

Self-Quiz

____ 1. The biological species concept states that ____________. [p.294]
   a. all living creatures belong to one species
   b. species are groups of interbreeding natural populations that are reproductively isolated from other groups
   c. reproduction may occur either sexually or asexually
   d. all organisms that have similar morphological characteristics belong to the same species
   e. species are determined by their genetic composition

____ 2. When something prevents gene flow between two populations or subpopulations, ____________ may occur. [p.294]
   a. genetic drift, natural selection, and mutation
   b. genetic divergence
   c. a buildup of differences in the separated gene pools of alleles
   d. all of the above

____ 3. When potential mates occupy different local habitats within the same area, it is termed ____________ isolation. [p.295]
   a. temporal
   b. behavioral
   c. mechanical
   d. ecological

_____ 4. If potential mates occupy overlapping ranges but reproduce at different times, they are exhibiting _______________ isolation. [p.295]
a. temporal
b. behavioral
c. mechanical
d. ecological

_____ 5. Of the following, _______________ is a postzygotic isolating mechanism. [p.295]
a. behavioral isolation
b. gametic mortality
c. temporal isolation
d. mechanical isolation
e. hybrid sterility

For questions 6–9, choose from the following answers:

a. parapatric speciation
b. sympatric speciation
c. allopatric speciation

_____ 6. When the Isthmus of Panama was formed, it provided contemporary scientists with an ideal natural laboratory for studying _______________. [p.296]

_____ 7. Polyploidy and cichlid fishes of crater lakes provide evidence for _______________. [p.298]

_____ 8. The identification of a hybrid zone between the ranges of eastern Baltimore orioles and western Bullock orioles is an example of _______________. [p.299]

_____ 9. Archipelagos are ideal locations for _______________. [p.297]

_____ 10. The branching speciation pattern revealed by the fossil record is called _______________. [p.300]
a. anagenesis
b. the gradual model
c. cladogenesis
d. the punctuation model

_____ 11. Species formation by many small changes over long time spans fits the _______________ model of speciation; rapid speciation with morphological changes compressed into a brief period of population divergence describes the _______________ model of speciation. [p.300]
a. anagenesis; punctuation
b. punctuation; gradual
c. gradual; cladogenesis
d. gradual; punctuation

_____ 12. Adaptive radiation is _______________. [p.300]
a. a burst of microevolutionary activity within a lineage
b. the formation of new species in a wide range of habitats
c. the spreading of lineages into unfilled adaptive zones
d. a lineage radiating into an adaptive zone when it has physical, ecological, or evolutionary access to it
e. all of the above

_____ 13. A catastrophe during which major groups disappear is called a _______________. [p.301]
a. mass extinction
b. adaptive radiation
c. background extinction
d. speciation extinction

_____ 14. The condition in which somatic cells have three or more of each type of chromosome characteristic of the species is called _______________. [p.298]
a. adaptive radiation
b. extinction
c. allopatric speciation
d. polyploidy

_____ 15. The changes in allele frequencies that are significant enough to mark the formation of a descendant species from the ancestral one are called _______________. [p.293]
a. the biological species concept
b. mass extinction
c. speciation
d. polyploidy

Chapter Objectives/Review Questions

1. Define speciation. [p.293]
2. State the basic principles of the biological species concept as stated by Ernst Mayr. [p.294]
3. __________ __________ is a buildup of differences in separated pools of alleles. [p.294]
4. The evolution of reproductive __________ __________ paves the way for genetic divergence and speciation. [p.294]
5. Briefly define the major categories of prezygotic and postzygotic isolating mechanisms. [p.295]
6. __________ speciation occurs when daughter species form gradually by divergence in the absence of gene flow between geographically separate populations. [p.296]
7. Explain why archipelagos are ideal locations for the formation of new species. [p.297]
8. In __________ speciation, daughter species arise, sometimes rapidly, from a small proportion of individuals within an existing population. [p.297]
9. Explain why sympatric speciation by polyploidy is a rapid method of speciation. [p.298]
10. When daughter species form from a small proportion of individuals along a common border between two populations, it is called __________ speciation. [p.299]
11. How does a subspecies differ from a species? [p.299]
12. Distinguish cladogenesis from anagenesis. [p.300]
13. Be able to interpret evolutionary tree diagrams and understand the symbolism used. [p.300]
14. Explain the difference between gradual and punctuation models of speciation. [p.300]
15. An adaptive radiation is a burst of __________ activity within a lineage that results in the formation of new species in a variety of habitats. [p.301]
16. Cite examples of adaptive zones. [p.301]
17. How does a key innovation differ from mass extinction? [p.301]

Integrating and Applying Key Concepts

1. *Systematics* is defined as the practice of describing, naming, and classifying living things; this includes the comparative study of organisms and all relationships among them. Plant systematists who work with flowering plants readily accept Ernst Mayr's definition of *biological species,* which was described in this chapter. In actual practice, however, systematists often must also work with a concept known as *morphological species.* For example, the statement is sometimes made that "two plant specimens belong to the same morphological species but not to the same biological species." Explain the meaning of this statement. What type of experimental evidence would be necessary as a basis for this statement? From your study of this chapter, can you suggest why the application of the term *morphological species* is sometimes necessary? What difficulties and inaccuracies, in terms of identifying a species, might the use of both species concepts present?
2. Paleontologists study the fossilized remains of ancient organisms. Why does the biological species concept usually not apply to their field of study? What challenges does this present to that scientific discipline?
3. The modern world frequently creates geographical barriers to gene flow. What challenges does this present to conservation biologists? How might they overcome these problems?

19

THE MACROEVOLUTIONARY PUZZLE

Interactive Exercises

Measuring Time [pp.304–305]

19.1. FOSSILS — EVIDENCE OF ANCIENT LIFE [pp.306–307]

19.2. FOCUS ON SCIENCE: *Dating Pieces of the Macroevolutionary Puzzle* [pp.308–309]

Selected Words: *geologic* time [p.304], *Homo sapiens* [p.304], *fossils* [p.306], *fossil record* [p.306], *trace* fossils [p.306], "modern" era [p.308], *unstable* [p.309], *extinction crisis* [p.309], *adaptive radiations* [p.309], *mass extinction* [p.309]

Boldfaced, Page-Referenced Terms

[p.304] asteroids ______________________________

____

[p.306] fossils ______________________________

____

[p.306] fossilization ______________________________

____

[p.306] stratification ____

____

[p.307] lineage ____

____

[p.308] geologic time scale ____

____

[p.308] macroevolution ____

____

[p.309] radiometric dating ____

____

[p.309] half-life ____

____

Matching

Choose the most appropriate answer for each question.

1. ____ fossils [p.306]
2. ____ stratification [p.306]
3. ____ asteroids [p.304]
4. ____ K–T boundary [p.304]
5. ____ fossil record [p.306]
6. ____ lineage [p.307]
7. ____ fossilization [p.306]

A. An iridium rich layer of Earth's crust that is evidence of a large asteroid impact
B. The process by which sedimentary rock layers are deposited, with the oldest on the bottom
C. The historical record of life
D. Rocky, metallic bodies that orbit the sun
E. A slow process by which organic material is mineralized and converted to a hard record of the organism
F. A line of descent
G. The buried, preserved remains of ancient organisms

Fill-in-the-Blanks

One of the biggest questions facing the early discoverers of fossils was how to determine the age of the fossils. Eventually, by examining stratified rock layers, geologists were able to construct a(n) (8) ____________ ____________ ____________ [p.308], or chronology of Earth's history. The large boundaries in this time scale were assigned according to abrupt transitions in the (9) ____________ ____________ [p.308]. The geologic time scale can be correlated with the process of (10) ____________ [p.308], that is, the major trends and patterns among lineages. However, the geologic time scale only provided the relative ages of fossils until the discovery of (11) ____________ ____________ [p.309]. Since the number of neutrons and protons in a radioactive element is (12) ____________ [p.309], the nucleus spontaneously (13) ____________ [p.309] until it reaches a more stable configuration. The (14) ____________-____________ [p.309] is the amount of time that it takes for half of a given quantity of a radioisotope to decay. For example, carbon-14 has a half-life of (15) ____________ [p.309] years.

Complete the Table

16. In the table below, first give the era and the period that correspond to the major biological event listed. Then, in the last column, indicate the order of the events, starting with the earliest. [pp.308–309]

Era	*Period*	*Biological Event*	*Order*
a.		Origin of mammals; gymnosperms dominant land plants	
b.		Origin of amphibians; first adaptive radiation of marine invertebrates	
c.		Origin of flowering plants	
d.		Origin of modern humans	
e.		Origin of eukaryotic cells	

Short Answer

17. Assume that a sample of organic material contains 4.0 grams of carbon-14. How many grams would be left after three half-lives? How many years would have elapsed? [p.309] ________________

____

19.3. EVIDENCE FROM BIOGEOGRAPHY [pp.310–311]

19.4. EVIDENCE FROM COMPARATIVE MORPHOLOGY [pp.312–313]

19.5. EVIDENCE FROM PATTERNS OF DEVELOPMENT [pp.314–315]

19.6. EVIDENCE FROM COMPARATIVE BIOCHEMISTRY [pp.316–317]

Selected Words: "supercontinent" [p.310], "hot spots" [p.311], *homo-* [p.312], *morpho-* [p.312], "stem" reptile [p.312], *analogos* [p.313], *divergence* [p.313], *convergence* [p.313], *rate* of development [p.314], "rubber grid" [p.315], *Alu* transposon [p.315]

Boldfaced, Page-Referenced Terms

[p.310] theory of uniformity ________________

____

[p.310] Pangea ________________

____

[p.310] plate tectonics theory ________________

____

[p.310] Gondwana ________________

____

[p.312] comparative morphology ________________

____

[p.312] homologous structures ______________________________

____

[p.312] morphological divergence ______________________________

____

[p.313] morphological convergence ______________________________

____

[p.313] analogous structures ______________________________

____

[p.315] transposons ______________________________

____

[p.316] nucleic acid hybridization ______________________________

____

[p.317] neutral mutations ______________________________

____

[p.317] molecular clock ______________________________

____

Matching

Choose the most appropriate answer for each.

1. ____ Gondwana [p.310]
2. ____ Pangea [p.310]
3. ____ theory of uniformity [p.310]
4. ____ plate tectonics theory [p.310]
5. ____ *Glossopteris* and *Lystrosaurus* [p.310]
6. ____ hot spots [p.310]

A. Fossilized species which contributed to the idea that Africa and South America were once joined
B. Areas of Earth's mantle where plumes of molten material are located close to the surface
C. The first of the supercontinents
D. The supercontinent that broke up to form the major continents evident today
E. Explains the repetitive action of mountain building and erosion over geologic time
F. The idea that the crust of the Earth is in constant motion

Choice

For questions 7–16, identify to which category of evidence each statement belongs.

a. biogeography [pp.310–311] b. comparative morphology [pp.312–313]
c. patterns of development [pp.314–315] d. comparative biochemistry [pp.316–317]

7. ____ Molecular clocks that record the accumulation of neutral mutations
8. ____ Homologous structures
9. ____ *Lystrosaurus* fossils in Africa and South America
10. ____ Embryonic studies of early embryo stages in vertebrates

11. _____ *Alu* transposon

12. _____ Comparison of cytochrome *c* protein sequences

13. _____ Studies of homologous structures in related organisms

14. _____ The movement of continents according to the theory of plate tectonics

15. _____ Studies of mutations in the DNA

16. _____ Examination of how unrelated organisms evolved similar structures over time

Fill-in-the-Blanks

The type of evolutionary evidence that focuses on changes in body structure and form is called (17) ____________ ____________ [p.312]. Frequently, such studies reveal a genetic basis for similarities in a particular body part. These structures are called (18) ____________ ____________ [p.312]. Since populations of the same species diverge (19) ____________ [p.312] when the (20) ____________ ____________ [p.312] ceases, the (21) ____________ [p.312] traits that characterize their species may also diverge. Such a change in the body form of a common ancestor is called (22) ____________ [p.312]. However, if lineages are not closely related, body parts with similar form or function may evolve (23) ____________ [p.313]. If comparable body parts were subjected to similar (24) ____________ [p.313] pressures, there may have been (25) ____________ [p.313] for similar modifications. This could result in (26) ____________ ____________ [p.313], or the independent evolution of similar body structures in remotely related organisms. Since these body parts may have similar functions, such as the wings of bats, birds, and insects, they are called (27) ____________ [p.313] structures.

Choice

For questions 28–35, use the following answers:

a. associated with the developmental program of vertebrates [pp.314–315]
b. an example of a nucleic acid comparison [p.316]
c. associated with the developmental program of flowering plants (larkspurs) [p.314]
d. an example of a protein comparison [p.316] e. associated with the molecular clock [p.317]

28. _____ Represented by changes in rates of development

29. _____ Neutral mutations that accumulate at predictable rates

30. _____ Proceed through similar stages of embryonic development

31. _____ A comparison of the primary structure of cytochrome *c*

32. _____ Action of the *Alu* transposon

33. _____ Patterns of nucleic acid hybridization

34. _____ Changes in the pattern of gene expression slow down development

35. _____ Adults are different as a result of changes in growth rates of organs and tissues following embryonic development

19.7. HOW DO WE INTERPRET THE EVIDENCE? [pp.318–319]

19.8. FOCUS ON SCIENCE: *Constructing a Cladogram* [pp.320–321]

19.9. ON INTERPRETING AND MISINTERPRETING THE PAST [pp.322 –323]

Selected Words: *classical* taxonomy [p.318], *cladistic* taxonomy [p.318], *clad-* [p.318], "single tribe" [p.318], "ancestral community" [p.319], *in-group* [p.320], *outgroup* [p.320], *ancestor* [p.321], *existing species* [p.321], "missing links" [p.322], *Archaeopteryx* [p.322]

Boldfaced, Page-Referenced Terms

[p.318] taxonomy ______________________________

____

[p.318] binomial system ______________________________

____

[p.318] genus ______________________________

____

[p.318] specific epithet ______________________________

____

[p.318] classification systems ______________________________

____

[p.318] higher taxa ______________________________

____

[p.318] phylogeny ______________________________

____

[p.318] derived trait ______________________________

____

[p.318] cladograms ______________________________

____

[p.318] monophyletic group ______________________________

____

[p.319] Monera ______________________________

____

[p.319] Protista ______________________________

____

[p.319] Fungi ______________________________

____

[p.319] Plantae ______________________________

____

[p.319] Animalia ____

____

[p.319] three-domain system ____

____

[p.319] six-kingdom classification scheme ____

____

[p.319] Archaebacteria ____

____

[p.319] Eubacteria ____

____

Matching

Choose the most appropriate answer for each. [p.318]

1. ____ monophyletic group
2. ____ higher taxa
3. ____ genus
4. ____ taxonomy
5. ____ cladograms
6. ____ phylogeny
7. ____ derived trait
8. ____ binomial system
9. ____ specific epithet

A. Evolutionary relationships among species, from the most ancestral forms to modern species
B. The field of biology that attempts to classify and name species
C. The use of a two-part Latin name to identify a species
D. All of the descendants from a single ancestral species in which a specific trait first evolved
E. Groupings of species into family, order, class, phylum, and kingdom
F. The first part of a species name
G. A novel feature that evolved once and is shared only by descendants of the organism in which it first evolved
H. Evolutionary tree diagrams
I. The second, most-specific part of a species name

Short Answer

10. Arrange the following jumbled taxa in proper order, with the most inclusive first: family, class, species, kingdom, genus, phylum (or division), and order. [p.318] ____

____

11. Explain the difference between classical and cladistic taxonomy. [p.318] ____

____

____

____

12. Briefly explain the difference between the three-domain and six-kingdom classification systems. [p.319]

___

___

___

Matching

Choose the single best answer for each. [p.319]

13. ___ Monera
14. ___ Protista
15. ___ Fungi
16. ___ Plantae
17. ___ Animalia

A. Multicelled heterotrophs that feed by extracellular digestion and absorption
B. Single-celled prokaryotes — some autotrophs, others heterotrophs
C. Diverse multicelled heterotrophs, including predators and parasites
D. Multicelled photosynthetic autotrophs
E. Diverse single-celled eukaryotes — some photosynthetic autotrophs, many heterotrophs

Cladogram Interpretation

For questions 18–24, study the following cladogram of seven taxa. The vertical bars on the stem of the cladogram represent shared derived traits. Various taxa are indicated by letters. [pp.320–321]

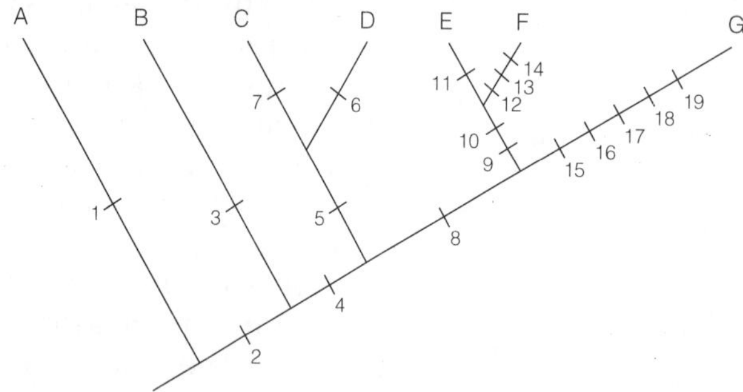

18. Give some examples of classes of traits that might be represented by the numbers on the cladogram. What types of information could be represented by the numbered traits? ___

___

___

___

19. Which derived trait is shared by taxa EFG? ___
20. Which derived trait is shared by taxa CDEFG? ___
21. What is the unique derived trait shared by taxa C and D? ___
22. What does it mean if some taxa are closer together on the cladogram than others? ___

___

23. Which taxon on the cladogram represents the outgroup condition? ___
24. Give an example of a monophyletic taxon. ___

___

Choice

For questions 25–29, choose from the following: [pp.322–323]

a. Archaeopteryx b. Rodhocetus and Basilosaurus c. both a and b

25. _____ Established link between aquatic mammals and hoofed land mammals
26. _____ A feathered reptile
27. _____ Transitional species
28. _____ Provided the initial proof of Darwin's theory of evolution by natural selection
29. _____ Fossilized remains indicate a gradual adaptation to aquatic life

---

Self-Quiz

_____ 1. The patterns, trends, and rates of change among lineages over geologic time are known as _______________. [p.308]
   a. taxonomy
   b. classification
   c. a binomial system
   d. macroevolution

For questions 2–3, choose from the following answers: [pp.312–313]
   a. morphological convergence
   b. morphological divergence

_____ 2. The wings of pterosaurs, birds, and bats serve as examples of _______________.

_____ 3. Penguins and porpoises serve as examples of _______________.

For questions 4–8, determine to which of the following forms of evolutionary evidence the statement belongs:
   a. comparative morphology [pp.312–313]
   b. biogeography [pp.310–311]
   c. patterns of development [pp.314–315]
   d. comparative biochemistry [pp.316–317]

_____ 4. Comparisons of mutation rates in DNA

_____ 5. Examinations of early embryonic similarities in vertebrates

_____ 6. Comparisons of protein structure between species

_____ 7. Studies of descent with modification in lines of descent

_____ 8. Studies of the distribution of similar species around the world

_____ 9. Early embryos of vertebrates strongly resemble one another *because* _______________. [pp.314–315]
   a. the genes that guide early embryonic development are the same (or similar) in all vertebrates
   b. the analogous structures they each contain are evidence of morphological convergence
   c. the homologous structures they each contain provide very strong evidence of morphological divergence
   d. nucleic acid hybridization techniques reveal many structural alterations that have resulted from gene mutations in early embryos

_____ 10. The significant contribution by Linnaeus was to develop _______________. [p.318]
   a. the idea that organisms should be placed in two kingdoms
   b. the binomial system
   c. the theory of evolution
   d. the idea that a species could have several common names

_____ 11. Phylogeny is _____________. [p.318]
a. the identification of organisms and the assignment of names to them
b. producing a "retrieval system" of several levels
c. an evolutionary history of organism(s), both living and extinct
d. the study of the adaptive responses of various organisms

_____ 12. Lineages sharing a common evolutionary heritage are termed _____________. [p.318]
a. phenetic
b. polyphyletic
c. monophyletic
d. homologous

_____ 13. All living organisms have eukaryotic cell structure except for members of the _____________. [p.319]
a. Animalia
b. Plantae
c. Fungi
d. Archae- and Eubacteria
e. Protista

Chapter Objectives/Review Questions

1. Give a generalized definition of *fossil;* cite examples. [p.306]
2. _________ begins with rapid burial in sediments or volcanic ash. [p.306]
3. What is the name given to the layering of sedimentary deposits? [p.306]
4. Are the oldest fossils found in the deepest layers of sedimentary rocks or nearer the surface? [p.306]
5. Describe the major biological events in each period/era of geologic time. [pp.308–309]
6. Explain the importance of radiometric dating to the study of fossils. [p.309]
7. Calculate the amount of radioactive material that would remain after a given number of half-lives. [p.309]
8. Explain the importance of the theories of uniformity and plate tectonics to the study of biogeography. [pp.310–311]
9. Explain the difference between morphological divergence and convergence and give an example. [pp.312–313]
10. Distinguish homologies from analogies as applied to organisms. [pp.312–313]
11. Explain why two species of larkspur have flowers with different shapes that attract different pollinators. [p.314]
12. Early in the vertebrate developmental program, _________ of different lineages proceed through strikingly similar stages. [p.314]
13. Explain why, despite similar embryonic stages, the adults of vertebrate species vary so widely in their physical characteristics. [pp.314–315]
14. What can analyses of protein primary structure tell us about evolutionary history? [p.316]
15. Some nucleic acid comparisons are based on _________ _________ hybridization. [p.316]
16. Define *neutral mutations* and discuss their use as molecular clocks. [p.317]
17. Describe the binomial system as developed by Linnaeus. [p.318]
18. Arrange the classification groupings correctly, from most to least inclusive. [p.318]
19. Explain the importance of derived traits in the interpretation of a cladogram. [p.318]
20. From a cladogram, be able to identify monophyletic groups, outgroups, and derived traits. [pp.318–321]
21. List the five kingdoms of life in the Whittaker system and give examples of the organisms in each. [p.319]
22. Explain the major units of the three-domain system. [p.319]
23. Explain the major difference between the six-kingdom classification system and Whittaker's five-kingdom system. [p.319]
24. Explain the importance of transitional species in providing evidence for macroevolution. [pp.322–323]

Integrating and Applying Key Concepts

1. Water crowfoot plants (family Ranunculaceae) often have two or more distinct leaf types within the same species and on the same plant. One type is capillary (hairlike), found submerged in the water or growing well above the water. The other type is laminate (flat and expanded) and is found floating on the water or submerged. In some *Ranunculus* species, three different leaf shapes form on a plant in a sequence. Suppose two taxonomists describe the same crowfoot plant as being two different and distinct species due to these two different leaf shapes. Can you think of difficulties that might arise when other researchers are required to refer to these "two species" in the course of their work?
2. Species have previously been defined as groups of interbreeding organisms that are reproductively isolated from other groups. What problems does this definition entail for scientists who study macroevolutionary processes?
3. Biologists have frequently proposed a "missing link" between primates and humans. What characteristics would this transitional species need to have?

20

THE ORIGIN AND EVOLUTION OF LIFE

Interactive Exercises

In the Beginning . . . [pp.326–327]

20.1. CONDITIONS ON THE EARLY EARTH [pp.328–329]

20.2. EMERGENCE OF THE FIRST LIVING CELLS [pp.330–331]

Selected Words: *nebula* [p.326], *complex* organic molecules [p.329], *metabolism* [p.330], "nanobes" [p.331]

Boldfaced, Page-Referenced Terms

[p.327] Big Bang ____________________

[p.328] crust ____________________

[p.328] mantle ____________________

[p.331] RNA world ____________________

[p.331] protocells ____________________

Choice

For questions 1–20, choose from the following probable stages in the chemical and physical evolution of life the one that the statement fits best.

a. formation of Earth [p.328] b. early Earth and the first atmosphere [p.328]
c. synthesis of organic compounds [pp.328–329] d. origin of agents of metabolism [p.330]
e. origin of self-replicating systems [pp.330–331] f. origin of the first plasma membranes [p.331]

1. ____ Most likely, the protocells were little more than membrane sacs protecting information-storing templates and various metabolic agents from the environment.

2. ____ Experiments indicate that RNA, enzymes, and coenzymes can be created in the laboratory.

3. ____ Four billion years ago, Earth was a thin-crusted inferno.

4. ____ Sunlight, lightning, or heat escaping from Earth's crust could have supplied the energy to drive their condensation into complex organic molecules.

5. ____ During the first 600 million years of Earth's history, enzymes, ATP, and other molecules could have assembled spontaneously at the same locations.

6. ____ Gaseous oxygen (O_2) was absent from the atmosphere.

7. ____ Sidney Fox heated amino acids under dry conditions to form protein chains, which he placed in hot water. The cooled chains self-assembled into small, stable spheres. The spheres were selectively permeable.

8. ____ Stanley Miller mixed hydrogen, methane, ammonia, and water in a reaction chamber that recirculated the mixture and bombarded it with a spark discharge. Within a week, amino acids and other small organic compounds had been formed.

9. ____ Without an oxygen-free atmosphere, the organic materials with which the story of life began never would have formed on their own. Free oxygen would have attacked them.

10. ____ Much of Earth's inner rocky material melted.

11. ____ Imagine an ancient sunlit estuary, rich in clay deposits. Countless aggregations of organic molecules stick to the clay. Some of these may be porphyrin molecules that are part of the chlorophylls and cytochromes.

12. ____ In other experiments, fatty acids and glycerol combined to form long-tail lipids under conditions that simulated evaporating tidepools. The lipids self-assembled into small, water-filled sacs, which in many ways were like cell membranes.

13. ____ The origin of DNA is unclear.

14. ____ Rocks collected from Mars, meteorites, and the moon contain chemical precursors that must have been present on the early Earth.

15. ____ This differentiation resulted in the formation of a crust of basalt, granite, and other types of low-density rock; a rocky region of intermediate density (the mantle); and a high-density, partially molten core of nickel and iron.

16. ____ When the crust finally cooled and solidified, water condensed into clouds and the rains began. For millions of years, runoff from rains stripped mineral salts and other compounds from Earth's parched rocks.

17. ____ The close association of enzymes, ATP, and other molecules would have promoted chemical interactions. Sunlight energy alone could have driven the spontaneous formation of RNA molecules.

18. ______ Even if amino acids did form in the early seas, they wouldn't have lasted long. In water, the favored direction of most spontaneous reactions is toward hydrolysis, not condensation.

19. ______ Between 4.6 and 4.5 billion years ago, the outer regions of the cloud cooled.

20. ______ At first, it probably consisted of gaseous hydrogen (H_2), nitrogen (N_2), carbon monoxide (CO), and carbon dioxide (CO_2).

20.3. ORIGIN OF PROKARYOTIC AND EUKARYOTIC CELLS [pp.332–333]

20.4. WHERE DID ORGANELLES COME FROM? [pp.334–335]

Selected Words: bacteria-eating *animals* [p.333], *endo-* [p.334], *symbiosis* [p.334], "mitochondrial code" [p.335]

Boldfaced, Page-Referenced Terms

[p.332] Archean ______________________________

____

[p.332] prokaryotic cells ______________________________

____

[p.332] eubacteria ______________________________

____

[p.332] archaebacteria ______________________________

____

[p.332] eukaryotic cells ______________________________

____

[p.332] stromatolites ______________________________

____

[p.333] Proterozoic ______________________________

____

[p.334] endosymbiosis ______________________________

____

[p.335] protistans ______________________________

____

Choice

For questions 1–15, choose from the following:

a. Archean eon [p.332] b. Proterozoic eon [p.333]

1. ______ By 2.5 billion years ago, the noncyclic pathway of photosynthesis had evolved in some eubacterial species.

2. ______ By 1.2 billion years ago, eukaryotes had originated.

3. _____ The first prokaryotic cells emerged.

4. _____ Oxygen, one of the by-products of photosynthesis, began to accumulate in the atmosphere.

5. _____ There was an absence of free oxygen.

6. _____ There was a divergence of the original prokaryotic lineage into two major evolutionary directions.

7. _____ About 800 million years ago, stromatolites began a dramatic decline.

8. _____ Between 3.5 and 3.2 billion years ago, the cyclic pathway of photosynthesis evolved in some species of eubacteria.

9. _____ Fermentation pathways were the most likely sources of energy.

10. _____ An oxygen-rich atmosphere brought the chemical origination of living cells to a halt.

11. _____ Food was available, predators were absent, and biological molecules were free from oxygen attacks.

12. _____ Aerobic respiration became the dominant energy-releasing pathway.

13. _____ Archaebacteria and eubacteria arose.

14. _____ The first cells may have originated in tidal flats or on the seafloor, in muddy sediments warmed by heat from volcanic vents.

15. _____ Along the shores of Rhodinia, the first supercontinent, early animals began using stromatolites as a food supply.

Fill-in-the-Blanks

The most important feature of eukaryotic cells is the abundance of membrane-bound (16) _______________ [p.334] in the cytoplasm. The origin of these structures remains a mystery. Some probably evolved gradually, through (17) _______________ [p.334] mutations and natural selection. Some species of prokaryotic cells do possess infoldings of the (18) _______________ [p.334] membrane, a site where enzymes and other metabolic agents are embedded. In prokaryotic cells that were ancestral to eukaryotic cells, similar membranous infoldings may have served as (19) _______________ [p.334] to the surface. These membranes may have evolved into (20) _______________ _______________ [p.334] and into the (21) _______________ [p.334] envelope. The advantage of such membranous enclosures may have been to protect (22) _______________ [p.334] and their products from foreign invader cells. A nuclear envelope would be favored because it would remove the cell's genes and enzymes of (23) _______________ [p.334] and (24) _______________ [p.334] out of the cytoplasm.

Accidental partnerships between different (25) _______________ [p.334] species may have formed countless times. Some partnerships resulted in the origin of (26) _______________ [p.334], (27) _______________ [p.334], and other organelles. This is the theory of (28) _______________ [p.343], developed in greatest detail by Lynn Margulis. In this case, one species spends its entire life cycle (29) _______________ [p.334] the host species.

According to one hypothesis, (30) _______________ [p.334] cells arose after the noncyclic pathway of photosynthesis emerged and (31) _______________ [p.335] had accumulated to significant levels.

(32) ______________ [p.343] transport systems had evolved in some bacteria to include extra cytochromes to donate electrons to (33) ______________ [p.335]. By 1.2 billion years ago or earlier, forerunners of eukaryotes were engulfing aerobic (34) ______________ [p.335] and perhaps forming endocytic vesicles around the food for delivery to the cytoplasm for digestion. Some aerobic bacteria resisted digestion and thrived in a new, protected, nutrient-rich environment. In time they released extra (35) ______________ [p.335]—which the hosts came to depend on for growth, greater activity, and assembly of other structures. The guest aerobic bacteria came to depend on the metabolic functions of their (36) ______________ [p.335] cell. The aerobic and anaerobic cells were now (37) ______________ [p.335] of independent existence. The guests had become (38) ______________ [p.335], supreme suppliers of (39) ______________ [p.335].

Mitochondria are bacteria sized and replicate their own (40) ______________ [p.335], dividing independently of the host cell's division process. The inner mitochondrial membrane is like a (41) ______________ [p.335] plasma membrane. Mitochondria have a few genetic code words with unique meanings; this is called the (42) ______________ [p.335] code. In addition, (43) ______________ [p.335] may be descended from aerobic eubacteria that engaged in oxygen-producing photosynthesis. Such cells may have been engulfed, resisted digestion, and provided their respiring (44) ______________ [p.335] cells with needed oxygen. Chloroplasts resemble some (45) ______________ [p.343] in metabolism and overall nucleic acid sequence. Chloroplasts have self-replicating (46) ______________ [p.335] and divide independently of the host cell's division.

New cells appeared on the evolutionary stage equipped with a nucleus, endomembranes, and mitochondria or chloroplasts (or both); they were the first eukaryotic cells, the first (47) ______________ [p.335].

Labeling

For questions 48–59, fill in the blanks, using the diagram on the next page. Choose from the following: animals, plants, fungi, cyanobacteria, chloroplasts, mitochondria, eukaryotes, eubacteria, archaebacteria, prokaryotes, and methanogens. [pp.332–333]

48. ______________
49. ______________
50. ______________
51. ______________
52. ______________
53. ______________
54. ______________
55. ______________
56. ______________
57. ______________
58. ______________
59. ______________

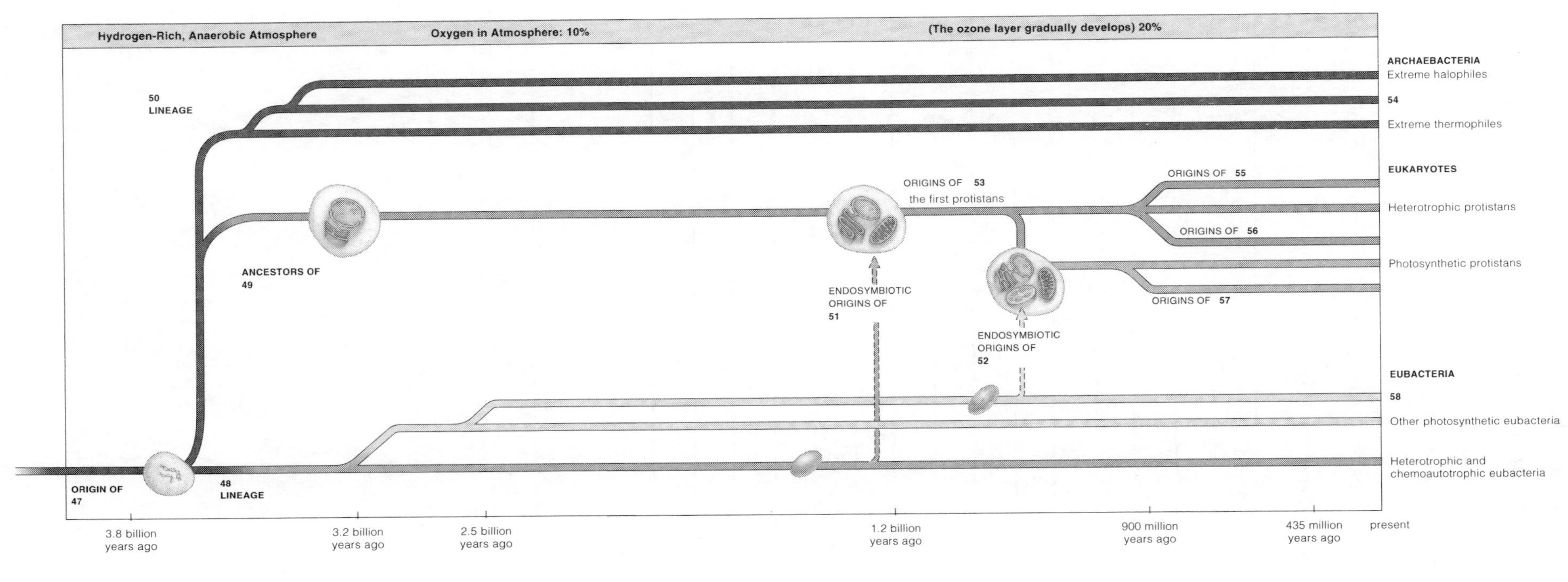

Hydrogen-Rich, Anaerobic Atmosphere
Oxygen in Atmosphere: 10%
(The ozone layer gradually develops) 20%
50
LINEAGE
ARCHAEBACTERIA
Extreme halophiles
54
Extreme thermophiles
ORIGINS OF 53
the first protistans
ORIGINS OF 55
EUKARYOTES
Heterotrophic protistans
ORIGINS OF 56
Photosynthetic protistans
ANCESTORS OF
49
ENDOSYMBIOTIC
ORIGINS OF
51
ENDOSYMBIOTIC
ORIGINS OF
52
ORIGINS OF 57
EUBACTERIA
58
Other photosynthetic eubacteria
Heterotrophic and
chemoautotrophic eubacteria
ORIGIN OF
47
48
LINEAGE
3.8 billion
years ago
3.2 billion
years ago
2.5 billion
years ago
1.2 billion
years ago
900 million
years ago
435 million
years ago
present

20.5. LIFE IN THE PALEOZOIC ERA [pp.336–337]

20.6. LIFE IN THE MESOZOIC ERA [pp.338–339]

20.7. FOCUS ON SCIENCE: *Horrendous End to Dominance* [p.340]

20.8. LIFE IN THE CENOZOIC ERA [pp.340–341]

Selected Words: "Cambrian explosion" [p.336], "The Great Dying" [p.337], "greenhouse" gases [p.338]

Boldfaced, Page-Referenced Terms

[p.336] Paleozoic ____

____

[p.336] Ediacarans ____

____

[p.338] Mesozoic ____

____

[p.338] gymnosperms ____

____

[p.338] angiosperms ____

____

[p.338] dinosaurs ____

____

[p.348] asteroids ____

____

[p.340] K–T asteroid impact theory ____

____

[p.340] global broiling hypothesis ____

____

[p.340] Cenozoic ____

____

Sequence

Refer to Figure 20.18 in the text. Study of the geologic record reveals that as the major events in the evolution of Earth and its organisms occurred, there were periodic major *extinctions* of organisms followed by major *radiations.* Using the following list, arrange the letters designating these extinctions and radiations according to the order in which they most likely occurred, from most recent to most ancient. [pp.342–343]

1. ____ A. Pangea; worldwide ocean forms, shallow seas squeezed out. Major radiations of reptiles, gymnosperms
2. ____ B. Land masses dispersed near equator. Simple marine communities only. Origin of animals with hard parts
3. ____ C. Mass extinction of many marine invertebrates, most fishes
4. ____ D. Pangea breakup begins. Rich marine communities. Major radiations of dinosaurs
5. ____ E. Asteroid impact? Mass extinction of all dinosaurs and many marine organisms
6. ____ F. Recovery, radiations of marine invertebrates, fishes, dinosaurs. Gymnosperms the dominant land plants. Origin of mammals
7. ____ G. Major glaciations. Modern humans emerge and begin what may be the greatest mass extinction of all time on land, starting with Ice Age hunters
8. ____ H. Unprecedented mountain building as continents rupture, drift, collide. Major climatic shifts; vast grasslands emerge. Major radiations of flowering plants, insects, birds, mammals. Origin of earliest human forms
9. ____ I. Oxygen accumulating in atmosphere; origin of aerobic metabolism
10. ____ J. Mass extinction of many marine species. Vast swamplands, early vascular plants. Radiation of fishes continues. Origin of amphibians
11. ____ K. Pangea breakup continues, broad inland seas form. Major radiations of marine invertebrates, fishes, insects, dinosaurs; origin of angiosperms (flowering plants)
12. ____ L. Asteroid impact? Mass extinction of many organisms in seas, some on land; dinosaurs, mammals survive
13. ____ M. Mass extinction. Nearly all species in seas and on land perish
14. ____ N. Recurring glaciations. Major radiations of insects, amphibians. Spore-bearing plants dominant; gymnosperms present. Origin of reptiles

Chronology of Events — Geologic Time

Refer to Figure 20.18 in the text. From the list of evolutionary events above (A–N), select those occurring in a particular era by circling the appropriate letters. [pp.342–343]

15. Cenozoic: A – B – C – D – E – F – G – H – I – J – K – L – M – N
16. Boundary of Cenozoic–Mesozoic: A – B – C – D – E – F – G – H – I – J – K – L – M – N
17. Mesozoic: A – B – C – D – E – F – G – H – I – J – K – L – M – N
18. Boundary of Mesozoic–Paleozoic: A – B – C – D – E – F – G – H – I – J – K – L – M – N
19. Paleozoic: A – B – C – D – E – F – G – H – I – J – K – L – M – N

Choice

For questions 20–31, choose from the following answers:

a. Paleozoic era [pp.336–337] b. Mesozoic era [pp.338–339] c. Cenozoic era [pp.340–341]

20. ____ The supercontinent Gondwana forms
21. ____ A group of animals called the Ediacarans are present in the world's oceans
22. ____ Evolution of the gymnosperm and angiosperm plants
23. ____ The global broiling hypothesis suggests that atmospheric temperatures may have increased rapidly during this time, possibly as the result of an asteroid impact
24. ____ The Cambrian explosion of biodiversity occurs
25. ____ The formation of modern mountain ranges such as the Andes and the Alps
26. ____ The first lobe-finned fish appears
27. ____ The mammals begin their domination
28. ____ The end of this era is marked by the "Great Dying" and the formation of Pangea
29. ____ The first dinosaurs appear
30. ____ The K–T impact theory marks the end of this era
31. ____ A mass extinction caused by human activity occurs

Self-Quiz

____ 1. Between 4.6 and 4.5 billion years ago, ______________. [pp.328–329]
   a. Earth was a thin-crusted inferno
   b. the outer regions of the cloud from which our solar system formed was cooling
   c. life originated
   d. the atmosphere was laden with oxygen

____ 2. More than 3.8 billion years ago, ______________. [pp.328–329]
   a. Earth was a thin-crusted inferno
   b. the outer regions of the cloud from which our solar system formed were cooling
   c. life originated
   d. the atmosphere was laden with oxygen

____ 3. The concept of endosymbiosis is used to explain which of the following? [pp.334–335]
   a. the origin of the archaebacteria
   b. the evolution of organelles such as mitochondria and chloroplasts
   c. the major trend in the evolution of the animal kingdom
   d. the formation of the first replicating proteins
   e. none of the above

For questions 4–11, choose from these answers:

a. Archean [p.332] b. Cenozoic [pp.340–341] c. Mesozoic [pp.338–339]
d. Paleozoic [pp.336–337] e. Proterozoic [p.332]

4. _____ The Alps, Andes, Himalayas, and Cascade Range were born during major reorganization of land masses early in the _____________ era.

5. _____ The composition of Earth's atmosphere changed during the _____________ eon, from one that was anaerobic to one that was aerobic.

6. _____ Invertebrates, primitive plants, and primitive vertebrates were the principal groups of organisms on Earth during the _____________ era.

7. _____ Before the close of the _____________ eon, the first photosynthetic bacteria had evolved.

8. _____ The _____________ era ended with the greatest of all extinctions, the Permian extinction.

9. _____ Late in the _____________ era, flowering plants arose and underwent a major radiation.

10. _____ The _____________ era included adaptive zones into which plant-eating mammals and their predators radiated.

11. _____ Dinosaurs and gymnosperms were the dominant forms of life during the _____________ era.

Chapter Objectives/Review Questions

1. Explain the concept of the Big Bang and the formation of the first stars. [p.327]
2. Describe the formation of early Earth before the formation of the first atmosphere. [p.328]
3. List the probable chemical constituents of Earth's first atmosphere. [p.328]
4. __________ oxygen was probably not present in early Earth's atmosphere. [p.328]
5. Describe experimental evidence provided by Stanley Miller (and others) that the formation of biological molecules from simple precursor molecules might have occurred on early Earth. [pp.328–329]
6. __________ might have been the medium on which condensation reactions yielding complex organic compounds occurred. [p.329]
7. During the first __________ years of Earth's history, enzymes, ATP, and other molecules could have assembled at the same location; their close association would have promoted chemical interactions—and the beginning of __________ pathways. [p.330]
8. Although simple, self-replicating systems of RNA, enzymes, and coenzymes have been created in the laboratory, the chemical ancestors of __________ and DNA remain unknown. [p.330]
9. Describe the experiments performed by Sidney Fox that aided understanding of the origin of plasma membranes. [p.331]
10. The first eon of the geologic time scale is now called the __________, "the beginning." [p.332]
11. List the two major evolutionary directions of the original prokaryotic lineage during the early Archean era. [p.332]
12. Explain the significance of stromatolites. [p.332]
13. Describe how the endosymbiosis theory may help explain the origin of eukaryotic cells. [pp.334–335].
14. Explain the importance of the evolution of the mitochondrion, nucleus, and chloroplast. [pp.334–335].
15. Generally discuss the important geological and biological events occurring throughout the Paleozoic, Mesozoic, and Cenozoic eras. [pp.336–341]
16. Explain the importance of the K–T asteroid impact theory and the global broiling hypothesis. [p.340]
17. List the three geologic eras and the two geologic eons in proper order, from oldest to youngest. [pp.342–343]

Integrating and Applying Key Concepts

1. As Earth becomes increasingly loaded with carbon dioxide and various industrial waste products, how do you think living forms on Earth will evolve to cope with these changes?
2. The evolution of life on this planet has been shaped by the dynamic nature of the Earth's surface. Scientists are currently examining the possibility that life may have existed, or may exist, on some other planets and moons of our solar system, yet most of these bodies have remained relatively stable for billions of years. What does this say about the possible evolution of life on those planets and moons? Do you think that they would have the same level of diversity as found on Earth? Why or why not?

21

PROKARYOTES AND VIRUSES

Interactive Exercises

The Unseen Multitudes [pp.346–347]

21.1. CHARACTERISTICS OF PROKARYOTIC CELLS [pp.348–349]

21.2. PROKARYOTIC GROWTH AND REPRODUCTION [pp.350–351]

Selected Words: nanometer [p.346], *Escherichia coli* [p.346], *photoautotrophic* [p.348], *chemoautotrophs* [p.348], *photoheterotrophs* [p.348], *chemoheterotrophic* [p.348], parasitic [p.348], saprobic [p.348], peptidoglycan [p.349], *gram-positive* [p.349], *gram-negative* [p.349], *Staphylococcus aureus* [p.349], *Bacillus cereus* [p.350], *Salmonella* [p.351], *Streptococcus* [p.351], sex pilus [p.351]

Boldfaced, Page-Referenced Terms

[p.346] microorganisms ______________________________

____

[p.346] pathogens ______________________________

____

[p.348] coccus, cocci ______________________________

____

[p.348] bacillus, bacilli ______________________________

____

[p.348] spirillum, spirilla ______________________________

[p.348] prokaryotic cells ______________________________

[p.349] cell wall ______________________________

[p.349] gram stain ______________________________

[p.349] glycocalyx ______________________________

[p.349] bacterial flagella (sing., flagellum) ______________________________

[p.349] pilus, pili ______________________________

[p.350] bacterial chromosome ______________________________

[p.350] prokaryotic fission ______________________________

[p.351] plasmid ______________________________

[p.351] bacterial conjugation ______________________________

Choice

Choose from the lettered terms; write the letter in the blank next to the best description. [p.348]

a. chemoautotrophic eubacteria b. chemoheterotrophic eubacteria
c. photoautotrophic eubacteria d. photoheterotrophic eubacteria

1. ____ Use CO_2 from the environment as their usual source of carbon atoms and use electrons, hydrogen, and energy released from chemical reactions to assemble chains of carbon (food storage)

2. ____ Use CO_2 and H_2O from the environment as sources of carbon, hydrogen, and oxygen atoms and use sunlight to power the assembly of food storage molecules

3. ____ Cannot use CO_2 from the environment to construct own carbon chains; instead obtain nutrients from the products, wastes, or remains of other organisms; can break down glucose to pyruvate and follow it with fermentation of some sort

4. ____ Cannot use CO_2 from the environment to construct own cellular molecules but can absorb sunlight and transfer some of that energy to the bonds of ATP; must obtain food molecules (carbon chains) produced by other organisms to construct their own molecules

Fill-in-the-Blanks

Bacteria are microscopic (5) ______________ [p.348] cells having one bacterial chromosome as well as, often, a number of smaller (6) ______________ [p.351]. The cells of nearly all bacterial species have a(n) (7) ______________ [p.349] around the plasma membrane and a(n) (8) ______________ [p.349] or slime layer surrounding the cell wall. Typically, the width or length of these cells falls between 1 and 10 (9) (choose one) □ millimeters □ nanometers □ centimeters □ micrometers [p.348]. Most bacteria reproduce by (10) ______________ ______________ [p.350]. Spherical bacteria are (11) ______________ [p.348], rod-shaped bacteria are (12) ______________ [p.348], and helical bacteria are (13) ______________ [p.348]. Gram- (14) ______________ [p.349] bacteria retain the purple stain when washed with alcohol.

21.3. PROKARYOTIC CLASSIFICATION [p.351]

21.4. MAJOR PROKARYOTIC GROUPS [p.352]

21.5. ARCHAEBACTERIA [pp.352–353]

21.6. EUBACTERIA — THE TRUE BACTERIA [pp.354–355]

Selected Words: *Sulfolobus* [p.352], *Thermus aquaticus* [p.352], *Methanococcus jannaschii* [p.353], *Methanosarcina* [p.353], anaerobic electron transport [p.353], *Halobacterium* [p.353], bacteriorhodopsin [p.353], peptidoglycan [p.354], cyanobacteria [p.354], *Anabaena* [p.354], *Lactobacillus* [p.354], *Azospirillum* [p.354], *Rhizobium* [p.354], *Deinococcus radiourans* [p.354], bioremediation [p.354], *Clostridium botulinum* [p.354], *botulism* [p.354], *Clostridium tetani* [p.354], *tetanus* [p.354], *Borrelia burgdorferi* [p.355], *Lyme disease* [p.355], *Rickettsia rickettsii* [p.355], *Rocky Mountain spotted fever* [p.355], magnetotactic bacteria [p.355], *collective* behavior [p.355], *Myxococcus xanthus* [p.355]

Boldfaced, Page-Referenced Terms

[p.351] numerical taxonomy ____

____

[p.351] eubacteria ____

____

[p.351] archaebacteria ____

____

[p.352] strain ____

____

[p.352] extreme thermophiles ____

____

[p.353] methanogens ____

____

[p.353] extreme halophiles ____

____

[p.354] heterocysts ____

____

[p.354] endospore ____

____

[p.355] fruiting bodies ____

____

Fill-in-the-Blanks

The (1) ______________ [p.352] have cell structure, metabolism, and nucleic acid sequences that are, in many respects, unique to the group; for example, none has a plasma membrane that contains (2) ______________ ______________ [p.352]. On the other hand, (3) ______________ [p.354] are far more common than the three rather unusual types of (1). The most common photoautotrophic bacteria are the (4) ______________ [p.354] (also called the blue-green algae). *Anabaena* and others of the (4) group produce oxygen during photosynthesis. Heterocysts are cells in *Anabaena* that carry out (5) ______________ - ______________ [p.354]. Many species of chemoautotrophic eubacteria affect the global cycling of nitrogen, (6) ______________ [p.354], and other nutrients. Sugarcane and corn plants benefit from a(n) (7) ______________ [p.354] -fixing spirochete, *Azospirillum*. Beans and other legumes benefit from the nitrogen-fixing activities of (8) ______________ [p.354], which dwells in their roots.

When environmental conditions become adverse, many bacteria form (9) ______________ [pp.354–355], which resist moisture loss, heat, irradiation, disinfectants, and even acids. Because they form endospores, some bacteria can be lethal to humans if they enter the food supply. If their endospores are not killed by high (10) ______________ [p.355] and high pressure, these endospores germinate, releasing aerobic bacteria that can live and reproduce in canned food while producing deadly toxins. Two examples of pathogenic (disease-causing) bacteria that form endospores harmful to humans are (11) ______________ ______________ [p.354] and (12) ______________ ______________ [p.354]. (13) ______________ ______________ [p.355] and (14) ______________ ______________ ______________ ______________ [p.355] are tick-borne diseases that are common in the United States; tick bites deliver (15) ______________ [p.355] from one host to another. Bacterial behavior depends on (16) ______________ ______________ [p.355], which change shape when they absorb or connect with chemical compounds. Cyanobacteria require (17) ______________ [recall p.348] as a source of energy to drive their metabolic activities. Most of the world's bacteria are (18) (choose one) ☐ producers ☐ consumers ☐ decomposers [p.354], so we think of them as "good" heterotrophs. (19) ______________ [p.362] are one of the major producers of antibiotics. *Escherichia coli,* which dwells in our gut, synthesizes vitamin (20) ______________ [p.354] and substances useful in digesting fats.

Matching

Match each of the items below with a lowercase letter designating its principal bacterial group and an uppercase letter denoting its best descriptor from the right-hand column.

a. archaebacteria b. chemoautotrophic eubacteria c. chemoheterotrophic eubacteria
d. photoautotrophic eubacteria e. photoheterotrophic eubacteria

21. ____, ____ *Anabaena* [pp.352,354]
22. ____, ____ *Bacillus, Clostridium* [pp.352,354–355]
23. ____, ____ *Escherichia coli* [pp.349,354,356]
24. ____, ____ *Halobacterium* [pp.354–355]
25. ____, ____ *Lactobacillus* [pp.352,354]
26. ____, ____ *Methanobacterium* [pp.354–355]
27. ____, ____ *Nitrobacter, Nitrosomonas* [p.354]
28. ____, ____ *Rhizobium, Agrobacterium* [pp.352,354]
29. ____, ____ *Rhodospirillum* [p.352]
30. ____, ____ *Salmonella* [pp.351–352]
31. ____, ____ *Spirochaeta, Treponema* [p.352]
32. ____, ____ *Staphylococcus, Streptococcus* [pp.349,352]
33. ____, ____ *Streptomyces, Actinomyces* [p.352]
34. ____, ____ *Thermoplasma, Sulfolobus* [pp.352–353]

A. Lives in anaerobic sediments of swamps and in animal gut; chemosynthetic; used in sewage treatment facilities
B. Purple; generally in anaerobic sediments of lakes or ponds; do not produce oxygen; do not use water as a source of electrons
C. Endospore-forming rods and cocci that live in the soil and in the animal gut; some major pathogens
D. Gram-positive cocci that live in the soil and in the skin and mucous membranes of animals; some major pathogens
E. Gram-positive, nonsporulating rods that ferment plant and animal material; some are important in dairy industry; others contaminate milk, cheese
F. In acidic soil, hot springs, hydrothermal vents on seafloor; may use sulfur as a source of electrons for ATP formation
G. Live in extremely salty water; have a unique form of photosynthesis
H. Gram-negative aerobic rods and cocci that live in soil or aquatic habitats or are parasites of animals and/or plants; some fix nitrogen
I. Nitrifying bacteria that live in the soil, fresh water, and marine habitats; play a major role in the nitrogen cycle
J. Gram-negative, anaerobic rod that inhabits the human colon, where it produces vitamin K
K. Major gram-negative pathogens of the human gut that cause specific types of food poisoning
L. Mostly in lakes and ponds; cyanobacteria; produce O_2 from water as an electron donor
M. Major producer of antibiotics; an actinomycete that lives in soil and some aquatic habitats
N. Helically coiled, motile parasites of animals; some are major pathogens

21.7. THE VIRUSES [pp.356–357]

21.8. VIRAL MULTIPLICATION CYCLES [pp.358–359]

21.9. CONNECTIONS: *Evolution and Infectious Diseases* [pp.360–361]

Selected Words: *helical, polyhedral, enveloped,* and *complex* viruses [p.356], *AIDS* [p.356], tobacco mosaic virus [p.357], *Herpes simplex,* type I [p.358], *cold sores* [p.358], *kuru* [p.359], *Creutzfeldt-Jakob disease* [p.359], scrapie [p.359], *mad cow disease* (BSE) [p.359], *contagious, sporadic,* and *endemic* diseases [p.360], *Ebola* virus [p.360], *monkeypox* [p.360], *Streptococcus pneumoniae* [p.361], *Salmonella enteridis* [p.361]

Boldfaced, Page-Referenced Terms

[p.356] virus ______________________________

____

[p.356] bacteriophages ______________________________

____

[p.358] lytic pathway ______________________________

____

[p.358] lysis ______________________________

____

[p.358] lysogenic pathway ______________________________

____

[p.359] viroids ______________________________

____

[p.359] prions ______________________________

____

[p.360] infection ______________________________

____

[p.360] disease ______________________________

____

[p.360] epidemic ______________________________

____

[p.360] pandemic ______________________________

____

[p.360] emerging pathogens ______________________________

____

Short Answer

1. a. State the principal characteristics of viruses. [p.356]

____

____

____

b. Describe the structure of viruses. [pp.356–359]

____

____

____

c. Distinguish among the ways viruses replicate themselves. [pp.358–359]

____

____

____

2. a. List five specific viruses that cause human illness. [pp.356–357]

____

____

____

b. Describe how each virus in (a) does its dirty work. [pp.356–360]

____

____

____

Fill-in-the-Blanks

A(n) (3) ______________ [p.356] is a noncellular, nonliving infectious agent; each agent consists of a central (4) ______________ ______________ [p.356] core surrounded by a protective (5) ______________ ______________ [p.356]. (6) ______________ [p.356] contain the blueprints for making more of themselves but cannot carry on metabolic activities. Chickenpox and shingles are two infections caused by DNA viruses from the (7) ______________ [p.357] category. (8) ______________ [p.357] are RNA viruses that infect animal cells, cause AIDS, and follow (9) ______________ [pp.358–359] pathways of replication. During a period of (10) ______________ [pp.358–359], viral genes remain inactive inside the host cell and any of its descendants. Naked strands or circles of RNA that lack a protein coat are called (11) ______________ [p.359], while pathogenic protein particles are called (12) ______________ [p.367]. (13) ______________ [recall p.346] are the usual units of measurement that microbiologists use to measure viruses, whereas bacteria and protistans are usually measured in terms of (14) ______________ [recall p.346]. A bacterium 86 micrometers in length is (15) ______________ [recall p.346] nanometers long.

Identification

Identify the virus that causes each of the following illnesses by writing its name in the blank preceding the disease. Then tell whether it is a DNA virus or an RNA virus. [p.357]

16. ________________, ______ Common cold
17. ________________, ______ AIDS, leukemia
18. ________________, ______ Cold sores, chickenpox

Matching

Match each item below with the correct lettered description.

19. ______ antibiotic [p.361, glossary]
20. ______ bacteriophage [p.356]
21. ______ endemic [p.360]
22. ______ epidemic [p.308]
23. ______ lysogenic pathway [p.358]
24. ______ lytic pathway [p.358]
25. ______ microorganism [recall p.346]
26. ______ pathogen [recall p.346]
27. ______ prion [p.359]
28. ______ sporadic [p.360]
29. ______ viroid [p.359]
30. ______ virus [p.356]

A. "Naked" RNA bits that resemble introns
B. Disease that breaks out irregularly; affects few organisms
C. Chemical substance that interferes with gene expression or other normal functions of bacteria
D. Disease that spreads abruptly through large portions of a population
E. Disease that occurs continuously but is localized to a relatively small portion of the population
F. Small proteins that are altered products of a gene; linked to eight degenerative diseases of the nervous system
G. Any organism too small to be seen without a microscope
H. A virus that infects a bacterium
I. Damage and destruction to host cells occurs quickly
J. Noncellular infectious agent that must take over a living cell in order to reproduce itself
K. Viral nucleic acid is integrated into the nucleic acid system of the host cell and replicated during this time
L. Any disease-causing organism or agent

Self-Quiz

Multiple Choice

______ 1. Which of the following diseases is *not* caused by a virus? [p.357]
   a. hepatitis B
   b. polio
   c. influenza
   d. syphilis

______ 2. Bacteriophages are ________________. [p.356]
   a. viruses that parasitize bacteria
   b. bacteria that parasitize viruses
   c. bacteria that phagocytize viruses
   d. composed of a protein core surrounded by a nucleic acid coat

Matching

Match all applicable letters with the appropriate terms. A letter may be used more than once, and a blank may contain more than one letter.

3. ____ *Anabaena, Nostoc* [pp.352,542]
4. ____ *Clostridium botulinum* [pp.352,354]
5. ____ *Escherichia coli* [pp.349,352–354]
6. ____ *Herpes simplex* [p.357]
7. ____ HIV [pp.356–357]
8. ____ *Lactobacillus* [pp.352,354]
9. ____ *Staphylococcus* [p.352]

A. Bacteria
B. Virus
C. Cyanobacteria
D. Gram-positive eubacteria
E. Cause cold sores and a type of venereal disease
F. Associated with AIDS

Matching

Match the following pictures with their names.

10. ____ [p.357]
11. ____ [p.356]
12. ____ [p.354]
13. ____ [p.356]
14. ____ [p.355]
15. ____ [p.346]

A. *Bacillus*
B. bacteriophage
C. *Clostridium tetani*
D. Cyanobacterium
E. *Influenza* virus
F. HIV

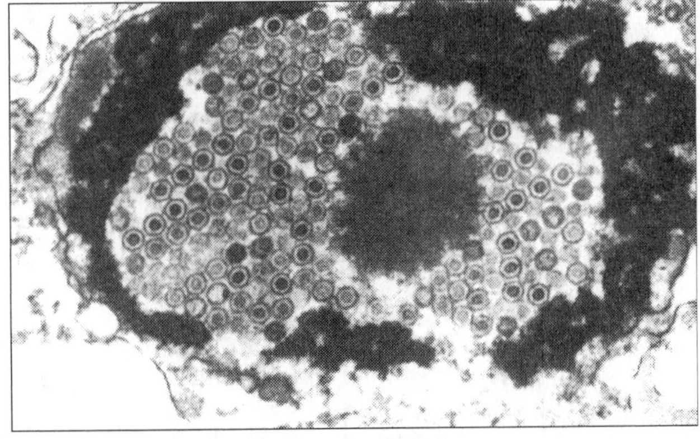

10.

11.

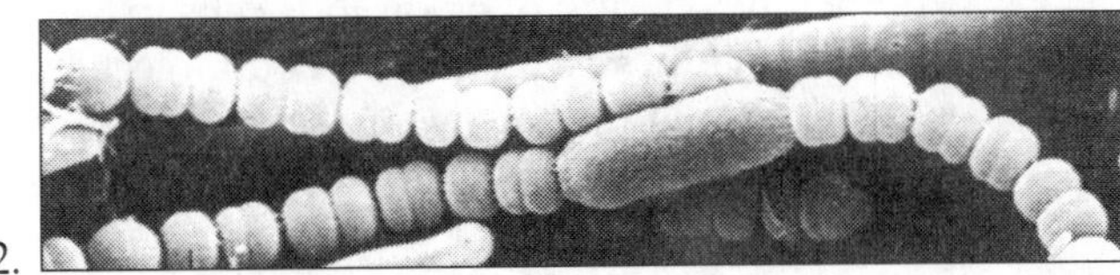

12.

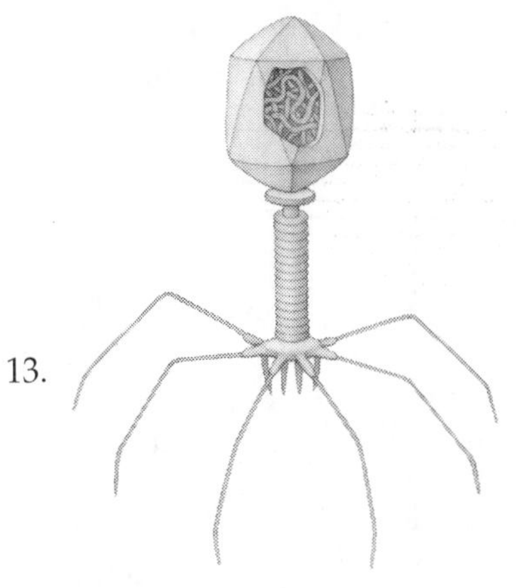

13.

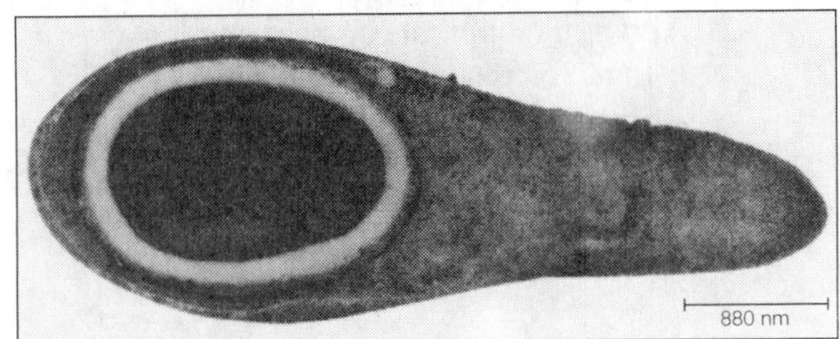

14.

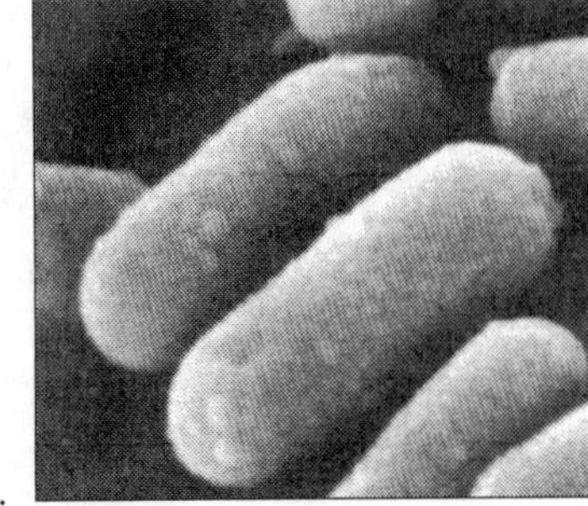

15.

Crossword Puzzle — Bacteria and Viruses

Across

2. __________ can assemble their own food molecules by using light energy to join together carbon, hydrogen, and oxygen atoms [p.354]
3. Prokaryotic __________ is the process most bacteria use to reproduce [p.350]
6. Disease causers [p.346]
7. Bacterial __________ transfers a plasmid from a donor to a recipient cell [p.351]
8. A chemical substance produced by *actinomycetes* and other microorganisms that kills or inhibits the growth of other microorganisms [p.354]
10. A whiplike organelle of locomotion [p.348]
12. Boy, young man
14. Group that includes blue-green algae [p.354]
16. Sticky mesh of polysaccharides, polypeptides, or both; alternate name of either a capsule or a slime layer [p.349]
17. __________ cells existed before the origin both of the nucleus and of eukaryotic cells [p.348]
19. Cannot synthesize their own food molecules; *can* use sunlight as an energy source to make ATP but cannot fix CO_2 from their surroundings; must obtain carbon compounds from other organisms [p.348]
20. Viruses that infect bacterial host cells [p.356]
23. Any organism that can be seen most clearly by using a microscope [p.346]
24. A short, filamentous protein that projects above a cell wall; helps tether a bacterium to another bacterium or to a surface [p.349]

25. A(n) ________ occurs when, over the entire world, a disease spreads through large portions of many populations for a limited period of time [p.360]

26. ________ are structures that resist heat, drying, boiling, and radiation; can give rise to new bacterial cells [pp.354–355]

27. A(n) ________ occurs when a disease abruptly spreads through large portions of a single population for a limited period of time [p.360]

28. Salt-loving archaebacterium [p.353]

29. Methane-producing archaebacterium [p.353]

30. Spherical bacterium [p.348]

Down

1. Rod-shaped bacterium [p.348]
2. Substance not in archaebacterial cell walls [p.354]
4. Helical [p.348]
5. ________ positive bacterial cell walls have an affinity for crystal violet stain [p.349]
9. ________ include Earth's major decomposers, nitrogen-fixers, and fermenters of milk [p.354]
11. The ________ pathway of bacteriophage multiplication assembles new viral particles very soon after infecting the host cell; no integration into host cell's DNA [p.358]
13. This group includes the most ancient of Earth's bacteria [p.352]
15. Bacteria that can synthesize their own food molecules by using energy released from specific chemical reactions [p.348]
18. Heat-tolerant bacterial type [p.352]
21. The ________ pathway of bacteriophage multiplication includes integrating viral DNA into its bacterial host's chromosome [p.358]
22. Small loops of DNA in addition to the main bacterial "chromosome" [p.351]

Chapter Objectives/Review Questions

1. Distinguish chemoautotrophs from photoautotrophs. [p.348]
2. Describe the principal body forms of monerans (inside and outside). [pp.348–349]
3. Explain how, with no nucleus or few, if any, membrane-bound organelles, bacteria reproduce themselves and obtain the energy to carry on metabolism. [pp.350–351]
4. State the ways in which archaebacteria differ from eubacteria. [pp.352–354]
5. Describe the shapes of various viral types and explain the ways in which viruses infect their hosts. [pp.356–357]
6. Distinguish between the lytic and lysogenic patterns of viral replication. [p.358]

Integrating and Applying Key Concepts

The textbook (Fig. 51.16) identifies natural gas as a nonrenewable fuel resource, yet there is a group of archaebacteria that produce methane, the burning of which can serve as a fuel for heating and/or cooking. Recall or imagine how these bacteria could be incorporated into a system to serve human societies by generating methane in a cycle that is renewable. Why did your text categorize natural gas as a nonrenewable resource? Is methane a constituent of natural gas? Why or why not?

22

PROTISTANS

Interactive Exercises

Confounding Critters at the Crossroads [pp.364–365]

22.1. CONNECTIONS: *An Emerging Evolutionary Road Map* [pp.366–367]

Selected Words: Walter Judd [p.366], *monophyletic* [p.366], eyespot [p.367], clade [p.367]

Boldfaced, Page-Referenced Terms

[p.364] protistans ______________________________

____

[p.366] euglenoids ______________________________

____

[p.367] pellicle ______________________________

____

Labeling

1. For each structure, indicate with a "P" if it is a prokaryotic (bacterial) characteristic, an "E" if it is a eukaryotic characteristic, and a "B" if it is a characteristic of both.

   a. _____ nucleus [p.365]

   b. _____ circular DNA [recall Ch.21]

   c. _____ mitosis/meiosis [p.365]

   d. _____ can be photoautotrophs [recall Ch.21, p.365]

   e. _____ mitochondria [p.365]

   f. _____ ribosomes [recall Ch.21, p.365]

   g. _____ endoplasmic reticulum [p.365]

Fill-in-the-Blanks

The vast majority of protistans are (2) ____________ [p.366] -celled, although nearly every lineage has (3) ____________ [p.366] members. Since the diverse members of the kingdom do not share a common ancestor and derived traits found in no other group, this kingdom is not considered to be (4) ____________ [p.366]. (5) ____________ [p.367], which are assigned to this kingdom, share many derived characteristics with plants and not with other protistans. Although they share the photopigments (6) ____________ [p.367] and (7) ____________ [p.367], euglenoids are not closely related to chlorophytes. Like some (8) ____________ ____________ [p.367], euglenoids have a flexible body covering called a(n) (9) ____________ [p.367]. Most euglenoids are autotrophs but some are (10) ____________ [p.367]; however, even the autotrophs often require organic material such as (11) ____________ [p.367]. Puzzles like these make the study of protistan taxonomy a challenge.

22.2. ANCIENT LINEAGES OF FLAGELLATED PROTOZOANS [p.368]

22.3. AMOEBOID PROTOZOANS [p.369]

22.4. THE CILIATES [pp.370–371]

22.5. THE SPOROZOANS [p.372]

22.6. FOCUS ON HEALTH: *Malaria and Night-Feeding Mosquitoes* [p.373]

Selected Words: *Trypanosoma brucei* [p.368], African sleeping sickness [p.368], *T. cruzi* [p.368], *Chagas disease* [p.368], *Trichomonas vaginalis* [p.368], *Giardia lamblia* [p.368], *giardiasis* [p.368], opportunistic parasites [p.369], foraminiferans [p.368], radiolarians [p.368], heliozoans [p.368], *Paramecium* [p.368], *micro*nucleus [p.371], *macro*nucleus [p.371], sporozoites [p.372], *cryptosporidiosis* [p.372], *Pneumocystis carinii* [p.372], *toxoplasmosis* [p.372], *Plasmodium* [p.372], malaria [p.373], merozoites [p.373], vaccines [p.373]

Boldfaced, Page-Referenced Terms

[p.368] kinetoplastids ____

____

[p.368] parabasilids ____

____

[p.368] diplomonads ____

____

[p.369] amoeboid protozoans ____

____

[p.369] pseudopods ____

____

[p.369] rhizopods ____

____

[p.369] actinopods ____

____

[p.369] plankton ____

____

[p.370] ciliates ____

____

[p.370] contractile vacuoles ____

____

[p.371] binary fission ____

____

[p.371] conjugation ____

____

Fill-in-the-Blanks

Examples of flagellated protozoans that are parasitic include the (1) ______________ [p.368], two species of which cause African sleeping sickness and Chagas disease, respectively. (2) ______________ ______________ is a sexually transmitted flagellate that can cause damage to the reproductive system. (3) ______________, caused by *Giardia lamblia,* is an intestinal problem that is often transmitted by drinking contaminated water. Amoebas move by sending out (4) ______________ [p.369], which surround food and engulf it. (5) ______________ [p.369] secrete a hard exterior covering of calcareous material that is peppered with tiny holes through which sticky, food-trapping pseudopods extend. Shelled actinopods include (6) ______________ [p.369] and (7) ______________ [p.369]. *Paramecium* is a ciliate that lives in (8) ______________ [p.378] environments and depends on (9) ______________ ______________ [p.370] for eliminating the excess water constantly flowing into the cell. *Paramecium* has a(n)(10) ______________ [p.370], a groove that opens to the external watery world. Once inside the cavity, food particles become enclosed in (11) ______________-______________ ______________ [p.370], where digestion takes place. (12) ______________ [p.372] is a famous sporozoan that causes malaria. When a particular (13) ______________ [p.373] draws blood from an infected individual, (14) ______________ [p.373] of the parasite fuse to form zygotes, which eventually develop within the mosquito.

Matching

Put as many letters in each blank as are applicable.

15. ________ *Amoeba proteus* [p.369]
16. ________ foraminiferans [p.369]
17. ________ *Paramecium* [p.370]
18. ________ *Plasmodium* [pp.372–373]
19. ________ *Toxoplasma* [p.372]
20. ________ *Trichomonas vaginalis* [p.368]
21. ________ *Trypanosoma brucei* [p.368]

A. Ciliophora
B. Rhizopods
C. Amoeboid protozoans
D. Flagellated protozoans
E. African sleeping sickness
F. Malaria
G. Can cause miscarriages
H. Sporozoans
I. Primary component of many ocean sediments
J. Ciliated protozoans
K. Transmitted by biting insects
L. Sexually transmitted

Matching

Match the following pictures with the names below.

22. _____ [p.369]
23. _____ [p.369]
24. _____ [p.369]
25. _____ [p.368]
26. _____ [p.370]

A. *Amoeba proteus*
B. Flagellated protozoans
C. Foraminiferans
D. Heliozoans
E. *Paramecium*

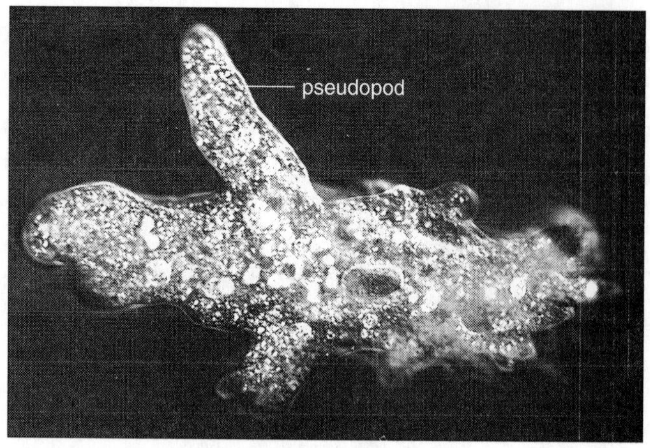

22.

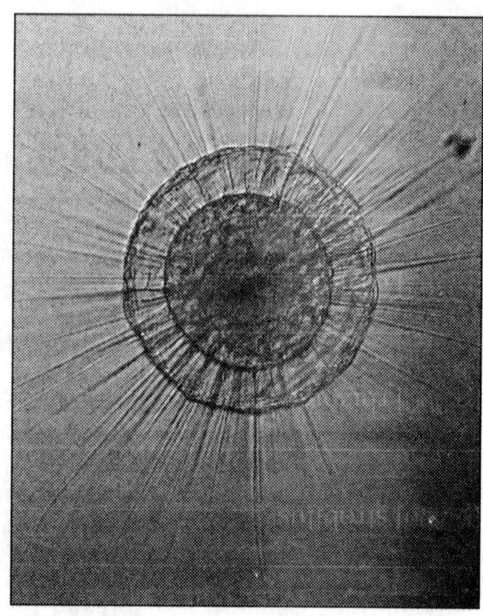

23.

24.

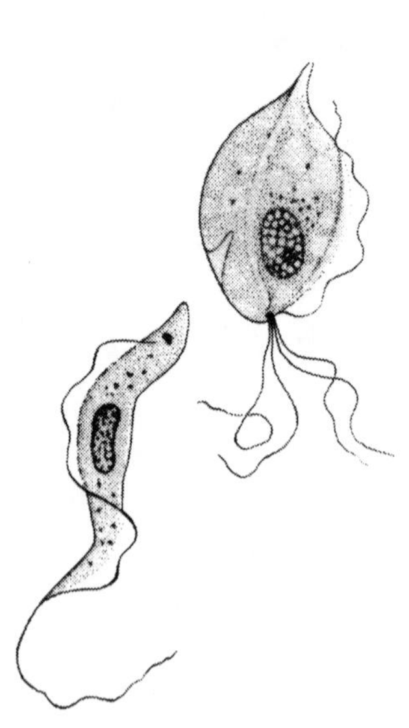

25.

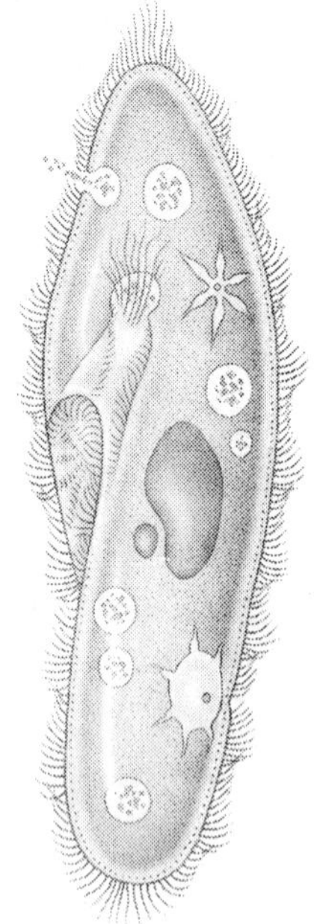

26.

22.7. THE CELL FROM HELL AND OTHER DINOFLAGELLATES [p.374]

22.8. OOMYCOTES — ANCIENT STRAMENOPILES [p.375]

22.9. PHOTOSYNTHETIC STRAMENOPILES — CHRYSOPHYTES AND BROWN ALGAE [pp.376–377]

22.10. GREEN ALGAE AND THEIR CLOSEST RELATIVES [pp.378–379]

22.11. RED ALGAE [p.380]

22.12. SLIME MOLDS [p.381]

Selected Words: phytoplankton [p.374], *Pfiesteria piscicida* [p.374], oomycotes [p.375], *Phytophthora infestans* [p.375], late blight [p.375], *P. ramorum* [p.375], *sudden oak death* [p.375], cankers [p.375], fucoxanthin [p.376], silica "shell" [p.376], Phaeophyta [p.376], *Saragassum* [p.376], giant kelps [p.376], stipes [p.376], blades [p.376], holdfasts [p.376], *Macrocystis* [pp.376–377], algins [p.377], Chlorophyta [p.378], cellulose [p.378], desmid [p.378], *Volvox* [p.378], *Chlamydomonas* [p.379], Spirogyra [p.379], Rhodophyta [p.380], *Porphyra* [p.380], phycobilins [p.380], cyanobacteria [p.380], agar [p.380], carrageenan [p.380], *nori* [p.380], *cellular* slime molds [p.381], *plasmodial* slime molds [p.381], *Dictyostelium discoideum* [p.381], cyclic AMP [p.381].

Boldfaced, Page-Referenced Terms

[p.374] dinoflagellates ______________________________

[p.374] algal bloom ______________________________

[p.374] red tide ______________________________

[p.375] stramenopiles ______________________________

[p.375] water molds ______________________________

[p.375] downy mildews ______________________________

[p.376] chrysophytes ______________________________

[p.376] golden algae ______________________________

[p.376] yellow-green algae ______________________________

[p.376] diatoms ______________________________

[p.376] coccolithophores ____________________

____

[p.376] brown algae ____________________

____

[p.378] green algae ____________________

____

[p.380] red algae ____________________

____

[p.381] slime molds ____________________

____

Fill-in-the-Blanks

Dinoflagellates are key (1) ____________ [p.374] among the marine phytoplankton. Tremendous increases in population called (2) ____________ ____________ [p.374] often lead to (3) ____________ ____________ [p.374], which can cause major fish kills. (4) ____________ ____________ [p.374], known as "the cell from hell," has caused major fish kills along the central Atlantic coast of the United States.

(5) ____________ [p.375] are ancient nonphotosynthetic stramenopiles. Two types are the (6) ____________ ____________ [p.375], which are mainly aquatic saprobes, and the (7) ____________ ____________ [p.375], which are major plant pathogens and include the species responsible for the mid-nineteenth-century potato famine in Ireland.

(8) ____________ [p.376] include 600 species of yellow-green algae, about 500 species of golden algae, and more than 5,600 existing species of (9) ____________ [p.376]. Except for yellow-green algae, photosynthetic chrysophytes contain xanthophylls and (10) ____________ [p.376]; these pigments mask the green color of chlorophyll in golden algae and diatoms. Diatom cells have thin, external, overlapping "shells" of (11) ____________ [p.376] that fit together like a pillbox. Roughly 270,000 metric tons of (12) ____________ ____________ [p.376] are extracted annually from a quarry near Lompoc, California, and are used to make abrasives, (13) ____________ [p.376] materials, and insulating materials. Related (14) ____________ ____________ [p.376] include the giant kelps, the largest, most complex protistans. Off the coast of California, giant kelp beds serve as productive (15) ____________ [p.377]. Kelp is also the commercial source of (16) ____________ [p.377], which serve as thickening and emulsifying agents. Green algae are thought to be ancestral to more complex plants, because they have the same types and proportions of (17) ____________ [p.378] pigments, have (18) ____________ [p.378] in their cell walls, and store their carbohydrates as (19) ____________ [p.378].

Red algae have chloroplasts very similar to (20) ____________ [p.380], which hints at an endosymbiotic origin. (21) ____________ [p.380] is extracted from the cell walls of some red algal species and is used as a gelling agent.

Labeling and Matching

Identify each indicated part of the following illustration by entering its name in the appropriate numbered blank. Choose from the following terms: *cytoplasmic fusion, asexual reproduction, resistant zygote, fertilization, zygote, meiosis and germination, gamete production, gametes meet.* Complete the exercise by matching from the list below, entering the correct letter in the parentheses following each label. [all from p.389; consult the illustration in the text on p.248]

22. ______________ ( )

23. ______________ ______________ ( )

24. ______________ and ______________ ( )

25. ______________ ______________ ( )

26. ______________ ______________ ( )

27. ______________ ______________ ( )

28. ______________ ______________ ( )

29. ______________ ______________ ( )

A. Fusion of two gametes of different mating types
B. A device for surviving unfavorable environmental conditions
C. More spore copies are produced
D. Fusion of two haploid nuclei
E. Haploid cells form smaller haploid gametes when nitrogen levels are low
F. Formed after fertilization
G. Two haploid gametes coming together
H. Reduction of the chromosome number

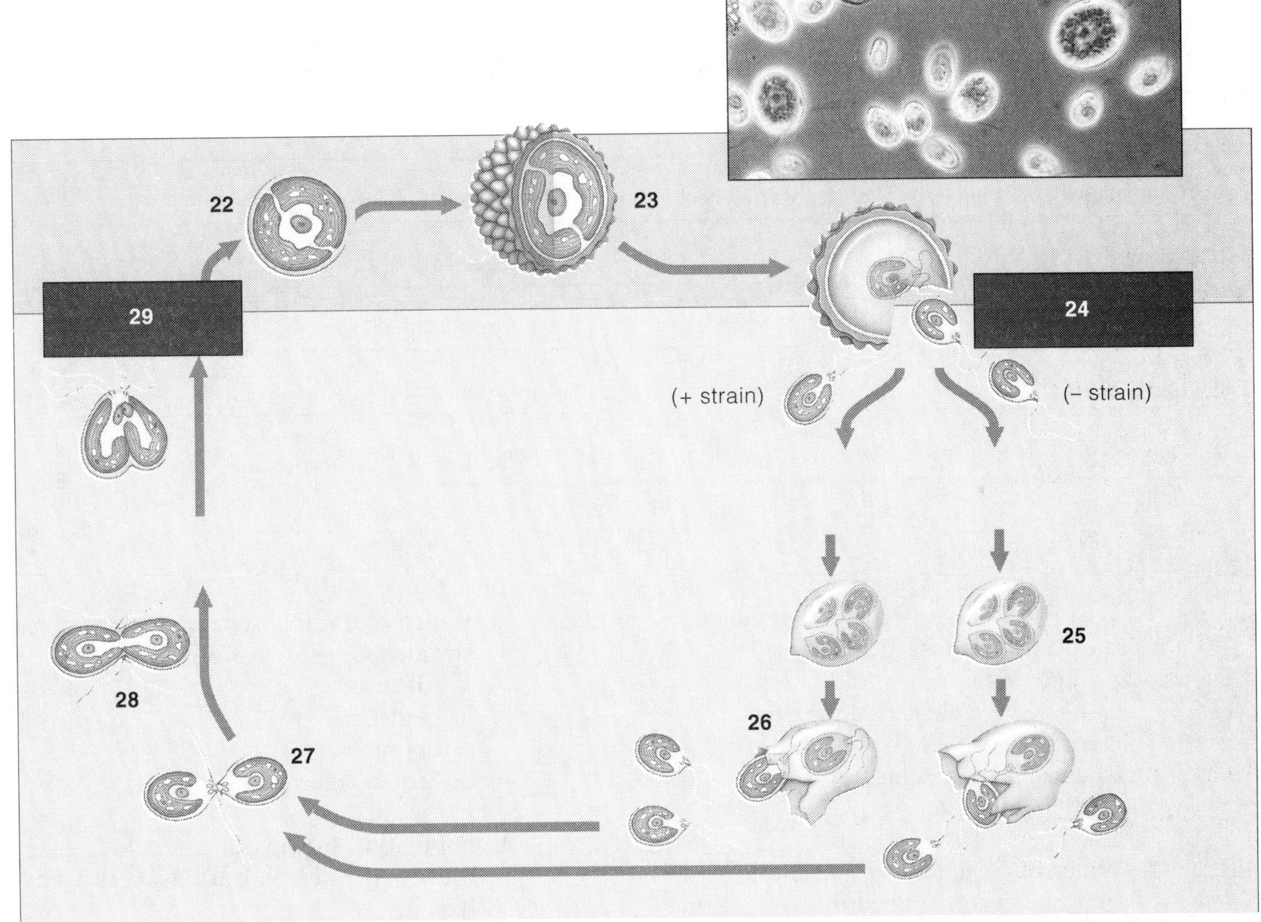

Complete the Table

30. Complete the following table.

Type of Alga	*Typical Pigments*	*Human Uses*	*Representatives*
Golden algae diatoms (Chrysophytes) [p.376]	a.	b.	c.
Brown algae (Phaeophyta) [pp.376–377]	d.	e.	f.
Green algae (Chlorophyta) [pp.378–379]	g.	h.	i.
Red algae (Rhodophyta) [p.380]	j.	k.	l.

More Choice

31. For each of the following groups, indicate with a "+" if members of the group have chloroplasts and a "−" if its members lack chloroplasts and the ability to carry on photosynthesis.

a. brown algae [p.377]	f. protozoans [pp.368–371]
b. chrysophytes [p.376]	g. red algae [p.380]
c. dinoflagellates [p.374]	h. slime molds [p.381]
d. euglenoids [p.366]	i. sporozoans [p.372]
e. green algae [pp.378–379]	j. water molds [p.375]

Self-Quiz

_____ 1. Which is *not* regarded as one of the major protistan lineages? [p.382]
  a. green algae
  b. blue-green algae (cyanobacteria)
  c. red algae
  d. ciliates, sporozoans, and dinoflagellates

_____ 2. Which of the following specialized structures is not correctly paired with a function? [p.370]
  a. gullet — ingestion
  b. cilia — food gathering
  c. contractile vacuole — digestion
  d. chloroplast — food production

_____ 3. Which of the following is not one of the stramenopiles? [pp.375–376]
  a. diatom
  b. euglena
  c. oomycote
  d. brown algae

_____ 4. Population blooms of _____________ cause red tides and extensive fish kills. [p.374]
  a. *Euglena*
  b. specific dinoflagellates
  c. diatoms
  d. *Plasmodium*

_____ 5. Which of the following is not a parasitic flagellated protozoan? [p.372]
a. cryptosporidium
b. *Giardia lamblia*
c. *Trichomonas vaginalis*
d. *Trypanosoma cruzi*

_____ 6. Which of the following protistans does *not* cause great misery to humans? [p.381]
a. *Dictyostelium discoideum*
b. *Giardia lamblia*
c. *Plasmodium*
d. *Trypanosoma brucei*

_____ 7. Which of the following is not a sporozoan? [p.370]
a. *Cryptosporidium*
b. *Pneumocystis carinii*
c. *Plasmodium*
d. *Paramecium*

_____ 8. Which of the following are saprobic decomposers? [pp.375,381]
a. slime molds
b. water molds
c. both a and b
d. neither a nor b

_____ 9. Red algae _______________. [p.380]
a. are primarily marine organisms
b. contain chlorophylls a and b
c. contain xanthophyll as their main accessory pigments
d. all of the above

_____ 10. Stemlike structure, leaflike blades, and gas-filled floats are found in the species of _______________. [pp.376–377]
a. red algae
b. brown algae
c. bryophytes
d. green algae

_____ 11. Because of similarities in pigmentation, cellulose walls, and starch storage, the _______________ algae are thought to be ancestral to more complex plants. [p.378]
a. red
b. brown
c. blue-green
d. green

Matching

Match all applicable letters with the appropriate terms. A letter may be used more than once, and a blank may contain more than one letter.

12. __________ *Amoeba proteus* [p.369]
13. __________ diatoms [p.376]
14. __________ *Dictyostelium* [p.381]
15. __________ foraminifera [p.369]
16. __________ *Pfiesteria piscicida* [p.374]
17. __________ *Paramecium* [pp.370–371]
18. __________ *Plasmodium* [pp.372–373]
19. __________ *Phytophthora infestans* [p.375]

A. Protista
B. Slime mold
C. Oomycote
D. Dinoflagellates
E. Obtain food by using pseudopodia
F. Causes malaria
G. A sporozoan
H. A ciliate
I. Live in "glass houses"
J. Live in hardened shells that have thousands of tiny holes through which pseudopodia protrude

Matching

Match the following pictures with their names.

20. ____ [p.376]
21. ____ [p.370]
22. ____ [p.369]
23. ____ [p.369]
24. ____ [p.367]

A. *Amoeba proteus*
B. diatoms
C. *Euglena*
D. foraminiferans
E. *Paramecium*

21.

20.

22.

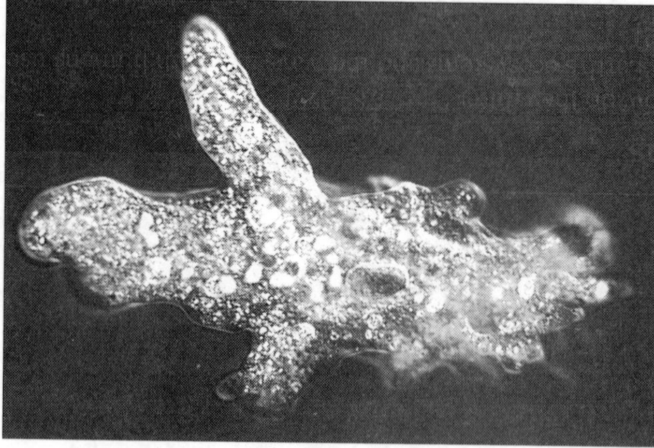

23.

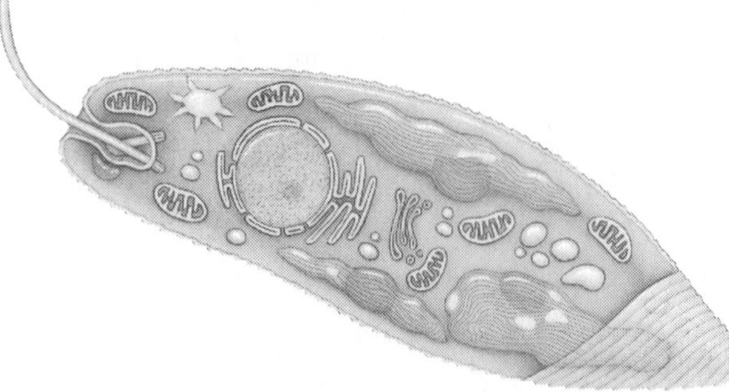

24.

Chapter Objectives/Review Questions

1. In what ways do green algae resemble plants? How do they differ? [pp.378–379]
2. Discuss the contributions that protistans make to Earth's ecosystems and the ways in which humans use protistans to make specific products. [pp.366–382]
3. State the principal characteristics of the amoebas, radiolarians, and foraminiferans. Indicate how they generally move from one place to another and how they obtain food. [p.369]
4. Two flagellated protozoans that cause human misery are ________ and ________. [p.368]
5. Characterize the sporozoan group, identify the group's most prominent representative, and describe the life cycle of that organism. [pp.372–373]
6. List the features common to most ciliated protozoans. [pp.370–371]
7. How do golden algae resemble diatoms? [p.376]
8. Explain what causes red tides. [p.374]
9. State the outstanding characteristics of the red, brown, and green algae. [pp.376–380]

Integrating and Applying Key Concepts

Explain why the taxonomy of protista has been so difficult. What are the currently accepted groupings within this kingdom?

23

PLANTS

Interactive Exercises

Pioneers in a New World [pp.384–385]

23.1. TRENDS IN PLANT EVOLUTION [pp.386–387]

Selected Words: *non*vascular plants [p.386], *seedless* vascular plants [p.386], *seed-bearing* vascular plants [p.386], *haploid (n)* [p.386], *heterospory* [p.387], *homo*spory [p.387]

Boldfaced, Page-Referenced Terms

[p.386] vascular plants ____

____

[p.386] bryophytes ____

____

[p.386] gymnosperms ____

____

[p.386] angiosperms ____

____

[p.386] root systems ______________________________

[p.386] shoot systems ______________________________

[p.386] lignin ______________________________

[p.386] xylem ______________________________

[p.386] phloem ______________________________

[p.386] cuticle ______________________________

[p.386] stomata ______________________________

[p.386] gametophytes ______________________________

[p.386] sporophyte ______________________________

[p.386] spores ______________________________

[p.387] pollen grains ______________________________

[p.387] seed ______________________________

Matching

Choose the most appropriate answer for each term. [p.386]

1. ____ seedless vascular plants
2. ____ angiosperms
3. ____ vascular plants
4. ____ bryophytes
5. ____ gymnosperms

A. Liverworts, hornworts, and mosses
B. Seed-bearing plants that include cycads, ginkgo, gnetophytes, and conifers
C. Whisk ferns, lycophytes, horsetails, and ferns
D. A group of plants that bear flowers and seeds
E. In general, a large number of plants having internal tissues that conduct water and solutes through the plant body

Complete the Table

6. As plants evolved, several key evolutionary developments occurred that solved the problems of living in new land environments. Complete the following table to summarize these events. Choose from the following: heterospory, xylem and phloem, seed, shoot systems, sporophytes, spores, lignin, gametophytes, root systems, stomata, pollen grains, and cuticle. [pp.386–387]

Evolutionary Adaptation	*Survival Problem Solved*
a.	Allows extensive growth of stems and branches; a very hard organic substance that strengthens cell walls, thus allowing erect plant parts to display leaves to sunlight
b.	Haploid cells produced by meiosis in sporophyte plants; later divide by mitosis to give rise to the gametophytes
c.	Condition in some seedless plant species and seed-bearing plants where two kinds of spores are produced
d.	Developed from one type of spore in gymnosperms and angiosperms; in turn develops into mature, sperm-bearing male gametophytes
e.	Consists of an embryo sporophyte, nutritive tissues, and a protective coat
f.	Provides a large surface area for rapidly taking up soil water and dissolved mineral ions; often anchors the plant
g.	Provides cellular pipelines to distribute water, dissolved ions, and solutes such as dissolved sugars to all living plant cells
h.	The haploid (n) phase that dominates the life cycles of algae
i.	The diploid ($2n$) phase that dominates the life cycles of most plants
j.	A waxy coat on many plant organs that helps conserve water on hot, dry days
k.	Tiny openings across the surfaces of stems and some leaves that help control the absorption of CO_2 and restrict evaporative water loss
l.	Consist of stems and leaves that function efficiently in the absorption of sunlight energy and CO_2 from the air

23.2. BRYOPHYTES [pp.388–389]

Boldfaced, Page-Referenced Terms

[p.406] mosses ____

____

[p.406] liverworts ____

____

[p.406] hornworts ____

____

[p.407] peat bogs ____

____

True/False

If the statement is true, write a "T" in the blank. If the statement is false, make it correct by changing the underlined word(s) and writing the correct word(s) in the answer blank. [pp.388–389]

_______________ 1. Mosses are especially sensitive to <u>water</u> pollution.

_______________ 2. Bryophytes have leaflike, stemlike, and rootlike parts although they <u>do not</u> contain xylem or phloem.

_______________ 3. Most bryophytes have <u>rhizomes</u>, elongated cells or threads that attach gametophytes to soil and serve as absorptive structures.

_______________ 4. Bryophytes are the simplest plants to exhibit a cuticle, cellular jackets around parts that produce sperm and eggs, and large gametophytes that do not depend on <u>sporophytes</u> for nutrition. [p.406]

_______________ 5. The true <u>liverworts</u> are the most common bryophytes.

_______________ 6. Following fertilization, zygotes give rise to <u>gametophytes</u>.

_______________ 7. Each <u>sporophyte</u> consists of a stalk and a jacketed structure in which spores will develop.

_______________ 8. <u>Club</u> moss is a bog moss whose large, dead cells in their leaflike parts soak up five times as much water as cotton does.

_______________ 9. Bryophyte sperm reach eggs by movement through <u>air</u>.

_______________ 10. Bryophytes are <u>vascular</u> plants with flagellated sperm.

Labeling

Each of the following numbers corresponds to a stage or process labeled in the diagram below. For each question, name the structure or process. In the case of structures, indicate within the parentheses whether it is diploid (2*n*) or haploid (*n*). [p.388]

11. ________________ ( )

12. ________________

13. ________________ ( )

14. ________________ ( )

15. ________________ ( )

16. ________________

17. ________________ ( )

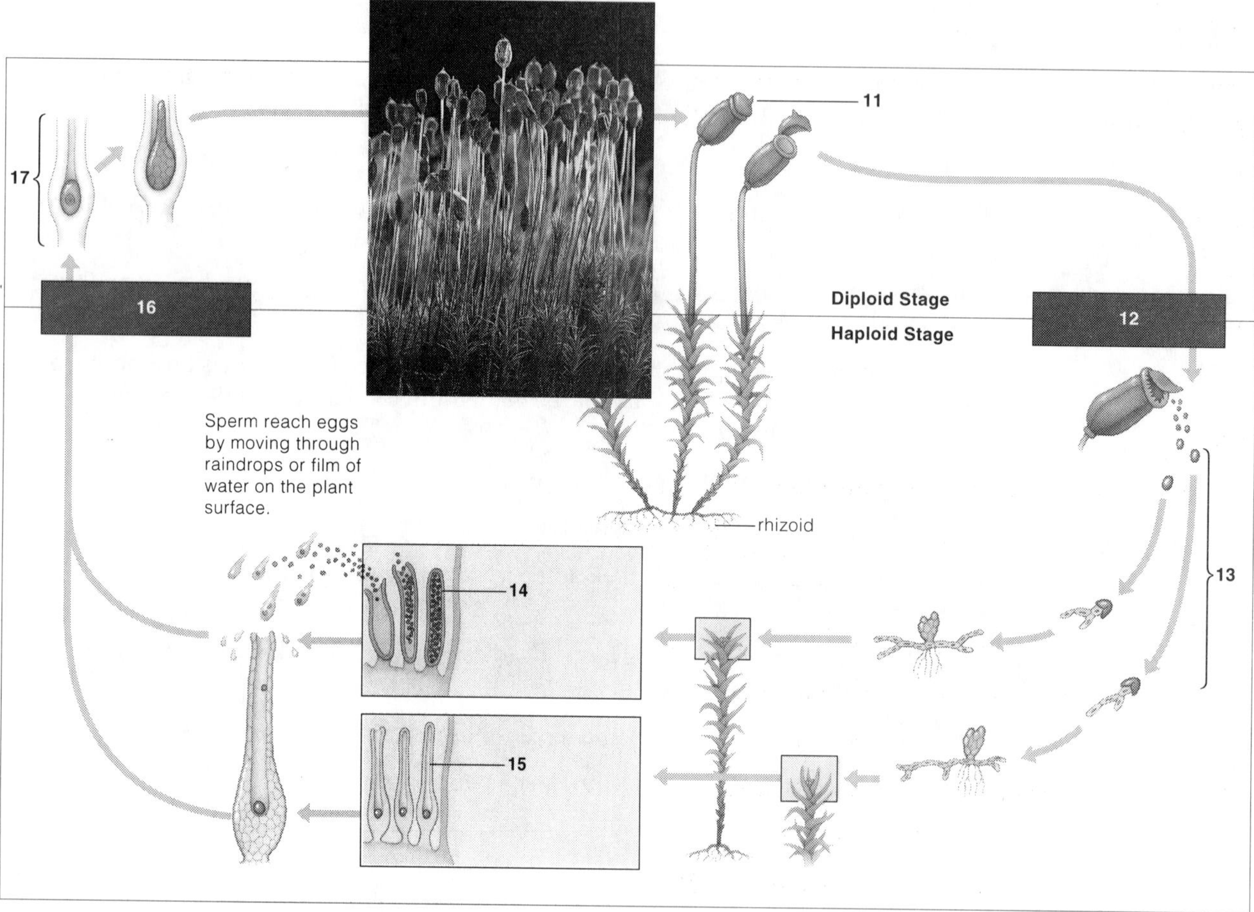

23.3. EXISTING SEEDLESS VASCULAR PLANTS [pp.390–391]

23.4. FOCUS ON THE ENVIRONMENT: *Ancient Carbon Treasures* [p.392]

Selected Words: "amphibians" of plant kingdom [p.390], *epiphyte* [p.391], "fossil fuels" [p.392]

Boldfaced, Page-Referenced Terms

[p.390] whisk ferns ______________________________

____

[p.390] lycophytes ______________________________

____

[p.390] horsetails ______________________________

____

[p.390] ferns ______________________________

____

[p.390] strobilus ______________________________

____

[p.390] rhizomes ______________________________

____

[p.392] coal ______________________________

____

Choice

For questions 1–20, choose from the following:

a. whisk ferns [p.390] b. lycophytes [p.390] c. horsetails [pp.390–391] d. ferns [p.391]
e. applies to a, b, c, and d [pp.390–391]

1. _____ A group in which only the genus *Equisetum* survives
2. _____ The club mosses belong to this group
3. _____ Seedless vascular plants
4. _____ An example is the genus *Psilotum*
5. _____ Rust-colored patches, the sori, are on the lower surface of their fronds
6. _____ Some tropical species are the size of trees
7. _____ The largest, most diverse group, commonly called the pterophytes
8. _____ Stems were used by pioneers of the American West to scrub cooking pots
9. _____ The sporophytes have no roots
10. _____ Mature leaves are usually divided into leaflets
11. _____ The sporophyte has vascular tissues
12. _____ When the spore chamber snaps open, spores catapult through the air

13. _____ Grow in mud soil of streambanks and in disturbed habitats, such as roadsides and vacant lots

14. _____ Most have vascularized rhizomes that give rise to roots and leaves

15. _____ The sporophyte is the larger, longer lived phase of the life cycle

16. _____ An example is the genus *Lycopodium*

17. _____ The young leaves are coiled into the shape of a fiddlehead

18. _____ A germinating spore develops into a small, green, heart-shaped gametophyte

19. _____ An example is the heterosporous genus *Selaginella*

20. _____ "Amphibians" of the plant kingdom; life cycles require water

Labeling

Each of the following numbers corresponds to a structure indicated in the diagram below. For each question, name the structure and indicate within the parentheses whether it is diploid (2*n*) or haploid (*n*). [p.391]

21. ________________ ( )

22. ________________ ( )

23. ________________ ( )

24. ________________ ( )

25. ________________ ( )

26. ________________ ( )

27. ________________ ( )

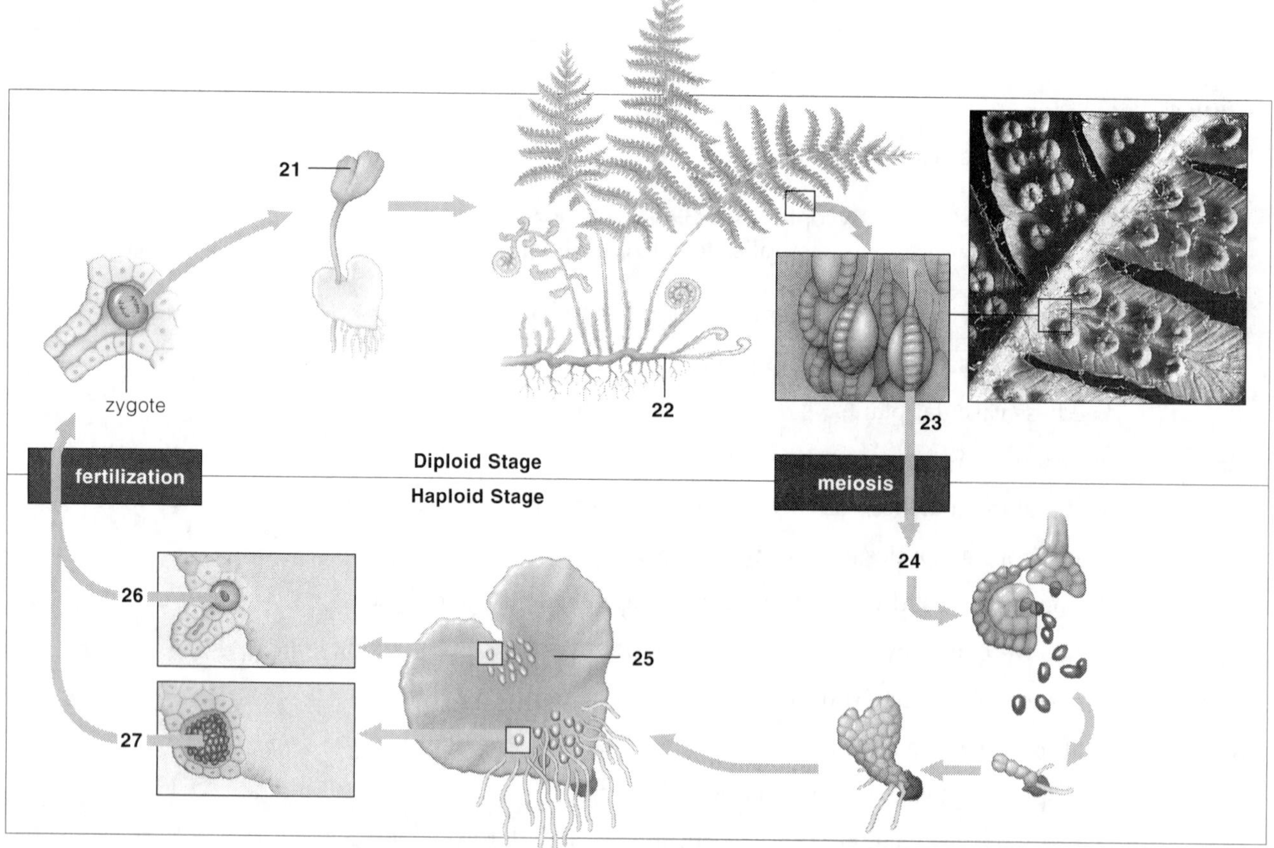

Fill-in-the-Blanks

About 300 million years ago, during the (28) ________________ [p.392], mild climates prevailed and swamp forests carpeted the wet lowlands. In this environment, the plants having (29) ________________ - ________________ [p.392] tissues had the competitive edge. For example, some of the (30) ________________ [p.392] trees were over 40 meters in height. Some horsetails were over (31) ________________ [p.392] meters tall. During this period of geologic time, the (32) ________________ ________________ [p.392] rose and fell over 50 times. Each time, the remains of these plants were buried in sediment. Over time these saturated, decayed remains were compressed into (33) ________________ [p.392]. As the sediments continued to accumulate, the increased heat and (34) ________________ [p.392] compressed the peat into great seams of (35) ________________ [p.392]. Coal is one of the premier (36) ________________ ________________ [p.392] and represents a(n) (37) ________________ [p.392] source of energy.

23.5. THE RISE OF THE SEED-BEARING PLANTS [p.393]

Selected Words: "tetrad scar" [p.393]

Boldfaced, Page-Referenced Terms

[p.393] microspores ____

____

[p.393] pollination ____

____

[p.393] megaspores ____

____

[p.393] ovules ____

____

[p.393] seed ferns ____

____

[p.393] progymnosperms ____

____

Matching

Choose the most appropriate answer for each term.

1. ____ microspores [p.393]
2. ____ tetrad scar [p.393]
3. ____ pollination [p.393]
4. ____ megaspore [p.393]
5. ____ ovule [p.393]
6. ____ seed ferns [p.393]
7. ____ progymnosperms [p.393]

A. The spore type that develops within ovules in seed-bearing plants
B. Female reproductive parts, which are seeds at maturity
C. The spore type that gives rise to pollen grains in seed-bearing plants
D. Formerly dominant seed-bearing plant group along with gymnosperms and the later angiosperms; replaced by cycads, conifers, and other gymnosperms
E. The name for the arrival of pollen grains on the female reproductive structures
F. A feature of meiotic division in seedless vascular plants that is absent in seed-bearing plants
G. The name given to the earliest plants to produce seedlike structures

23.6. GYMNOSPERMS—PLANTS WITH "NAKED" SEEDS [pp.394–395]

23.7. A CLOSER LOOK AT THE CONIFERS [pp.396–397]

Selected Words: gymnos [p.394], *sperma* [p.394], *evergreen* [p.394], *deciduous* [p.394], "male" plants [p.394], "female" plants [p.394], "tree farms" [p.397]

Boldfaced, Page-Referenced Terms

[p.394] conifers ____________________

____

[p.394] cones ____________________

____

[p.394] cycads ____________________

____

[p.395] ginkgos ____________________

____

[p.395] gnetophytes ____________________

____

[p.397] deforestation ____________________

____

Choice

For questions 1–12, choose from the following:

a. cycads [p.394] b. ginkgos [p.395] c. gnetophytes [p.395] d. conifers [pp.394,396–397]
e. all gymnosperms (includes a, b, c, d) [pp.394–397]

1. _____ Fleshy-coated seeds of female trees produce an awful stench
2. _____ Includes pines and redwoods
3. _____ Seeds and a flour made from the trunk are edible after removal of poisonous alkaloids
4. _____ Only a single species survives, *Ginkgo biloba*
5. _____ Includes the genera *Welwitschia, Gnetum,* and *Ephedra*
6. _____ Form pollen-bearing cones and seed-bearing cones on separate plants; leaves superficially resemble those of palm trees
7. _____ Seeds are mature ovules
8. _____ The favored male trees are now widely planted; they have attractive, fan-shaped leaves and are resistant to insects, disease, and air pollutants
9. _____ Their ovules and seeds are not covered; they are borne on surfaces of spore-producing reproductive structures
10. _____ The oldest of the trees belong to this group
11. _____ Most species are evergreen trees and shrubs with needlelike or scalelike leaves
12. _____ Includes conifers, cycads, ginkgos, and gnetophytes

Labeling

Each of the following numbers corresponds to a structure labeled in the diagram below. For each question, name the structure and indicate within the parentheses whether it is diploid (2*n*) or haploid (*n*). [p.396]

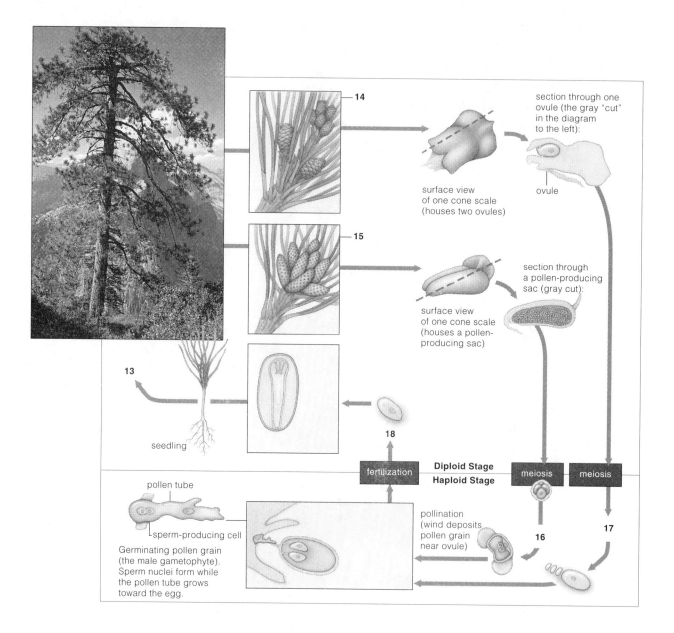

13. ________________ ( )

14. ________________ ( )

15. ________________ ( )

16. ________________ ( )

17. ________________ ( )

18. ________________ ( )

14

section through one ovule (the gray "cut" in the diagram to the left):

surface view of one cone scale (houses two ovules)

ovule

15

section through a pollen-producing sac (gray cut):

surface view of one cone scale (houses a pollen-producing sac)

13

seedling

18

fertilization

Diploid Stage

Haploid Stage

meiosis

meiosis

pollen tube

sperm-producing cell

Germinating pollen grain (the male gametophyte). Sperm nuclei form while the pollen tube grows toward the egg.

pollination (wind deposits pollen grain near ovule)

16

17

Short Answer

19. What is a tree farm? ______________________________

____

____

____

20. Define *deforestation.* ______________________________

____

____

____

23.8. ANGIOSPERMS — THE FLOWERING, SEED-BEARING PLANTS [pp.398–399]

23.9. SEED PLANTS AND PEOPLE [pp.400–401]

Selected Words: "double" fertilization [p.399], "vegetables" [p.400]

Boldfaced, Page-Referenced Terms

[p.398] flowers ______________________________

____

[p.398] pollinators ______________________________

____

[p.398] magnoliids ______________________________

____

[p.398] monocots ______________________________

____

[p.398] eudicots ______________________________

____

Matching

Choose the most appropriate answer for each term.

1. _____ examples of monocot plants [p.398]
2. _____ flowers [p.398]
3. _____ pollinators [p.398]
4. _____ double fertilization [p.399]
5. _____ examples of magnoliid plants [p.398]
6. _____ examples of eudicot plants [p.398]

A. Orchids, palms, lilies, and grasses, including rye, sugarcane, corn, rice, and wheat
B. Unique angiosperm reproductive structures
C. Insects, bats, birds, and other animals that withdraw nectar or pollen from a flower and, in so doing, transfer pollen to its female reproductive parts
D. Among all plants, unique to flowering plant life cycles; one sperm fertilizes the egg, the other sperm fertilizes a cell that gives rise to endosperm
E. Most herbaceous plants, such as cabbages and daisies; most flowering shrubs and trees, such as oaks and apple trees; water lilies and cacti
F. Black pepper plants, avocados, nutmeg, and magnolias

Labeling

Label each of the following structures indicated in the diagram of a monocot life cycle. [p.399]

7. ______________ ______________
8. ______________
9. ______________
10. ______________
11. ______________
12. ______________ ______________
13. ______________
14. ______________ ______________
15. ______________
16. ______________
17. ______________

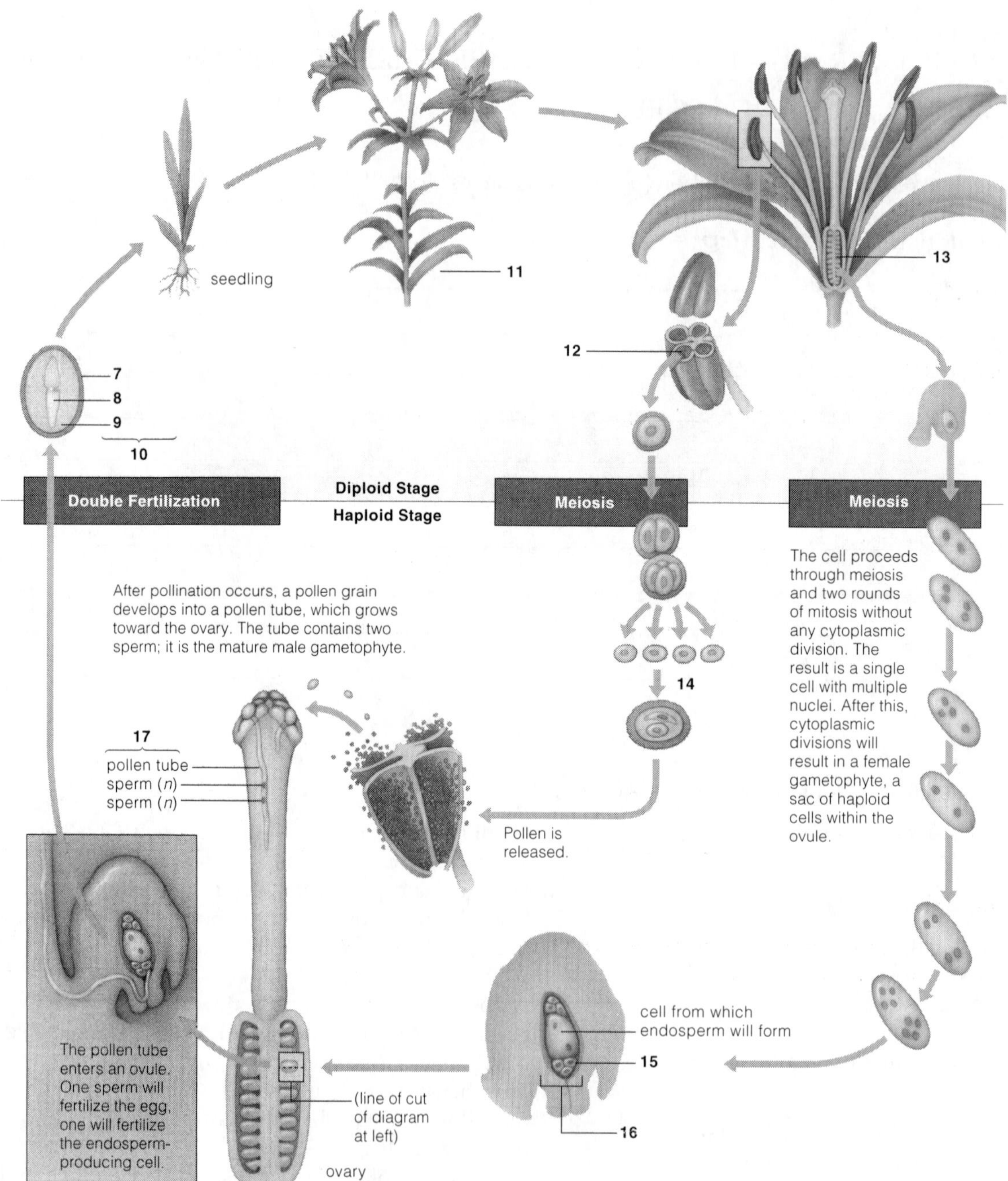

seedling
11
13
12
7
8
9
10
Double Fertilization
Diploid Stage
Haploid Stage
Meiosis
Meiosis
The cell proceeds through meiosis and two rounds of mitosis without any cytoplasmic division. The result is a single cell with multiple nuclei. After this, cytoplasmic divisions will result in a female gametophyte, a sac of haploid cells within the ovule.
After pollination occurs, a pollen grain develops into a pollen tube, which grows toward the ovary. The tube contains two sperm; it is the mature male gametophyte.
14
17
pollen tube
sperm (n)
sperm (n)
Pollen is released.
The pollen tube enters an ovule. One sperm will fertilize the egg, one will fertilize the endosperm-producing cell.
cell from which endosperm will form
15
16
(line of cut of diagram at left)
ovary

Matching

Match each term with its contribution to human society. [pp.400–401]

18. ____ *Agave*
19. ____ *Digitalis purpurea*
20. ____ *Secale*
21. ____ *Triticum*
22. ____ *Theobroma*
23. ____ *Saccharum officinarum*
24. ____ *Aloe vera*
25. ____ *Nicotiana*
26. ____ *Cannabis sativa*
27. ____ *Hyoscyamus niger*
28. ____ neem tree

A. Tobacco
B. Common bread wheat
C. Leaf extracts kill nematodes, insects, and mites but not the natural predators of these common pests
D. Source of marijuana and other mind-altering substances
E. Sugarcane
F. Extracts stabilize the heartbeat and blood circulation
G. Used to make twine and ropes from century plants
H. Henbane; source of belladonna and other alkaloids
I. Wheat and rye
J. Cocoa butter or chocolate
K. Juices soothe sun-damaged skin

Self-Quiz

Complete the Table

1. Complete the following table comparing the plant groups studied in this chapter.

Plant Group	*Dominant Generation*	*Vascular Tissue Present?*	*Seeds Present?*
a. Bryophytes [pp.388–389]			
b. Lycophytes [p.390]			
c. Horsetails [pp.390–391]			
d. Ferns [p.391]			
e. Gymnosperms [pp.394–395]			
f. Angiosperms [pp.398–399]			

Multiple Choice

____ 2. The plants around you today are descendants of ancient species of ________. [p.384]
a. brown algae
b. green algae
c. bryophytes
d. red algae

____ 3. The ________ represents a major trend in the evolution of plants. [pp.386–387]
a. evolution of roots, stems, and leaves
b. shift from haploid to diploid dominance
c. development of xylem and phloem
d. development of cuticles and stomata
e. all of the above

____ 4. Existing nonvascular plants do *not* include ____________. [p.388]
a. horsetails
b. mosses
c. liverworts
d. hornworts

____ 5. Plants possessing xylem and phloem are called ____________ plants. [p.386]
a. gametophyte
b. nonvascular
c. vascular
d. seedless

____ 6. Bryophytes ____________. [p.388]
a. have vascular systems that enable them to live on land
b. include lycophytes, horsetails, and ferns
c. have true roots but no stems
d. include mosses, liverworts, and hornworts

____ 7. ____________ are *not* seedless vascular plants. [p.390]
a. Lycophytes
b. Gymnosperms
c. Horsetails
d. Whisk ferns
e. Ferns

____ 8. In horsetails, lycophytes, and ferns, ____________. [p.391]
a. spores give rise to gametophytes
b. the dominant plant body is a gametophyte
c. the sporophyte bears sperm- and egg-producing structures
d. the dominant stage is haploid
e. all of the above

____ 9. ____________ are seed plants. [pp.393–395]
a. Cycads
b. Conifers
c. Angiosperms
d. Ginkgos
e. all of the above

____ 10. Flowers and double fertilization are characteristics of ____________. [pp.398–399]
a. gymnosperms
b. bryophytes
c. ferns
d. angiosperms
e. all of the above

____ 11. Monocots and dicots are groups of ____________. [p.398]
a. gymnosperms
b. club mosses
c. angiosperms
d. horsetails

____ 12. Which of the following is not matched correctly with its contribution to human society? [pp.400–401]
a. *Saccharum officinarum*—sugar
b. *Aloe vera*—skin protectant
c. *Triticum*—wheat
d. *Agave*—cancer cure

Chapter Objectives/Review Questions

1. Every plant you see today is a descendant of ancient species of ________ algae that lived near the water's edge or made it onto land. [p.384]
2. Most of the members of the plant kingdom are ________ plants, with internal tissues that conduct water and solutes through roots, stems, and leaves. [p.386]
3. State the general functions of the root systems and shoot systems of vascular plants. [p.386]
4. Explain the significance of lignin in the evolution of plants [p.386]
5. Distinguish between xylem and phloem. [p.386]
6. Explain the significance of cuticle and stomata in plant evolution. [p.386]
7. Give the reasons why diploid dominance allowed plants to successfully exploit the land environment. [pp.404–405]

8. List the differences between a sporophyte and a gametophyte. [p.386]
9. Plants evolved two spore types (heterospory): One spore type develops into pollen grains that become mature sperm-bearing male ________; the other spore type develops into female ________, where eggs form and later become fertilized. [p.387]
10. The combination of a plant embryo, nutritive tissues, and protective tissues constitutes a(n) ________. [p.387]
11. Mosses, liverworts, and hornworts belong to a plant group called the ________. [p.388]
12. Describe and state the functions of rhizoids. [p.388]
13. What group of plants first displayed cuticles, cellular jackets around the parts that produce sperm and eggs, and large gametophytes that retain sporophytes? [p.388]
14. The remains of peat mosses accumulate into compressed, excessively moist mats called ________ ________. [p.389]
15. Understand the major aspects of the bryophyte life cycle. [p.388]
16. List the four groups of seedless vascular plants. [p.390]
17. The ________ is the larger, longer lived phase of the seedless vascular plants. [p.391]
18. Understand the key characteristics of ferns, lycophytes, whisk ferns, and horsetails. [pp.390–391]
19. Some sporophytes of club mosses have nonphotosynthetic, cone-shaped clusters of leaves known as ________ that bear spore-producing structures. [p.390]
20. Understand the major aspects of a fern life cycle. [p.391]
21. ________ are underground, branching, short, mostly horizontal absorptive stems. [p.391]
22. Explain the meaning of the general term *epiphyte.* [p.391]
23. Describe the process by which coal is formed. [p.392]
24. Distinguish between microspores and megaspores. [p.393]
25. Define the term *pollination.* [p.393]
26. What is the evolutionary significance of the progymnosperms? [p.393]
27. Conifers, cycads, ginkgos, and gnetophytes are all members of the ________ lineage. [p.394]
28. Briefly characterize plants known as cycads, ginkgos, and gnetophytes. [pp.394–395]
29. Give reasons why *Ginkgo biloba* is a unique plant. [p.395]
30. *Gnetum, Ephedra,* and *Welwitschia* represent genera of ________. [p.395]
31. Describe the life cycle of *Pinus,* a somewhat typical gymnosperm. [p.396]
32. ________ is the removal of all trees from large tracts of land, as by clear-cutting. [p.397]
33. Only angiosperms produce reproductive structures known as ________. [p.398]
34. Most flowering plants coevolved with ________ such as insects, bats, birds, and other animals that withdraw nectar or pollen from a flower. [p.398]
35. Name and cite examples of the three major classes of flowering plants. [p.398]
36. Define *double fertilization.* [p.399]
37. Understand the key aspects of a monocot life cycle. [p.399]
38. Understand the development of seeds and fruits in angiosperms. [p.399]
39. List human uses for the following plants: *Agave,* neem tree, *Digitalis, Secale, Triticum, Theobroma cacao, Saccharum officinarum, Aloe vera, Nicotiana, Cannabis sativa,* and *Hyoscyamus.*

Integrating and Applying Key Concepts

1. Why is coal said to be a nonrenewable source of energy? According to the geologic conditions under which coal formed, how long would it take to replenish the world's supply by natural means?
2. What is the importance of gymnosperms to the world economy? Given the life cycle of gymnosperms, why are these trees more susceptible to the negative aspects of deforestation than angiosperm tree species?

24

FUNGI

Interactive Exercises

Ode to the Fungus among Us [pp.404–405]

24.1. CHARACTERISTICS OF FUNGI [p.406]

Selected Words: "fungus-root" [p.404], "fungi" [p.406], "imperfect fungi" [p.406]

Boldfaced, Page-Referenced Terms

[p.404] symbiosis ____________________

[p.404] mutualism ____________________

[p.404] lichen ____________________

[p.404] mycorrhiza ____________________

[p.404] decomposers ____________________

[p.404] extracellular digestion and absorption ____________________ [p.406] fungi ______________________________

[p.406] saprobes ______________________________

[p.406] parasites ______________________________

[p.406] zygomycetes ______________________________

[p.406] sac fungi ______________________________

[p.406] club fungi ______________________________

[p.406] spores ______________________________

[p.406] mycelium ______________________________

[p.406] hypha ______________________________

Fill-in-the-Blanks

(1) ______________ [p.404] refers to species that live closely together. In some cases, called (2) ______________ [p.404], their interaction either benefits both partners or benefits one without harming the other. An example of this are the (3) ______________ [p.404], a vegetative body in which a fungus has become intertwined with one or more photosynthetic organisms. Fungi that enter into mutualistic interactions with young tree roots form a(n) (4) ______________ [p.404], which means "fungus-root." Plants benefit from fungi because the latter act as (5) ______________ [p.404], which break down organic compounds in their surroundings. The method by which fungi ingest nutrients is through (6) ______________ ______________ ______________ ______________ [p.404], which also may provide some nutrients to plants.

Matching

Choose the most appropriate answer for each term. [all from p.406]

7. _____ parasites
8. _____ mycelium
9. _____ fungi
10. _____ hypha
11. _____ zygomycetes, sac fungi, club fungi
12. _____ spores
13. _____ saprobes

A. Reproductive cells or multicelled structures of fungi
B. Represent major lineages of fungal evolution
C. A mesh of branching fungal filaments that grows over and into organic matter, secretes digestive enzymes, and functions in food absorption
D. Fungi that obtain nutrients from nonliving organic matter and so cause its decay
E. Fungi that extract nutrients from tissues of a living host
F. Each filament in a mycelium; consists of cells of interconnecting cytoplasm and chitin-reinforced walls
G. A richly diverse group of heterotrophs that are premier decomposers

24.2. CONSIDER THE CLUB FUNGI [pp.406–407]

Selected Words: *dikaryotic* mycelium [p.407]

Boldfaced, Page-Referenced Terms

[p.407] mushrooms ____

____

[p.407] basidiospores ____

____

Fill-in-the-Blank

The club fungi represent a diverse group of organisms. Some saprobic species are (1) ____________ [p.406] of litter on and in the soil. Others are (2) ____________ [p.406] of young roots of trees. Still others, such as the (3) ____________ ____________ [p.406], are parasites of important crops such as wheat and corn. Also included in this group is the common mushroom, whose species name is (4) ____________ ____________ [p.406], which is frequently found in grocery stores. The largest organism on the planet, the (5) ____________ ____________ [p.407], found in the soils of Oregon, belongs to this group of fungi as well.

Labeling

The numbered items in the following illustration of the club fungi life cycle represent missing information. For each item, give the correct name of the structure or process. For each structure, indicate whether it is haploid (*n*) or diploid (2*n*) in the parentheses. [p.407]

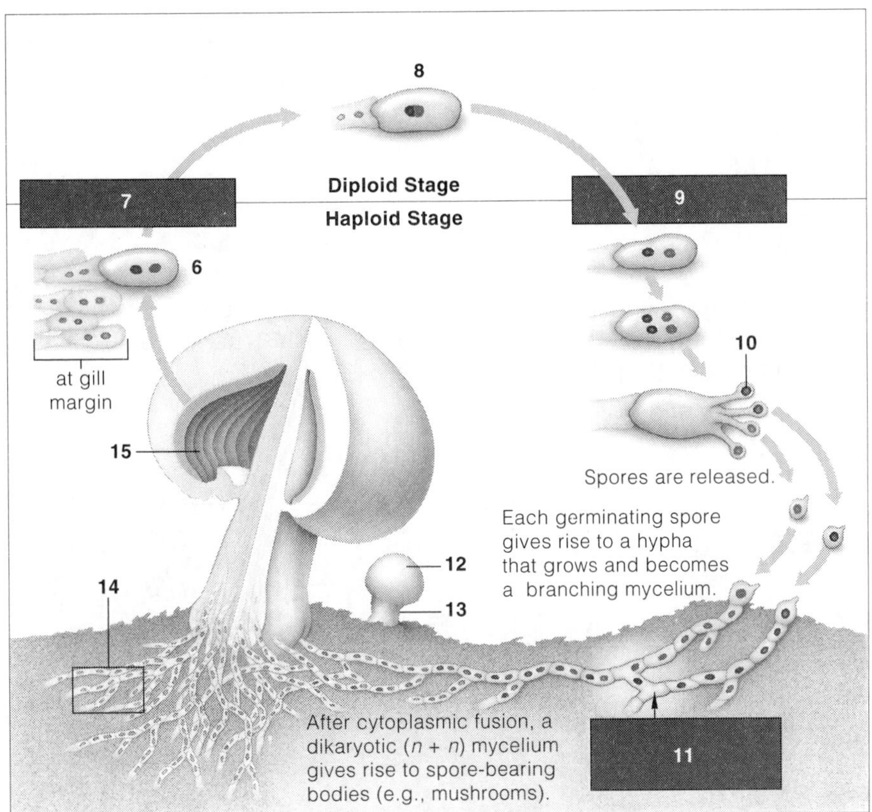

8
7
Diploid Stage
Haploid Stage
9
6
at gill margin
10
15
Spores are released.
Each germinating spore gives rise to a hypha that grows and becomes a branching mycelium.
12
13
14
After cytoplasmic fusion, a dikaryotic (n + n) mycelium gives rise to spore-bearing bodies (e.g., mushrooms).
11

6. ______________ ( )

7. ______________ ______________

8. ______________ ( )

9. ______________

10. ______________ ( )

11. ______________ ______________

12. ______________ ( )

13. ______________ ( )

14. ______________ ______________ ( )

15. ______________ ( )

24.3. SPORES AND MORE SPORES [pp.408–409]

Selected Words: "conidia" [p.409], "yeast" [p.409]

Boldfaced, Page-Referenced Terms

[p.408] zygospore ____

____

[p.408] ascospores ____

____

Labeling

The numbered items in the following illustration (*Rhizopus* life cycle) represent missing information. For each item, give the correct name of the structure or process. For each structure, indicate whether it is haploid (n) or diploid ($2n$) in the parentheses. [p.408]

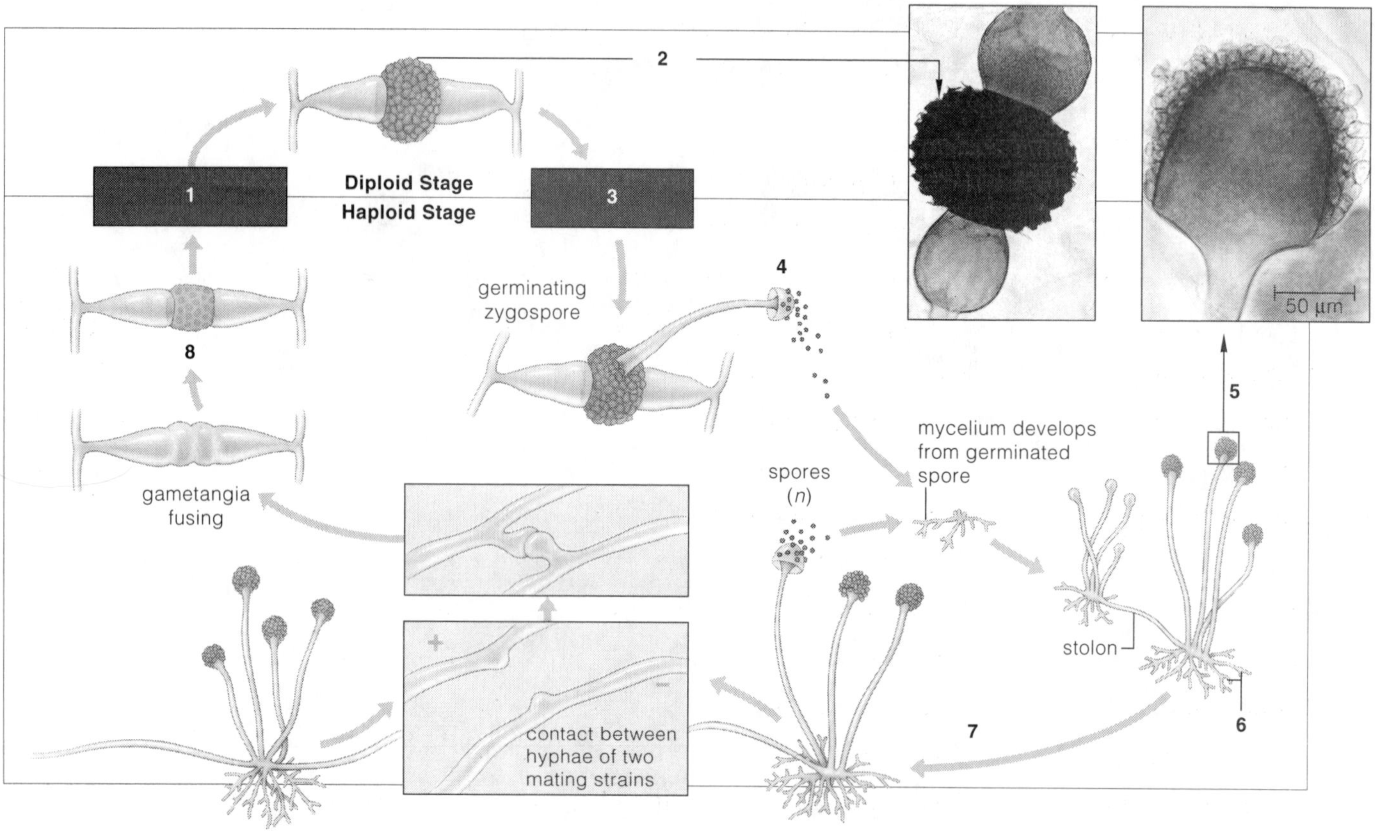

1. ______________ ______________

2. ______________ ( )

3. ______________

4. ______________ ( )

5. ______________ ______________ ( )

6. ______________ ( )

7. ______________ ______________

8. ______________ ( )

Matching

Choose the most appropriate answer for each term.

9. ____ truffles and morels [p.409]
10. ____ *Saccharomyces cerevisiae* [p.409]
11. ____ *Rhizopus stolonifer* [p.408]
12. ____ *Candida albicans* [p.409]
13. ____ *Penicillium* [p.409]
14. ____ *Aspergillis* [p.409]
15. ____ *N. crassa* [p.409]
16. ____ *Arthrobotrys dactyloides* [p.409]

A. "Flavor" Camembert and Roquefort cheeses; produce antibiotics
B. Has uses in genetic research
C. A predatory species that feeds on roundworms
D. Causes vexing infections in humans
E. Highly prized edibles
F. Make citric acid for candies and soft drinks; ferment soybeans for soy sauce
G. Involved in the commercial production of alcoholic beverages
H. Bread mold

24.4. THE SYMBIONTS REVISITED [pp.410–411]

24.5. FOCUS ON SCIENCE: *A Look at the Unloved Few* [p.412]

Selected Words: *myco*biont [p.410], *photo*biont [p.410], *ecto*mycorrhiza [p.411], *endo*mycorrhizae [p.411], *histoplasmosis* [p.412], *ergotism* [p.412], "blind staggers" [p.412]

Fill-in-the-Blanks

(1) ______________ [p.410] refers to species that live in close ecological association. In cases of symbiosis called (2) ______________ [p.410], interaction either benefits both partners or benefits one without harming the other. A(n) (3) ______________ [p.410] is a mutualistic interaction between a fungus and one or more photosynthetic species. The fungal part of a lichen is known as the (4) ______________ [p.410]; the photosynthetic component is the (5) ______________ [p.410]. Of about 13,500 known types of lichens, nearly half incorporate (6) ______________ [p.410] fungi. A lichen forms after the tip of a fungal (7) ______________ [p.410] binds with a suitable host cell. Both lose their wall, and their (8) ______________ [p.410] fuses or else the hypha induces the host cell to cup around it. The (9) ______________ [p.410] and the (10) ______________ [p.410] grow and multiply together. The lichen commonly has distinct (11) ______________ [p.410]. The overall pattern of growth may be leaflike, flattened, pendulous, or erect. Lichens typically colonize places that are too (12) ______________ [p.410] for most organisms. Almost always, the (13) ______________ [p.411] is the largest component of the lichen. The fungus benefits by having a long-term source of nutrients that it absorbs from cells of the (14) ______________ [p.411]. Nutrient withdrawals affect the (15) ______________ [p.410] growth a bit, but the lichen may help (16) ______________ [p.411] it. If more than one fungus is present in the lichen, it may be a mycobiont, a(n) (17) ______________ [p.411], or even an opportunist that is using the lichen as a substrate.

Fungi are also mutualists with young tree roots, as (18) ______________ [p.411]. Without them, plants cannot grow as (19) ______________ [p.411]. In (20) ______________ [p.411], hyphae form a dense net around living cells in the roots but do not penetrate them. Ectomycorrhizae are common in (21) ______________ [p.411] forests and help trees withstand seasonal changes in temperature and rainfall. About 5,000 fungal species, mostly (22) ______________ [p.411] fungi, enter into such associations. The more common (23) ______________ [p.411] form in about 80 percent of all vascular plants. These fungal hyphae (24) ______________ [p.411] plant cells, as they do in lichens. Fewer than 200 species of (25) ______________ [p.411] serve as the fungal partner. Their hyphae branch extensively, forming tree-shaped (26) ______________ [p.411] structures in plant cells. Hyphae also extend several centimeters into the (27) ______________ [p.411].

Matching

For each of the following, choose the most appropriate answer. [p.413]

28. ____ ergotism
29. ____ histoplasmosis
30. ____ "blind staggers"
31. ____ *Cryphonectria parasitica*
32. ____ *Aspergillus, Cladosporium, Stachybotrys*
33. ____ *Claviceps purpurea*

A. Household molds that are responsible for many common infections
B. A lung disease caused by the fungus *Ajellomycetes capsulatus*
C. The parasitic fungi that are responsible for eliminating chestnut trees
D. A disease that causes vomiting, diarrhea, and hallucinations
E. A form of ergotism that affected the armies of Peter the Great
F. The species responsible for the disease ergotism

Self-Quiz

Labeling and Matching

In the blank corresponding to each of the following illustrations (1–8), identify the organism by its common name (or its scientific name if a common name is unavailable). Then match each organism with the appropriate item from the lettered list (may be used more than once) by entering the correct letter in the parentheses.

A. Sac fungi B. Zygomycetes C. Club fungi D. Imperfect fungi E. Lichen

1. ______________ ______________ ( ) [p.405]

2. ______________ ( ) [p.413]

3. ______________ ( ) [p.406]

4. ______________ ( ) [p.409]

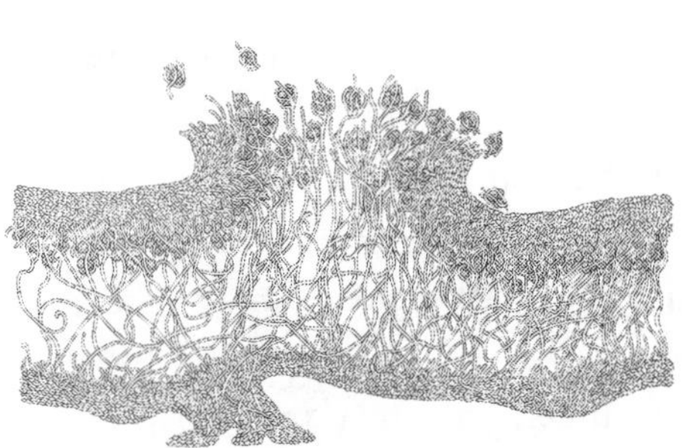

5. ______________ ( ) [p.410]

6. ______________ ( ) [p.410]

7. ______________ ______________

______________ ( ) [p.405]

8. ______________ ______________

______________ ( ) [p.409]

Multiple Choice

_____ 9. Most true fungi send out cellular filaments called _______________ . [p.406]
a. mycelia
b. hyphae
c. mycorrhizae
d. asci

_____ 10. Heterotrophic species of fungi can be _______________ . [pp.406,410]
a. saprobic
b. parasitic
c. mutualistic
d. all of the above

Choice

For questions 11–20, choose from the following:

a. club fungi [pp.406–407] b. imperfect fungi [p.409]
c. sac fungi [pp.408–409] d. zygomycetes [p.408]

_____ 11. The group that includes *Rhizopus stolonifer,* the notorious black bread mold, is the _______________.

_____ 12. The group that includes delectable morels and truffles but also includes bakers' and brewers' yeasts is the _______________.

_____ 13. The group that includes the honey mushroom, the largest and oldest organism on the planet, is the _______________.

_____ 14. The group that includes the commercial mushroom, *Agaricus brunnescens,* as well as the death cap mushroom, *Amanita phalloides,* is the _______________.

_____ 15. The group that includes *Penicillium,* which has a variety of species that produce penicillin and substances that flavor Camembert and Roquefort cheeses, is the _______________.

_____ 16. The group whose spore-producing structures (asci) are shaped like flasks, globes, and shallow cups is the _______________.

_____ 17. The group that forms a thin, clear covering around the zygote, the zygospore, is the _______________.

_____ 18. The group whose spore-producing structures are club shaped is the _______________.

_____ 19. The groups that are symbiotic with young roots of shrubs and trees in mycorrhizal associations are _______________ and _______________.

_____ 20. A group of puzzling kinds of fungi whose members are lumped together but do not constitute a formal taxonomic group is the _______________.

Chapter Objectives/Review Questions

1. Define *symbiosis* and explain how mutualism is a form of symbiosis. [p.404]
2. Explain the similarities and differences between a lichen and a mycorrhiza. [p.404]
3. Explain the process by which fungi absorb their nutrients. [p.404]
4. Explain the difference between a saprobe and a parasite. [p.406]
5. List the common names for the four groups of fungi. [p.406]
6. Distinguish among the meanings of the following terms: *hypha, spore,* and *mycelium.* [p.406]
7. Describe the diverse appearances of the fungi classified as club fungi. [pp.406–407]
8. Explain what the term *mushroom* represents in the fungal life cycle. [p.407]
9. Review the generalized life cycle of a club fungus. [p.407]
10. What is the importance of the club fungi *Armillaria ostoyae?* [p.407]

11. Zygomycetes form sexual spores by way of ________. [p.408]
12. Review the life cycle of *Rhizopus.* [p.408]
13. Most sac fungi produce sexual spores called ________. [p.408]
14. What are some typical shapes of the reproductive structures enclosing the asci? [pp.408–409]
15. Give the economic importance of the genus *Aspergillus.* [p.409]
16. What is the difference between *Saccharomyces cerevisiae* and *Candida albicans?* [p.409]
17. Why are some fungi assigned to the group called the imperfect fungi? [p.409]
18. Define *mutualism* and explain why a lichen fits that definition. [p.410]
19. Distinguish the mycobiont from the photobiont. [p.410]
20. Describe the fungus–plant root association known as mycorrhizae. [p.411]
21. Distinguish ectomycorrhizae from endomycorrhizae. [p.411]
22. What is the effect of pollution on mycorrhizae? [p.411]
23. Give the name of the fungus that causes the disease known as ergotism; describe the symptoms of ergotism. [p.412]
24. Name the cause and describe the symptoms of histoplasmosis. [p.412]

Integrating and Applying Key Concepts

1. Suppose humans acquired a few well-placed fungal genes that caused them to reproduce in the manner of a typical fungus. Try to imagine the behavioral changes that humans would likely undergo. Would their food supplies necessarily be different? table manners? stages of their life cycle? courtship patterns? habitat? Would the natural limits to population increase be the same? Would their body structure change? Would there necessarily have to be separate sexes? Compose a descriptive science-fiction tale about two mutants who find each other and set up housekeeping together.
2. What does the fact that the classification *imperfect fungi* exists tell you about fungal classification? Why does this group make identification of species according to the biological species concept (Ch. 19) difficult?

25

ANIMALS: THE INVERTEBRATES

Interactive Exercises

Madeleine's Limbs [pp.414–415]

25.1. OVERVIEW OF THE ANIMAL KINGDOM [pp.416–417]

25.2. PUZZLES ABOUT ORIGINS [p.418]

25.3. SPONGES — SUCCESS IN SIMPLICITY [pp.418–419]

Selected Words: Burgess Shale [p.414], *anterior* end [p.417], *posterior* end [p.417], *dorsal* surface [p.417], *ventral* surface [p.417], pseudocoel [p.417], *Paramecium* [p.418], *Volvox* [p.418], *Trichoplax adhaerens* [p.418], spicules [p.418], *Euplectella* [p.419], gemmules [p.419]

Boldfaced, Page-Referenced Terms

[p.416] animals ______________________________

[p.416] ectoderm ______________________________

[p.416] endoderm ______________________________

[p.416] mesoderm ______________________________

[p.416] vertebrates ______________________________

[p.416] invertebrates ______________________________

[p.416] radial symmetry ______________________________

[p.417] bilateral symmetry ______________________________

[p.417] cephalization ______________________________

[p.417] gut ______________________________

[p.417] coelom ______________________________

[p.418] placozoan ______________________________

[p.418] sponges ______________________________

[p.418] collar cells ______________________________

[p.419] larva (plural, larvae) ______________________________

Matching

Choose the most appropriate answer for each term.

1. ____ animals [p.416]
2. ____ ventral surface [p.417]
3. ____ ectoderm, endoderm, mesoderm [p.417]
4. ____ anterior end [p.417]
5. ____ vertebrates [p.416]
6. ____ invertebrates [p.416]
7. ____ radial symmetry [p.416]
8. ____ bilateral symmetry [p.417]
9. ____ dorsal surface [p.417]
10. ____ gut [p.417]
11. ____ coelom [p.417]
12. ____ thoracic cavity [p.417]
13. ____ abdominal cavity [p.417]
14. ____ posterior end [p.417]
15. ____ segmentation [p.417]
16. ____ cephalization [p.417]

A. All animals whose ancestors evolved before the vertebral column did
B. An evolutionary process whereby sensory structures and nerve cells became concentrated in a head
C. The back surface
D. Animal body cavity lined with a peritoneum—found in most bilateral animals; some worms lack this cavity, other worms have a false cavity
E. Head end
F. Animals having body parts arranged regularly around a central axis, like spokes of a bike wheel
G. Upper coelom cavity holding a heart and lungs
H. Primary tissue layers that give rise to all adult animal tissues and organs
I. Surface opposite the dorsal surface
J. Region inside animal body in which food is digested
K. Animals having right and left halves that are mirror images of each other
L. Series of animal body units that may or may not be similar to one another
M. Tail end
N. Lower coelom cavity holding a stomach, intestines, and other organs
O. Multicellular organisms with tissues forming organs and organ systems; diploid body cells; heterotrophic; aerobic respiration; sexual reproduction, sometimes asexual; most are motile in some part of the life cycle; the life cycle shows embryonic development
P. Animals with a vertebral column

Complete the Table

17. Complete the following table by filling in the appropriate phylum or representative group name. [p.416]

Phylum	*Some Representative*	*Number of Known Species*
a.	*Trichoplax;* simplest animal	1
b. Porifera		8,000
c.	Hydrozoans, jellyfishes, corals, sea anemones	11,000
d. Platyhelminthes		15,000
e.	Pinworms, hookworms	20,000
f.	Tiny body with crown of cilia, great internal complexity; "wheel animals"	1,800
g. Mollusca		110,000
h.	Leeches, earthworms, polychaetes	15,000
i. Arthropoda		1,000,000
j.	Sea stars, sea urchins	6,000

Choice

For questions 18–27, concerning animal origins, choose from the following: [p.418]

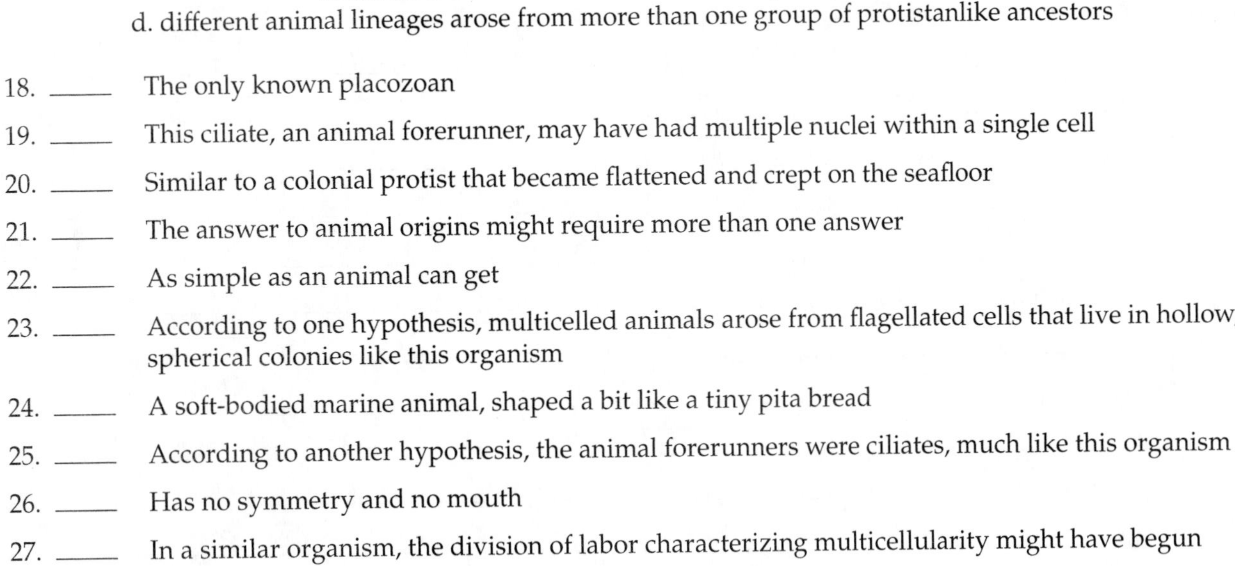

a. *Paramecium* b. *Volvox* c. *Trichoplax adhaerens*
d. different animal lineages arose from more than one group of protistanlike ancestors

18. ____ The only known placozoan
19. ____ This ciliate, an animal forerunner, may have had multiple nuclei within a single cell
20. ____ Similar to a colonial protist that became flattened and crept on the seafloor
21. ____ The answer to animal origins might require more than one answer
22. ____ As simple as an animal can get
23. ____ According to one hypothesis, multicelled animals arose from flagellated cells that live in hollow, spherical colonies like this organism
24. ____ A soft-bodied marine animal, shaped a bit like a tiny pita bread
25. ____ According to another hypothesis, the animal forerunners were ciliates, much like this organism
26. ____ Has no symmetry and no mouth
27. ____ In a similar organism, the division of labor characterizing multicellularity might have begun

Matching

Choose the most appropriate answer for each term.

28. ____ fragmentation [p.419]
29. ____ sponge phylum [p.418]
30. ____ adult [p.419]
31. ____ larva [p.419]
32. ____ amoeboid cells [pp.418–419]
33. ____ sponge skeletal elements [p.418]
34. ____ *Trichoplax* [p.418]
35. ____ microvilli [p.419]
36. ____ collar cells [pp.418–419]
37. ____ pathway of water flow [p.418]
38. ____ gemmules [p.419]

A. Flagellated cells that absorb and move water through a sponge as well as engulf food
B. Reside in a gelatinous substance between inner and outer cell linings
C. An organism whose two cell layers resemble those of a sponge
D. Sexually mature form of a species
E. Form the "collars" of collar cells
F. Clusters of sponge cells capable of germinating and establishing new colonies
G. Microscopic pores and chambers
H. Random chunks of sponge tissue break off and grow into more sponges
I. Spongin fibers, glasslike spicules of silica or calcium carbonate or both
J. Sexually immature form preceding the adult
K. Porifera

25.4. CNIDARIANS — TISSUES EMERGE [pp.420–421]

Selected Words: scyphozoans [p.420], anthozoans [p.420], hydrozoans [p.420] *Hydra* [p.420], *Chironix* [p.420], gastrodermis [p.420], mesoglea [p.420], *Obelia* [p.421], *Physalia* [p.421]

Boldfaced, Page-Referenced Terms

[p.420] cnidarians ____

____

[p.420] nematocysts ____

____

[p.420] medusa ____

____

[p.420] polyp ____

____

[p.420] epithelium (plural, epithelia) ____

____

[p.420] nerve cells ____

____

[p.420] contractile cells ____

____

[p.421] hydrostatic skeleton ____

____

[p.421] gonads ____

[p.421] planula ____

Fill-in-the-Blanks

All members of the phylum (1) ______________ [p.420] are radial animals; they include jellyfishes, sea anemones, corals, and animals such as *Hydra*. Most of these animals live in the sea, and they alone produce (2) ______________ [p.420], capsules capable of discharging threads that entangle or pierce prey. Cnidarians have two common body plans, the (3) ______________ [p.420], which looks like a bell or an upside-down saucer, and the (4) ______________ [p.420], which has a tubelike body with a tentacle-fringed mouth at one end. The saclike cnidarian gut processes food with its (5) ______________ [p.420], a sheetlike lining with glandular cells that secrete digestive enzymes. A(n) (6) ______________ [p.420] lines the rest of the body's surfaces. Each of these linings is referred to as a(n) (7) ______________ [p.420], a tissue with a free surface that faces the environment or some type of fluid inside the body. Cnidaria epithelia house (8) ______________ [p.420] cells, which receive signals from receptors which sense changes in the surroundings and send signals to (9) ______________ [p.420] cells that can carry out suitable responses. The nerve cells interact as a(n) (10) "______________ [p.420] net", a simple nervous tissue, to control movement and changes in shape. The (11) ______________ [p.420] is a layer of secreted gelatinous material that lies between the epidermis and gastrodermis. (12) ______________ [pp.420–421] contain enough mesoglea to impart buoyancy and to serve as a firm yet deformable skeleton against which contractile cells can act. Any fluid-filled cavity or cell mass against which contractile cells can act is a(n) (13) ______________ [p.421] skeleton. The contractile cells of most polyps, which have little (14) ______________ [p.421], act against water in their gut.

Although some cnidarian species have only a polyp or a medusa stage in the life cycle, many cnidarians, such as *Obelia* and *Physalia*, exhibit (15) ______________ [p.421] body forms, with the medusa being the sexual form. They have simple (16) ______________ [p.421] that rupture and release gametes. Zygotes formed at fertilization develop into (17) ______________ [p.421], a kind of swimming or creeping larva that usually possesses ciliated epidermal cells. In time, a mouth opens at one end, transforming the larva into a polyp or a medusa, and the cycle begins anew.

Another example of cnidarian diversity is *Physalia*, informally called the (18) ______________ [p.421] man-of-war. The (19) ______________ [p.421] in the nematocysts of this infamous hydrozoan poses a danger to bathers and fishermen as well as to prey organisms (fish). Although *Physalia* lives mainly in warm waters, currents sometimes move it up to the (20) ______________ [p.421] coasts of North America and Europe. A blue, gas-filled float that develops from the (21) ______________ [p.421] keeps the colony near the water's surface, where winds move it about. Under the float, groups of polyps and medusae interact as (22) "________________" [p.421] in feeding, reproduction, defense, and other specialized tasks.

The (23) ________________ [p.421] forming corals and other colonial anthozoans are a fine example of variation on the basic cnidarian body plan. The colonies consist of polyps that have secreted (24) ________________ [p.421] reinforced skeletons, which interconnect with one another. Over time, the (25) ________________ [p.421] accumulate and so become the main building materials for reefs such as the Great Barrier Reef. Reef-building corals receive (26) ________________ [p.421] inputs from the changing tides and from (27) ________________ [p. 421] mutualists living in their tissues.

Labeling

Identify each indicated part of the following illustration: [p.421]

28. ________________ ________________

29. ________________ ________________

30. ________________ ________________

31. ________________

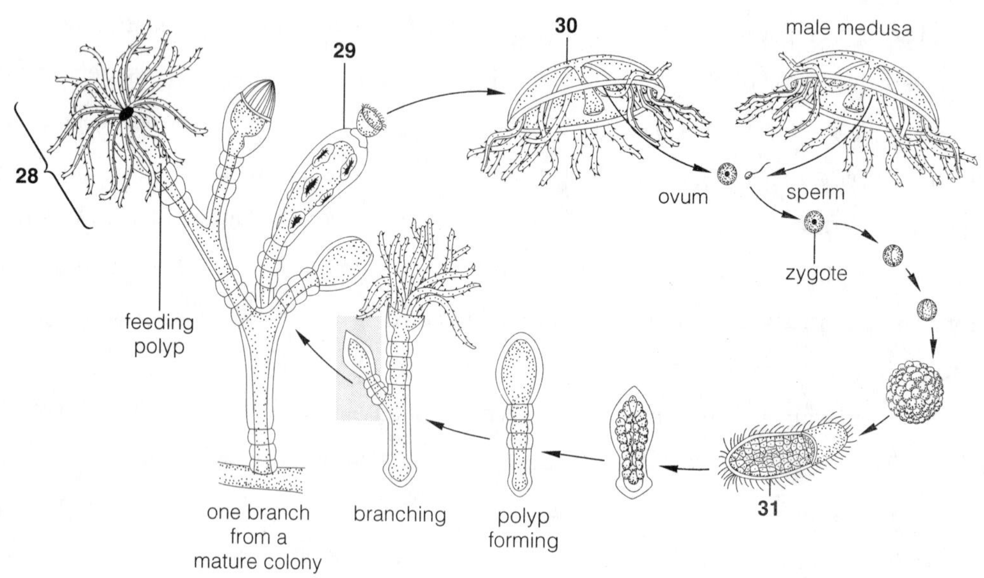

25.5. ACOELOMATE ANIMALS — AND THE SIMPLEST ORGAN SYSTEMS

[pp.422–423]

Selected Words: *organ-system* level of construction [p.422], *definitive* host [p.422], *intermediate* host [p.422], Platyhelminthes [p.422], *transverse* fission [p. 422], scolex [p.423]

Boldfaced, Page-Referenced Terms

[p.422] organ ____

[p.422] organ system ____

[p.422] flatworms ____

[p.422] pharynx ____

[p.422] hermaphrodites ____

[p.423] proglottids ____

Choice

For questions 1–12, choose from the following. Letters can be used more than once, and blanks can have more than one letter.

a. turbellarians b. flukes c. tapeworms

1. ____ Parasitic worms [pp.422–423]
2. ____ Possess a scolex [p.423]
3. ____ Ancestral forms probably had a gut but later lost it during their evolution in animal intestines [p.423]
4. ____ Water-regulating systems have one or more tiny, branched tubes called protonephridia [p.423]
5. ____ Only planarians and a few others live in freshwater habitats [p.422]
6. ____ Their life cycles have sexual and asexual phases and at least two kinds of hosts [p.423]
7. ____ Flame cells, each with a tuft of cilia, that drive out excess water into the surroundings [p.422]
8. ____ Thrive in predigested food in vertebrate intestines [p.423]
9. ____ After division, each half regenerates the missing parts [p.422]
10. ____ Proglottids are new units of the body that bud just behind the head [p.423]
11. ____ Possess a structure equipped with suckers, hooks, or both [p.423]
12. ____ Older proglottids store fertilized eggs; they break off and leave the body in feces [p.423]

Labeling

Identify the parts of the animal shown dissected in the accompanying drawings. [p.422]

13. ________________ ________________
14. ________________
15. ________________
16. ________________ ________________
17. ________________
18. ________________

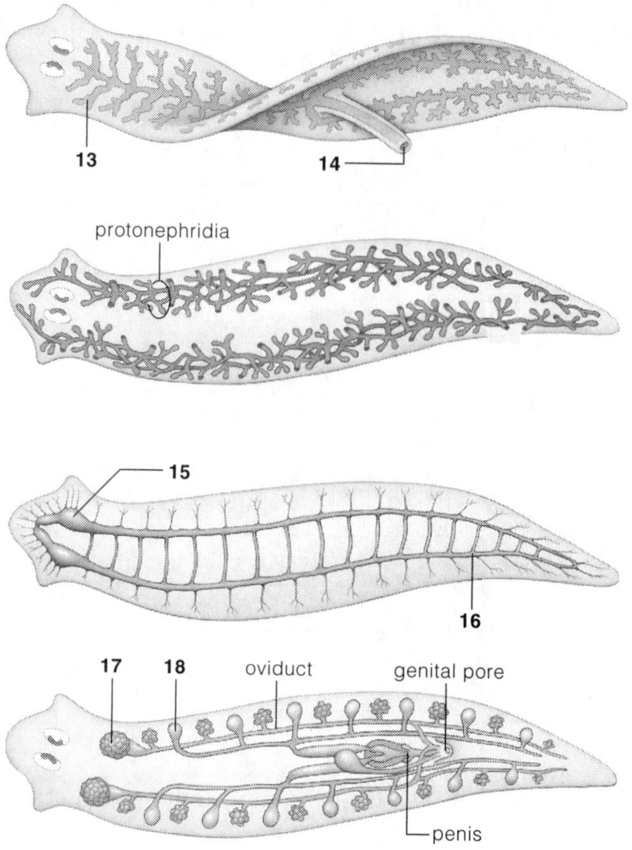

Short Answer

Answer questions 19–22 with reference to the accompanying drawing of a dissected animal. [p.422]

19. What is the common name (or genus) of the animal dissected? ________________
20. Is the animal parasitic? ________________
21. Is the animal hermaphroditic? ________________
22. What is the coelom type exhibited by this animal? ________________

25.6. ROUNDWORMS [p.423]

25.7. FOCUS ON HEALTH: A Rogue's Gallery of Worms [pp.424–425]

25.8. ROTIFERS [p.425]

Selected Words: Nematoda [p.423], *Caenorhabditis elegans* [p.423], *schistosomiasis* [p.424], *Schistosoma japonicum* [p.424], *Enterobius vermicularis* [p.424], *Taenia saginata* [p.424], *Trichinella spiralis* [p.425], *Wuchereria bancrofti* [p.425], *elephantiasis* [p.425], *Philodina roseola* [p.425]

Boldfaced, Page-Referenced Terms

[p.423] roundworms ____

____

[p.423] cuticles ____

____

[p.425] rotifers ____

____

Short Answer

Answer questions 1–3 with reference to the following drawing of a dissected animal. [p.423]

1. What is the common name of the animal dissected? ___
2. Is the animal hermaphroditic? ___
3. What is the coelom type exhibited by this animal? ___

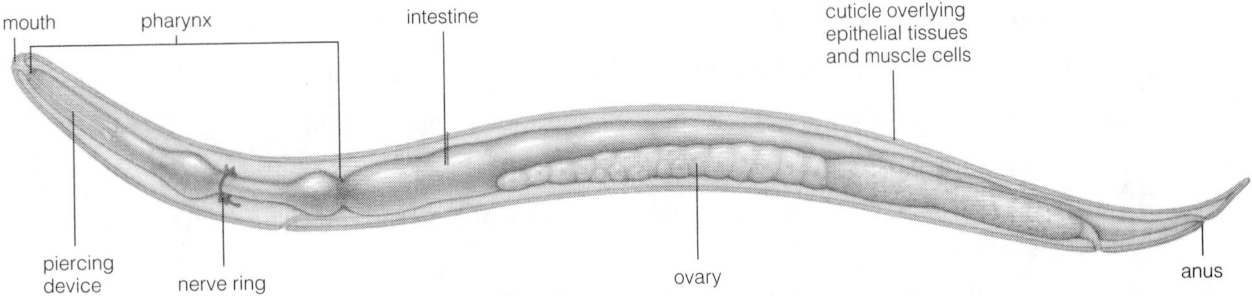

Answer questions 4–8 with reference to the following drawing of a dissected animal. [p.425]

4. What is the common name of this animal? ___
5. What type of symmetry does this animal possess? ___
6. Is this animal cephalized? ___
7. What type of coelom does this animal possess? ___
8. Why is this animal said to be a "wheel animal"? ___

___

___

___

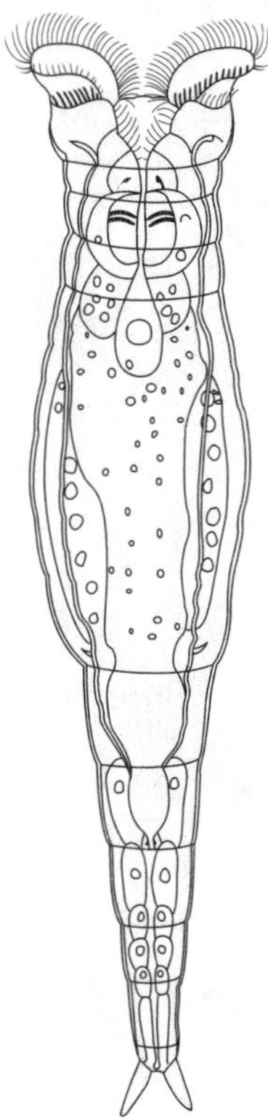

Choice

For questions 9–26, choose from the following:

a. roundworms b. rotifers

9. ______ All but about 5 percent live in fresh water, such as lakes and ponds, and even within films of water on mosses and other plants. [p.425]
10. ______ At night, the centimeter-long females migrate to the host's anal region to lay eggs; itching leads to scratching, which transfers eggs. [p.424]
11. ______ Hookworms, pinworms, and other types parasitize plants and animals. [pp.423–424]
12. ______ These are probably the most abundant of all multicelled animals alive today. [p.423]
13. ______ All have a crown of cilia used in swimming and wafting food to the mouth. [p.425]
14. ______ All have a bilateral, cylindrical body, usually tapered at both ends and protected by a tough cuticle. [p.423]
15. ______ Parasitic forms can do extensive damage to their hosts, which include humans, cats, dogs, cows, and sheep, as well as valued crop plants. [p.423]
16. ______ Most species are less than a millimeter long, yet they have a pharynx, an esophagus, digestive glands, a stomach, usually an intestine and anus, and protonephridia. [p.425]
17. ______ Adult forms become lodged in the body's lymph nodes; elephantiasis occurs, an enlargement of the legs and other body regions due to blockage of lymph flow. [p.425]
18. ______ They cause infection when a juvenile form penetrates the bare skin; the parasite then travels the bloodstream to the lungs. [p.424]
19. ______ Some have "eyes." [p.425]
20. ______ Two "toes" exude substances that attach free-living species to substrates. [p.425]
21. ______ A mosquito is *Wuchereria*'s intermediate host. [p.424]
22. ______ Its rhythmic motions reminded early microscopists of a turning wheel. [p.425]
23. ______ Humans become infected by *Trichinella spiralis* mostly by eating insufficiently cooked meat from pigs or certain game animals. [pp.424–425]
24. ______ They eat bacteria and microscopic algae. [p.425]
25. ______ *Enterobius vermicularis* lives in the large intestine of humans. [p.424]
26. ______ Thousands of scavenging types may occupy a handful of rich topsoil. [p.423]

Fill-in-the-Blanks

The numbered items in the following illustration represent missing information; fill in the corresponding answer blanks to complete the narrative of the life cycle of the blood fluke. [p.424]

The life cycle of the Southeast Asian blood fluke (*Schistosoma japonicum*) requires a human primary host standing in water in which the fluke larvae can swim. This life cycle also requires an aquatic snail as an intermediate host. Flukes reproduce sexually, producing (27) ________________ that mature in a human body. (27) leave the body in feces, then hatch into ciliated, swimming (28) ________________ that burrow into a(n) (29) ________________ and multiply asexually. In time, many fork-tailed (30) ________________ develop. These leave the snail and swim until they contact (31) ________________ skin. They bore inward and migrate to thin-walled intestinal veins, and the cycle begins anew. In infected humans, white blood cells that defend the body attack the masses of fluke eggs, and grainy masses form in tissues. In time, the liver, spleen, bladder, and kidneys deteriorate.

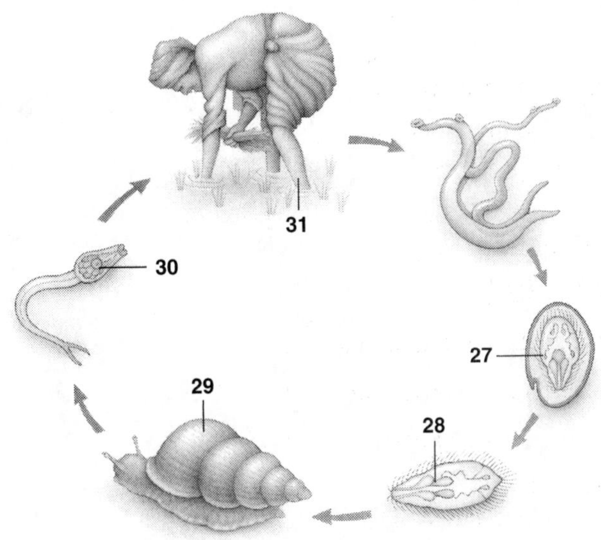

Fill-in-the-Blanks

The numbered items in the following illustration represent missing information; fill in the corresponding answer blanks to complete the narrative of the life cycle of the beef tapeworm. [p.424]

(32) ________________, each with the inverted scolex of a future tapeworm, become encysted in (33) ________________ host tissues (such as skeletal muscle). A(n) (34) ________________, a definitive host, eats infected and undercooked beef containing tapeworm cysts. The scolex of a larva turns inside out, attaches to the wall of the host's (35) ________________ ________________, and begins to absorb host nutrients. Many (36) ________________ form, by budding; each of these segments becomes sexually mature and has both male and female reproductive (37) ________________. Ripe (38) ________________ containing fertilized eggs leave the host in (39) ________________, which may contaminate water and vegetation. Inside each fertilized egg, an embryonic (40) ________________ form develops. Cattle may ingest embryonated eggs or ripe proglottids and so become (41) ________________ hosts.

a. 32 each with inverted scolex of future tapeworm become encysted in **33** host tissues (e.g. skeletal muscle).

b. A **34** , a definitive host, eats infected, undercooked beef (mainly skeletal muscle).

c. Scolex of larva turns inside out, attaches to small **35** wall. Larva absorbs host nutrients.

d. Many **36** form by budding.

e. Each sexually mature proglottid has female and male **37** . Ripe **38** containing fertilized eggs leave host in **39** which may contaminate water and vegetation.

f. Inside each fertilized egg, an embryonic **40** form develops. Cattle may ingest embryonated eggs or ripe proglottids, and so become **41** hosts.

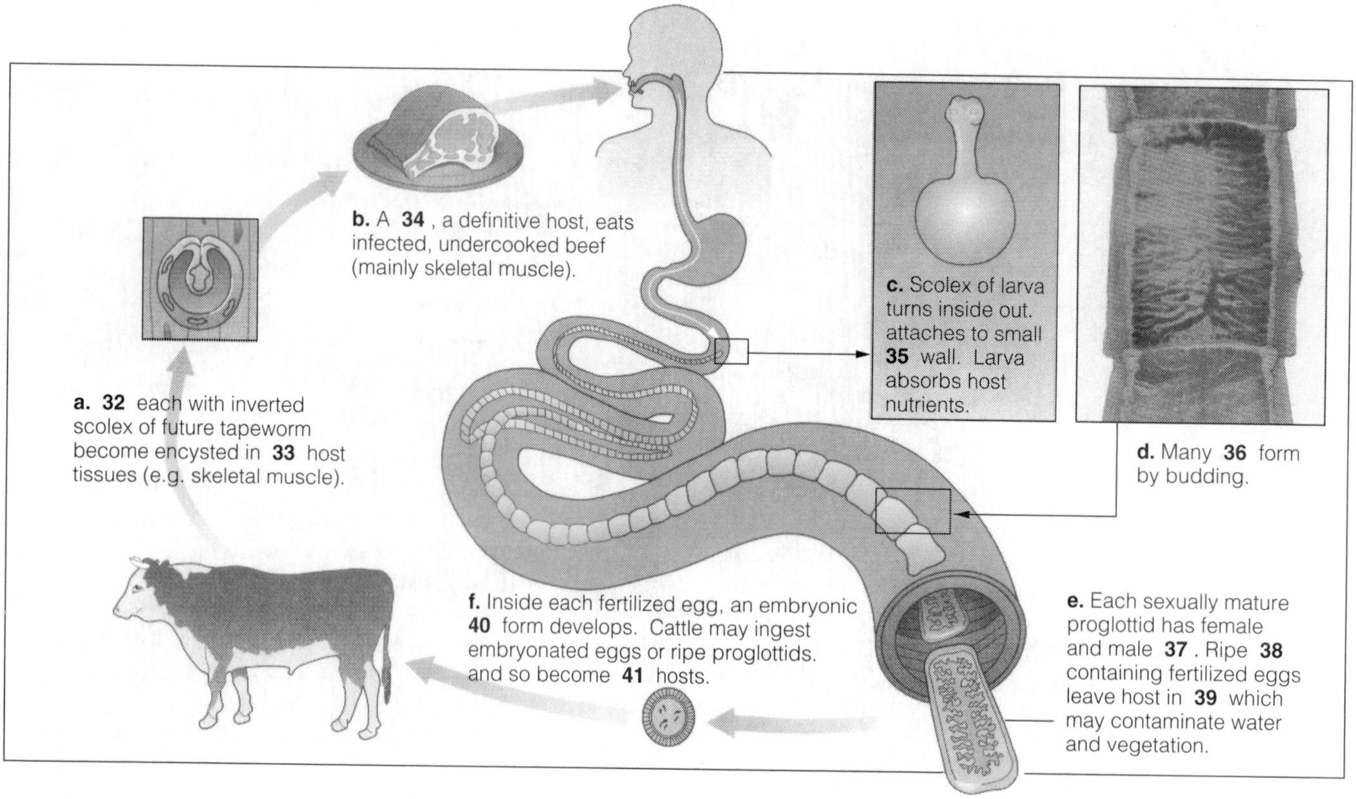

25.9. A MAJOR DIVERGENCE [p.426]

25.10. A SAMPLING OF MOLLUSKS [pp.426–427]

25.11. EVOLUTIONARY EXPERIMENTS WITH MOLLUSCAN BODY PLANS [pp.428–429]

Selected Words: *spiral* cleavage [p.426], *radial* cleavage [p.426], *molluscus* [p.426], radula [p.426], chitons [pp.426–427], gastropods [pp.426–427], bivalves [pp.426–427], cephalopods [pp.426–427], *Aplysia* [p.428], *jet propulsion* [p.429]

Boldfaced, Page-Referenced Terms

[p.426] protostomes ______________________________

____

[p.426] deuterostomes ______________________________

____

[p.426] mollusks ______________________________

____

[p.426] mantle ______________________________

____

[p.428] torsion ______________________________

____

Choice

For questions 1–10, choose from the following: [p.426]

a. protostomes b. deuterostomes

1. _____ The first external opening in these embryos becomes the anus; the second becomes the mouth
2. _____ Animals having a developmental pattern in which the early cell divisions are parallel and perpendicular to the axis
3. _____ A coelom arises from spaces in the mesoderm
4. _____ Radial cleavage
5. _____ The first external opening in these embryos becomes the mouth
6. _____ A coelom forms from outpouchings of the gut wall
7. _____ Spiral cleavage
8. _____ Animal having a developmental pattern in which early cell divisions are at oblique angles relative to the genetically prescribed body axis
9. _____ Echinoderms and chordates
10. _____ Mollusks, annelids, and arthropods

Matching

Identify the animals pictured below by matching each with the appropriate description.

11. ______ [pp.426,428] Animal A
12. ______ [p.429] Animal B
13. ______ [p.429] Animal C

I. Bivalve
II. Cephalopod
III. Gastropod

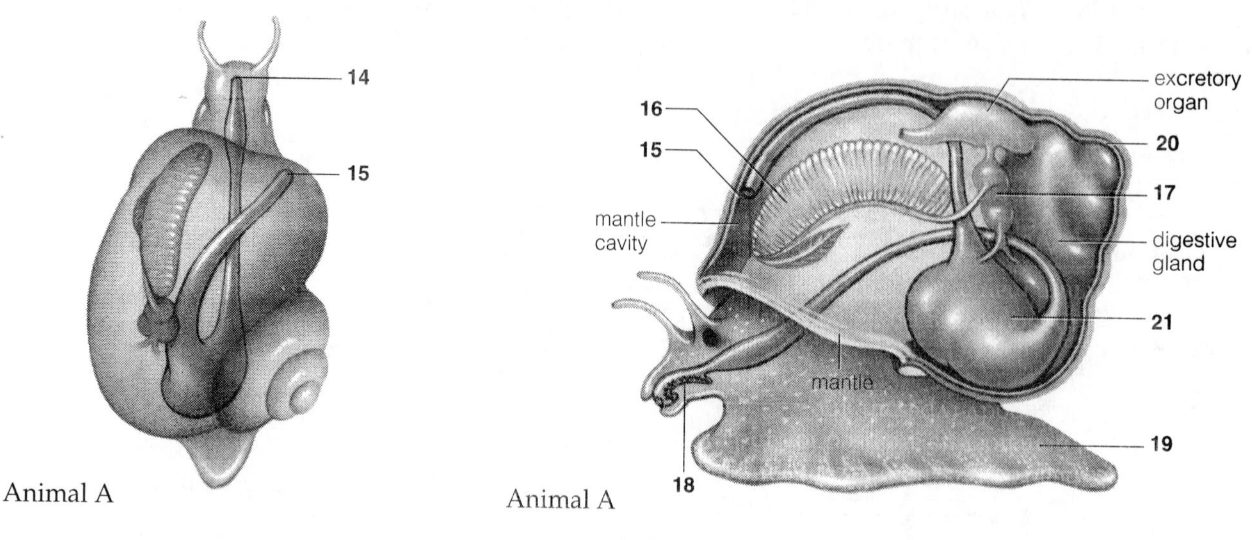

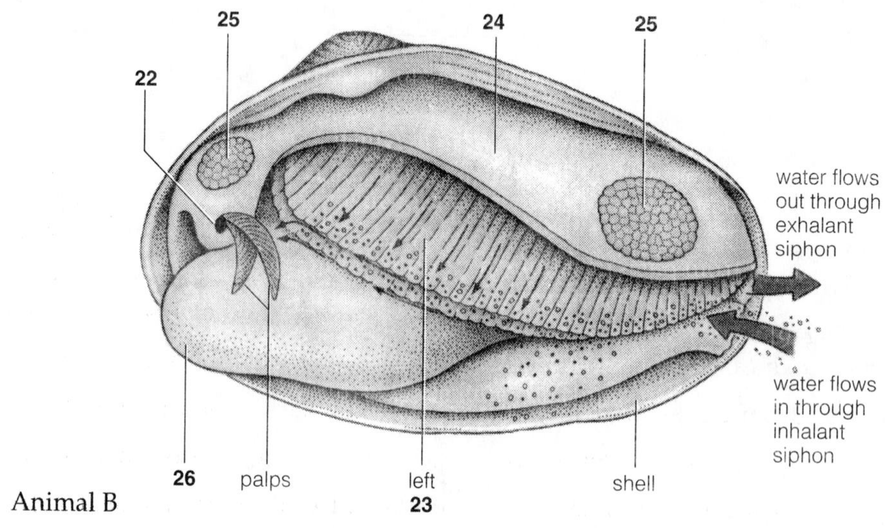

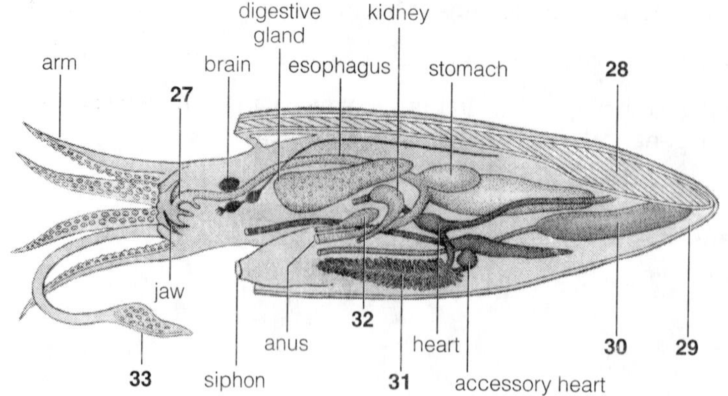

Labeling

Identify each numbered part in the preceding drawings by writing its name in the appropriate blank.

14. ______________ [p.428]
15. ______________ [pp.426,428]
16. ______________ [p.426]
17. ______________ [p.426]
18. ______________ [p.425]
19. ______________ [p.426]
20. ______________ [p.426]
21. ______________ [p.426]
22. ______________ [p.429]
23. ______________ [p.429]
24. ______________ [p.429]
25. ______________ [p.429]
26. ______________ [p.429]
27. ______________ [p.429]
28. ______________ ______________ [p.429]
29. ______________ [p.429]
30. ______________ ______________ [p.429]
31. ______________ [p.429]
32. ______________ ______________ [p.429]
33. ______________ [p.429]

Fill-in-the-Blanks [p.426]

A(n) (34) ______________ is a bilateral animal having a small coelom and a fleshy soft body. Most have a(n) (35) ______________ of calcium carbonate and protein, secreted from cells of a tissue that drapes like a skirt over the body mass. This tissue, the (36) ______________, is unique to mollusks. Special respiratory organs, the (37) ______________, contain thin-walled leaflets for gas exchange. Most mollusks have a fleshy (38) ______________. Many have a(n) (39) ______________, a tonguelike, toothed organ that by rhythmic protractions and retractions rasps small algae and other food from substrates and draws it into the mouth. Mollusks with a well-developed head have (40) ______________ and (41) ______________; but not all have a head. There is great diversity in the phylum, and here we review four classes: chitons, gastropods, bivalves, and cephalopods.

Choice

For questions 42–65, choose from the following classes of the phylum Mollusca:

a. chitons b. gastropods c. bivalves d. cephalopods

42. ____ Class with the swiftest invertebrates, the squids [p.427]
43. ____ The "belly foots" [p.426]
44. ____ The largest class, with 90,000 species [p.426]
45. ____ Possess a dorsal shell divided into eight plates [p.427]
46. ____ Class with the smartest invertebrates, the octopuses [p.427]
47. ____ So named because their soft foot spreads out as they crawl [pp.426–427]
48. ____ Animals having a "two-valved shell" [p.427]
49. ____ Class with the largest invertebrates, the giant squids [p.427]
50. ____ Coiling compacts the organs into a mass that can be balanced above the body, rather like a backpack [p.427]
51. ____ Class with the most complex invertebrates in the world, the octopuses and squids, with a memory and a capacity for learning [p.427]
52. ____ Highly active predators of the seas [p.427]
53. ____ Many have spirally coiled or conical shells [p.427]
54. ____ The only mollusks with a closed circulatory system [p.429]
55. ____ Includes clams, scallops, oysters, and mussels [p.427]
56. ____ Includes aquatic snails, land snails, and sea slugs [p.426]
57. ____ Move rapidly with a system of jet propulsion [p.429]
58. ____ Water is drawn into the mantle cavity through one siphon and leaves through the other, carrying wastes [p.429]
59. ____ Most can discharge dark fluid from an ink sac, perhaps to confuse predators [p.429]
60. ____ The anus dumps wastes near the mouth [p.428]
61. ____ Being highly active, they have great demands for oxygen [p.429]
62. ____ The only class of mollusks where torsion operates [p.428]
63. ____ Class containing the chambered nautilus [p.429]
64. ____ The shells of many are lined with iridescent mother-of-pearl [p.427]
65. ____ As the embryo develops, a cavity between the mantle and shell twists 180 degrees counterclockwise, as does nearly all of the visceral mass [p.428]

25.12. ANNELIDS — SEGMENTS GALORE [pp.430–431]

Selected Words: polychaetes [p.430], *setae* or *chaetae* [p.430], oligochaetes [p.430], *Hirudo medicinalis* [p.430], ganglion [p.431]

Boldfaced, Page-Referenced Terms

[p.430] annelids ____

____

[p.431] nephridia (singular, nephridium) ______________________________

____

[p.431] brain ______________________________

____

[p.431] nerve cords ______________________________

____

Matching

Choose the appropriate answer for each term.

1. ____ cuticle [p.431]
2. ____ annelids [p.430]
3. ____ earthworms [pp.430–431]
4. ____ marine polychaetes [p.430]
5. ____ brain [p.431]
6. ____ nephridia [p.431]
7. ____ nerve cords [p.431]
8. ____ setae or chaetae [p.430]
9. ____ advantage of segmentation [p.430]
10. ____ hydrostatic skeleton [p.431]
11. ____ leeches [p.430]
12. ____ earthworm locomotion [p.431]
13. ____ ganglion [p.431]

A. Fluid-cushioned coelomic chambers
B. Oligochaete scavengers with a closed circulatory system and few bristles per segment
C. Paired, each a bundle of extensions of nerve cell bodies leading away from the brain
D. Possess many bristles per segment
E. Muscle contraction with protraction and retraction of segment bristles
F. Secreted wrapping around the body surface of most annelids that permits respiratory exchange
G. A rudimentary aggregation of nerve cell bodies that integrate sensory input and muscle responses for the whole body
H. Lack bristles
I. Different body parts can evolve separately and specialize in different tasks
J. Except for leeches, chitin-reinforced bristles on each side of the body on nearly all segments
K. The term means "ringed forms"
L. An enlargement of the nerve cord in each segment of an earthworm
M. Regulates volume and composition of body fluids; often begins with a funnel-shaped structure in each segment

Short Answer

14. Describe how an earthworm burrows into the soil. ______________________________

____

____

____

Labeling

Identify each indicated part of the following illustrations.

15. ______________ [p.431]

16. ______________ [p.431]

17. ______________ ______________ [p.431]

18. ______________ [p.431]

19. ______________ [p.431]

20. ______________ ______________ [p.431]

21. ______________ [p.431]

22. ______________ [p.431]

23. ______________ [p.431]

24. ______________ [p.431]

25. ______________ ______________ [p.431]

26. Name this animal. ______________ [p.431]

27. Name this animal's phylum. ______________ [p.430]

28. Name two distinguishing characteristics of this group. ______________

______________ [pp.430–431]

29. Is this animal segmented? ( ) yes ( ) no [pp.430–431]

30. Symmetry of adult: ( ) radial ( ) bilateral [p.430]

31. Does this animal have a true coelom? ( ) yes ( ) no [p.430]

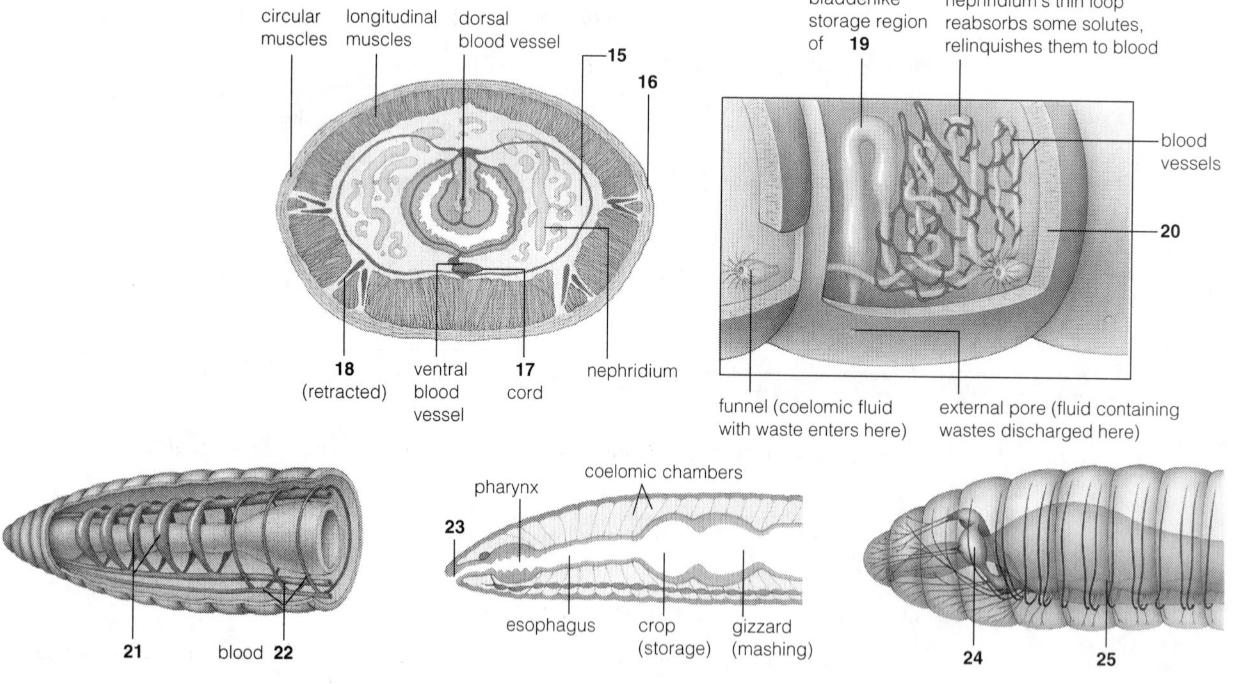

25.13. ARTHROPODS — THE MOST SUCCESSFUL ORGANISMS ON EARTH [p.432]

Selected Words: chelicerates [p.432], crustaceans [p.432], uniramians [p.432], *juvenile* [p.432], *division of labor* [p.432]

Boldfaced, Page-Referenced Terms

[p.432] arthropods ______________________________

____

[p.432] exoskeleton ______________________________

____

[p.432] molting ______________________________

____

[p.432] metamorphosis ______________________________

____

Choice

For questions 1–13, choose from the following six adaptations that contributed to the success of arthropods: [p.432]

a. hardened exoskeletons b. fused and modified segments c. jointed appendages
d. respiratory structures e. specialized sensory structures f. division of labor

1. _____ Intricate eyes and other sensory organs that contributed to arthropod success
2. _____ The new individual is a *juvenile,* a miniaturized form of the adult that simply changes in size and proportion until reaching sexual maturity
3. _____ A cuticle of chitin, proteins, and surface waxes that may be impregnated with calcium carbonate
4. _____ In most of their existing descendants, however, the serial repetitions of the body wall and organs are masked, for many fused-together, modified segments perform more specialized functions
5. _____ Metamorphosis from embryo to adult forms
6. _____ Might have evolved as defenses against predation
7. _____ Immature stages, such as caterpillars, specialize in feeding and growing in size
8. _____ Gills of aquatic arthropods
9. _____ In the ancestors of insects, different segments became combined into a head, a thorax, and an abdomen
10. _____ A jointed exoskeleton was a key innovation that led to appendages as diverse as wings, antennae, and legs
11. _____ Numerous species have a wide angle of vision and can process visual information from many directions
12. _____ Air-conducting tubes evolved among insects and other land dwellers
13. _____ Their waxy surfaces restrict evaporative water loss and can support a body deprived of water's buoyancy

Short Answer

14. Why are the arthropods said to be the most biologically successful organisms on Earth? [p.432] ________

____

____

____

25.14. A LOOK AT SPIDERS AND THEIR KIN [p.433]

25.15. A LOOK AT THE CRUSTACEANS [pp.434–435]

25.16. HOW MANY LEGS? [p.435]

Selected Words: arachnids [p.433], *open* circulatory system [p.433], book lungs [p.433], cephalothorax [p.434], copepods [p.434], *Scutigera* [p.435]

Boldfaced, Page-Referenced Terms

[p.435] millipedes ____

____

[p.435] centipedes ____

____

Matching

Select the most appropriate answer for each term. [p.433]

1. ____ ticks
2. ____ arachnid forebody appendages
3. ____ arachnids
4. ____ arachnid hindbody appendages
5. ____ chelicerates
6. ____ spider, internal organs
7. ____ efficient predatory arachnids

A. Scorpions and spiders that sting, bite, and may subdue prey with venom
B. An open circulatory system and book lungs
C. A group including mites, horseshoe crabs, sea spiders, spiders, ticks, and chigger mites
D. Spin out silk thread for webs and egg cases
E. Some transmit bacterial agents of Rocky Mountain spotted fever or Lyme disease to humans
F. A familiar group including scorpions, spiders, ticks, and chigger mites
G. Four pairs of legs, a pair of pedipalps that have mainly sensory functions, and a pair of chelicerae that can inflict wounds and discharge venom

Dichotomous Choice

Circle one of two possible answers given between parentheses in each statement.

8. Nearly all arthropods possess (strong claws/an exoskeleton). [p.434]
9. The giant crustaceans are (lobsters and crabs/barnacles and pillbugs). [p.434]
10. The simplest crustaceans have many pairs of (different/similar) appendages along their length. [p.434]
11. (Barnacles/Lobsters and crabs) have strong claws that collect food, intimidate other animals, and sometimes dig burrows. [p.434]
12. (Barnacles/Lobsters and crabs) have feathery appendages that comb microscopic bits of food from the water. [p.434]

13. (Barnacles/Copepods) are the most numerous animals in aquatic habitats, maybe even in the world. [p.435]
14. Of all arthropods, only (lobsters and crabs/barnacles) have a calcified "shell." [p.435]
15. Adult (barnacles/copepods) cement themselves to wharf pilings, rocks, and similar surfaces. [p.435]
16. As is true of other arthropods, crustaceans undergo a series of (rapid feedings/molts) and so shed the exoskeleton during their life cycle. [p.435]
17. As (millipedes/centipedes) develop, pairs of segments fuse, so each segment in the cylindrical body ends up with two pairs of legs. [p.435]
18. Adult (millipedes/centipedes) have a flattened body, are fast moving, and have a pair of walking legs on every segment except two. [p.435]
19. (Millipedes/Centipedes) mainly scavenge for decaying vegetation in soil and forest litter. [p.435]
20. (Millipedes/Centipedes) are fast-moving, aggressive predators, outfitted with fangs and venom glands. [p.435]

Labeling

Identify each numbered part of the animal pictured at right. [p.434]

21. ______________
22. ______________
23. ______________
24. ______________
25. ______________
26. ______________

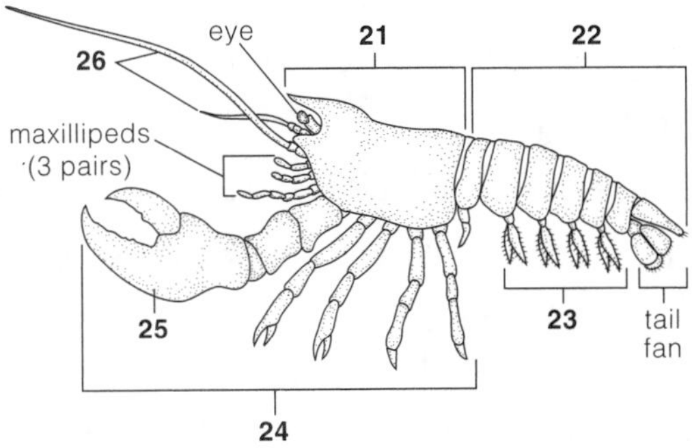

Answer questions 27–32 for the animal pictured above. [p.434]

27. Name the animal pictured. ______________
28. Name the subgroup of arthropods to which this animal belongs. ______________
29. Name two distinguishing characteristics of this group. ______________

____

30. Is this animal segmented? ( ) yes ( ) no
31. Symmetry of adult: ( ) radial ( ) bilateral
32. Does this animal have a true coelom? ( ) yes ( ) no

Identify each numbered body part in this illustration, then answer question 38. [p.433]

33. ______________ ______________
34. ______________
35. ______________
36. ______________
37. ______________ ______________
38. Name the subgroup of arthropods to which the animal belongs. ______________

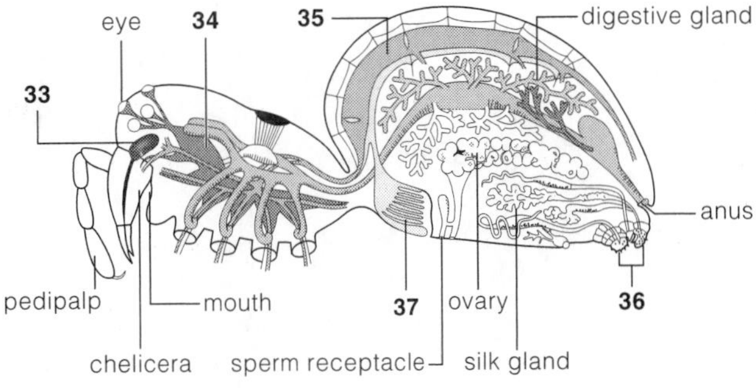

25.17. A LOOK AT INSECT DIVERSITY [pp.436–437]

25.18. FOCUS ON HEALTH: *Unwelcome Arthropods* [pp.438–439]

Selected Words: *incomplete* metamorphosis [p.436], *complete* metamorphosis [p.436], *Loxosceles* [p.438], *Ixodes* [p.438], Lyme disease [p.438], *Ixodes dammini* [p.438], *Borrelia burgdorferi* [p.438], *Centruroides sculpuratus* [p.439], *Diabrotica virgifera* [p.439]

Boldfaced, Page-Referenced Terms

[p.436] Malpighian tubules ______________________________

____

[p.436] nymphs ______________________________

____

[p.436] pupae ______________________________

____

Matching

Choose the most appropriate answer for each term.

1. ____ complete metamorphosis [p.436]
2. ____ the most successful species of insects [p.436]
3. ____ incomplete metamorphosis [p.436]
4. ____ insect success based on aggressive competition with humans [pp.436–437]
5. ____ Malpighian tubules [p.436]
6. ____ metamorphosis [p.436]
7. ____ a variation of insect headparts [p.436]
8. ____ insect activities considered beneficial to humans [p.437]
9. ____ insect life cycle stages [p.436]
10. ____ shared insect adaptations [p.436]

A. Postembryonic resumption of growth and transformation into an adult form
B. Destruction of crops, stored food, wool, paper, and timber; drawing blood from humans and their pets; transmitting pathogenic microorganisms
C. Head, thorax, and abdomen; paired sensory antennae and mouthparts; three pairs of legs and two pairs of wings
D. Chewing, sponging up, siphoning, piercing, and sucking
E. Involves gradual, partial change from the first immature form until the last molt
F. Winged insects, also the only winged invertebrates
G. Pollination of flowering plants and crop plants; parasitizing or attacking plants humans would rather do without
H. Larva–nymph–pupa–adult
I. Structures that collect nitrogen-containing wastes from blood and convert them to harmless crystals of uric acid that are eliminated with feces
J. Tissues of immature forms are destroyed and replaced before emergence of the adult

Choice

For questions 11–20, choose from the following:

a. spiders b. ticks c. scorpions d. beetles

11. _____ Corn rootworm [p.439]
12. _____ Lyme disease [pp.438–439]
13. _____ The poisonous brown recluse [p.438]
14. _____ *Centruroides sculpturatus* from Arizona, the most dangerous species in the United States [p.439]
15. _____ About five people die each year in the United States as a result of black widow bites [p.438]
16. _____ *Ixodes dammini* [p.438]
17. _____ Possess large, prey-seizing pincers and a venom-dispensing stinger at the tip of a narrowed, jointed abdomen [p.439]
18. _____ Rocky Mountain spotted fever, scrub typhus, tularemia, babesiasis, and encephalitis [p.439]
19. _____ Cucurbitacin is being used effectively against corn rootworms [p.439]
20. _____ Do not jump or fly; they crawl onto grasses and shrubs, then onto animals that brush past [p.439]

25.19. THE PUZZLING ECHINODERMS [pp.440–441]

Selected Words: *Echinodermata* [p.440], crinoids [p.440]

Boldfaced, Page-Referenced Terms

[p.440] echinoderms ____

____

[p.441] water-vascular system ____

____

Fill-in-the-Blanks

The second lineage of coelomate animals is referred to as the (1) ______________ [p.440]. The major invertebrate phylum of this lineage is the (2) ______________ [p.440]. The body wall of all echinoderms has protective spines, spicules, or plates made rigid with (3) ______________ ______________ [p.440]. Most echinoderms also have a well-developed internal (4) ______________ [p.440], which is composed of calcium carbonate and other substances secreted from specialized cells. Oddly, adult echinoderms have (5) ______________ [p.440] symmetry, with some bilateral features; but many produce larvae with (6) ______________ symmetry [p.440]. Adult echinoderms have no (7) ______________ [p.440], but a decentralized (8) ______________ [p.440] system allows them to respond to information about food, predators, and so forth coming from different directions. For example, any (9) ______________ [p.441] of a sea star that senses the shell of a tasty scallop can become the leader, directing the remainder of the body to move in a direction suitable for prey capture.

The (10) ________________ [p.441] feet of sea stars are used for walking, burrowing, clinging to a rock, or gripping a meal of clam or snail. These "feet" are parts of a(n) (11) ________________- [p.441] vascular system unique to echinoderms. Each foot contains a(n) (12) ________________ [p.441], which acts like a rubber bulb on a medicine dropper as it contracts and forces fluid into a foot that then lengthens. Tube feet change shape constantly as (13) ________________ [p.441] action redistributes fluid through the water-vascular system. Some sea stars are able simply to swallow their prey (14) ________________ [p.441]; others can push part of their stomach outside the mouth and around their prey, then start (15) ________________ [p.441] their meal even before swallowing it. Coarse, indigestible remnants are regurgitated back through the mouth. Their small (16) ________________ [p.441] is of no help in getting rid of empty clam or snail shells.

Labeling

Identify each indicated part of the two following illustrations. [p.441]

17. ________________ ________________
18. ________________ ________________
19. ________________
20. ________________
21. ________________
22. ________________ ________________
23. ________________
24. ________________ ________________

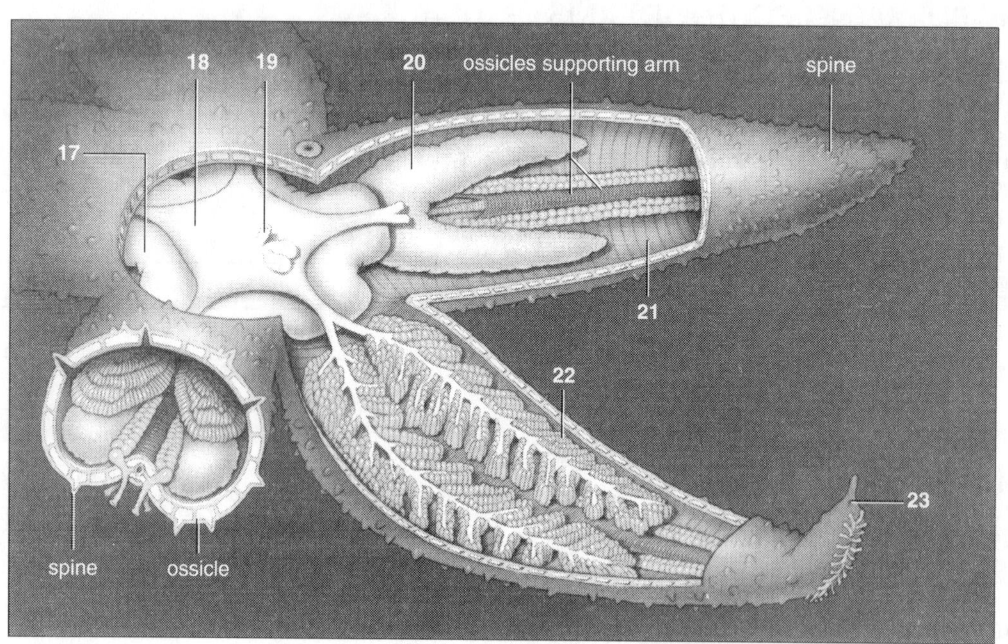

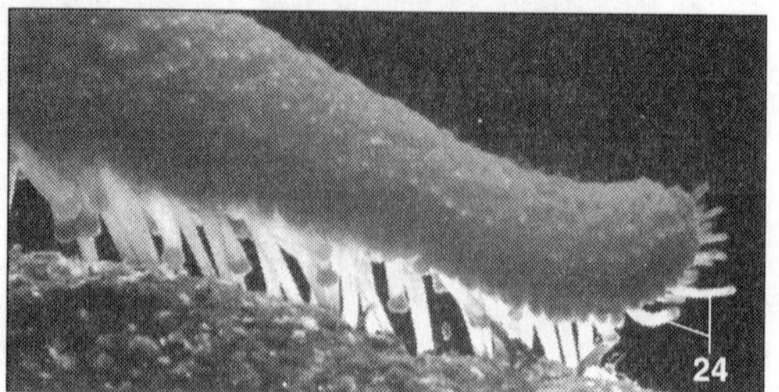

Short Answer

Answer questions 25–26 for the animal pictured above.

25. What is the animal shown? ________________ [p.441]
26. Is it a protostome ( ) or a deuterostome ( )? [p.440]

Identifying

27. Name the system shown at the right. ________________ - ________________ system [p.441]

28. Identify each creature below by its common name. Write the name in the blank below each picture, a–d.

29. Name the phylum of the animals pictured below. ________________ [p.440]

30. Name two distinguishing characteristics of this group. ____________________________________ [pp.440–441]

31. Symmetry of adults represented below is: ( ) radial ( ) bilateral. [p.440]

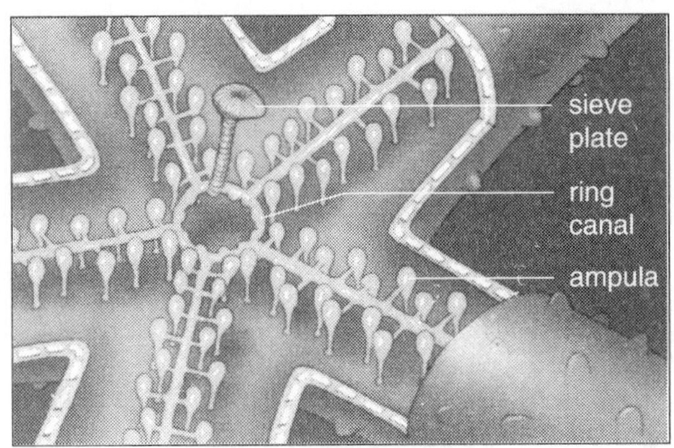

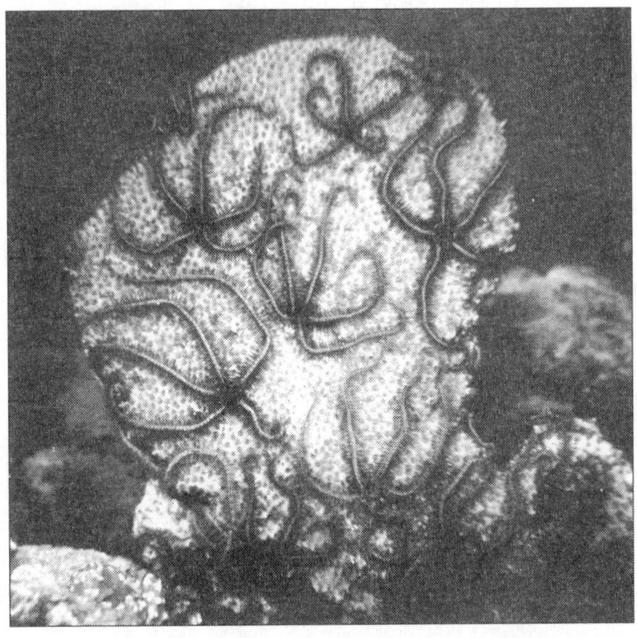

a. ____________ ____________ [p.440]

b. ____________ ____________ [p.440]

c. ____________ ____________ [p.440]

d. ____________ ____________ [p.440]

Self-Quiz

_____ 1. Which of the following is *not* true of sponges? They have no _______________. [pp.418–419]
a. distinct cell types
b. nerve cells
c. muscles
d. gut

_____ 2. Which of the following is *not* a protostome? [p.440]
a. earthworm
b. crayfish or lobster
c. sea star
d. squid

_____ 3. Deuterostomes undergo _______________ cleavage; protostomes undergo _________ cleavage. [p.426]
a. radial; spiral
b. radial; radial
c. spiral; radial
d. spiral; spiral

_____ 4. Bilateral symmetry is characteristic of _______________. [p.422]
a. cnidarians
b. sponges
c. jellyfish
d. flatworms

_____ 5. Flukes and tapeworms are parasitic _______________. [pp.422–423]
a. leeches
b. flatworms
c. jellyfish
d. roundworms

_____ 6. Insects include _______________. [pp.436–437]
a. spiders, mites, ticks
b. centipedes and millipedes
c. termites, aphids, and beetles
d. all of the above

_____ 7. The simplest animals to exhibit the organ/system level of construction are the _______________. [p.422]
a. sponges
b. arthropods
c. cnidarians
d. flatworms

_____ 8. Torsion is a process characteristic of _______________. [p.428]
a. chitons
b. bivalves
c. gastropods
d. cephalopods
e. echinoderms

_____ 9. The _______________ body plan is characterized by bilateral symmetry, a flattened body, cephalization, a digestive system with a pharynx for feeding, and hermaphroditism. [p.422]
a. annelid
b. roundworm
c. echinoderm
d. flatworm

_____ 10. The _______________ have bilateral symmetry, cylindrical bodies tapered on both ends, a tough protective cuticle, and a false coelom and present the simplest example of a complete digestive system. [p.424]
a. roundworms
b. cnidarians
c. flatworms
d. echinoderms

_____ 11. A _______________ with a peritoneum separates the gut and body wall of most bilateral animals. [p.426]
a. coelom
b. mesoderm
c. mantle
d. water-vascular system

_____ 12. A complete digestive tract with a mouth and an anus is not seen in _______________. [p.444]
a. annelids
b. flatworms
c. mollusks
d. roundworms

Matching

Match each of the following phyla with its corresponding characteristics (letters a–i) and representatives (A–M). A phylum may match with more than one letter from the group of representatives.

13. _____, _____ Annelida [pp.430–431]
14. _____, _____ Arthropoda [pp.432–439]
15. _____, _____ Cnidaria [pp.420–421]
16. _____, _____ Echinodermata [pp.440–441]
17. _____, _____ Mollusca [pp.426–429]
18. _____, _____ Nematoda [pp.423–425]
19. _____, _____ Platyhelminthes [pp.422–423]
20. _____, _____ Porifera [pp.418–419]
21. _____, _____ Rotifera [p.426]

a. choanocytes (= collar cells) + spicules
b. jointed legs + an exoskeleton
c. pseudocoelomate + wheel organ + soft body
d. soft body + mantle; may or may not have radula or shell
e. bilateral symmetry + blind-sac gut
f. radial symmetry + blind-sac gut; stinging cells
g. body compartmentalized into repetitive segments; coelom containing nephridia (= primitive kidneys)
h. tube feet + calcium carbonate structures in skin
i. complete gut + bilateral symmetry + cuticle; includes many parasitic species, some of which are harmful to humans

A. Small animals with a crown of cilia and two "exuding toes"
B. Corals, sea anemones, and *Hydra*
C. Tapeworms and planaria
D. Insects
E. Jellyfish and the Portuguese man-of-war
F. Sand dollars and starfishes
G. Earthworms and leeches
H. Lobsters, shrimp, and crayfish
I. Organisms with spicules and collar cells
J. Scorpions and millipedes
K. Octopuses and oysters
L. Flukes
M. Hookworm, trichina worm

Chapter Objectives/Review Questions

1. List the six general characteristics that define an "animal." [p.416]
2. __________ cells are the forerunners of the primary tissue layers: the ectoderm, endoderm, and, in most species, mesoderm. [p.416]
3. What is the primary characteristic that separates vertebrates from invertebrates? [p.416]
4. Distinguish radial symmetry from bilateral symmetry, and generally describe various animal gut types. [pp.416–417]
5. Give meanings for the following terms relating to aspects of an animal body: *anterior* and *posterior* ends, *dorsal* and *ventral* surfaces, and *cephalization*. [p.417]
6. The __________ is a tubular or saclike region in the body in which food is digested, then absorbed into the internal environment. [p.417]
7. List two benefits that the development of a coelom brings to an animal. [p.417]
8. Define *pseudocoel* and describe the types of animals to which this term is applied. [p.417]
9. What is meant by a *segmented animal?* [p.417]
10. *Trichoplax* is as simple as an animal can get, having only __________ distinct layers of cells. [p.418]
11. List the characteristics that distinguish sponges from other animal groups. [pp.418–419]
12. Describe the processes involved in sponge reproduction, both asexual and sexual. [p.419]
13. State what nematocysts are used for and explain how they function. [p.420]
14. Two cnidarian body types are the __________ and the __________ [p.420]

15. Describe the structure typical of a cnidarian, using terms such as *epithelium, nerve cells, contractile cells, nerve net, mesoglea,* and *hydrostatic skeleton.* [pp.420–421]
16. Describe the life cycle of *Obelia.* [p.421]
17. Define *organ-system* level of construction and relate it to flatworms. [p.422]
18. List the three main types of flatworms and briefly describe each; name the groups that are parasitic. [pp.422–423]
19. Describe the body plan of roundworms, comparing its various systems with those of the flatworm body plan. [p.423]
20. Southeast Asian blood flukes, tapeworms, and pinworms are examples of ________ parasites. [pp.424–425]
21. Describe the size, structure, and environment of the rotifers. [p.426]
22. Define, by their characteristics, *protostome* and *deuterostome* lineages, and cite examples of animal groups belonging to each lineage. [p.426]
23. List and generally describe the major groups of mollusks and their members. [p.427]
24. Define *mantle* and tell what role it plays in the molluscan body. [p.426]
25. Describe the process of torsion that occurs only in gastropods. [p.428]
26. Explain why cephalopods came to have such well-developed sensory and motor systems and are able to learn. [p.429]
27. Describe the advantages of segmentation, and tell how this relates to the development of specialized internal organs. [p.430]
28. Generally describe the external and internal structure of a typical annelid, the earthworm. [pp.430–431]
29. List four different lineages of arthropods; briefly describe each. [p.432]
30. List the six arthropod adaptations that led to their success. [p.432]
31. List the groups of organisms known as the familiar chelicerates. [p.433]
32. Describe the characteristics and significance of the arachnid lifestyle. [p.433]
33. Name the most obvious characteristic shared by most crustaceans. [p.434]
34. Name some common types of crustaceans. [pp.434–435]
35. Compare and contrast the structural characteristics of millipedes and centipedes. [p.435]
36. Name the characteristics that all insects share, even though they may appear very dissimilar. [p.436]
37. The development of many insects proceeds through very different postembryonic stages and then to an adult form by a process known as ________ . [p.436]
38. List members of the following groups that are unwelcome to humans: spiders, mites, scorpions, and beetles; tell why each is unwelcome. [pp.438–439]
39. The major invertebrate members of the ________ lineage are the echinoderms; list their major characteristics. [pp.440–441]
40. List five examples of the animals known as echinoderms. [pp.440–441]
41. Describe how locomotion and eating occur in sea stars. [p.441]

Integrating and Applying Key Concepts

Study Figure 25.41 on p.442 of the text to verify that most highly evolved invertebrate animals have bilateral symmetry, a complete gut, a true coelom, and segmented bodies. Why do you suppose that having a true coelom and a segmented body is considered more highly evolved than lacking a coelom or possessing a false coelom and having an unsegmented body? Cite evidence in the chapter that would support or not support the information found in Figure 25.41.

26

ANIMALS: THE VERTEBRATES

Interactive Exercises

So You Think the Platypus Is a Stretch [pp.444–445]

26.1. THE CHORDATE HERITAGE [pp.446–447]

26.2. CONNECTIONS: TRENDS IN VERTEBRATE EVOLUTION [p.448]

26.3. EXISTING JAWLESS FISHES [p.449]

26.4. EXISTING JAWED FISHES [pp.450–451]

Selected Words: duck-billed platypus [p.444], salps [p.444], tunicates [p.446], sea squirts [p.446], lancelets [p.446], dentin [p.446], jawless fish [p.448], scales [p.450], teleosts [p.450], coelacanths [p.451]

Boldfaced, Page-Referenced Terms

[p.446] chordates ______

[p.446] notochord ______

[p.446] nerve cord ______

[p.446] pharynx ______

[p.446] vertebrates ______

[p.446] urochordates ______

[p.446] filter feeders ______

[p.446] gill slits ______

[p.446] cephalochordates ______

[p.447] craniates ______

[p.447] ostracoderms ______

[p.447] jaws ______

[p.447] placoderms ______

[p.448] bone tissue ______

[p.448] vertebrae ______

[p.448] fins ______

[p.448] gills ______

[p.448] lungs ____

____

[p.450] swim bladder ____

____

[p.450] cartilaginous fishes ____

____

[p.450] ray-finned fishes ____

____

[p.451] lobe-finned fishes ____

____

[p.451] lungfishes ____

____

[p.451] tetrapods ____

____

Fill-in-the-Blanks

Each animal is an assortment of (1) ______________ [p.445]; many have been conserved from remote ancestors, while others are unique to the animal's branch on the animal family tree. Four major features distinguish the embryos of chordates from those of all other animals: a tubular dorsal (2) ______________ ______________ [p.446], a pharynx with (3) ______________ ______________ [p.446] in its wall, a(n) (4) ______________ [p.446], and a tail that extends past the anus during at least part of its life. In some chordates, the (5) ______________ [p.446] chordates, the notochord is *not* replaced by a vertebral column of separate, bony segments; in others, the (6) ______________ [p.446], it is. Invertebrate chordates living today are represented by tunicates and (7) ______________ [p.446], which obtain their food by (8) ______________ - ______________ [p.447]; they draw in plankton-laden water through the mouth and pass it over sheets of mucus, which trap the particulate food before the water exits through the (9) ______________ ______________ [p.447] in the pharynx. (10) ______________ [p.446] are among the most primitive of all living chordates; when they are tiny, they look and swim like (11) ______________ [p.446]. A rod of stiffened tissue, the (12) ______________ [p.446], cooperates with muscles to act like a torsion bar propelling the larva forward. Most (10) remain attached to rocks or hard substrates in marine habitats after their larvae undergo (13) ______________ [p.446].

The ancestors of the vertebrate line may have been mutated forms of their closest relatives, the (14) ______________ [p.447], in which the notochord was replaced by a bony (15) ______________ ______________ [p.448]. The vertebral column was the foundation for fast-moving (16) ______________ [p.448], some of which were ancestral to all other vertebrates. The evolution of (17) ______________ [pp.447–448] intensified the competition for prey as well as the competition to avoid being preyed on;

animals in which mutations expanded the nerve cord into a(n) (18) ________________ [p.448] that enabled the animal to compete effectively survived more frequently than their duller witted fellows and passed along their genes to the next generation. (19) ________________ [p.448] fins, equipped with skeletal supports, set the stage for the development of legs, arms, and wings in later groups.

As the ancestors of land vertebrates began spending less time immersed and more time exposed to air, use of gills declined and (20) ________________ [p.448] evolved; more elaborate and efficient (21) ________________ [p.448] systems evolved along with more complex and efficient lungs.

Labeling

22. Name the organism in diagram A. ________________ [p.447]

Name the numbered structures in the same diagram. [p.447]

23. ________________ with ________________ ________________

24. ________________

25. ________________

26. ________________ ________________ ________________ ________________

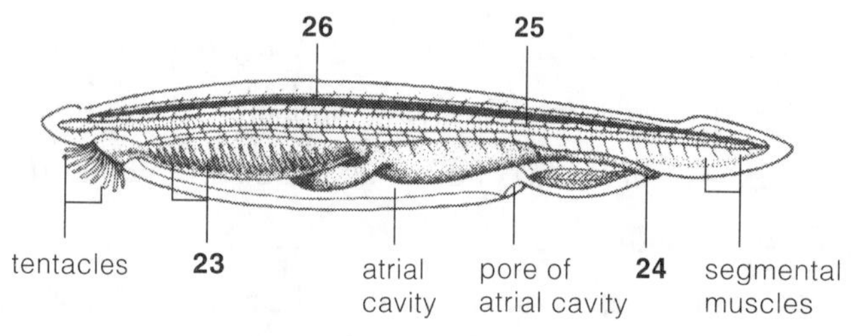

A

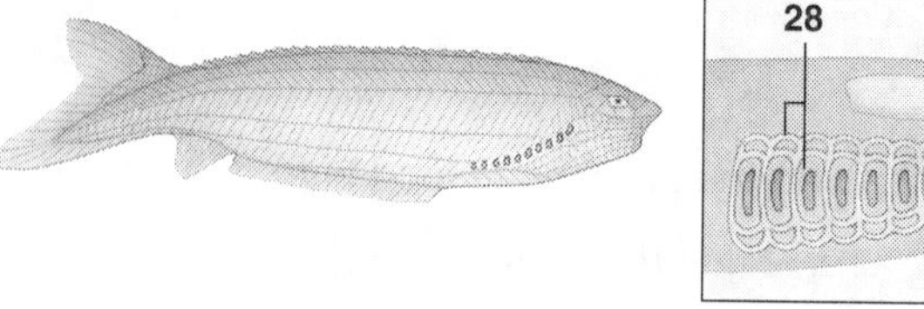

B

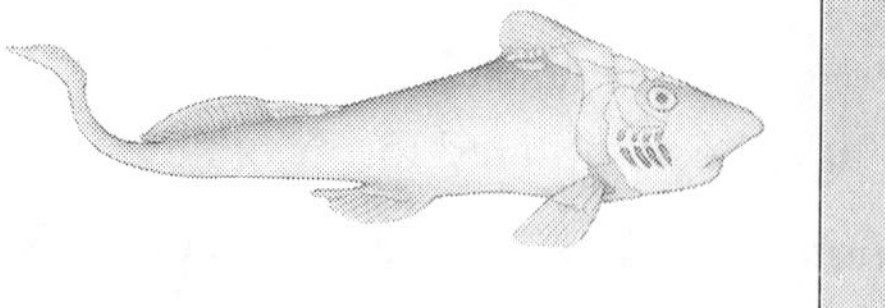

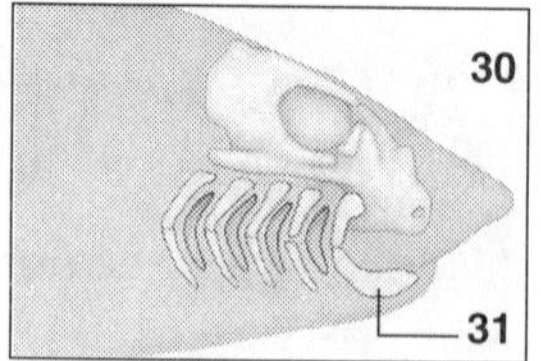

C

Short Answer

Answer the following questions about diagrams B and C in the blanks provided. [p.448]

27. Name the creature whose head is shown in illustration B. ____________________

28. Name this structure. ____________ ____________

29. Name this structure. ____________ ____________

30. Name the creature shown in illustration C. ____________________

31. What structures occupy the front of its head space? ____________________

Fill-in-the-Blanks

Existing jawless fishes include the scavenging (32) ____________ [p.449] and the often-parasitic (33) ____________ [p.449]. To protect themselves, (34) ____________ [p.449] can release up to a gallon of sticky (35) ____________ [p.449]. Aggressive (36) ____________ [p.449] have invaded the fresh water of the Great Lakes and have threatened many native fish species. Cartilaginous fishes include about 850 species of rays, skates, (37) ____________ [p.450], and chimaeras. They have conspicuous fins and five to seven (38) ____________ ____________ [p.450] on both sides of the pharynx. Ninety-six percent of the existing species of fishes are (39) ____________ [p.450]. Their ancestors arose during the Silurian period, perhaps as early as 450 million years ago, and soon gave rise to three lineages: the (40) ____________ - ____________ [p.450] fishes, the (41) ____________ - ____________ [p.450] fishes, and the (42) ____________ [p.450]. Bony fishes (Osteichthyes) have lunglike structures called (43) ____________ ____________ [p.450], which increase buoyancy in the water. (44) ____________ [p.451] are the only living representatives of lobe-finned fishes. These lobe fins have bones that are similar to the (45) ____________ [p.451] of tetrapods.

Matching

Match the numbered item with its letter. [all from p.451]

46. _____
47. _____
48. _____
49. _____
50. _____
51. _____
52. _____
53. _____
54. _____
55. _____
56. _____
57. _____

muscle segments
fin supports
46
olfactory bulb
47
liver
gall bladder
48
49
52
51
kidney
50

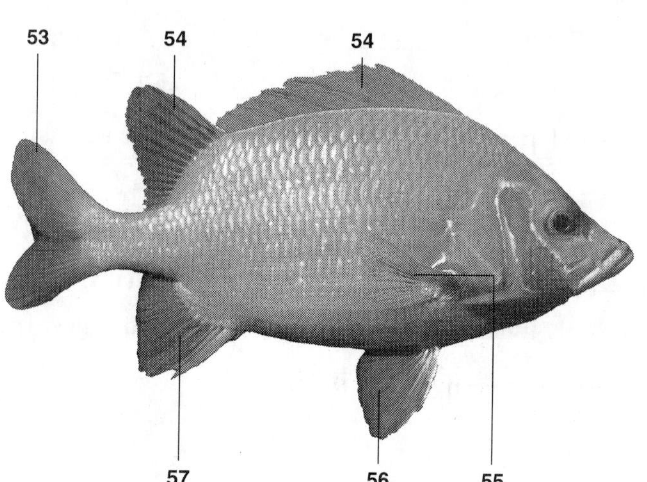

A. Anal fin
B. Anus
C. Brain
D. Caudal fin
E. Dorsal fins
F. Heart
G. Intestine
H. Pectoral fin (paired)
I. Pelvic fin (paired)
J. Stomach
K. Swim bladder
L. Urinary bladder

Analysis and Short Answer

Figure 26.10 in the text mentions features that distinguish the amphibian–lungfish lineage from the lineage that leads to sturgeons and other bony fishes. At least one of those features may have developed near (5) in the evolutionary tree on the next page. Exercises 58–68 refer to this evolutionary diagram.

58. What single feature do the lampreys, hagfishes, and extinct ostracoderms have in common that is different from the placoderms? [p.449] ____________________

59. How did ostracoderms feed? [p.449] ____________________

60. A mutation in ostracoderm stock led to the development of what feature in all organisms that descended from (1)? [p.449] ____________________

61. A genetic event at (2) led to the development of an endoskeleton made of what? [p.450] ____________________

____

62. Mutations at (3) led to an endoskeleton made of what? [p.450] ____________________

63. Mutations at (4) led to which spectacularly diverse fishes that have delicate fins originating from the dermis? [pp.450–451] ____________________

64. Mutations at (5) led to which fishes whose fins incorporate fleshy extensions from the body? [pp.451–452] ____________________

65. Which branch, (4) or (5), gave rise to the amphibians? [p.452] ____________________

66. Which branch gave rise to the modern bony fishes? [p.450] ____________________

67. In which period did three distinctly different lineages of *bony* fishes first appear in the fossil record? [p.450] ____________________

68. Approximately how many million years ago did the fork in the evolutionary path that led to the amphibians occur? [estimated from the following evolutionary diagram] ____________________

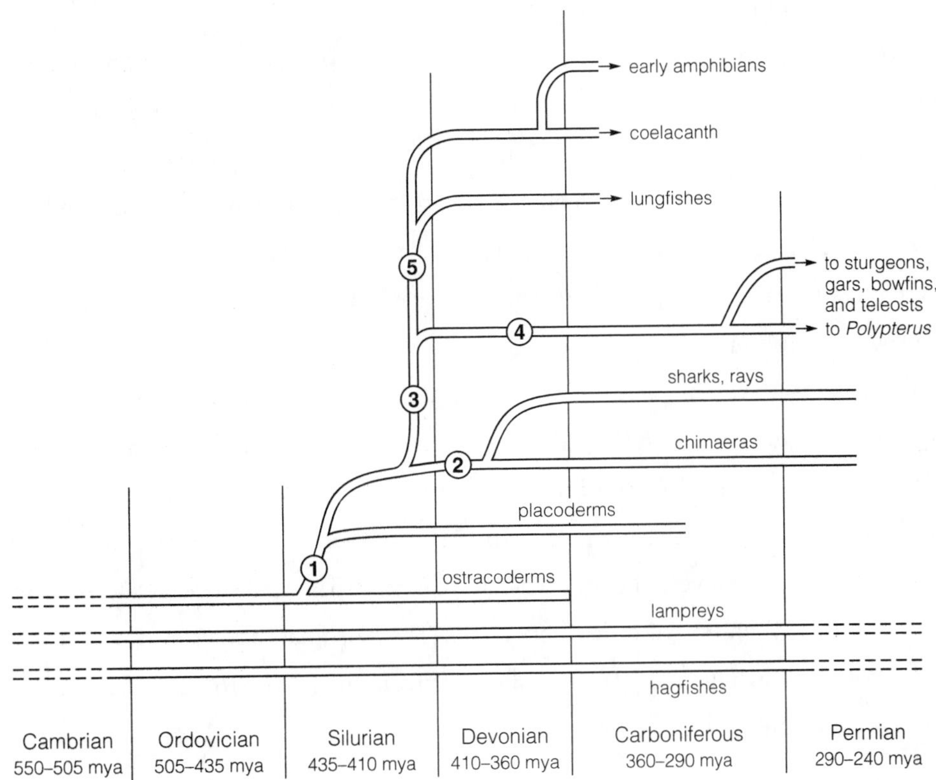

26.5. AMPHIBIANS [pp.452–453]

26.6. THE RISE OF AMNIOTES [pp.454–455]

26.7. A SAMPLING OF EXISTING REPTILES [pp.456–457]

Selected Words: *Acanthostega* [p.452], *Ichthyostega* [p.452], frog [p.452], toad [p.452], salamander [p.452], caecilian [p.452], *Ribelroia* [p.453], Carboniferous [p.454], *synapsids* [p.454], *sauropsids* [p.454], *Lystrosaurus* [p.454], *Maiasaura* [pp.454–455], Triassic [p.455], dinosaurs [p.455], turtles [p.456], lizards [p.456], snakes [p.456], tuataras [p.457], crocodilians [p.457]

Boldfaced, Page-Referenced Terms

[p.452] amphibian ____________________

____

[p.454] reptiles ____

____

[p.454] amniotes ____

____

Fill-in-the-Blanks

Natural selection acting on lobe-finned fishes during the Devonian period favored the evolution of ever more efficient (1) ______________ [p.452] used in gas exchange and stronger (2) ______________ [p.452] used in locomotion. Without the buoyancy of water, an animal traveling over land must support its own weight against the pull of gravity. The (3) ______________ [p.452] of early amphibians underwent dramatic modifications that involved evaluating incoming signals related to vision, hearing, and (4) ______________ [p.452]. (5) ______________ [p.452] systems became more efficient in order to rapidly move (6) ______________ [p.452] to cells. This and increased gas exchange allowed for greater production of (7) ______________ [p.452] to support more active lifestyles. The Devonian period also brought humid, forested swamps with an abundance of aquatic invertebrates and (8) ______________ [p.452]—ideal prey for amphibians.

There are three groups of existing amphibians: (9) ______________ [p.452], frogs and toads, and caecilians. Amphibians require free-standing (10) ______________ [p.452] or at least a moist habitat in which to (11) ______________ [p.452]. Amphibian skin must also be kept moist, since it acts as an additional (12) ______________ ______________ [p.452].

(13) ______________ [p.454] were the first true land animals. They produced water-conserving (14) ______________ [p.454] with hard or leathery shells. They also had dry, (15) ______________ ______________ [p.454], which helped prevent water loss from the body, (16) ______________ [p.454] fertilization, and very efficient (17) ______________ [p.454].

(For questions 18–26, consult Figure 26.14 in the main text and the figure for the Analysis and Short Answer exercise on page 323 of this Study Guide.)

Although amphibians originated during Devonian times, ancestral "stem" reptiles appeared during the (18) ______________ [p.455] period, about 340 million years ago. Reptilian groups living today that have existed on Earth longest are the (19) ______________ [p.455]; their ancestral path diverged from that of the so-called stem reptiles during the (20) ______________ [p.455] period. Crocodilian ancestors appeared in the late (21) ______________ [p.455] period, about 220 million years ago. Snake ancestry diverged from lizard stocks during the late (22) ______________ [p.455] period, about 140 million years ago. Modern (23) ______________ [pp.455,457] are more closely related to extinct dinosaurs and crocodiles than to any other existing vertebrates; they, too, can adjust (24) ______________ [p.457]. Mammals have descended from therapsids, which in turn are descended from the (25) ______________ [p.455] group of reptiles, which diverged earlier from the stem reptile group during the (26) ______________ [p.455] period, approximately 320 million years ago.

Sequence

In blanks 27–33, arrange the following groups in sequence, from earliest to latest, according to their appearance in the fossil record: [all from p.455]

A. Birds B. Crocodilians C. Dinosaurs D. Early ancestors of mammals (= therapsid reptiles) E. Early ancestors of turtles (= anapsid reptiles) F. Snakes G. Stem reptiles

27. ____ 28. ____ 29. ____ 30. ____ 31. ____ 32. ____ 33. ____

Analysis and Short Answer

34. This chapter mentions features that distinguish, say, the mammals from all the other groups on the right-hand side of the evolutionary diagram below. It follows that, at some time during the evolution of mammals, mutations that produced those features must have appeared. Consult the diagram below and imagine what sort of mutation(s) occurred at the numbered places. [diagram adapted from text Figure 26.14]

a. What may have occurred at (1)? ____________________

b. What may have occurred at (2)? ____________________

c. What may have occurred at (3)? ____________________

d. What may have occurred at (4)? ____________________

e. What may have occurred at (5)? ____________________

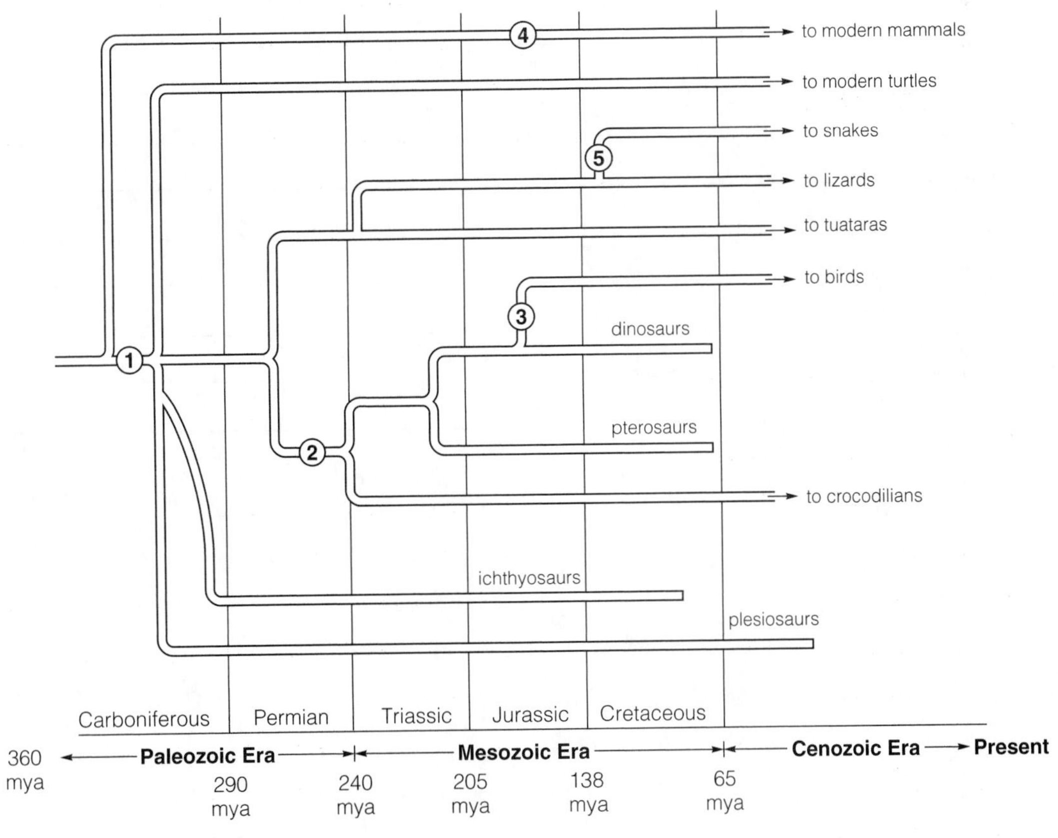

26.8. BIRDS [pp.458–459]

26.9. THE RISE OF MAMMALS [pp.460–461]

26.10. A PORTFOLIO OF EXISTING MAMMALS [pp.462–463]

Selected Words: feathers [p.458], *Archaeopteryx* [p.458], sternum [p.459], incisors, canines, premolars, and molars [p.460], therapsids [p.461], therians [p.461], egg-laying mammals [p.461], pouched mammals [p.461], placental mammals [p.461], spiny anteater [p.462], duck-billed platypus [p.462], koala [p.462], Tasmanian devil [p.462], uterus [p.463]

Boldfaced, Page-Referenced Terms

[p.458] birds ______________________________

____

[p.459] migration ______________________________

____

[p.460] mammals ______________________________

____

[p.460] dentition ______________________________

____

[p.461] monotremes ______________________________

____

[p.461] marsupials ______________________________

____

[p.461] eutherians ______________________________

____

[p.462] convergent evolution ______________________________

____

[p.463] placenta ______________________________

____

Labeling

Label the structures pictured at the right.

1. ______________ [p.458]
2. ______________ [p.458]
3. ______________ ______________ [p.458]

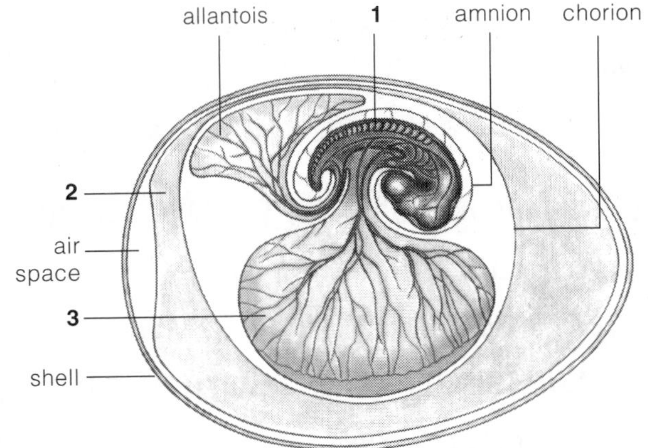

Fill-in-the-Blanks

Birds descended from (4) ______________ [p.458] that ran around on two legs some 160 million years ago. Almost 9,000 species of birds show an amazing amount of variation in body structure. The smallest known adult bird weighs only (5) ______________ [p.458] grams. The largest existing bird is the (6) ______________ [p.459], which weighs about 150 kilograms; it can sprint fast but cannot fly. All birds have (7) ______________ [p.458] that insulate and help get the bird aloft. Generally, birds have a greatly enlarged (8) ______________ [p.459] to which flight muscles are attached. Bird bones contain (9) ______________ ______________ [p.459], which decrease weight, and birds have air sacs attached to the lungs, which increase gas exchange. Birds also lay (10) ______________ [p.458] eggs and generally show (11) ______________ [p.458] behavior. Like their closest relatives, birds have (12) ______________ [p.458] on their legs.

Flight demands high (13) ______________ [p.458] rates, which require an abundant supply of (14) ______________ [p.458] pumped to all body parts by means of a large, durable, (15) ______________-chambered [p.458] heart.

Most mammals have (16) ______________ [p.460] as a means of insulation, and mammalian mothers suckle their young with milk produced by (17) ______________ ______________ [p.460]. Mammals differ from reptiles and birds in (18) ______________ [p.460] (the type, number, and size of teeth). Most mammalian females produce one (19) ______________ [p.462] per developing offspring inside the uterus.

The ancestors of today's mammals diverged from small, hairless reptiles called (20) ______________ [p.461] more than 200 million years ago, during the Triassic. Through mutation and natural selection during Jurassic times, they developed hair and major changes in body form, jaws, and teeth; mammals called (21) ______________ [p.461] had evolved. They coexisted with the many groups of dominant (22) ______________ [p.461] through the Cretaceous. When the (22)s became extinct, diverse adaptive zones awaited exploitation by the three principal lineages: those that lay eggs (examples are the

(23) ______________ [p.462] and the spiny anteater); those that are (24) ______________ [p.461] (examples are the opossum and the kangaroo); and those that are (25) ______________ [p.461] mammals, which, during the subsequent 65 million years, have blossomed into 4,500 known mammalian species. (26) ______________ ______________ [p.462] has occurred as evolutionarily distinct, geographically isolated lineages evolved in similar ways in similar habitats.

26.11. CONNECTIONS: TRENDS IN PRIMATE EVOLUTION [pp.464–465]

26.12. FROM EARLY PRIMATES TO HOMINIDS [pp.466–467]

26.13. EMERGENCE OF EARLY HUMANS [pp.468–469]

26.14. EMERGENCE OF MODERN HUMANS [p.470]

26.15. FOCUS ON SCIENCE: *Out of Africa — Once, Twice, or . . .* [p.471]

Selected Words: prosimians [p.464], tarsioids [p.464], anthropoids [p.464], arboreal [p.464], savannas [p.464], prehensile [p.464], opposable [p.464], Paleocene [p.466], Eocene [p.466], *Plesiadapis* [p.466], *Aegyptopithecus* [p.466], dryopiths [p.466], Miocene [p.466], *Sahelanthropus tchadensis* [p.466], "Lucy" [p.467], *Australopithecus* [p.467], gracile [p.467], robust [p.467], *Homo habilis* [p.468], Olduvai Gorge [p.469], *Homo erectus* [p.470], *Homo sapiens* [p.470], cultural [p.470], "races" [p.471]

Boldfaced, Page-Referenced Terms

[p.464] primates ____

____

[p.464] hominoids ____

____

[p.464] hominids ____

____

[p.464] bipedalism ____

____

[p.464] culture ____

____

[p.467] australopiths ____

____

[p.468] humans ____

____

[p.471] multiregional model ____

____

[p.471] African emergence model ____

____

Fill-in-the-Blanks

Within the mammals, a group arose, called (1) ______________ [p.464], that included prosimians, tarsioids, and (2) ______________ [p.464], the latter group including (3) ______________ [p.464], apes, and humans. The evolution from early prosimians to hominids required several adaptations. Excellent (4) ______________ [p.464] vision with (5) ______________ - ______________ [p.464] eyes allowed them to interpret and respond quickly and to rely less on the sense of (6) ______________ [p.464]. The (7) ______________ [p.464] and (8) ______________ [p.464] movements of the hands allowed hominids to (9) ______________ [p.464], which eventually led to the ability to make (10) ______________ [p.464]. The pelvic girdle and vertebral column were adapted for (11) ______________ ______________ [p.464], while (12) ______________ [p.464] and jaws were modified for a mixed diet. As the (13) ______________ [p.464] expanded, more complex (14) ______________ [p.464] occurred, which eventually led to the development of (15) ______________ [p.465] and (16) ______________ [p.465].

In central (17) ______________ [p.466], 6 or 7 million years ago, hominids were becoming distinct from the apes. During the Miocene and Pliocene, many different "southern apes," or (18) ______________ [p.467], evolved. Between 2.4 and 1.6 million years ago, the earliest forms of (19) ______________ [p.468] were living in Africa. About (20) ______________ [p.470] million years ago, *Homo erectus* developed. This group migrated out of Africa into both (21) ______________ [p.470] and (22) ______________ [p.470]. *H. erectus* had a larger (23) ______________ [p.470] and was a more creative (24) ______________ [p.470] than its ancestors. By (25) ______________ [p.470] years ago, modern humans, (26) ______________ ______________ [p.470], appeared. As they spread, earlier lineages, like the cold-adapted (27) ______________ [p.470] of Europe, disappeared.

One debate in human evolution concerns the way in which *H. sapiens* arose. Human fossils older than (28) ______________ [p.471] years have not been found on any continent except Africa. According to the (29) ______________ ______________ [p.471] model, modern humans arose in sub-Saharan Africa and migrated out, replacing archaic (30) ______________ ______________ [p.471] populations from prior migrations. This hypothesis is supported by the discovery of the oldest (31) ______________ ______________ [p.471] fossils in Africa. According to the (32) ______________ [p.471] model, the *H. erectus* population that migrated earlier evolved separately into (33) ______________ ______________ [p.471].

Identify each numbered part of the illustration below [p.469].

34. ________________

35. ________________

36. ________________

37. ________________

38. ________________

39. ________________ ________________

40. ________________ ________________

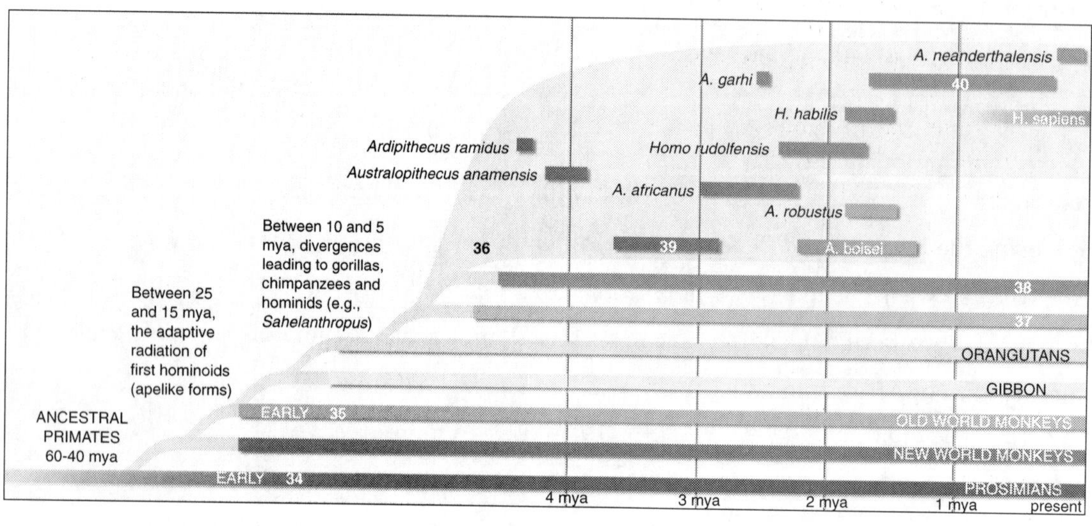

Self-Quiz

_____ 1. Filter-feeding chordates rely on ________________, which have cilia that create water currents and mucous sheets that capture nutrients suspended in the water. [p.447]
a. notochords
b. differentially permeable membranes
c. filiform tongues
d. gill slits

_____ 2. In true fishes, the gills serve primarily ________________ function. [p.448]
a. a gas-exchange
b. a feeding
c. a water-elimination
d. both a feeding and a gas-exchange

_____ 3. The feeding and defense behavior of true fishes selected for highly developed ________________. [p.448]
a. parapodia
b. notochords
c. sense organs
d. gill slits

_____ 4. Which of the following is an important adaptation to life on land? [p.454]
a. amniote egg
b. lungs
c. internal fertilization
d. all of the above

_____ 5. Which of the following is *not* considered to have been a key characteristic in early primate evolution? [pp.464–465]
a. eyes adapted for discerning color and shape in a three-dimensional field
b. body and limbs adapted for tree climbing
c. bipedalism and increased cranial capacity
d. eyes adapted for discerning movement in a three-dimensional field

_____ 6. Primitive primates generally live _____________. [p.464]
a. in tropical and subtropical forest canopies
b. in temperate savanna and grassland habitats
c. near rivers, lakes, and streams in the East African Rift Valley
d. in caves where there are abundant supplies of insects

_____ 7. The earliest fossils of *Homo habilis*, *Homo erectus*, and *Homo sapiens* date from approximately _____________, _____________, and _____________ years ago, respectively. [pp.468–469]
a. 5, 4, and 2 million
b. 2.5, 2, and 0.1 million
c. 25 million, 2 million, and 100,000
d. 5 million, 4 million, and 20,000

_____ 8. The hominid evolutionary line stems from a divergence (fork in a phylogenetic tree) from the ape line that apparently occurred _____________. [p.469]
a. somewhere between 6 million and 4 million years ago
b. about 3 million years ago
c. during the Pliocene epoch
d. less than 2 million years ago

_____ 9. _____________ was an Oligocene anthropoid that probably predated the divergence leading to Old World monkeys and the apes, with dentition more like that of dryopiths and less like that of the Paleocene primates with rodentlike teeth. [p.466]
a. *Aegyptopithecus*
b. *Australopithecus*
c. *Homo erectus*
d. *Plesiadapis*

_____ 10. Donald Johanson, from the University of California at Berkeley, discovered Lucy (named for the Beatles tune), who was a(n) _____________. [p.467]
a. dryopith
b. australopith
c. member of *Homo*
d. prosimian

_____ 11. A hominid of Europe and Asia that became extinct nearly 30,000 years ago was _____________. [p.470]
a. a dryopith
b. *Australopithecus*
c. *Homo erectus*
d. Neandertal

Matching

Choose the most appropriate answer for each.

12. ____ anthropoids [p.464]
13. ____ australopiths [p.467]
14. ____ Cenozoic [recall geologic time scale]
15. ____ hominids [p.464]
16. ____ hominoids [p.464]
17. ____ Miocene [p.466]
18. ____ primates [p.464]
19. ____ prosimians [p.464]

A. A group that includes apes and humans
B. Organisms in a suborder that includes New World and Old World monkeys, apes, and humans
C. An era that began 65 to 63 million years ago; characterized by the evolution of birds, mammals, and flowering plants
D. A group that includes humans and their most recent ancestors
E. An epoch of the Cenozoic era lasting from 25 million to 5 million years ago; characterized by the appearance of primitive apes, whales, and grazing animals of the grasslands
F. Organisms in a suborder that includes tree shrews, lemurs, and others
G. A group that includes prosimians, tarsioids, and anthropoids
H. Bipedal organisms living from about 4 million to 1 million years ago, with essentially human bodies and ape-shaped heads; brains no larger than those of chimpanzees

Matching

Match the following groups and classes with the corresponding characteristics (a–i) and representatives (A–I).

20. ____, ____ Amphibians [p.452]
21. ____, ____ Birds [p.458]
22. ____, ____ Bony fishes [pp.450–451]
23. ____, ____ Cartilaginous fishes [p.450]
24. ____, ____ Cephalochordates [pp.446–447]
25. ____, ____ Jawless fishes [p.449]
26. ____, ____ Mammals [p.460]
27. ____, ____ Reptiles [p.454]
28. ____, ____ Urochordates [p.446]

a. hair + vertebrae
b. feathers + hollow bones
c. jawless + cartilaginous skeleton (in existing species)
d. two pairs of limbs (usually) + glandular skin + "jelly"-covered eggs
e. amniote eggs + scaly skin + bony skeleton
f. invertebrate + sessile adult that cannot swim
g. jaws + cartilaginous skeleton + vertebrae
h. in adult, notochord stretches from head to tail; mostly burrowed-in; adult can swim
i. bony skeleton + skin covered with scales and mucus

A. lancelet
B. loons, penguins, and eagles
C. tunicates, sea squirts
D. sharks and manta rays
E. lampreys and hagfishes (and ostracoderms)
F. true eels and sea horses
G. lizards and turtles
H. caecilians and salamanders
I. platypuses and opossums

Identification

Provide the common name and the major chordate group for each creature pictured as follows.

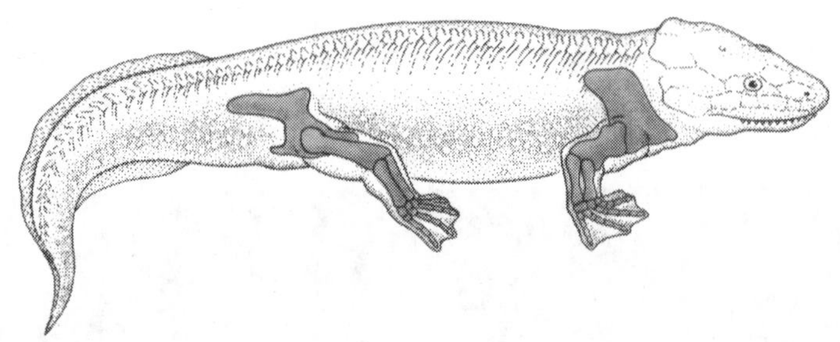

29. [p.452] ______________ ______________, ______________

30. [p.463] ______________ ______________, ______________

31. [p.451] ______________ ______________, ______________ ______________

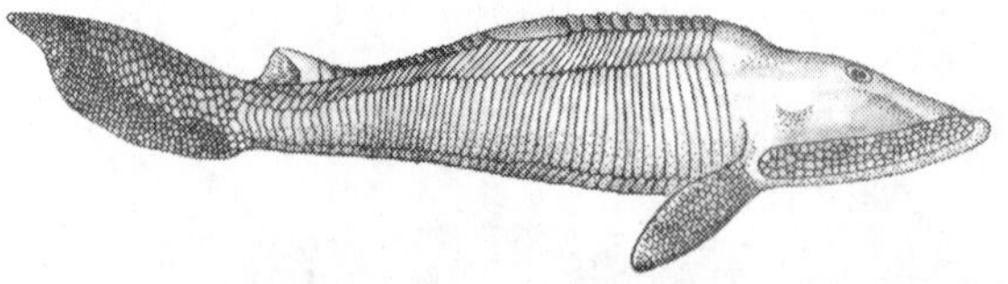

32. [p.447] ________________, ________________ ________________

33. [p.459] ________________, ________________

34. [p.456] ________________, ________________

35. [p.450] ________________, ________________ ________________

36. [p.451] ________________, ________________ ________________

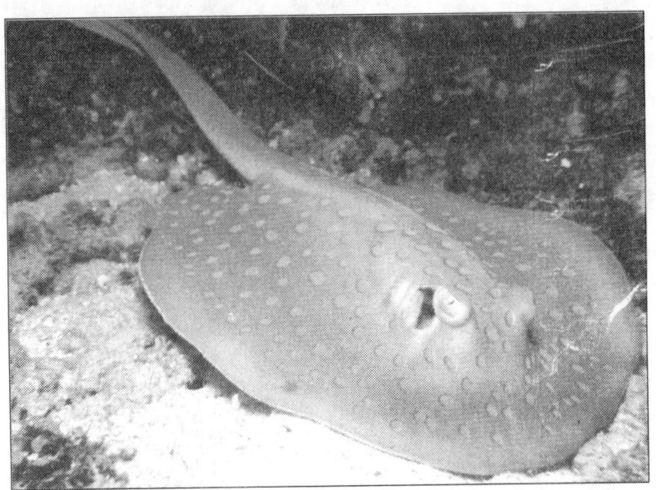

37. [p.450] ________________ ________________, ________________

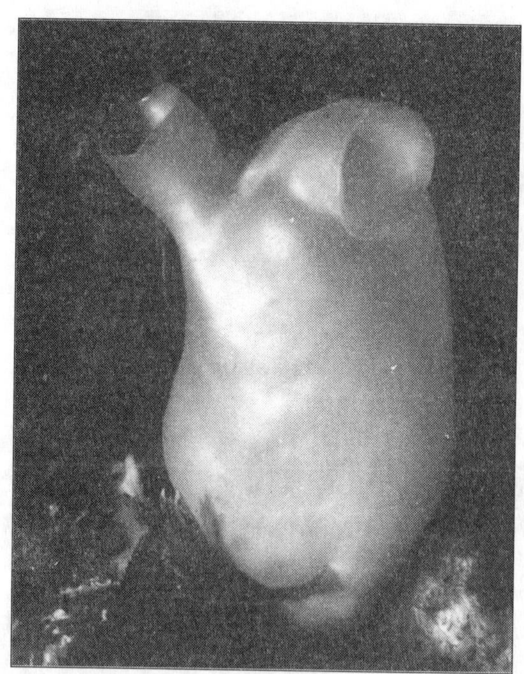

38. [p.446] ________________, ________________

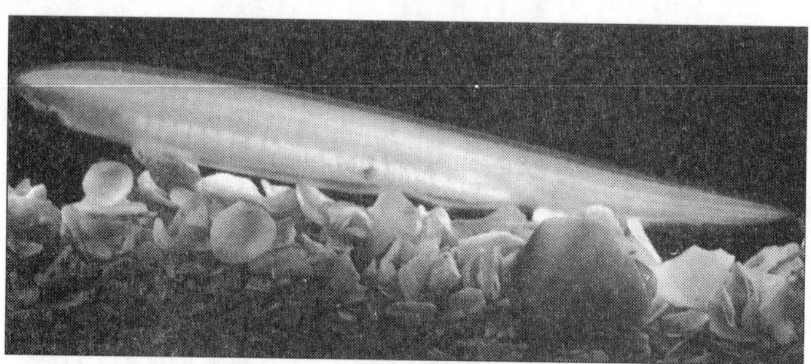

39. [p.447] ________________, ________________

Chapter Objectives/Review Questions

1. Name each class of vertebrates and give the distinguishing features of each. [pp.446–460]
2. List four characteristics found only in chordates. [p.446]
3. Describe the adaptations that sustain the sessile or sedentary lifestyle seen in primitive chordates such as tunicates and lancelets. [pp.446–447]
4. State what sort of changes occurred in the primitive chordate body plan that could have promoted the emergence of vertebrates. [pp.448–449]
5. Describe the differences between primitive and advanced fishes in terms of skeleton, jaws, special senses, and brain. [pp.447–451]
6. Describe the changes that enabled aquatic fishes to give rise to land dwellers. [pp.451–452]
7. Discuss the effects that increased parental nurture of offspring in birds and mammals has had on courtship behavior and reproductive physiology. [pp.458–462]
8. Where do tarsier survivors dwell on Earth today? [p.464]
9. Beginning with those most closely related to humans, list the main groups of primates in order by decreasing closeness of relationship to humans. [p.469]
10. Five key characters of primate evolution are ________, ________, ________, ________, and ________. [pp.464–465]
11. Describe the general physical features and behavioral patterns attributed to early primates. [p.466]
12. State which anatomical features underwent the greatest changes along the evolutionary line from early anthropoids to humans. [pp.464–467]
13. Explain how you think *Homo sapiens* arose. Make sure your theory incorporates existing paleontological (fossil), biochemical, and morphological data. [pp.468–471]

Integrating and Applying Key Concepts

Both birds and mammals have four-chambered hearts and high metabolic rates, and both regulate their body temperatures efficiently. Since both groups evolved from reptiles, one might think that these same traits would have developed in ancestral reptiles. Data suggest that most reptiles have a heart that is intermediate between three and four chambers, lower metabolic rates, and body temperatures that are not well regulated but tend to rise and fall in accord with the environmental temperature. If the three traits mentioned in the first sentence had developed in reptilian groups, how might their lives have been different?

Suppose someone told you that sometime between 12 to 6 million years ago, dryopiths were forced by larger predatory members of the cat family to flee the forests and move to estuarine, riverine, and sea coastal habitats, where they could take refuge in the nearby water to evade the tigers. Those that, through mutations, became naked, developed an upright stance, developed subcutaneous fat deposits as insulation, and developed a bridged nose that had advantages in watery habitats (features which dryopiths that remained inland never developed) survived and expanded their populations. As time went on, predation by the big cats and competition with other animals for available food caused most of the terrestrial dryopiths to become extinct; but the water-habitat varieties survived as scattered remnant populations, adapting to easily available shellfish and fish, wild rice and oats, and various tubers, nuts, and fruits. It was in these aquatic habitats that the first food-getting tools (baskets, nets, and pebble tools) were developed as well as the first words that signified different kinds of food. How does such a story fit with current speculations about and evidence for human origins? How could such a story be shown to be true or false?

Crossword Puzzle—Primate Evolution

1 2 3 4 5 6 7 8 9 10 11 12 13 14 15 16 17 18 19

ACROSS

1. A group that includes prosimians, tarsioids, and anthropoids [p.464]
3. A group that includes apes and humans [p.464]
7. Small, daisylike flowers
10. A kind of tall, showy flower; often lavender, purple, or yellow
11. A cheek tooth that crushes and grinds [p.460]
12. All the behavior patterns of a social group passed by means of learning and language from generation to generation [p.465]
13. Pay great homage to; worship
14. Forest apes that lived 13 million years ago in Africa, Europe, and southern Asia [p.466]
17. Pointed teeth that enable the tearing of flesh [p.460]
18. *Homo* ________ was the first out-of-Africa hominid to control and use fire for heating and cooking [p.470]
19. A group of primates intermediate between lemurs and monkeys [p.464]

DOWN

2. Southern ape–humans, the fossils of which have dates from 3.7 to 1.25 million years ago [p.467]
4. A vertebrate with hair [p.460]
5. A flat chisel or conelike tooth that nips or cuts food [p.460]
6. A group that includes modern humans and their direct-line ancestors since divergence from the ape line [p.460]
8. The specific (species) epithet of modern humans [p.470]
9. Habitual two-legged method of locomotion [p.464]
15. Early hominids had shorter, flatter ________ than chimpanzees [p.466]
16. Hard structures that can provide clues about what an animal typically eats [p.460]

27

BIODIVERSITY IN PERSPECTIVE

Interactive Exercises

The Human Touch [pp.474–475]

27.1. ON MASS EXTINCTIONS AND SLOW RECOVERIES [pp.476–477]

27.2. THE NEWLY ENDANGERED SPECIES [pp.478–479]

27.3. FOCUS ON THE ENVIRONMENT: *Case Study: The Once and Future Reefs* [pp.480–481]

27.4. FOCUS ON SCIENCE: *Rachel's Warning* [p.482]

Selected Words: *endemic* [p.478], *exotic* species [p.479], *fringing* reefs [p.480], *barrier* reefs [p.480], *atolls* [p.480]

Boldfaced, Page-Referenced Terms

[p.478] endangered species ____

____

[p.478] habitat loss ____

____

[p.479] habitat fragmentation ____

____

[p.479] habitat islands ____

____

[p.479] indicator species ____

____

[p.480] coral reefs ____

____

Complete the Table

1. Refer to the section in the text called *The Human Touch.* For each of the following items initially found on Easter Island, indicate the primary reason for its decline. [pp.474–475]

Labeling

Resource	*Reason for Decline*
a. fertile soil	
b. forests	
c. native animals	
d. human population	

For questions 2–5, give a possible cause for each of the mass extinction events indicated in the figure below. [p.476]

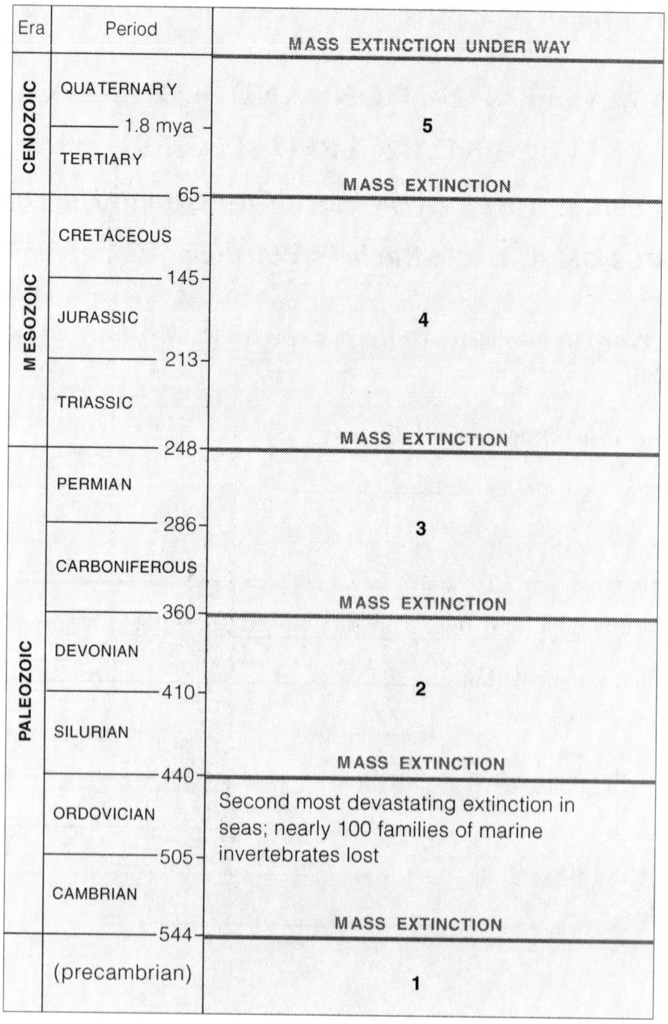

Era	Period		
			MASS EXTINCTION UNDER WAY
CENOZOIC	QUATERNARY		
		1.8 mya	5
	TERTIARY		MASS EXTINCTION
		65	
MESOZOIC	CRETACEOUS		
		145	
	JURASSIC		4
		213	
	TRIASSIC		MASS EXTINCTION
		248	
PALEOZOIC	PERMIAN		
		286	3
	CARBONIFEROUS		MASS EXTINCTION
		360	
	DEVONIAN		
		410	2
	SILURIAN		MASS EXTINCTION
		440	
	ORDOVICIAN		Second most devastating extinction in seas; nearly 100 families of marine invertebrates lost
		505	
	CAMBRIAN		MASS EXTINCTION
		544	
	(precambrian)		1

2. ______________________________

3. ______________________________

4. ______________________________

5. ______________________________

Matching

Choose the most appropriate answer for each question.

6. _____ habitat loss [p.478]
7. _____ endemic species [p.478]
8. _____ exotic species [p.479]
9. _____ habitat fragmentation [p.479]
10. _____ indicator species [p.479]
11. _____ endangered species [p.478]
12. _____ habitat islands [p.479]

A. A species that originates in one geographic area and is found nowhere else
B. The breaking up of a species habitat into smaller pieces
C. Introduced species that are not endemic to an area
D. Any species that is extremely vulnerable to extinction
E. The physical reduction in places for a species to live
F. A species that can be used to warn of changes in habitat
G. Small areas of suitable habitat located within vast areas of unsuitable habitat

Fill-in-the-Blanks

(13) ______________ ______________ [p.480] are wave-resistant formations consisting of the accumulated remains of countless (14) ______________ [p.480] organisms. The photosynthetic organisms called (15) ______________ [p.480] often live as symbionts in the tissues of the reef-building corals. The coral provides (16) ______________ [p.480] for the dinoflagellates, which in turn provide (17) ______________ [p.480] and recycle wastes for the coral. However, when (18) ______________ [p.480], the coral expels its symbionts. Abnormal, widespread (19) ______________ [p.480] of corals began in the 1980s, when sea surface temperatures increased. Human activities, such as (20) ______________ ______________ ______________, (21) ______________ ______________, (22) ______________, and (23) ______________ [all p.480], also destroy coral reefs.

In 1962, (24) ______________ ______________ [p.481] published her book *Silent Spring*, which documented the harmful effects of pesticides on wildlife. She suggested that indiscriminate use of (25) ______________ [p.481] could endanger human populations as well as kill robins, catbirds, doves, jays, and wrens. She died from cancer without knowing that she had been instrumental in starting the (26) ______________ [p.481] movement.

27.5. CONSERVATION BIOLOGY [pp.482–483]

27.6. RECONCILING BIODIVERSITY WITH HUMAN DEMANDS [pp.484–485]

Selected Word: *biodiversity* [p.482]

Boldfaced, Page-Referenced Terms

[p.482] conservation biology ____

____

[p.482] hot spots ____

____

[p.483] ecoregion ____

____

[p.484] strip logging ____

____

[p.485] riparian zone ____

____

Fill-in-the-Blanks

In the next 50 years, Earth's human population size may reach (1) ______________ billion [p.482], with the majority of growth occurring in the (2) ______________ [p.482] countries, where scientific knowledge of their biological treasure trove is small and (3) ______________ [p.482] is greatest. Awareness of the impending extinction crisis gave rise to the field of pure and applied research known as (4) ______________ ______________ [p.482]. Scientists in this field perform a systematic (5) ______________ [p.482] and description of the full range of biological diversity; they try to understand its evolutionary and ecological origins, and to identify (6) ______________ [p.482] that might maintain populations at sufficient levels to keep Earth's ecosystems functioning well, with enough abundance to sustain the burgeoning human populations. (7) ______________ ______________ [p.482] are habitats with the greatest number of endemic species found nowhere else that are in the greatest danger of extinction due to human activities. These are part of a greater (8) ______________ [p.483], a broad region defined by climate, geography, and producer species. There are currently (9) ______________ [p.483] regions defined as crucial to biodiversity by the World Wildlife Fund. Many medicines and other chemical products are derived from plants. Quite possibly, large (10) ______________ [p.483] companies would be willing to pay royalties to countries which have large numbers of different plant species that might serve as sources of medicines or herbal tonics.

The richest sources of biodiversity on land are the (11) ______________ ______________ ______________ [p.484]. In an attempt to provide wood from these forests in a profitable, (12) ______________ [p.484] way, Gary Hartshorn has proposed the method of (13) ______________

_______________ [p.484] in portions of forests that are sloped and have a number of streams. The idea is to clear a narrow (14) _______________ [p.484] that follows the land's (15) _______________ [p.484], using the upper part to build a (16) _______________ [p.484] to haul away the logs taken from the lower part of the corridor. After a few years, (17) _______________ [p.484] seeded from the native trees start to grow in the original corridor, and another corridor is cleared above the road. Precious (18) _______________ [p.484] leached from the exposed soil trickle down into the first corridor and are taken up by the young saplings, which benefit by more rapid (19) _______________ [p.484] and begin a profitable logging cycle that the region can sustain over time.

Plants associated with (20) _______________ [p.485] zones along streams or rivers afford a line of defense against damage by (21) _______________ [p.485] because they "drink up" water from spring (22) _______________ [p.485] and summer storms. In the western part of the United States, these zones shelter 67 to 75 percent of the (23) _______________ [p.485] species that must spend at least part of their life cycles there. Unfortunately, compared with wild (24) _______________ [p.485], cattle drink a lot more water, so they tend to congregate in these zones, trampling and feeding on the grasses and tender shrubs until they are gone and leaving bare banks that erode easily. Developing feeding and (25) _______________ [p.485] sites away from these zones and providing supplemental feed at different grazing areas can restore riparian zones and sustain endemic wildlife diversity by providing natural sources of food, shelter, and shade.

Self-Quiz

_____ 1. All the following statements are apparently true of Easter Island *except:* [pp.474–475]

a. It was settled by voyagers from Hawaii around A.D. 350, possibly after being blown off course by a series of storms.
b. Dense palm forests, hau hau trees, toromino shrubs, and grasses were abundant on the island.
c. The settlers planted crops of taro, bananas, sugarcane, and sweet potatoes.
d. Edible seafood species vanished from the protected waters around the island, and so fishermen had to build larger boats to sail far out on the open sea.
e. The islanders ate rats and each other; the population vanished sometime after 1724, when Captain James Cook visited Easter Island.

_____ 2. The most important lesson for humans worldwide that can be understood from the story of Easter Island is _______________. [p.475]

a. that you shouldn't cut down the last tree on the island, because then you won't be able to make a boat with which to escape to another place
b. that preserving biodiversity generally sustains societies and allows them to enjoy abundant resources as long as they keep their human population in check
c. that arable land grows richer with nutrients as time passes if intelligent planting methods are used
d. that by 1400, at least 10,000 humans were living on an island that consisted of approximately 165 square kilometers (64 square miles) of surface land
e. that it resulted in humans cannibalizing other humans

_____ 3. Human population worldwide is expected to reach _____________ billion by 2050, and human demands for food, materials, and living space are threatening biodiversity around the world. [p.475]
a. 3
b. 6
c. 9
d. 12
e. 15

_____ 4. Conservation biology includes all the following *except* _____________. [p.482]
a. identifying methods to maintain and use biodiversity for the benefit of humans
b. surveying the full range of biodiversity in the world ecoregions
c. encouraging the use of genetically altered crops planted as monocultures to reduce competition with native varieties of plants
d. analyzing the evolutionary and ecological origins of diversity

For questions 5–8, choose from these possible answers:

a. Strip logging b. Habitat fragmentation c. Indicator species
d. Endemic species e. Relocating watering and feeding stations

_____ 5. can play a major role in the restoration of riparian zones. [p.485]

_____ 6. can play a major role in the sustainable use of tropical rain forests. [p.484]

_____ 7. can provide signs of changes in a habitat and potential loss of biodiversity. [p.479]

_____ 8. can increase the vulnerability of a species to the effects of natural disasters. [p.479]

_____ 9. All of the following are true of coral reefs *except:* [p.480]
a. Coral reefs generally develop between latitudes 25° North and South.
b. The surface temperatures of the tropical and subtropical oceans have increased since the 1980s.
c. Some unscrupulous reef fishermen use dynamite and/or sodium cyanide to make reef animals come out of their hiding places.
d. Since the 1980s, Indonesian fisherman have been causing widespread bleaching of living corals by squirting laundry bleach on them to prepare them for sale to pet stores and gift shops.
e. Reef formations off Florida's Key Largo have declined by one-third, mostly since 1970.

Chapter Objectives/Review Questions

1. Explain how poor use of resources such as soil, native foods, and forests contributed to the downfall of the civilization on Easter Island. [pp.474–475]
2. Explain how it can be that 99 percent of all species that have ever lived are extinct, but that biodiversity is greater now than it has ever been. [p.476].
3. List some of the historical causes of mass extinctions. [p.476]
4. Describe the supposed relationship between the dodo and the tree *Calvaria major* on the island of Mauritius. [p.487]
5. Earth's sixth major extinction event is underway. State its principal cause and then list four secondary causes brought about by the principal cause. [pp.478–479]
6. Explain why the coral reef is a biodiversity hot spot. List several of the threats to coral reefs. [pp.480–481]
7. What was Rachel Carson's contribution to science? [p.482]
8. Define the science of conservation biology. [p.482]

9. How do bioeconomics play a role in protecting biodiversity? [p.483]
10. How is strip logging an example of sustainable development? [p.484]
11. What is a riparian zone and how does it relate to biodiversity? [pp.484–485]

Integrating and Applying Key Concepts

1. Think about your life, and identify five specific things that you could do to encourage biodiversity in ecosystems, some of them far away from your home and some of them very close. Are there any streams, rivers, or forests undergoing changes for the worse near where you live? How could you find information about what these places were like before your region was settled by the human ancestors of its current inhabitants? What can you do to help these places maintain high biodiversity? Why should you bother to put your personal energy into such activities?
2. You have inherited a large sum of money and want to invest it for your retirement, yet you also want your investment to be environmentally sound. What characteristics would you look for in a company that might satisfy both of your purposes? What types of activities could a company perform that would make it more friendly to sustainable development? Could these also be profitable for the company?

28

HOW PLANTS AND ANIMALS WORK

Interactive Exercises

On High-Flying Geese and Edelweiss [pp.488–489]

28.1. LEVELS OF STRUCTURAL ORGANIZATION [pp.490–491]

28.2. THE NATURE OF ADAPTATION [pp.492–493]

Selected Words: *quantitative* terms [p.490], *qualitative* terms [p.490], *internal* environment [p.491], "adaptation" [p.492], short-term adaptations [p.492]

Boldfaced, Page-Referenced Terms

[p.489] anatomy ______________________________

____

[p.489] physiology ______________________________

____

[p.490] division of labor ______________________________

____

[p.490] tissue ____

____

[p.490] organ ____

____

[p.490] organ system ____

____

[p.490] growth ____

____

[p.490] development ____

____

[p.491] internal environment ____

____

[p.491] homeostasis ____

____

[p.492] adaptation ____

____

Matching

For each of the following, choose the most appropriate answer.

1. ____ physiology [p.489]
2. ____ growth [p.490]
3. ____ development [p.490]
4. ____ anatomy [p.489]
5. ____ division of labor [p.490]
6. ____ homeostasis [p.491]

A. Increase in the size, number, and volume of an organism's cells
B. The form of an organism
C. The idea that different parts of the body have different functions but contribute toward the whole organism
D. The maintenance of internal body conditions within tolerable limits
E. Successive stages in the formation of specialized tissues, organs, and organ systems
F. Study of processes and patterns by which an organism survives and reproduces

Complete the Table

7. For each of the definitions below, indicate whether the term applies to tissues, organs, or organ systems. [p.490]

Level of Organization	*Definition*
a.	A group of two or more tissues that are organized in specific patterns or proportions
b.	Community of cells that interact in one or more tasks
c.	A group of organs that interact physically and/or chemically

Short Answer

8. Explain the importance of maintaining a stable internal environment for a cell. [p.491]

____

____

____

Fill-in-the-Blanks

(9) ____________ [p.492] is a term that has different meanings in different contexts. (10) ____________-____________ [p.492] adaptations occur in response to the environment and last only as long as the (11) ____________ [p.492]. A long-term adaptation represents some (12) ____________ [p.492] aspect of form, function, behavior, or development that improves the odds for (13) ____________ [p.492] and (14) ____________ [p.492] in a given environment. It is the (15) ____________ [p.492] of macroevolution.

28.3. MECHANISMS OF HOMEOSTASIS IN ANIMALS [pp.494–495]

28.4. DOES THE CONCEPT OF HOMEOSTASIS APPLY TO PLANTS? [pp.496–497]

Selected Words: "set points" [p.494], *negative* feedback [p.494], *intensify* change [p.495], "sleep" position [p.497]

Boldfaced, Page-Referenced Terms

[p.494] interstitial fluid ____

____

[p.494] plasma ____

____

[p.494] sensory receptors ____

____

[p.494] stimulus ______________________________

____

[p.494] integrator ______________________________

____

[p.494] effectors ______________________________

____

[p.494] negative feedback mechanism ______________________________

____

[p.495] positive feedback mechanisms ______________________________

____

[p.496] compartmentalization ______________________________

____

[p.497] circadian rhythm ______________________________

____

Matching

Choose the most appropriate answer for each of the following.

1. ____ stimulus [p.494]
2. ____ integrator [p.494]
3. ____ circadian rhythm [p.497]
4. ____ plasma [p.494]
5. ____ compartmentalization [p.496]
6. ____ interstitial fluid [p.494]
7. ____ sensory receptors [p.494]
8. ____ effectors [p.494]

A. A biological activity that is repeated in cycles
B. The location that processes information regarding stimuli and signal responses
C. A plant response that involves sealing off infected areas from the remainder of the plant
D. The specific form of energy detected by a receptor
E. The fluid that fills the spaces between the cells and tissues
F. The portion of the body that carries out the response
G. The fluid portion of the blood
H. The cells that are responsible for detecting various forms of energy, such as heat or pressure

Choice

Determine the type of feedback mechanism that each of the following represents. [pp.494–495]

a. positive feedback   b. negative feedback

9. ____ Body temperature in mammals
10. ____ Sexual intercourse in humans
11. ____ Secretion of oxytocin during childbirth
12. ____ The response of the system cancels or counteracts the effect of the original stimulus
13. ____ The response is an intensification of the effect of the original stimulus

Choice

Choose whether each of the following represents a plant or an animal mechanism for regulating the internal environment.

a. plant [pp.496–497] b. animal [pp.494–495]

14. _____ Compartmentalization of infected cells
15. _____ Regulation of the chemical content of the interstitial fluid
16. _____ Comparison of sensory input to a "set-point" value in the brain
17. _____ Structural adaptations for water retention in sand-dwelling organisms

28.5. COMMUNICATION AMONG CELLS, TISSUES, AND ORGANS [pp.498–499]

28.6. RECURRING CHALLENGES TO SURVIVAL [pp.500–501]

Selected Words: *organ-identity* genes [p.498], *leafy* [p.499], *multiple sclerosis* [p.499]

Boldfaced, Page-Referenced Terms

[p.498] hormones ____________________

____

[p.498] ABC model ____________________

____

[p.499] myelin ____________________

____

[p.500] diffusion ____________________

____

[p.500] surface-to-volume ratio ____________________

____

[p.501] active transport ____________________

____

[p.501] habitat ____________________

____

Fill-in-the-Blanks

The molecular mechanisms by which cells (1) "__________" [p.498] to one another evolved among the (2) __________ [p.498] species. Many of these mechanisms persist in the (3) __________ [p.498] organisms. In many cases, they involve three events — the (4) __________ [p.498] of a receptor, the (5) __________ [p.498] of a signal into a molecular form, and then the (6) __________ [p.498] response.

In plants, the growth and development of new parts are under the control of (7) ______________ [p.498], (8) ______________ [p.498] molecules, and (9) ______________ [p.498] cues. In plant cells (10) ______________ [p.498] are the main signals for plant communication. Genes that control how to make floral organs are called (11) ______________-______________ [p.498] genes. An example is the (12) ______________ [p.498] model, in which three groups of genes act as master switches for floral development. These groups encode factors governing (13) ______________-______________ [p.499] for the products that make sepals and petals.

An example of signal reception, transduction, and response in animals may be found in the vertebrate nervous system. In the neuron, parts of each cell act as (14) ______________ [p.499] input zones. Other parts called (15) ______________ [p.499], act as output zones. (16) ______________ [p.499] signals are released from output zones and (17) ______________ [p.499] into other cells. Some of these axons are covered with a sheath, called (18) ______________ [p.499], that serves to insulate the cell. In the disease known as (19) ______________ ______________ [p.499], this sheath may become damaged, resulting in a gradual loss of brain and spinal cord function.

Matching

For each of the following, choose the most appropriate answer.

20. _____ surface–volume ratio [p.500]
21. _____ habitat [p.501]
22. _____ active transport [p.501]
23. _____ diffusion [p.500]

A. The movement of particles against a concentration gradient
B. The net movement of molecules from an area of high concentration to one of low concentration
C. The place where the individuals of a species normally live
D. The physical property that determines the size and shape of cells

Self-Quiz

_____ 1. Which of the following terms is used to indicate the process by which an organism goes through successive stages in the formation of specialized tissues and organs? [p.490]
a. anatomy
b. physiology
c. growth
d. development

_____ 2. The action of an organism to keep the operating conditions of its internal environment within specified limits is called ______________. [p.491]
a. development
b. homeostasis
c. adaptation
d. diffusion

_____ 3. In animals, a mechanism for controlling homeostasis in which the output of a metabolic reaction is used to slow or reverse a pathway is called __________. [p.494]
   a. positive feedback mechanism
   b. adaptation
   c. negative feedback mechanism
   d. physiology

_____ 4. Which of the following processes expends energy to move molecules from an area of low concentration to one of high concentration? [p.501]
   a. diffusion
   b. anatomy
   c. compartmentalization
   d. active transport

_____ 5. Which of the following is true regarding the process of adaptation? [p.492]
   a. It is an outcome of macroevolution.
   b. It represents a heritable trait.
   c. It may involve some aspect of an organism's form, function, behavior, or development.
   d. It improves the odds of surviving and reproducing in a given environment.
   e. All of the above are correct.

Chapter Objectives/Review Questions

1. Explain the difference between the terms *anatomy* and *physiology.* [p.489]
2. Explain the differences in organization between a tissue, an organ, and an organ system. [p.490]
3. Define the terms *growth* and *development.* [p.491]
4. Define *homeostasis* in relation to the internal environment of an organism. [p.491]
5. Explain the relationship between adaptation and macroevolution. [p.492]
6. Understand how salt-tolerant tomatoes and polar bears represent the process of adaptation. [pp.492–493]
7. List some of the limitations of establishing a direct relationship between a specific adaptation and environmental change. [p.493]
8. Describe the difference between plasma and interstitial fluid. [p.494]
9. Explain the action of sensory receptors using the terms *stimulus, integrators,* and *effectors.* [p.494]
10. Explain the difference between a negative and a positive feedback mechanism and give an example of each. [pp.494–495]
11. Give examples of how plants may maintain homeostasis. [pp.496–497]
12. Define the term *circadian rhythm.* [p.497]
13. Give an example of communication between cells in both plants and animals. [pp.498–499]
14. Explain how the surface-to-volume ratio defines the physical size of a cell. [p.500]
15. Explain how a cell may use diffusion and active transport to maintain an internal environment. [pp.500–501]
16. Define *habitat.* [p.501]

Integrating and Applying Key Concepts

1. Choose a species of plant or animal that lives in an extreme environment. Explain how this species may have adapted to this habitat over time. What changes must have been made to the process of homeostasis in the organism?

29

PLANT TISSUES

Interactive Exercises

Plants versus the Volcano [pp.504–505]

29.1. OVERVIEW OF THE PLANT BODY [pp.506–507]

Selected Words: *Lycopersicon* [p.506], apical meristems [p.507], protoderm [p.507], ground meristem [p.507], procambium [p.507], primary growth [p.507], periderm [p.507], secondary growth [p.507]

Boldfaced, Page-Referenced Terms

[p.506] magnoliids ______

[p.506] eudicots ______

[p.506] monocots ______

[p.506] shoots ______

[p.506] roots ______

[p.506] ground tissue system ______

[p.506] vascular tissue system ______

[p.506] dermal tissue system ______

[p.507] meristems ______

[p.507] vascular cambium ______

[p.507] cork cambium ______

Labeling and Matching

Identify each part of the accompanying illustration. Choose from the following: dermal tissues (epidermis), root system, ground tissues, shoot system, and vascular tissues. Complete the exercise by matching and entering the letter of the proper description in the parentheses following each label. [p.506]

1. ______________ ______________ ( )
2. ______________ ______________ ( )
3. ______________ ______________ ( )
4. ______________ ______________ ( )
5. ______________ ______________ ( )

A. Typically consists of stems, leaves, and flowers (reproductive shoots)
B. Typically grows below ground, anchors aboveground parts, absorbs soil water and minerals, stores and releases food, and anchors the aboveground parts
C. Tissues that make up the bulk of the plant body
D. Covers and protects the plant's surfaces
E. Two conducting tissues that distribute water and solutes through the plant body

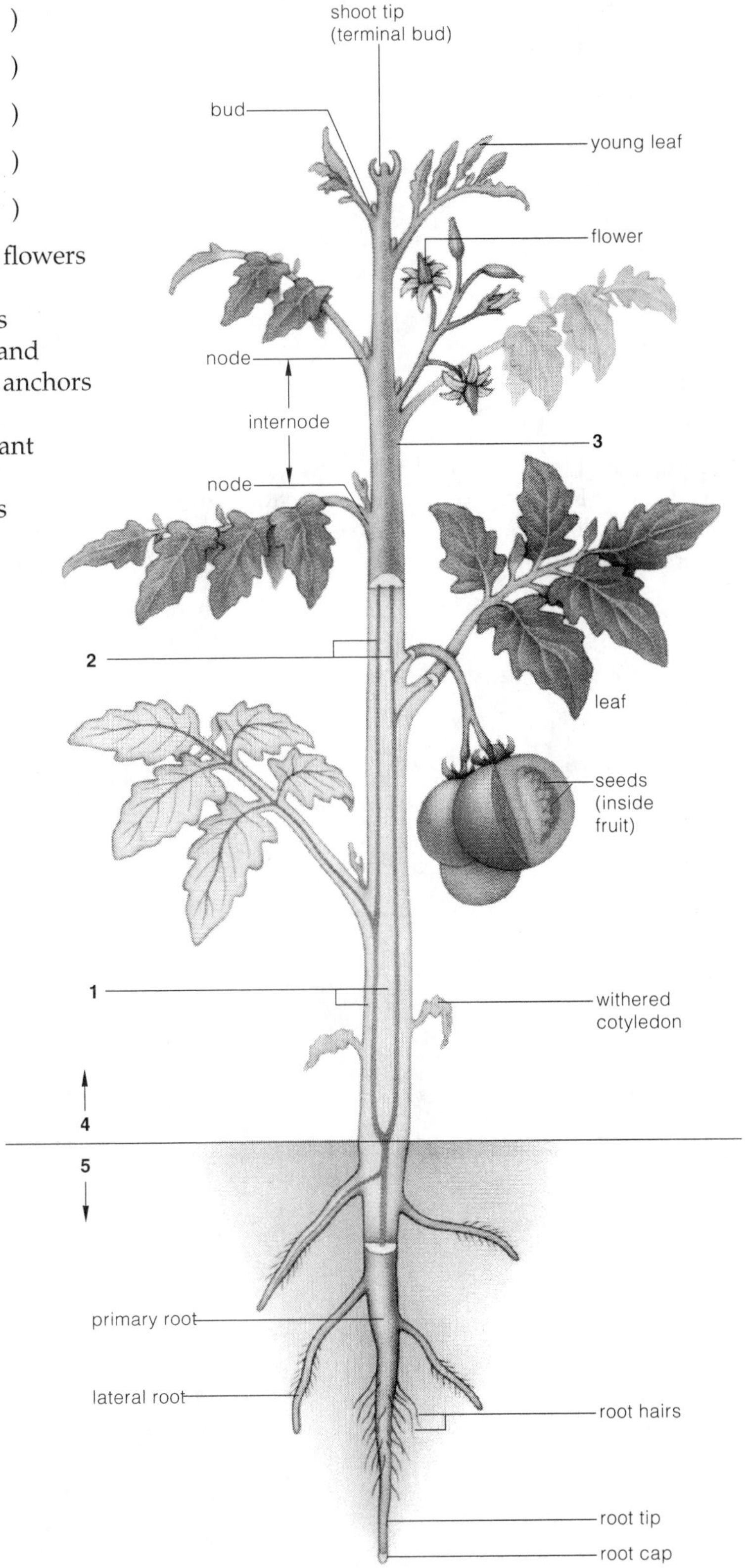

shoot tip (terminal bud)
bud
young leaf
flower
node
internode
node
3
2
leaf
seeds (inside fruit)
1
withered cotyledon
4
5
primary root
lateral root
root hairs
root tip
root cap

Matching

Choose the most appropriate answer for each term. [all from p.507]

6. ____ secondary growth
7. ____ vascular cambium and cork cambium
8. ____ meristems
9. ____ apical meristems
10. ____ primary growth
11. ____ primary meristematic tissue

A. Represented by a lengthening of stems and roots originating from cell divisions at apical meristems
B. Localized regions of dividing cells
C. Cell populations forming from apical meristems; include protoderm, ground meristem, and procambium
D. Lateral meristems giving rise to, respectively, secondary vascular tissues and a sturdier plant covering that replaces epidermis
E. Represented by a thickening of stems and roots caused by activity of the lateral meristems
F. Located in the dome-shaped tips of all shoots and roots; responsible for lengthening those organs

Labeling

It is important to understand the terms that identify the thin sections (slices) of plant organs and tissues that are prepared for study. Label each of the following three diagrams. Choose from the following: radial section (cut along the radius of the organ), transverse or cross section (cuts perpendicular to the long axis of the organ), and tangential section (cuts made at right angles to the radius of the organ). *Note:* The dark area indicates the slice. [p.507]

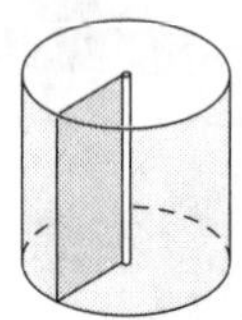

12. ______________ section

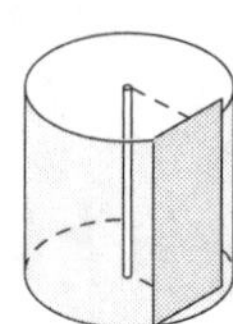

13. ______________ section

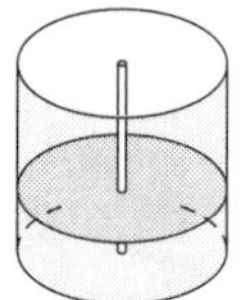

14. ______________ section

Labeling and Matching

Identify each part of the accompanying illustration. Choose from the following: shoot apical meristem, root apical meristem, root primary meristematic tissue, shoot primary meristematic tissue, and lateral meristems. Complete the exercise by matching and entering the letter of the proper description in the parentheses after each label. [all from p.507]

15. ________ ________ ________ ( )
16. ________ ________ ________ ________ ( )
17. ________ ________ ________ ________ ( )
18. ________ ________ ________ ( )
19. ________ ________ ( )

A. Near all root tips; also gives rise to three primary meristematic tissues from which the root's primary tissue systems develop (lengthening)
B. These embryonic descendants of apical meristem (protoderm, ground meristem, and procambium) divide, grow, and differentiate into a shoot's primary tissue systems
C. Found only inside the older stems and roots of woody plants; sources of secondary growth (increases in diameter)
D. A region of embryonic cells near the dome-shaped tip of all shoots; the source of primary growth (lengthening)
E. Embryonic descendants of the root apical meristem (protoderm, ground meristem, and procambium) that divide, grow, and differentiate into the root's primary tissue systems

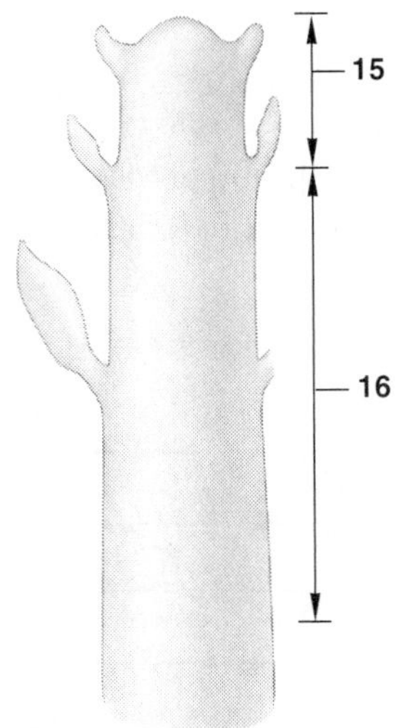

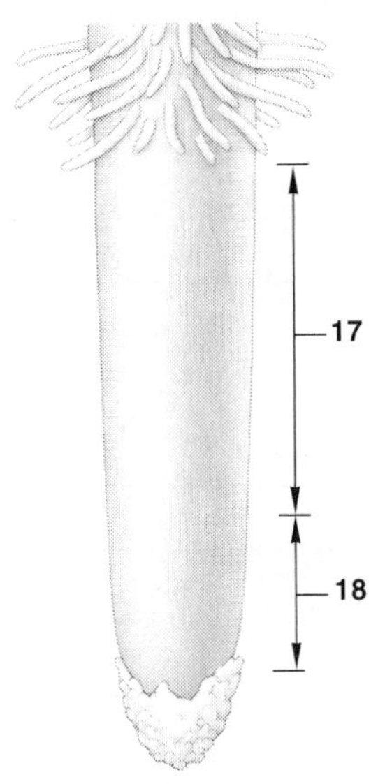

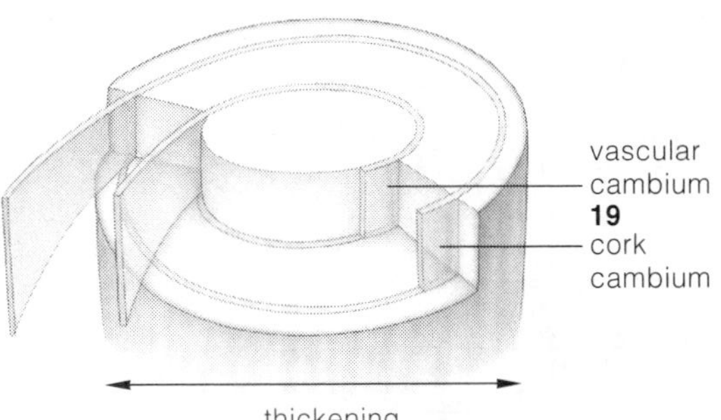

Complete the Table

20. The cells of the primary meristematic tissues and the lateral meristems give rise to tissues, either primary or secondary. Complete the following table, which summarizes the activity of these meristems. [p.507]

Meristem	*Tissue(s)*	*Primary or Secondary*
a. Protoderm		
b. Ground meristem		
c. Procambium		
d. Vascular cambium		
e. Cork cambium		

29.2. TYPES OF PLANT TISSUES [pp.508–509]

Selected Words: mesophyll [p.508], lignin [p.508], *fibers* [p.508], *sclereids* [p.508], *vessel members* [p.508], *tracheids* [p.508], *sieve-tube members* [p.508], *companion cells* [p.508], waxes [p.509], cutin [p.509], guard cells [p.509], periderm [p.509], suberin [p.509]

Boldfaced, Page-Referenced Terms

[p.508] parenchyma ______________________________

____

[p.508] collenchyma ______________________________

____

[p.508] sclerenchyma ______________________________

____

[p.508] xylem ______________________________

____

[p.508] phloem ______________________________

____

[p.509] epidermis ______________________________

____

[p.509] cuticle ______________________________

____

[p.509] stoma (plural, stomata) ______________________________

____

Labeling

Label the following illustrations as either *collenchyma* (uneven wall thickenings, often appearing like thickened corners); *parenchyma* (thin-walled with intercellular spaces); or *sclerenchyma* (very thick walls in cross section; includes fibers and sclereids such as stone cells). [all from p.508]

1. ________________

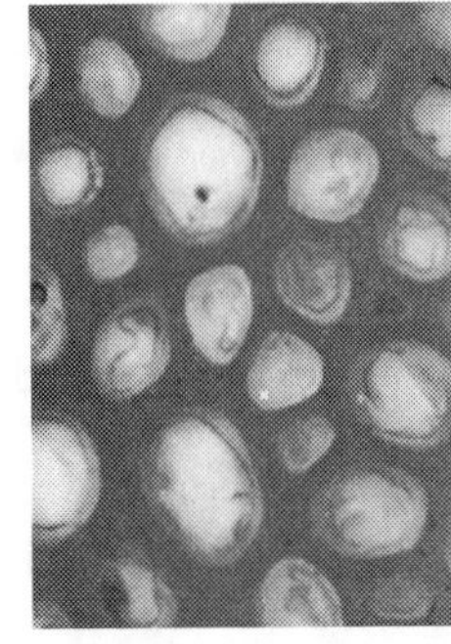

2. ________________

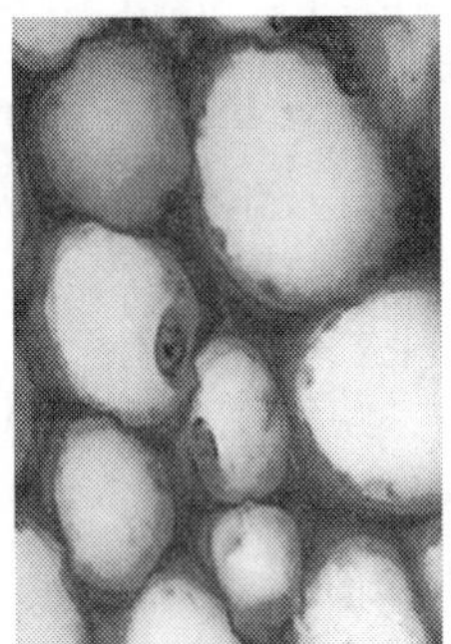

3. ________________

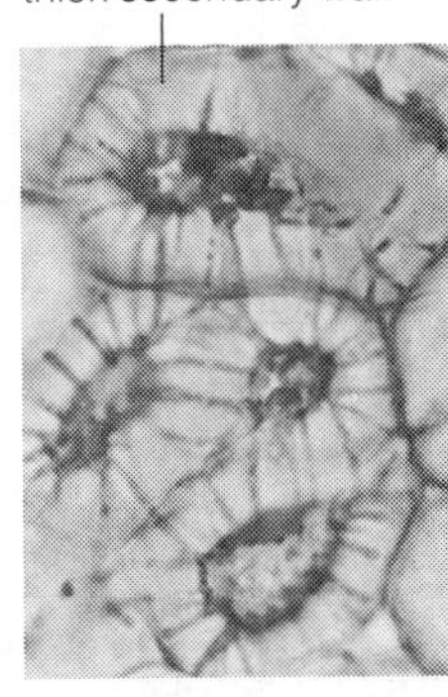

4. ________________

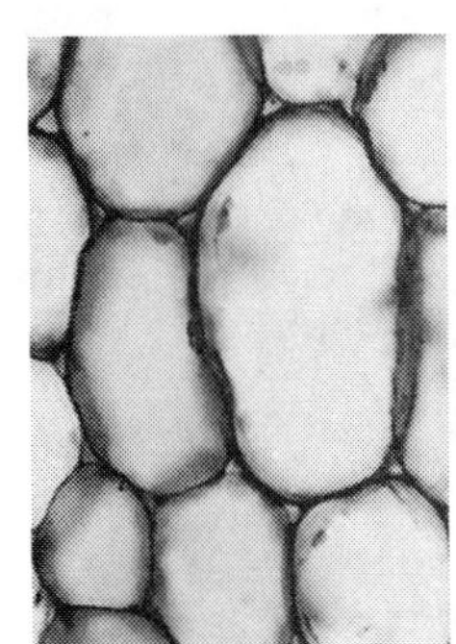

5. ________________

Choice

For questions 6–15, choose from the following: [p.508]

a. parenchyma b. collenchyma c. sclerenchyma

6. ____ Patches or cylinders of this tissue are found near the surface of lengthening stems
7. ____ Cells have thick, lignin-impregnated walls
8. ____ Cells are alive at maturity and retain the capacity to divide
9. ____ Some types specialize in storage, secretion, and other tasks
10. ____ Cells are mostly elongated, with unevenly thickened walls
11. ____ Provides flexible support for primary tissues
12. ____ Supports mature plant parts and often protects seeds
13. ____ Cells are thin walled, pliable, and many sided
14. ____ Mesophyll is specialized for photosynthesis
15. ____ Fibers and sclereids [p.502]

Labeling

Identify the cell types and cellular structures in the accompanying illustrations by entering the correct name in the appropriate blank. Complete the exercise by identifying the complex tissue (xylem or phloem) in which that cell or structure is found. [all from p.509]

16. ______________ (______________)
17. ______________ (______________)
18. ______________ (______________)
19. ______________ (______________)
20. ______________ (______________)
21. ______________ (______________)
22. ______________ (______________)
23. ______________ (______________)
24. ______________ (______________)

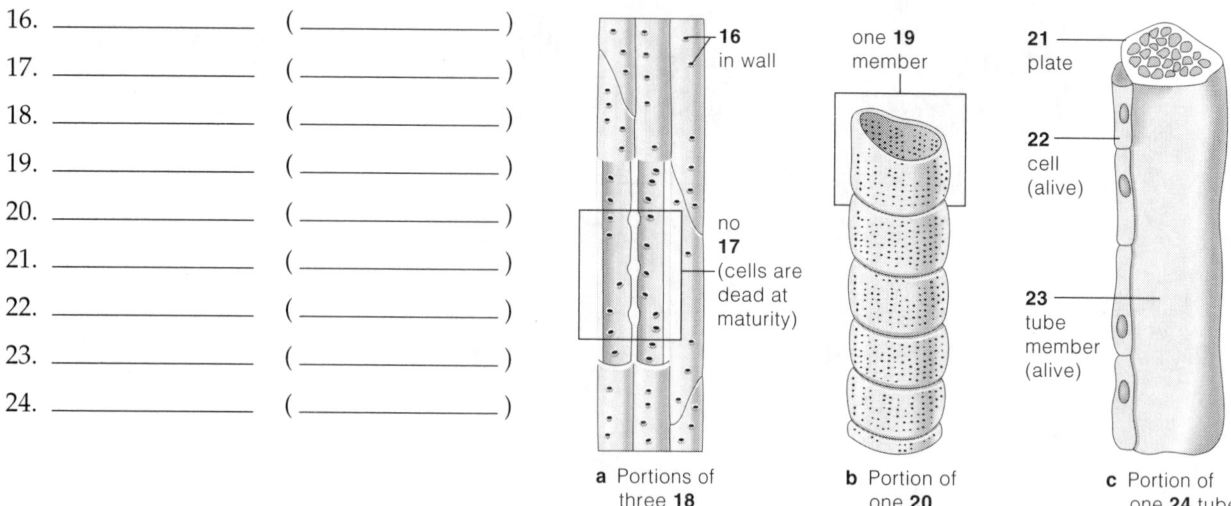

a Portions of three **18**

b Portion of one **20**

c Portion of one **24** tube

Complete the Table

25. The two types of vascular tissue, xylem and phloem, occur in strands called *vascular bundles* within the plant body (each of these complex tissues also has fibers and parenchyma). Complete the following table, which summarizes important information about these vascular tissues. [p.509]

Vascular Tissue	*Major Conducting Cells*	*Cells Alive?*	*Function(s)*
a. Xylem			
b. Phloem			

26. Two types of dermal tissue cover the plant body. Complete the following table, which summarizes information about the dermal tissues. [p.509]

Dermal Tissue	*Primary or Secondary Plant Body*	*Function(s)*
a. Epidermis		
b. Periderm		

27. There are two classes of flowering plants, monocots and dicots. Complete the following table, which summarizes information about the two groups of flowering plants. [p.509]

Class	*Number of Cotyledons*	*Number of Floral Parts*	*Leaf Venation*	*Pollen Grains*	*Vascular Bundles*
a. Monocots					
b. Dicots					

29.3. PRIMARY STRUCTURE OF SHOOTS [pp.510–511]

Selected Words: bud scale [p.510], *Coleus* [p.510], *Medicago* [p.511], *Zea mays* [p.511]

Boldfaced, Page-Referenced Terms

[p.510] bud ____

____

[p.510] vascular bundles ____

____

[p.510] cortex ____

____

[p.510] pith ____

____

Labeling

Name the structures numbered in the following illustrations of shoot primary development. [p.510]

1. ________________
2. ________________ ________________
3. ________________
4. ________________
5. ________________
6. ________________
7. ________________
8. ________________
9. ________________ ________________
10. ________________
11. ________________
12. ________________
13. ________________ ________________
14. ________________ ________________

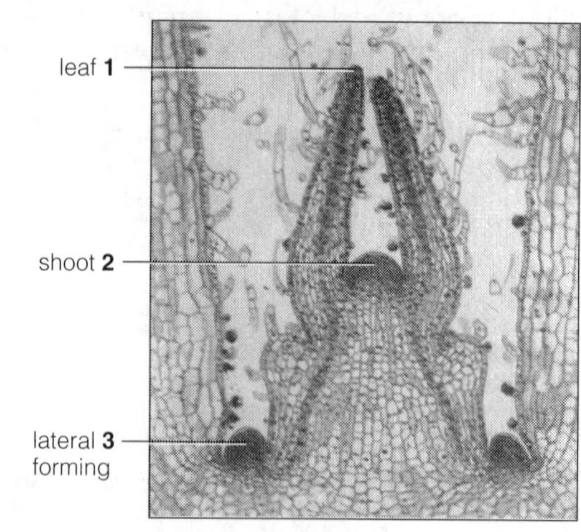

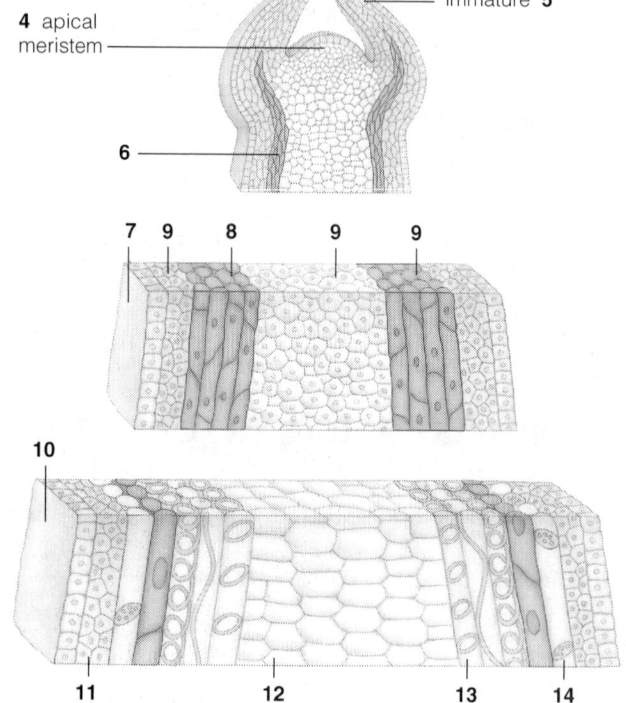

Labeling

Name the structures numbered in the following illustrations of the monocot stem. Choose from the following: air space, epidermis, sclerenchyma cells, xylem vessel, vascular bundle, sieve tube, ground tissue, xylem, and companion cell. [p.511]

15. ______________

16. ______________ ______________

17. ______________ ______________

18. ______________ ______________

19. ______________ ______________

20. ______________ ______________

21. ______________ ______________

22. ______________ ______________

23. ______________ is the tissue in which no. 20 is located.

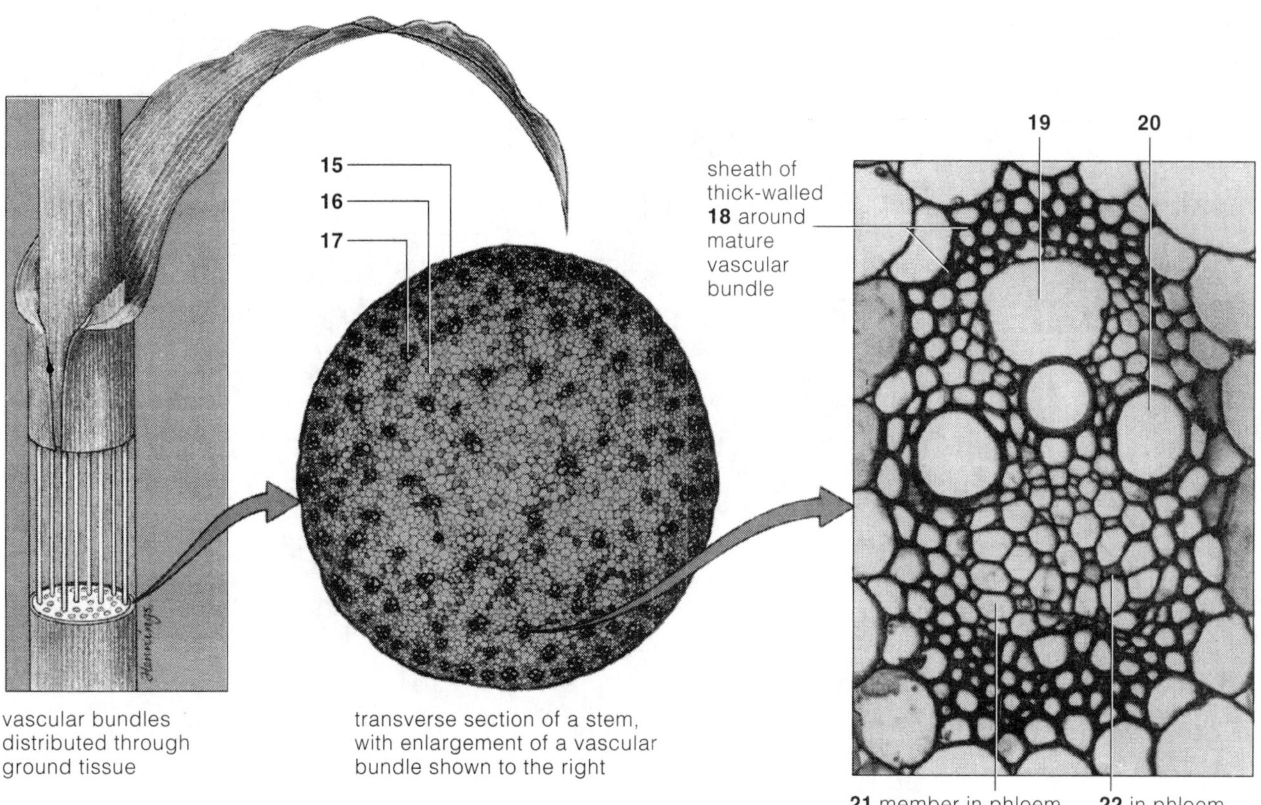

Labeling

Name the structures numbered in the following illustrations of the herbaceous dicot stem. Choose from the following: sieve tube, phloem fibers, vascular bundle, xylem vessel, cortex, meristematic cells, pith, and epidermis. [p.511]

24. ________________

25. ________________

26. ________________ ________________

27. ________________

28. ________________ ________________

29. ________________ ________________

30. ________________ ________________

31. ________________ ________________

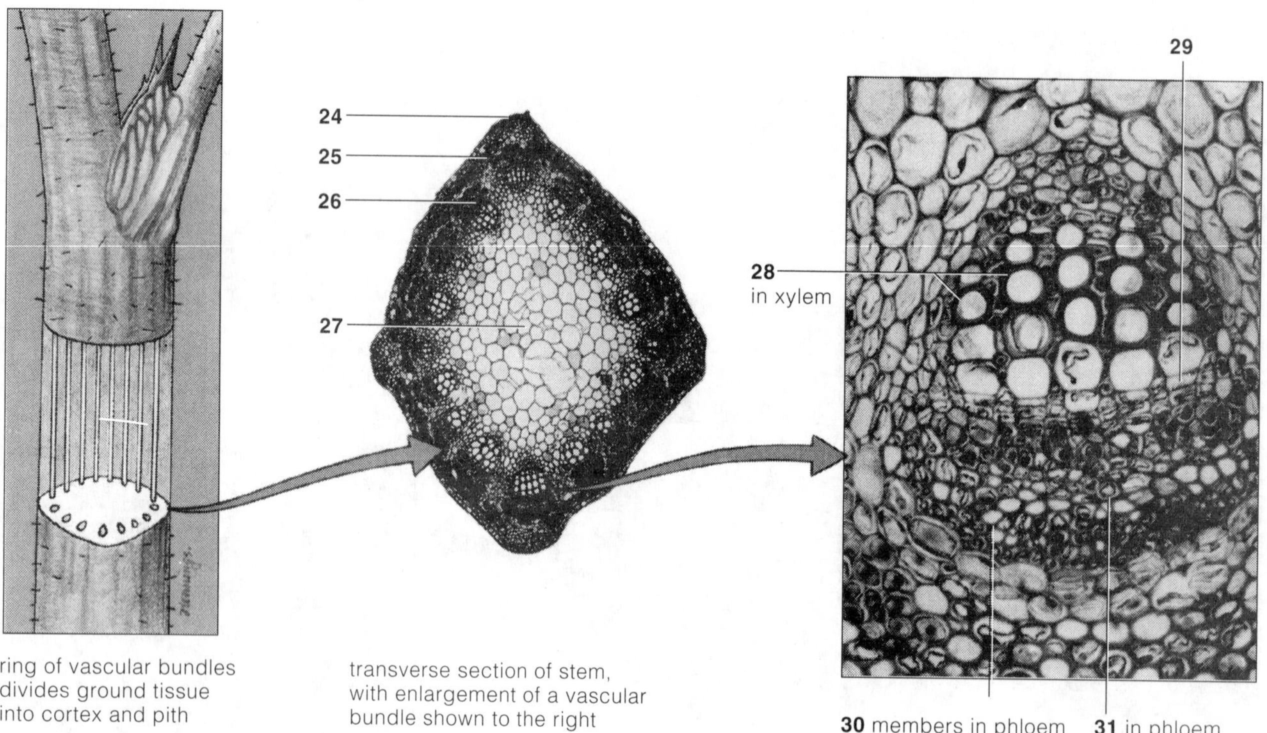

ring of vascular bundles divides ground tissue into cortex and pith

transverse section of stem, with enlargement of a vascular bundle shown to the right

29.4. A CLOSER LOOK AT LEAVES [pp.512–513]

Selected Words: *deciduous* [p.512], *evergreen* [p.512], blade [p.512], petiole [p.512], *simple* leaves [p.512], *compound* leaves [p.512], *palisade* mesophyll [p.513], *spongy* mesophyll [p.513]

Boldfaced, Page-Referenced Terms

[p.512] leaf ______________________________

____

[p.513] mesophyll ______________________________

____

[p.513] veins ______________________________

____

Labeling

Name the structures numbered in the accompanying illustrations of leaf development and leaf forms. [p.512]

1. ______________
2. ______________
3. ______________ ______________
4. ______________
5. ______________
6. ______________

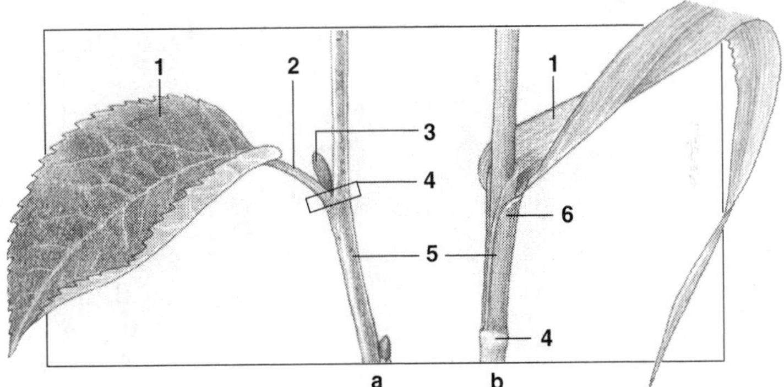

Dichotomous Choice

Circle one of two possible answers given between parentheses in each statement. [p.512]

7. The leaf illustrated on the left above is a (monocot/dicot).
8. The leaf illustrated on the right above is a (monocot/dicot).

Labeling and Matching

Identify each numbered part of the accompanying illustration. Complete the exercise by matching and entering the letter of the corresponding function description in the parentheses following each label. [p.513]

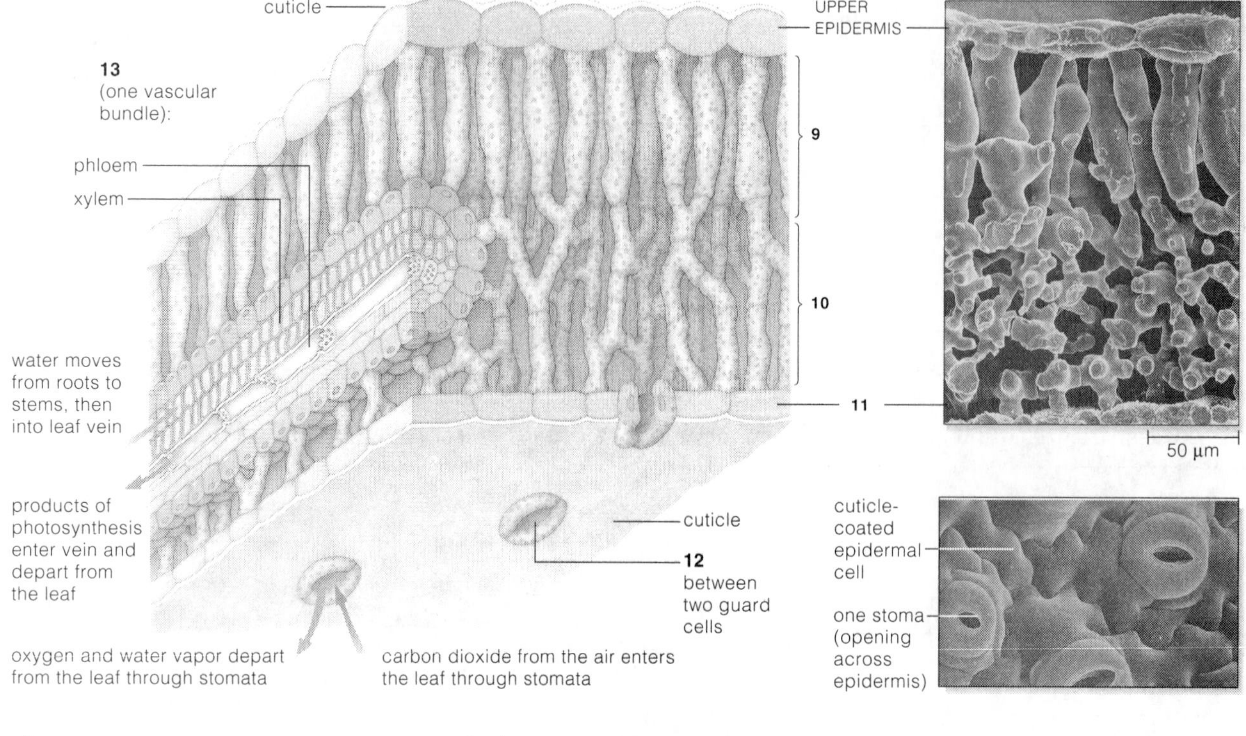

9. ______________ ______________ ( )

10. ______________ ______________ ( )

11. ______________ ______________ ( )

12. ______________ ( )

13. ______________ ______________ ( )

A. Lowermost cuticle-covered cell layer

B. Loosely packed photosynthetic parenchyma cells just above the lower epidermal layer

C. Allows movement of oxygen and water vapor out of leaves and allows carbon dioxide to enter

D. Photosynthetic parenchyma cells just beneath the upper epidermis

E. Move water and solutes to photosynthetic cells and carry products away from them

29.5. PRIMARY STRUCTURE OF ROOTS [pp.514–515]

Selected Words: *adventitious* [p.514], root cap [p.514], mucigel [p.514], endodermis [p.515], pericycle [p.515]

Boldfaced, Page-Referenced Terms

[p.514] primary root ____

[p.514] lateral roots ____

[p.514] taproot system ____

[p.514] fibrous root system ____

[p.514] root hairs ____

[p.514] vascular cylinder ____

Labeling and Matching

Identify each numbered part of the accompanying illustration. Complete the exercise by matching and entering the letter of the corresponding description in the parentheses following each label. Some choices will be used more than once. [all from p.514]

1. ______________ ______________ ( )
2. ______________ ( )
3. ______________ ( )
4. ______________ ( )
5. ______________ ( )
6. ______________ ______________ ( )
7. ______________ ______________ ( )
8. ______________ ( )
9. ______________ ( )
10. ______________ ______________ ( )

A. Dome-shaped cell mass produced by the apical meristem
B. Part of the vascular cylinder; gives rise to lateral roots
C. Part of the vascular cylinder; transports photosynthetic products
D. Ground tissue region surrounding the vascular cylinder
E. The absorptive interface with the root's environment
F. The region of apical and primary meristems
G. Innermost part of the root cortex; helps control water and mineral movement into the vascular column
H. Epidermal cell extensions; greatly increases the surface available for taking up water and solutes

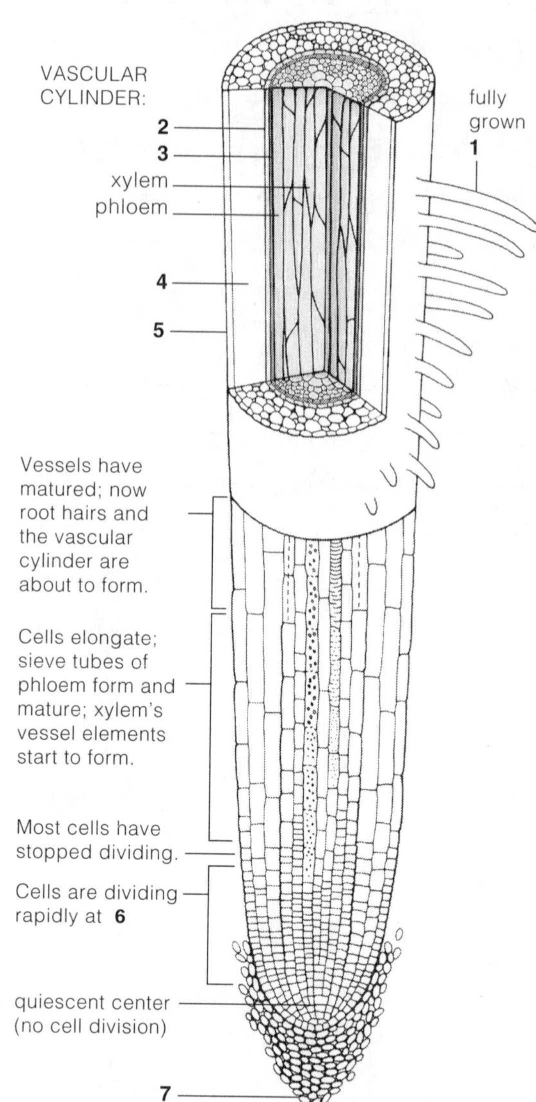

VASCULAR CYLINDER:
2
3
xylem
phloem
4
5
fully grown
1
Vessels have matured; now root hairs and the vascular cylinder are about to form.
Cells elongate; sieve tubes of phloem form and mature; xylem's vessel elements start to form.
Most cells have stopped dividing.
Cells are dividing rapidly at 6
quiescent center (no cell division)
7

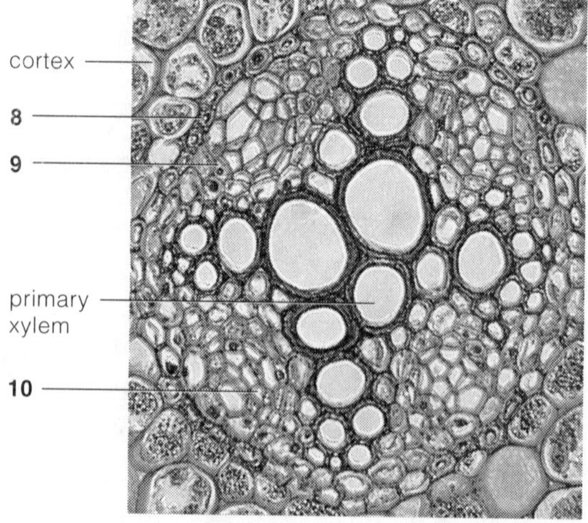

cortex
8
9
primary xylem
10

Vascular Cylinder

Fill-in-the-Blanks

The first part of a seedling to emerge from the seed coat is the (11) ______________ [p.514] root. In most dicot seedlings, the primary root increases in diameter while it grows downward. Later, (12) ______________ [p.514] roots begin forming in internal tissues at an angle perpendicular to the primary root's axis, then erupt through epidermis. Oak trees, carrots, poppies, and dandelions are examples of plants whose primary root and lateral branchings represent a(n) (13) ______________ [p.514] system. In monocots such as grasses, the primary root is short lived; in its place, numerous (14) ______________ [p.514] roots arise from the stem of the young plant. Such roots and their branches are somewhat alike in length and diameter and form a(n) (15) ______________ [p.514] root system.

Some root epidermal cells send out absorptive extensions called root (16) ______________ [p.514]. Vascular tissues form a(n) (17) ______________ [p.514] cylinder, a central column inside the root consisting of primary xylem and phloem and one or more layers of parenchyma called the (18) ______________ [p.515]. Ground tissues surrounding the cylinder are called the root (19) ______________ [p.515]. A monocot's vascular cylinder divides the ground tissue system into cortex and (20) ______________ [p.515] regions. There are many (21) ______________ [p.509] spaces in between cells of the ground tissue system, and (22) ______________ [p.515] can easily diffuse through them. Water entering the root moves from cell to cell until it reaches the (23) ______________ [p.515], the innermost cell layer of the root cortex. Abutting walls of its cells are waterproof, so they force incoming water to pass through the cytoplasm. This arrangement helps (24) ______________ [p.15] the movement of water and dissolved substances into the vascular cylinder. Just inside the endodermis is the (25) ______________ [p.515]. This part of the vascular cylinder gives rise to (26) ______________ [p.515] roots, which then erupt through the (27) ______________ [p.515] and epidermis.

29.6. ACCUMULATED SECONDARY GROWTH — THE WOODY PLANTS [pp.516–517]

29.7. A LOOK AT WOOD AND BARK [pp.518–519]

Selected Words: "nonwoody" [p.516], herbaceous [p.516], *woody* plants [p.516], fusiform initials [p.516], ray initials [p.516], *inner* face [p.517], *outer* face [p.517], secondary xylem [p.517], *early* wood [p.519], *late* wood [p.519], "tree rings" [p.519]

Boldfaced, Page-Referenced Terms

[p.516] annuals ____

____

[p.516] biennials ____

____

[p.516] perennials ____

____

[p.518] bark ____

____

[p.518] periderm ____

____

[p.518] cork ____

____

[p.518] heartwood ____

____

[p.518] sapwood ____

____

[p.519] growth rings ____

____

[p.519] hardwood ____

____

[p.519] softwood ____

____

Matching

Choose the most appropriate answer for each term.

1. _____ tree rings [p.519]
2. _____ heartwood [p.518]
3. _____ sapwood [p.518]
4. _____ early wood [p.519]
5. _____ nonwoody plants [p.516]
6. _____ woody plants [p.516]
7. _____ softwood [p.519]
8. _____ bark [p.518]
9. _____ cork [p.518]
10. _____ late wood [p.519]
11. _____ biennial [p.516]
12. _____ annuals [p.516]
13. _____ hardwood [p.519]
14. _____ perennials [p.516]

A. Produced by cells of the cork cambium
B. Includes all tissues external to the vascular cambium
C. Alternating bands of early and late wood, which reflect light differently; show secondary growth during two or more growing seasons
D. Complete the life cycle in a single growing season and are generally herbaceous
E. Another term for herbaceous plants
F. The first xylem cells produced at the start of the growing season; tend to have large diameters and thin walls
G. Produced by conifers that lack fibers and vessels in their xylem
H. Term applied to the wood of dicot trees that evolved in temperate and tropical regions; possess vessels, tracheids, and fibers in their xylem
I. Produced by some monocots and many dicots; they put on extensive secondary growth over two or more growing seasons
J. Formed in dry summer; has xylem cells with smaller diameters and thicker walls
K. A dumping ground at the center of older stems and roots for metabolic wastes such as resins, oils, gums, and tannins
L. Plants such as carrots that live for two growing seasons
M. Secondary growth located between heartwood and the vascular cambium; wet, usually pale
N. Plants that add secondary growth during two or more growing seasons

Labeling and Matching

Identify each numbered part of the accompanying illustration. Complete the exercise by matching and entering the letter of the corresponding description in the parentheses following each label. [p.519]

15. ________________ ( )

16. ________________ ________________ ( )

17. ________________ ________________ ( )

18. ________________ ( )

19. year ________________ ( )

20. and 21. years ________________

and ________________ ( )

22. ________________ ________________ ( )

A. All secondary growth
B. Produced later in the growing season; vessels have smaller diameters with thick walls
C. Includes the primary growth and some secondary growth
D. Includes all tissues external to the vascular cambium
E. Large-diameter conducting cell in the xylem
F. Produced early in the growing season; vessels have large diameters and thin walls
G. Produces secondary vascular tissues, xylem, and phloem

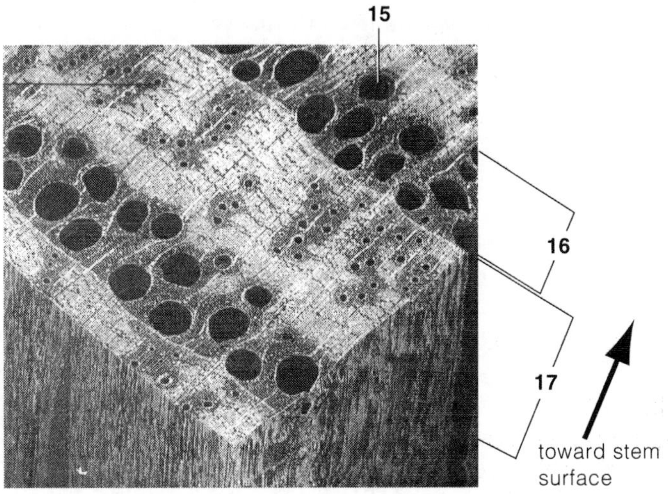

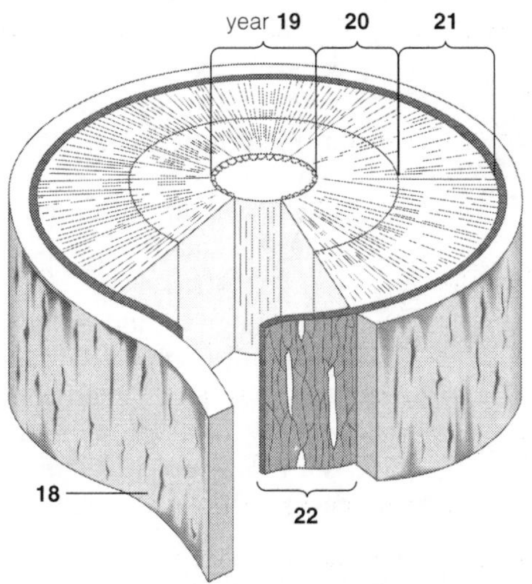

Self-Quiz

____ 1. ________________ covers and protects the plant's surfaces. [p.506]
a. Ground tissue
b. Dermal tissue
c. Vascular tissue
d. Pericycle

____ 2. Which of the following is *not* considered a type of simple tissue? [pp.508–509]
a. Epidermis
b. Parenchyma
c. Collenchyma
d. Sclerenchyma

____ 3. Of the following cell types, which one does *not* appear in vascular tissues? [p.508]
a. vessel members
b. cork cells
c. tracheids
d. sieve-tube members
e. companion cells

____ 4. The ________________ is a leaflike structure that is part of the embryo; monocot embryos have one, dicot embryos have two. [p.509]
a. shoot tip
b. root tip
c. cotyledon
d. apical meristem

____ 5. Each part of the stem where one or more leaves are attached is a(n) ________. [p.510]
a. node
b. internode
c. vascular bundle
d. cotyledon

____ 6. Which of the following structures is *not* considered meristematic? [p.507]
a. vascular cambium
b. lateral meristem
c. cork cambium
d. endodermis

____ 7. Which of the following statements about monocots is *false?* [p.509]
a. They are usually herbaceous.
b. They develop one cotyledon in their seeds.
c. Their vascular bundles are scattered throughout the ground tissue of their stems.
d. They have a single central vascular cylinder in their stems.

____ 8. New plants grow and older plant parts lengthen through cell divisions at ________ meristems present at root and shoot tips; older roots and stems of woody plants increase in diameter through cell divisions at ________ meristems. [p.507]
a. lateral; lateral
b. lateral; apical
c. apical; apical
d. apical; lateral

____ 9. Vascular bundles called ________ form a network through a leaf blade. [p.513]
a. xylem
b. phloem
c. veins
d. cuticles
e. vessels

____ 10. A primary root and its lateral branchings represent a(n) ________ system. [p.514]
a. lateral root
b. adventitious root
c. taproot
d. branch root

____ 11. Plants whose vegetative growth and seed formation continue year after year are ________ plants. [p.516]
a. annual
b. perennial
c. nonwoody
d. herbaceous

____ 12. The ________ layer of a root divides to produce lateral roots. [p.515]
a. endodermis
b. pericycle
c. xylem
d. cortex
e. phloem

Chapter Objectives/Review Questions

1. The aboveground parts of flowering plants are called ________; the plants' descending parts are called ________. [p.506]
2. Distinguish among the ground tissue system, vascular tissue system, and dermal tissue system. [p.506]
3. Plants grow at localized regions of self-perpetuating embryonic cells called ________. [p.507]
4. Lengthening of stems and roots originates at ________ meristems and all dividing tissues derived from them; this is called ________ growth. [p.507]
5. Cell populations of protoderm, ground meristem, and procambium are derived from the apical meristem and are known as ________ meristematic tissues. [p.507]
6. Increases in the thickness of a plant originate at ________ meristems. [p.507]
7. Describe the role of vascular cambium and cork cambium in producing secondary tissues of the plant body. [p.507]
8. Visually identify and generally describe the simple tissues called parenchyma, collenchyma, and sclerenchyma. [p.508]

9. Fibers and sclereids are both types of _________ cells. [p.508]
10. Name the cell wall compound that was necessary for the evolution of rigid and erect land plants. [p.508]
11. _________ tissue conducts soil water and dissolved minerals, as well as mechanically supporting the plant. [p.508]
12. _________ tissue transports sugars and other solutes. [p.508]
13. Name and describe the functions of the conducting cells in xylem and phloem. [p.508]
14. All surfaces of primary plant parts are covered and protected by a dermal tissue system called _________ and a surface coating called a(n) _________ . [p.509]
15. What is the general function of guard cells and stomata found within the epidermis of young stems and leaves? [p.509]
16. The cork cells of _________ replace the epidermis of stems and roots showing secondary growth. [p.509]
17. Distinguish between monocots and dicots by listing their characteristics and citing examples of each group. [p.509]
18. A _________ is an undeveloped shoot of mostly meristematic tissue, often protected by scales; it gives rise to new stems, leaves, and flowers. [p.510]
19. The primary xylem and phloem develop as vascular _________ . [p.510]
20. Distinguish between the stem's cortex and its pith. [p.510]
21. Visually distinguish between monocot stems and dicot stems, as seen in cross section. [p.511]
22. Describe the principal difference between deciduous and evergreen plants. [p.512]
23. How does the simple leaf type differ from compound leaves? [p.512]
24. Describe the structure (cells and layers) and major functions of leaf epidermis, mesophyll, and vein tissue. [p.513]
25. The _________ root is the first to poke through the coat of a germinating seed; later, _________ roots erupt through the epidermis. [p.514]
26. How does a taproot system differ from a fibrous root system? [p.514]
27. Define the term *adventitious*. [p.514]
28. Describe the origin and function of root hairs. [p.514]
29. A(n) _________ _________ consists of primary xylem and primary phloem and one or more layers of parenchyma cells called the pericycle. [pp.514–515]
30. Describe the passage of soil water through root epidermis to the xylem of the vascular cylinder; include the role of the endodermis. [p.515]
31. Define these categories of flowering plants: annuals, biennials, and perennials. [p.516]
32. Distinguish a woody plant from a nonwoody plant in terms of secondary growth. [p.516]
33. Each growing season, new tissues that increase the girth of woody plants originate at their _________ meristems, also called _________ . [p.516]
34. Describe the formation of cork and bark. [p.518]
35. Distinguish heartwood from sapwood and early wood from late wood. [pp.518–519]
36. Explain the origin of the annual growth layers (tree rings) seen in a cross section of a tree trunk. [p.519]
37. Hardwood trees possess _________ and _________ , but softwood trees lack these cells. [p.519]

Integrating and Applying Key Concepts

Imagine (and then describe) the specific behavioral restrictions that might be imposed if the human body resembled the plant body in having (1) open growth with apical meristematic regions, (2) stomata in the epidermis, (3) cells with chloroplasts, (4) excess carbohydrates stored primarily as starch rather than as fat, and (5) dependence on the soil as a source of water and inorganic compounds.

30

PLANT NUTRITION AND TRANSPORT

Interactive Exercises

Flies for Dinner [pp.522–523]

30.1. PLANT NUTRIENTS AND THEIR AVAILABILITY IN SOILS [pp.524–525]

Selected Words: *macro*nutrients [p.524], *micro*nutrients [p.524], *profile* properties [p.525]

Boldfaced, Page-Referenced Terms

[p.522] carnivorous plants ______________________________

____

[p.522] plant physiology ______________________________

____

[p.524] nutrient ______________________________

____

[p.524] soil ______________________________

____

[p.524] humus ______________________________

____

[p.525] loams ____

____

[p.525] topsoil ____

____

[p.525] leaching ____

____

[p.525] erosion ____

____

Matching

Choose the most appropriate answer for each term.

1. ____ soil [p.524]
2. ____ humus [p.524]
3. ____ loams [p.525]
4. ____ profile properties [p.525]
5. ____ topsoil [p.525]
6. ____ nutrients [p.524]
7. ____ macronutrients [p.524]
8. ____ micronutrients [p.524]
9. ____ leaching [p.525]
10. ____ erosion [p.525]

A. Elements essential for a given organism because, directly or indirectly, they have roles in metabolism that cannot be fulfilled by any other element
B. The layered characteristics of soils, which are in different stages of development in different places
C. The decomposing organic material in soil
D. The movement of land under the force of wind, running water, and ice
E. Elements other than the macronutrients that are essential for plant growth
F. Consists of particles of minerals mixed with variable amounts of decomposing organic material
G. Refers to the removal of some of the nutrients in soil as water percolates through it
H. Uppermost part of the soil that is highly variable in depth (A horizon); the most essential layer for plant growth
I. Soils having more or less equal proportions of sand, silt, and clay; best for plant growth
J. Nine of the essential elements required for plant growth

Fill-in-the-Blanks

The three essential elements that plants use as their main metabolic building blocks are oxygen, carbon, and (11) ______________ [p.524]. Besides these elements, plants depend on the uptake of at least (12) ______________ (number) [p.524] others. These are usually dissolved in soil in (13) ______________ [p.524] forms. Plants give up hydrogen ions to the clay in exchange for weakly bound elements such as (14) ______________ [p.524] and K^+. Nine essential elements are (15) ______________ [p.524] and are required in amounts above 0.5 percent of the plant's dry weight. The other elements are (16) ______________ [p.524]; they make up only traces of the plant's dry weight.

Complete the Table

17. Thirteen essential elements are available to plants as mineral ions. Complete the following table, which summarizes information about these important plant nutrients. [p.524]

Mineral or Element	*Macro- or Micronutrient*	*Known Functions*
a.		Component of most proteins, two vitamins
b.		Activation of enzymes; role in maintaining water–solute balance
c.		Component of nucleic acids, phospholipids, ATP
d.		Component of proteins, nucleic acids, coenzymes, chlorophyll
e.		Roles in cementing cell walls, regulation of many cell functions
f.		Component of several enzymes
g.		Roles in chlorophyll synthesis, electron transport
h.		Role in root, shoot growth; role in photolysis
i.		Activation of enzymes; component of chlorophyll
j.		Role in chlorophyll synthesis; coenzyme activity
k.		Role in formation of auxin, chloroplasts, and starch; enzyme component
l.		Component of enzyme used in nitrogen metabolism
m.		Roles in flowering, germination, fruiting, cell division, nitrogen metabolism

30.2. HOW DO ROOTS ABSORB WATER AND MINERAL IONS? [pp.526–527]

Boldfaced, Page-Referenced Terms

[p.526] vascular cylinder ____

[p.526] endodermis ____

[p.526] Casparian strip ____

[p.526] exodermis ____

[p.526] root hairs ____

[p.527] mutualism ____

[p.527] nitrogen fixation ____

[p.527] root nodules ____

[p.527] mycorrhizae ____

Labeling and Matching

Identify each numbered part of the following illustrations that deal with the control of nutrient uptake by plant roots. Choose from the following: epidermis, water movement, cytoplasm, root hair, vascular cylinder, endodermal cell wall, exodermis, endodermis, cortex, and Casparian strip. Complete the exercise by matching from the following list and entering the correct letter in the parentheses after each label. [pp.526–527]

1. ______________ ( )
2. ______________ ______________ ( )
3. ______________ ( )
4. ______________ ______________ ( )
5. ______________ ( )
6. ______________ ( )
7. ______________ ( )
8. ______________ ______________ ( )
9. ______________ ______________ ( )
10. ______________ ______________ ______________ ( )

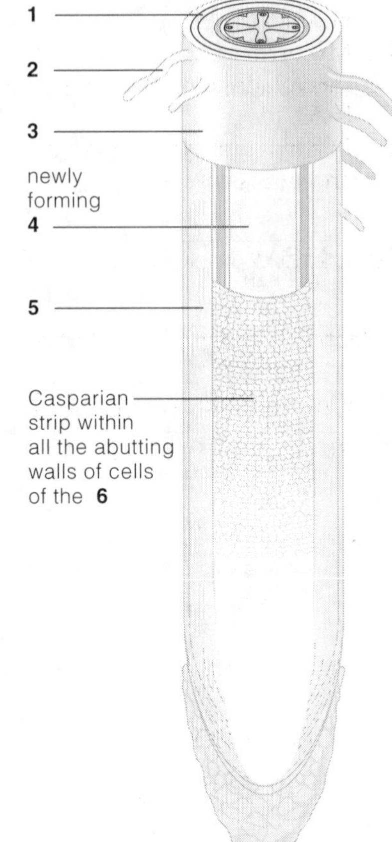

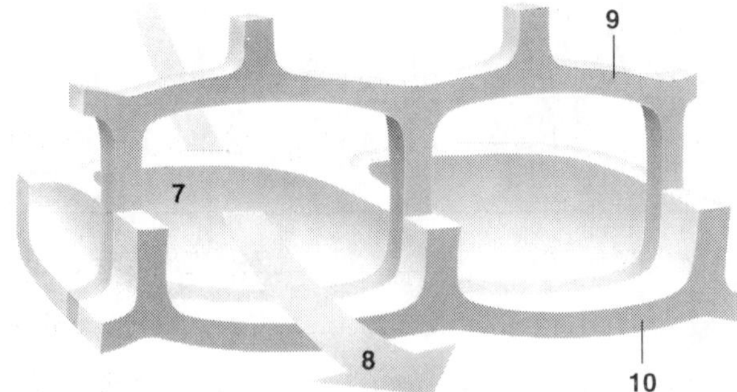

A. Cellular area through which water and dissolved nutrients must move, because of Casparian strips
B. A layer of cortex cells just inside the epidermis; also equipped with Casparian strips
C. Cellular area of a root between the exodermis (if present) and endodermis
D. Waxy band acting as an impermeable barrier between the walls of abutting endodermal cells; forces water and dissolved nutrients through the cytoplasm of endodermal cells
E. Tiny extensions of the root epidermal cells that greatly increase absorptive capacity of a root
F. Specific location of the waxy bands known as Casparian strips
G. Substance whose diffusion occurs through the cytoplasm of endodermal cells because of Casparian strips
H. Outermost layer of root cells
I. A cylindrical layer of cells that wraps around the vascular column
J. Tissues include the xylem, phloem, and pericycle

Dichotomous Choice

Circle one of two possible answers given between parentheses in each statement.

11. A two-way flow of benefits between species is a symbiotic interaction known as (mutualism/parasitism). [p.527]
12. (Gaseous nitrogen/Nitrogen "fixed" by bacteria) represents the chemical form of nitrogen plants can use in their metabolism. [p.527]
13. Nitrogen-fixing bacteria reside in localized swellings on legume plants known as (root hairs/root nodules). [p.527]
14. Mycorrhizae represent symbiotic relationships in which fungi and the roots they cover both benefit; in this relationship the fungus receives (sugars and nitrogen-containing compounds/scarce minerals). [p.527]
15. In a mycorrhizal interaction between a young root and a fungus, the root benefits by obtaining (sugars and nitrogen-containing compounds/scarce minerals that the fungus is better able to absorb). [p.527]

30.3. HOW IS WATER TRANSPORTED THROUGH PLANTS? [pp.528–529]

Selected Words: *cohesion* [p.528], *tension* [p.528]

Boldfaced, Page-Referenced Terms

[p.528] transpiration ____

____

[p.528] xylem ____

____

[p.528] tracheids ____

____

[p.528] vessel members ____

____

[p.528] cohesion–tension theory ____

____

Fill-in-the-Blanks

The cohesion–tension theory explains (1) ______________ [p.528] transport in plants. It travels pipelines in the (2) ______________ [p.522], which are formed by water-conducting cells called (3) ______________ [p.528] and (4) ______________ ______________ [p.528]. The process begins with the drying power of air, which causes (5) ______________ [p.528], the evaporation of water from plant parts exposed to air. The collective strength of (6) ______________ [p.528] bonds between water molecules in the narrow, tubular xylem cells imparts (7) ______________ [p.528]. This provides unbroken, fluid columns of water. The xylem water columns, as they are pulled upward, are under (8) ______________ [p.528]. This force extends from the veins inside leaves, down through the stems, and on into the (9) ______________ [p.529], where water is being absorbed. As long as water molecules continue to escape from the plant, the continuous tension inside the (10) ______________ [p.529] permits more molecules to be pulled upward from the roots to replace them.

30.4. HOW DO STEMS AND LEAVES CONSERVE WATER? [pp.530–531]

Selected Words: *inward* diffusion [p.530], *outward* diffusion [p.530]

Boldfaced, Page-Referenced Terms

[p.530] turgor pressure ____________________

____

[p.530] cuticle ____________________

____

[p.530] cutin ____________________

____

[p.530] stomata ____________________

____

[p.530] guard cells ____________________

____

[p.531] CAM plants ____________________

____

True/False

If the statement is true, write a "T" in the blank. If the statement is false, make it correct by changing the underlined word(s) and writing the correct word(s) in the answer blank.

_______________ 1. Of the water moving into a leaf, <u>2 percent</u> or more is lost by evaporation into the surrounding air. [p.530]

_______________ 2. When evaporation exceeds water uptake by roots, plant tissues <u>wilt</u> and water-dependent activities are seriously disrupted. [p.530]

_______________ 3. Epidermal plant cells secrete a translucent, water-impermeable layer called the <u>cuticle</u>, which coats the cells wall regions exposed to air. [p.530]

_______________ 4. Plant epidermal layers are peppered with tiny openings, called <u>guard cells</u>, through which water leaves the plant and carbon dioxide enters. [p.530]

_______________ 5. When a pair of <u>guard cells</u> swells with turgor pressure, the opening between them opens. [p.530]

_______________ 6. In most plants, stomata remain <u>open</u> during the daylight photosynthetic period; water is lost from plants, but they accumulate carbon dioxide. [p.530]

_______________ 7. Stomata stay <u>closed</u> at night in most plants. [p.530]

_______________ 8. Photosynthesis starts when the sun comes up; as the morning progresses, carbon dioxide levels <u>increase</u> in cells, including guard cells. [p.530]

_______________ 9. As the morning progresses, a drop in carbon dioxide level within guard cells triggers an inward active transport of potassium ions; water follows by osmosis and the fluid pressure opens the <u>cuticle</u>. [p.531]

_______________ 10. When the sun goes down and photosynthesis stops, carbon dioxide levels rise; stomata close when <u>potassium</u>, then water, moves out of the guard cells. [p.531]

____________ 11. CAM plants such as cacti and other succulents open stomata during the day when they fix carbon dioxide by way of a special C4 metabolic pathway. [p.531]

____________ 12. CAM plants use carbon dioxide in photosynthesis the following night when stomata are closed. [p.531]

30.5. HOW ARE ORGANIC COMPOUNDS DISTRIBUTED THROUGH PLANTS?

[pp.532–533]

Selected Words: *source* [p.532], *sink* [p.532]

Boldfaced, Page-Referenced Terms

[p.532] phloem ________________________

____

[p.532] sieve tubes ________________________

____

[p.532] companion cells ________________________

____

[p.532] translocation ________________________

____

[p.532] pressure flow theory ________________________

____

Fill-in-the-Blanks

Sucrose and other organic products resulting from (1) ____________ [p.532] are used throughout the plant. Most plant cells store their carbohydrates as (2) ____________ [p.532] in plastids. The (3) ____________ [p.532] and fats, which cells synthesize from carbohydrates and the amino acids distributed to them, are stored in many (4) ____________ [p.532]. Avocados and some other fruits also accumulate (5) ____________ [p.532]. (6) ____________ [p.532] molecules are too large to cross cell membranes and too insoluble to be transported to other regions of the plant body. Overall, (7) ____________ [p.532] are too large and fats are too insoluble to be transported from the storage sites. Plant cells convert storage forms of organic compounds to (8) ____________ [p.532] of smaller size that are more easily transported through the phloem. For example, the cells degrade starch to glucose monomers. When one of these monomers combines with fructose, the result is (9) ____________ [p.532], an easily transportable sugar. Experiments with phloem-embedded aphid mouthparts revealed that (10) ____________ [p.532], an easily transportable sugar, was the most abundant carbohydrate being forced out of those tubes.

Matching

Choose the most appropriate answer for each term.

11. ____ translocation [p.532]
12. ____ sieve-tube members [p.532]
13. ____ companion cells [p.532]
14. ____ source [p.532]
15. ____ sink [p.532]
16. ____ pressure flow theory [p.533]

A. Any region where organic compounds are being loaded into the sieve tubes
B. Nonconducting cells adjacent to sieve-tube members that supply energy to load sucrose at the source
C. Any region of the plant where organic compounds are being unloaded from the sieve-tube system and used or stored
D. Process occurring in phloem that distributes sucrose and other organic compounds through the plant (apparently under pressure)
E. Internal pressure builds up at the source end of a sieve-tube system and pushes the solute-rich solution toward a sink, where the solutes are removed
F. Passive conduits for translocation within vascular bundles; water and organic compounds flow rapidly through large pores on their end walls

---

Self-Quiz

____ 1. For plants, the essential elements include ____________. [p.524]
a. oxygen, carbon, and nitrogen
b. oxygen, hydrogen, and nitrogen
c. oxygen, carbon, and hydrogen
d. carbon, nitrogen, and hydrogen

____ 2. Macronutrients are the nine dissolved mineral ions that ____________. [p.524]
a. are found in major concentrations in the Earth's surface
b. occur in only small traces in plant tissues
c. are required in amounts above 0.5 percent of the plant's dry weight
d. can function only without the presence of micronutrients

____ 3. Gaseous nitrogen is converted to a plant-usable form by ____________. [p.527]
a. root nodules
b. mycorrhizae
c. nitrogen-fixing bacteria
d. Venus flytraps

____ 4. ____________ prevent(s) inward-moving water from moving past the abutting walls of the root endodermal cells. [p.526]
a. Cytoplasm
b. Plasma membranes
c. Osmosis
d. Casparian strips

____ 5. Most of the water moving into a leaf is lost through ____________. [p.528]
a. osmotic gradients being established
b. transpiration
c. pressure flow forces
d. translocation

____ 6. Stomata remain ____________ during daylight, when photosynthesis occurs, but remain ____________ during the night, when carbon dioxide accumulates through aerobic respiration. [p.530]
a. open; open
b. closed; open
c. closed; closed
d. open; closed

_____ 7. By control of _______________ levels inside the guard cells of stomata, the activity of stomata is controlled when leaves are losing more water than roots can absorb. [p.531]
   a. oxygen
   b. potassium
   c. carbon dioxide
   d. oxygen

_____ 8. Without _______________, plants would rapidly wilt and die during hot, dry spells. [p.530]
   a. a cuticle
   b. mycorrhizae
   c. phloem
   d. cotyledons

_____ 9. The _______________ theory of water transport states that hydrogen bonding allows water molecules to maintain a continuous fluid column as water is pulled from roots to leaves. [p.528]
   a. pressure flow
   b. cohesion–tension
   c. evaporation
   d. nitrogen-fixation

_____ 10. Leaves represent _____________ regions; growing leaves, stems, fruits, seeds, and roots represent _____________ regions. [p.532]
   a. source; source
   b. sink; source
   c. source; sink
   d. sink; sink

Chapter Objectives/Review Questions

1. Explain how carnivorous plants such as Venus flytraps and bladderworts satisfy their nutritional needs. [pp.522–523]
2. _____________ _____________ is the study of adaptations by which plants function in their environment. [p.522]
3. Define the terms *soil, humus,* and *loam.* [pp.524–525]
4. Explain what is meant by a soil profile, and name the soil layer that is the most essential for plant growth. [p.525]
5. Define the term *nutrients* in regard to plant nutrition. [p.526]
6. Name the three elements considered essential for plant nutrition. [p.526]
7. Distinguish between macronutrients and micronutrients in relation to their role in plant nutrition. [p.524]
8. Distinguish between leaching and erosion. [p.525]
9. Trace the path of water and mineral ions into roots, and name the structure and function of the plant structures involved. [p.526]
10. Differentiate between the endodermis and exodermis of the root cortex. [p.526]
11. Explain the importance of the Casparian strip. [p.526]
12. Explain why root hairs are so valuable in root absorption. [p.526]
13. Define the term *mutualism.* [p.527]
14. Describe the roles of root nodules and mycorrhizae in plant nutrition. [p.527]
15. The evaporation of water from leaves as well as from stems and other plant parts is known as _____________. [p.528]
16. Explain how the cohesion–tension theory describes how water moves in a plant. [p.528]
17. In a plant's vascular tissues, water moves through pipelines called _____________. [p.528]
18. Define *turgor pressure* and explain its role in controlling water loss at stomata. [p.525]
19. Explain the importance of the cuticle and cutin in plant physiology. [p.530]
20. Describe the operation of the guard cells and stomata during daylight hours. [pp.530–531]
21. Describe the operation of the guard cells and stomata during evening hours. [pp.530–531]
22. Describe the mechanisms by which CAM plants conserve water. [p.531]
23. Describe the role of phloem sieve-tube members and companion cells in the movement of organic compounds. [p.532]

24. ________ is the main form in which sugars are transported through most plants. [p.532]
25. According to the ________ ________ theory, pressure builds up at the source end of a sieve-tube system and pushes solutes toward a sink, where they are removed. [pp.532–533]

Integrating and Applying Key Concepts

1. How do you think maple syrup is made from maple trees? Which specific systems of the plant are involved, and why are maple trees tapped only at certain times of the year?

31

PLANT REPRODUCTION

Interactive Exercises

A Coevolutionary Tale [pp.536–537]

31.1. REPRODUCTIVE STRUCTURES OF FLOWERING PLANTS [pp.538–539]

31.2. FOCUS ON HEALTH: *Pollen Sets Me Sneezing* [p.539]

Selected Words: *Angraecum sesquipedale* [p.537], sepals [p.538], petals [p.538], calyx [p.538], corolla [p.538], anther [p.539], stigma [p.539], style [p.539], *perfect* flowers [p.539], *imperfect* flowers [p.539], *allergic rhinitis* [p.539], hypersensitivity [p.539]

Boldfaced, Page-Referenced Terms

[p.536] flower ______________________________

____

[p.536] coevolution ______________________________

____

[p.536] pollinator ______________________________

____

[p.538] sporophyte ______________________________

____

[p.538] gametophytes ______________________________

____

[p.539] stamens ______________________________

____

[p.539] pollen grains ______________________________

____

[p.539] carpels ______________________________

____

[p.539] ovary ______________________________

____

Labeling and Matching

Identify each numbered part of the accompanying illustration. Complete the exercise by matching and entering the letter of the proper description in the parentheses following each label. [p.538]

1. ______________ ( )
2. ______________ ( )
3. ______________ ( )
4. ______________ ( )
5. ______________ ( )
6. ______________ ( )

A. An event that produces a young sporophyte
B. A reproductive shoot produced by the sporophyte
C. The "plant"; a vegetative body that develops from a zygote
D. Cellular division event occurring within flowers to produce spores
E. Produces haploid eggs by mitosis
F. Produces haploid sperm by mitosis

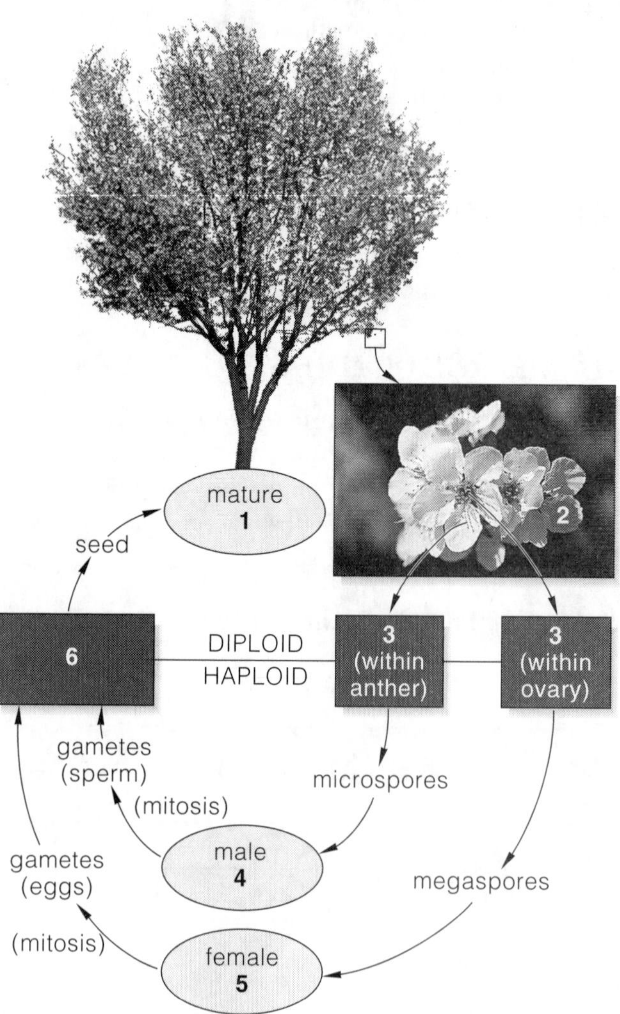

Labeling

Identify each numbered part of the accompanying illustration. [p.538]

7. ______________
8. ______________
9. ______________
10. ______________
11. ______________
12. ______________
13. ______________
14. ______________
15. ______________
16. ______________
17. ______________

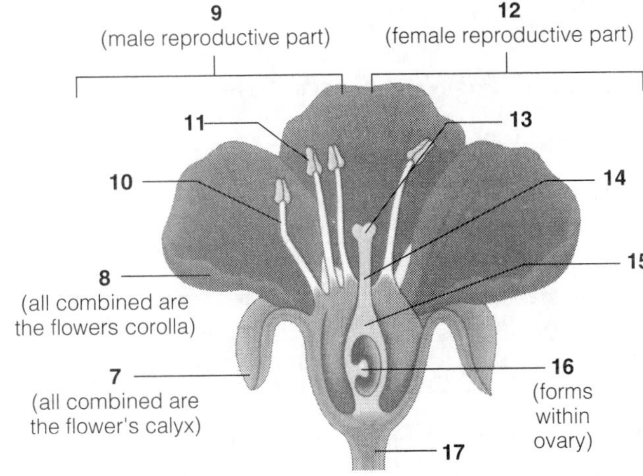

Matching

Choose the most appropriate answer for each term.

18. _____ sepals [p.538]
19. _____ petals [p.538]
20. _____ stamens [p.539]
21. _____ ovule [p.538]
22. _____ pollen grain [p.539]
23. _____ carpel [p.539]
24. _____ ovary [p.539]
25. _____ perfect flowers [p.539]
26. _____ imperfect flowers [p.539]

A. Have both male and female parts
B. Mature haploid spore whose contents develop into a male gametophyte
C. Collectively, the flower's "corolla"
D. Have male or female parts, but not both
E. Female reproductive part; includes stigma, style, and ovary
F. Structure inside the ovary; matures to become a seed
G. Found just inside the flower's corolla; the male reproductive parts
H. Lower portion of the carpel where egg formation, fertilization, and seed development occur
I. Outermost leaflike whorl of floral organs; collectively, the calyx

31.3. A NEW GENERATION BEGINS [pp.540–541]

Selected Words: sperm nuclei [p.540], embryo sac [p.540], endosperm mother cell [p.540], pollen tube [p.541], zygote [p.541], triploid [p.541]

Boldfaced, Page-Referenced Terms

[p.540] microspores ____

____

[p.540] ovule ____

____

[p.540] integuments ____

____

[p.540] megaspores ____

____

[p.540] endosperm ____

____

[p.540] pollination ____

____

[p.541] double fertilization ____

____

Fill-in-the-Blanks

The numbered items in the illustration on the next page represent missing information; complete the blanks in the following narrative to supply that information. [p.540]

Within each (1) ____________, mitotic divisions produce four masses of spore-forming cells, each mass forming within a(n) (2) ____________ ____________. Each one of these diploid cells is known as a(n) (3) ____________ ____________ cell and undergoes (4) ____________ to produce four haploid (5) ____________. Mitosis within each haploid microspore results in a two-celled haploid body, the immature male gametophyte. One of these cells will give rise to a(n) (6) ____________ ____________; the other cell will develop into a(n) (7) ____________ - ____________ cell. Mature microspores are eventually released from the pollen sacs of the anther as (8) ____________. Pollination occurs, and after the pollen lands on a(n) (9) ____________ of a carpel, the pollen tube develops from one of the cells in the pollen grain; the other cell within the pollen grain divides to form two sperm cells. As the pollen tube grows through the carpel tissues, it contains the two sperm cells and a tube nucleus. The pollen tube with its two sperm cells and the tube nucleus is known as the mature (10) ____________ ____________.

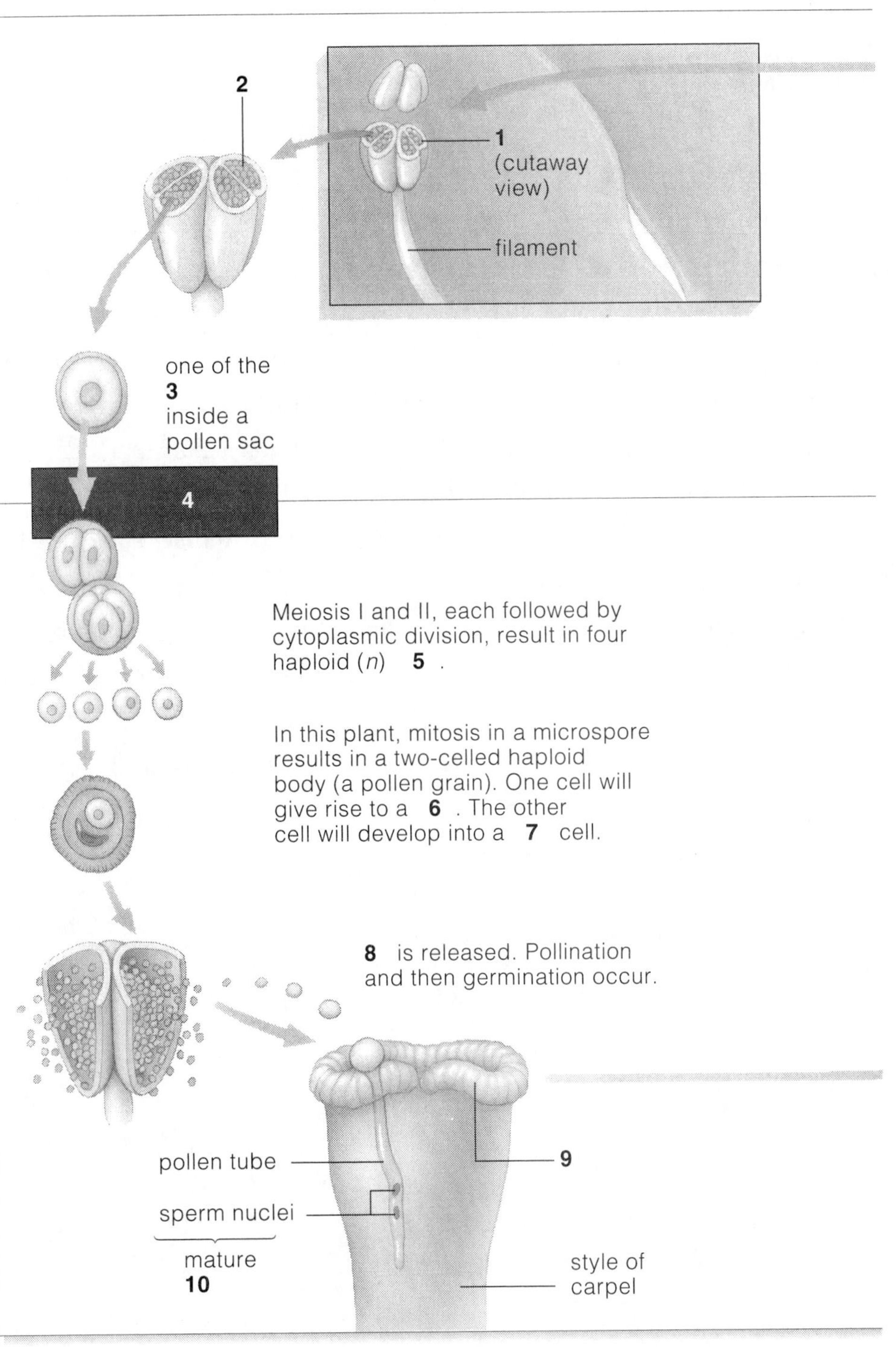
2
1
(cutaway
view)
filament
one of the
3
inside a
pollen sac
4
Meiosis I and II, each followed by
cytoplasmic division, result in four
haploid (n) 5 .
In this plant, mitosis in a microspore
results in a two-celled haploid
body (a pollen grain). One cell will
give rise to a 6 . The other
cell will develop into a 7 cell.
8 is released. Pollination
and then germination occur.
pollen tube
sperm nuclei
mature
10
9
style of
carpel

Fill-in-the-Blanks

The numbered items in the illustration on the next page represent missing information; fill in the numbered blanks to complete the following narrative. [pp.540–541]

In the carpel of the flower, one or more dome-shaped, diploid tissue masses develop on the inner wall of the ovary. Each mass is the beginning of a(n) (11) ________________ . A tissue forms inside a domed mass as it grows, and one or two protective layers called (12) ________________ form around it. Inside each mass, a diploid cell (the endosperm mother cell) divides by (13) ________________ to form four haploid spores known as (14) ________________ . Commonly, all but one (15) ________________ disintegrates. The remaining (16) ________________ undergoes (17) ________________ three times without cytoplasmic division. At first, this structure is a cell with (18) ________________ haploid nuclei. Cytoplasmic division results in a seven-cell (19) ________________ ________________ , which represents the mature (20) ________________ ________________ . Six of those cells have a single nucleus, but one cell has (21) ________________ (number) nuclei ($2n$) and represents the (22) ________________ mother cell ($n + n$). Another haploid cell within the embryo sac is the (23) ________________ . Following (24) ________________ ________________ with one sperm, the $n + n$ cell will help form the $3n$ (25) ________________ , a nutritive tissue for the forthcoming embryo. The other sperm involved in this unique fertilization process fertilizes the haploid egg; this combination forms the diploid (26) ________________ . Thus the ovule is transformed into a(n) (27) ________________ that is composed of three parts, a seed (28) ________________ , an embryo, and nourishment for the embryo, the endosperm.

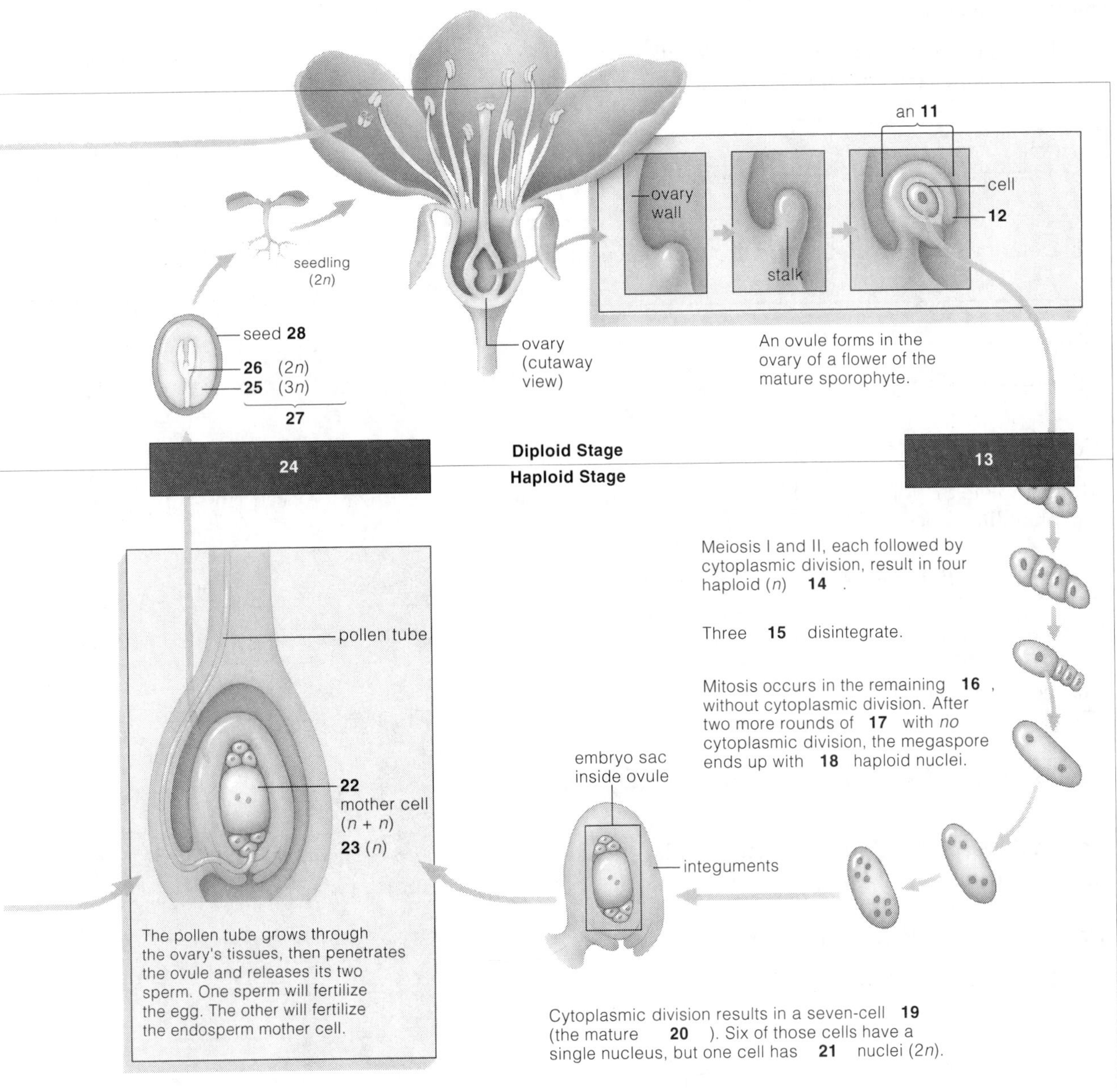
an 11
cell
12
ovary wall
stalk
seedling (2n)
seed 28
26 (2n)
25 (3n)
27
ovary (cutaway view)
An ovule forms in the ovary of a flower of the mature sporophyte.
24
Diploid Stage
Haploid Stage
13
Meiosis I and II, each followed by cytoplasmic division, result in four haploid (n) 14 .
Three 15 disintegrate.
Mitosis occurs in the remaining 16 , without cytoplasmic division. After two more rounds of 17 with *no* cytoplasmic division, the megaspore ends up with 18 haploid nuclei.
pollen tube
22
mother cell (n + n)
23 (n)
embryo sac inside ovule
integuments
The pollen tube grows through the ovary's tissues, then penetrates the ovule and releases its two sperm. One sperm will fertilize the egg. The other will fertilize the endosperm mother cell.
Cytoplasmic division results in a seven-cell 19 (the mature 20). Six of those cells have a single nucleus, but one cell has 21 nuclei (2n).

Short Answer

29. What guides the growth of the pollen tube down through the female floral tissues toward the chamber holding the egg? [pp.540–541] ______________________________

____

____

____

30. Describe the site of double fertilization, an event known only in flowering plants. [p.541] __________

____

____

____

Complete the Table

31. Complete the following table to summarize the unique double fertilization occurring only in flowering plant life cycles. [p.541]

Double Fertilization Products	*Origin*	*Produces?*	*Function*
a. Zygote ($2n$) nucleus			
b. Endosperm ($3n$) nucleus			

31.4. FROM ZYGOTE TO SEEDS AND FRUITS [pp.542–543]

Selected Words: *simple* fruits [pp.542–543], *aggregate* fruits [pp.542–543], *multiple* fruits [pp.542–543], accessory fruit [pp.542–543], endocarp [p.543], mesocarp [p.543], exocarp [p.543]

Boldfaced, Page-Referenced Terms

[p.542] fruit ______________________________

____

[p.542] cotyledons ______________________________

____

[p.543] seed ______________________________

____

[p.543] pericarp ______________________________

____

Complete the Table

1. Complete the following table, which summarizes concepts associated with seeds and fruits.

Structure	*Origin*
[p.542] a. Cotyledons	
[p.543] b. Seeds	
[p.543] c. Seed coat	
[pp.542–543] d. Fruit	

Labeling

Identify each numbered part of the accompanying illustration. [p.542]

2. ________________ - ________________
3. ________________ - ________________
4. ________________
5. ________________
6. ________________ - ________________

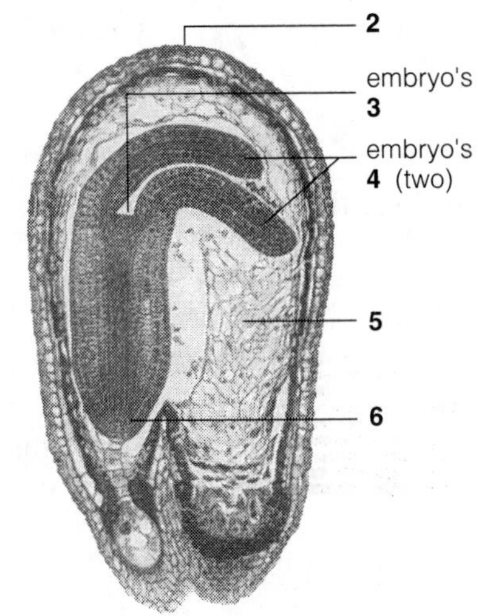

Fill-in-the-Blanks

Following fertilization, the newly formed zygote initiates a course of (7) ________________ [p.542] cell divisions that lead to a mature embryo (8) ________________ [p.542]. The embryo develops as part of a(n) (9) ________________ [p.542] and is accompanied by the formation of a(n) (10) ________________ [p.542], a mature ovary.

By the time a *Capsella* embryo is near maturity, two (11) ________________ [p.542], or seed leaves, have begun to develop from two lobes of meristematic tissue. Dicot embryos, such as *Capsella*, have (12) ________________ (number) [p.542] cotyledon(s), and monocot embryos have (13) ________________ [p.542] (number) cotyledon(s). Like most dicots, the *Capsella* embryo has rather thick cotyledons that absorb nutrients from the (14) ________________ [p.542] of the seed and store them in its cotyledons. In corn, wheat, and most other monocots, endosperm is not tapped until the seed (15) ________________ [p.542]. Digestive enzymes become stockpiled inside the (16) ________________ [p.542] cotyledons of monocot embryos.

When the (17) ______________ [p.542] do become active, nutrients stored in the endosperm will be released and transferred to the growing (18) ______________ [p.542].

From the time a zygote forms until an embryo matures, a parent sporophyte plant transfers nutrients to tissues of the (19) ______________ [p.543]. Food reserves accumulate in (20) ______________ [p.543] or cotyledons. Eventually, the ovule separates from the (21) ______________ [p.543] wall, and its integuments thicken and harden into a seed (22) ______________ [p.543]. The embryo, food reserves, and seed coat are a self-contained package—a(n) (23) ______________ [p.543], which is defined as a mature (24) ______________ [p.543]. While seeds are forming, changes occur in other floral parts and (25) ______________ [p.543] begin to form. There are three categories of these: (26) ______________ [p.543] (either dry or fleshy, and derived from a single ovary), (27) ______________ [p.543] (from many separate ovaries of a single flower), and (28) ______________ [p.543] (from many separate ovaries of a single flower, all attached to the same receptacle). An apple is a(n) (29) ______________ [p.543] fruit, composed mainly of an enlarged receptacle and a calyx. Botanists often refer to three divisions of fleshy fruits: The (30) ______________ [p.543] is the innermost portion around a seed or seeds; (31) ______________ [p.543] is the fleshy portion; and (32) ______________ [p.543] is the skin. Together, the three fruit regions are called a(n) (33) ______________ [p.543].

Choice

For questions 34–41, choose from the following fruit types. [p.542]

a. simple dry dehiscent b. simple dry indehiscent c. simple fleshy fruits
d. aggregate fruits e. multiple fruits f. accessory fruits

34. _____ pineapple
35. _____ nuts, grains
36. _____ raspberries
37. _____ grapes, tomatoes
38. _____ apples, pears, strawberries
39. _____ sunflowers and carrots
40. _____ oranges
41. _____ peaches, cherries, and other drupes

31.5. DISPERSAL OF FRUITS AND SEEDS [p.545]

31.6. FOCUS ON SCIENCE: *Why So Many Flowers and So Few Fruits?* [p.546]

Selected Words: *seed dispersal* [p.545], *Theobroma cacao* [p.545], giant saguaro [p.546]

Choice

For questions 1–10, choose from the following [p.544]:

a. wind-dispersed fruits b. fruits dispersed by animals c. water-dispersed fruits

1. _____ heavy wax coats
2. _____ coconut palms
3. _____ seed coats assaulted by digestive enzymes to assist in releasing embryos
4. _____ maples
5. _____ orchids
6. _____ air sacs
7. _____ hooks, spines, hairs, and sticky surfaces
8. _____ cacao
9. _____ dandelions
10. _____ cockleburs, bur clover, and bedstraw

Short Answer

11. Explain why, although it might not appear so, the following statement may actually reflect an adaptation for reproductive success in terms of Darwinian evolutionary theory: "Although giant saguaro cactus plants produce many flowers on each plant, perhaps less than 50 percent of these flowers set fruit with viable seeds." [p.545] ____________________

____

____

____

31.7. ASEXUAL REPRODUCTION OF FLOWERING PLANTS [pp.546–547]

Selected Words: *Populus tremuloides* [p.546], cuttings [p.546], grafted [p.546], *Daucus carota* [p.546]

Boldfaced, Page-Referenced Terms

[p.546] vegetative growth ____________________

____

[p.546] parthenogenesis ____________________

____

[p.546] tissue culture propagation ____________________

____

Matching

Match the following asexual reproductive modes of flowering plants. [p.546]

1. ____ corm
2. ____ bulb
3. ____ parthenogenesis
4. ____ runner
5. ____ vegetative propagation on modified stems
6. ____ rhizome
7. ____ tuber
8. ____ tissue culture propagation (induced propagation)
9. ____ vegetative growth

A. In a general sense, new plants develop from tissues or organs that drop or separate from parent plants
B. New shoots arise from axillary buds (enlarged tips of slender underground rhizomes)
C. New plants arise from cells in parent plant that were not irreversibly differentiated; a laboratory technique
D. New plants arise at nodes of underground horizontal stem
E. New plant arises from axillary bud on short, thick, vertical underground stem
F. New plants arise at nodes on an aboveground horizontal stem
G. New plant arises from an axillary bud on a short underground stem
H. Involves asexual reproduction using runners, rhizomes, corms, tubers, and bulbs
I. Embryo develops without nucleus or cellular fusion

Matching

Choose the most appropriate example for each modified stem. [p.546]

10. ____ bulb
11. ____ rhizome
12. ____ tuber
13. ____ runner
14. ____ corm

A. Potato
B. Strawberry
C. Gladiolus
D. Onion, lily
E. Bermuda grass

Self-Quiz

____ 1. The joint evolution of flowers and their pollinators is known as ______________ . [p.536]
   a. adaptation
   b. coevolution
   c. joint evolution
   d. covert evolution

____ 2. A stamen is ______________ . [p.539]
   a. composed of a stigma
   b. the mature male gametophyte
   c. the site where microspores are produced
   d. part of the vegetative phase of an angiosperm

____ 3. The portion of the carpel that contains an ovule is the ______________ . [p.539]
   a. stigma
   b. anther
   c. style
   d. ovary

____ 4. The phase in the life cycle of plants that gives rise to spores is known as the ______________ . [p.538]
   a. gametophyte
   b. embryo
   c. sporophyte
   d. seed

_____ 5. A gametophyte is _______________ . [p.538]
a. a gamete-producing plant
b. haploid
c. both a and b
d. the plant produced by the fusion of gametes

_____ 6. A characteristic of a seed is that it _______________ . [p.543]
a. contains an embryo sporophyte
b. represents an arrested growth stage
c. is covered by hardened and thickened integuments
d. all of these

_____ 7. An immature fruit is a(n) __________ and an immature seed is a(n) __________ . [p.543]
a. ovary; megaspore
b. ovary; ovule
c. megaspore; ovule
d. ovule; ovary

_____ 8. In flowering plants, one sperm nucleus fuses with that of an egg, and a zygote forms that develops into an embryo. Another sperm fuses with __________ . [p.541]
a. a primary endosperm cell to produce three cells, each with one nucleus
b. a primary endosperm cell to produce one cell with one triploid nucleus
c. both nuclei of the endosperm mother cell, forming a primary endosperm cell with a single triploid nucleus
d. one of the smaller megaspores to produce what will eventually become the seed coat

_____ 9. "Simple, aggregate, multiple, and accessory" refer to types of _______________ . [p.543]
a. carpels
b. seeds
c. fruits
d. ovaries

_____ 10. "When a leaf falls or is torn away from a jade plant, a new plant can develop from the leaf, from meristematic tissue." This statement refers to _______________ . [p.546]
a. parthenogenesis
b. runners
c. tissue culture propagation
d. vegetative propagation

Chapter Objectives/Review Questions

1. __________ refers to two or more species jointly evolving as an outcome of close ecological interactions. [p.536]
2. Describe the role of a pollinator. [p.536]
3. Distinguish between sporophytes and gametophytes. [p.538]
4. Identify the various parts of a typical flower and state their functions. [pp.538–539]
5. Walled microspores form in pollen sacs and develop into __________ __________ . [p.539]
6. Distinguish between a flower that is *perfect* and one that is *imperfect.* [p.539]
7. Describe the condition called allergic rhinitis. [p.539]
8. Relate the sequence of events and structures involved that give rise to microspores and megaspores. [p.540]

9. ________ is the transfer of pollen grains to a receptive stigma. [p.540]
10. What structures represent the male gametophyte and female gametophyte in flowering plants? List the contents of each. [pp.540–541]
11. The endosperm mother cell in the embryo sac is composed of two haploid ________. [p.541]
12. Describe the double fertilization that occurs uniquely in the flowering plant life cycle. [p.541]
13. After pollination and double fertilization, a(n) ________ and nutritive tissue form in the ________, which becomes a seed. [p.541]
14. How is endosperm formed? What is the function of endosperm? [p.541]
15. Describe the formation of the embryo sporophyte; describe the origin and formation of seeds and fruits. [p.542]
16. Review the general types of fruits produced by flowering plants. [pp.542–543]
17. Seeds and fruits are structurally adapted for ________ by air currents, water currents, and many kinds of animals. [p.544]
18. Define the term *vegetative growth.* [p.546]
19. Distinguish among parthenogenesis, vegetative propagation, and tissue culture propagation; give an example of each. [p.546]
20. List representative plant examples of a runner, a rhizome, a corm, a tuber, and a bulb. [p.546]

Integrating and Applying Key Concepts

In terms of botanical morphology, a flower is interpreted as "a reproductive shoot bearing organs." After studying floral organs in this chapter and leaf structure in Chapter 29 of the text, can you think of any comparable structural evidence that might have led botanists to arrive at this conclusion?

32

PLANT GROWTH AND DEVELOPMENT

Interactive Exercises

Foolish Seedlings and Gorgeous Grapes [pp.550–551]

32.1. PATTERNS OF EARLY GROWTH AND DEVELOPMENT—AN OVERVIEW [pp.552–553]

Selected Words: *bakane* "Foolish seedling" [p.550]

Boldfaced, Page-Referenced Terms

[p.550] gibberellin ____

____

[p.550] hormones ____

____

[p.552] germination ____

____

[p.552] imbibition ____

____

Fill-in-the-Blanks

Before or after seed dispersal from the parent plant, the growth of the (1) ______________ [p.552] idles. For seeds, (2) ______________ [p.552] is the resumption of growth by an immature stage in the life cycle after a period of arrested development. Germination depends on (3) ______________ [p.552] factors, such as temperature, moisture, and oxygen level in soil, together with the number of seasonal daylight hours available. By a process known as (4) ______________ [p.552], water molecules move into the seed. As more water moves in, the seed swells and its coat (5) ______________ [p.552]. Once the seed coat splits, more oxygen reaches the embryo, and (6) ______________ [p.552] respiration moves into high gear. The embryo's (7) ______________ [p.552] cells begin to divide rapidly. In general, the (8) ______________ [p.552] meristem is the first to be activated. Its meristematic descendants divide, elongate, and give rise to the (9) ______________ ______________ [p.552]. When this structure breaks through the seed coat, (10) ______________ [p.552] is over.

For both monocots and dicots, the patterns of germination, growth, and development that unfold have a(n) (11) ______________ [p.552] basis; they are dictated by the plant's (12) ______________ [p.552]. All cells in the new plant arise from the same cell, the (13) ______________ [p.552]. Thus all cells inherit the same (14) ______________ [p.552]. Unequal (15) ______________ [p.552] divisions between daughter cells lead to differences in their (16) ______________ [p.553] mechanisms. Activities in daughter cells start to vary as a result of (17) ______________ [p.553] gene expression. As an example, genes governing the synthesis of growth-stimulating (18) ______________ [p.553] are activated in some cells but not others. (19) ______________ [p.553] among genes, hormones, and the environment govern how an individual plant grows and develops.

Labeling

Identify each numbered part of the accompanying illustration. [p.553]

20. ______________
21. ______________
22. ______________
23. ______________
24. ______________
25. ______________
26. ______________
27. ______________
28. ______________
29. ______________
30. ______________
31. ______________
32. ______________
33. ______________
34. ______________
35. ______________
36. ______________
37. ______________

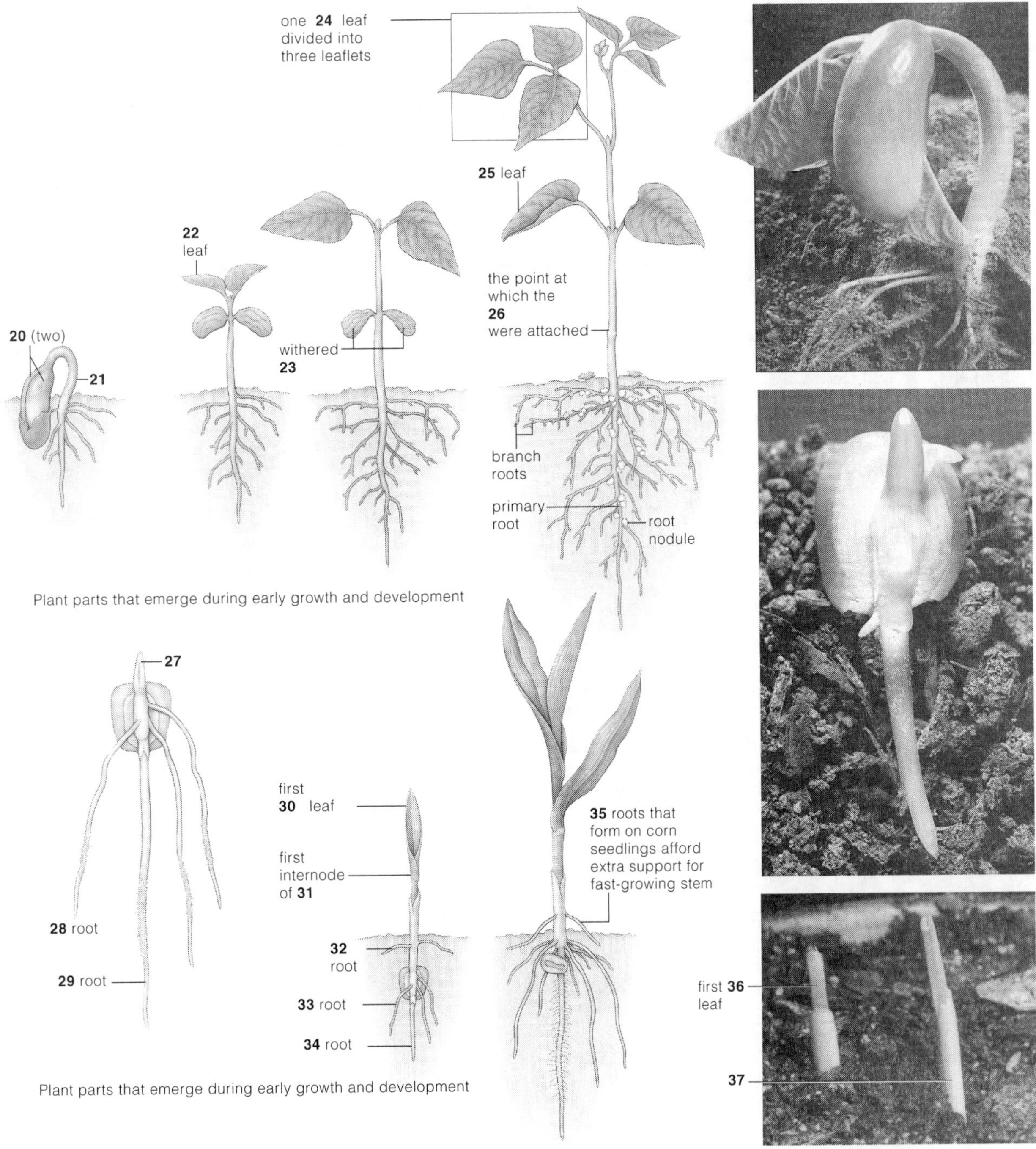

Plant parts that emerge during early growth and development

Plant parts that emerge during early growth and development

32.2. WHAT THE MAJOR PLANT HORMONES DO [pp.554–555]

32.3. ADJUSTING THE DIRECTION AND RATES OF GROWTH [pp.556–557]

Selected Words: *turgor* pressure [p.554], *apical dominance* [p.555], *Agent Orange* [p.555], *trope* [p.556], *auxin* [p.556], *auxein* [p.556], *thigma* [p.557]

Boldfaced, Page-Referenced Terms

[p.554] growth ____

____

[p.554] development ____

____

[p.555] auxins ____

____

[p.555] coleoptile ____

____

[p.555] herbicide ____

____

[p.555] cytokinins ____

____

[p.555] abscisic acid ____

____

[p.555] ethylene ____

____

[p.556] gravitropism ____

____

[p.556] statoliths ____

____

[p.556] phototropism ____

____

[p.557] flavoprotein ____

____

[p.557] thigmotropism ____

____

Choice

For questions 1–14, choose from the following: [all from pp.554–555]

a. auxins b. gibberellins c. cytokinins d. abscisic acid e. ethylene

1. _____ Ancient Chinese burned incense to hurry fruit ripening
2. _____ Natural and synthetic versions are used to prolong the shelf life of cut flowers, lettuces, mushrooms, and other vegetables
3. _____ In orchards, trees are sprayed with IAA to thin overcrowded seedlings in the spring
4. _____ Inhibits cell growth, induces bud dormancy, and prevents seeds from germinating prematurely; causes stomata to close when a plant is water stressed
5. _____ IAA, the most pervasive naturally occurring compound of its type
6. _____ Oranges and other citrus fruits are exposed to this to brighten the color of their rind before being displayed in the market
7. _____ Influences stem lengthening; influences plant responses to gravity and light and promotes coleoptile lengthening
8. _____ Promotes stem lengthening; helps buds and seeds end dormancy and resume growth in the spring
9. _____ Used to prevent premature fruit drop so that all fruit can be picked at the same time
10. _____ Stimulates cell division
11. _____ Used by food distributors to ripen green tomatoes and other fruit after delivery to grocery stores
12. _____ Most abundant in root and shoot meristems and in the tissues of maturing fruit
13. _____ Induces various aging responses, including fruit ripening and leaf drop
14. _____ Some synthetic forms, such as Agent Orange, serve as herbicides

Matching

Choose the most appropriate answer for each term.

15. _____ coleoptile [p.555]
16. _____ hormone [p.548]
17. _____ apical dominance [p.555]
18. _____ herbicide [p.555]
19. _____ growth [p.554]
20. _____ gibberellins [pp.554–555]
21. _____ IAA [p.555]
22. _____ development [p.554]
23. _____ target cell [p.554]
24. _____ turgor pressure [p.554]

A. The pressure of the water against the cell wall within a cell
B. The emergence of specialized, morphologically different body parts; measured in qualitative terms
C. Compounds that promote cell lengthening by influencing the orientation of microtubules
D. Hormonal effect that blocks lateral bud growth
E. An increase in the number, size, and volume of cells; measured in quantitative terms
F. A signaling molecule released from one cell that travels to target cells and stimulates or inhibits gene activity
G. The most important naturally occurring auxin
H. Any synthetic auxin compound used to kill some plants but not others
I. A thin sheath around the primary shoot of grass seedlings, such as corn plants
J. The location where a plant hormone acts as a signal, causing a specific change in metabolism or gene expression

Choice

For questions 25–34, choose from the following:

a. phototropism [pp.556–557] b. gravitropism [p.556]
c. thigmotropism [p.557] d. mechanical stress [p.557]

25. ____ More intense sunlight on one side of a plant — stems curve toward the light
26. ____ Vines climbing around a fencepost as they grow upward
27. ____ Plants grown outdoors have shorter stems than plants grown in a greenhouse
28. ____ A root turned on its side will curve downward
29. ____ Briefly shaking a plant daily inhibits the growth of the entire plant
30. ____ If a potted seedling is turned on its side; then the growing stem will curve upward
31. ____ Leaves turn until their flat surfaces face light
32. ____ Flavoprotein may be a central component in this response
33. ____ The tendrils of plants may sometimes be involved in this response
34. ____ Statoliths are generally involved in this type of response

32.4. HOW DO PLANTS KNOW WHEN TO FLOWER? [pp.558–559]

Selected Words: *circadian* rhythm [p.558], *night length* [p.558], "short-day" [p.559], "long-day" [p.559]

Boldfaced, Page-Referenced Terms

[p.558] biological clocks ________________________________

____

[p.558] phytochrome ________________________________

____

[p.558] photoperiodism ________________________________

____

[p.558] long-day plants ________________________________

____

[p.558] short-day plants ________________________________

____

[p.558] day-neutral plants ________________________________

____

Matching

Choose the most appropriate answer for each term.

1. ____ photoperiodism [p.558]
2. ____ Pr and Pfr [p.558]
3. ____ day-neutral plants [p.558]
4. ____ phytochrome activation [p.558]
5. ____ "long-day" plants [p.558]
6. ____ circadian rhythms [p.558]
7. ____ night length [pp.558–559]
8. ____ biological clocks [p.558]
9. ____ rhythmic leaf movements [p.558]
10. ____ phytochromes [p.558]

A. Biological activities that recur in cycles of 24 hours or so
B. Flower in spring when daylength exceeds a critical value
C. Internal time-measuring mechanisms with roles in adjusting daily activities
D. Any biological response to a change in the relative length of daylight and darkness in the 24-hour cycle; active Pfr may be an alarm button for this process
E. A blue-green pigment that absorbs red or far-red wavelengths, with different results
F. An example of a circadian rhythm in bean plants
G. The abbreviations for the active and inactive forms of phytochrome, respectively
H. Flower when mature enough to do so
I. The actual environmental cue that signals flower production
J. May induce plant cells to take up free calcium ions or induce certain plant cell organelles to release them

Labeling

Identify each numbered part of the accompanying illustration. [p.558]

11. ________
12. ________
13. ________
14. ________
15. ________

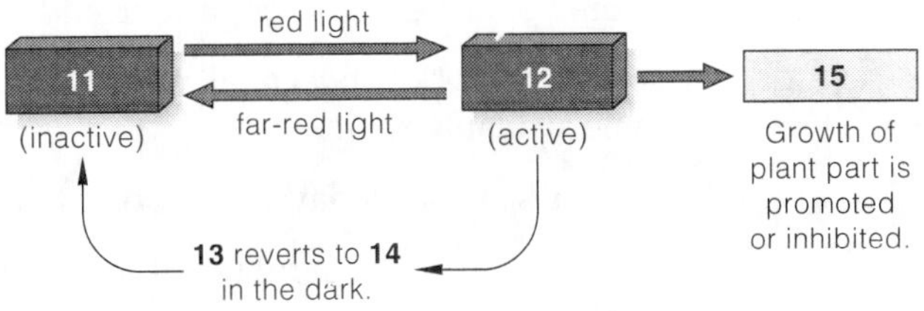

Fill-in-the-Blanks

(16) ____________ [p.558] plants flower in the spring, when daylength exceeds a critical value. (17) ____________ [p.558] plants flower in late summer and early autumn, when daylength is shorter than a critical value. The flowering of (18) ____________ - ____________ [p.558] plants is dependent on the maturity of the plant. While the term *daylength* is frequently used to describe the plants, it is actually the length of the (19) ____________ [p.558] that determines the flowering time in long-day and short-day plants. All plants respond to the wavelengths that predominate at (20) ____________ [p.559] and (21) ____________ [p.559]. Spinach is a short-day plant because it will not flower and produce seeds unless it is exposed to (22) ____________ hours of darkness for two weeks. To keep chrysanthemum plants from blooming, growers expose the plant to (23) ____________ [p.559] during the night to break up the darkness.

32.5. LIFE CYCLES END, AND TURN AGAIN [pp.560–561]

32.6. GROWING CROPS AND A CHEMICAL ARMS RACE [p.562]

Selected Words: "daylength" [p.561], *vernalis* [p.561], *2,4-D* [p.562], *Atrazine* [p.562], *DDT* [p.562], *malathion* [p.562], *herbicides* [p.562], *insecticides* [p.562], *fungicides* [p.562]

Boldfaced, Page-Referenced Terms

[p.560] abscission ____

____

[p.560] senescence ____

____

[p.560] dormancy ____

____

[p.561] vernalization ____

____

Choice

For questions 1–12, choose from the following:

a. senescence [p.560] b. abscission [p.560] c. vernalization [p.561]
d. entering dormancy [p.560] e. breaking dormancy [p.561]

1. ____ Dropping of leaves or other parts from a plant
2. ____ A process at work between fall and spring; temperatures become milder and rain and nutrients become available again
3. ____ Strong cues are short days; long, cold nights; and dry, nitrogen-deficient soil
4. ____ The process proceeds in tissues at the base of leaves, flowers, fruits, or other plant parts; a special zone is involved
5. ____ The sum total of processes leading to the death of plant parts or the whole plant
6. ____ A recurring cue for this process is a decrease in daylength
7. ____ Unless buds of some biennials and perennials are exposed to low winter temperatures, flowers will not form on their stems when spring rolls around
8. ____ The formation of ethylene in cells near the breakpoints may trigger the process
9. ____ Keeping germinating seeds of winter rye at near-freezing temperature to induce flowering the following summer
10. ____ Many perennial and biennial plants start to shut down growth as autumn approaches and days grow shorter
11. ____ Can be postponed when gardeners remove flower buds from plants to maintain vegetative growth
12. ____ The buds of the plant will resume growth after the convergence of precise environmental cues

Matching

Match each of the following terms with its most appropriate description. [p.562]

13. _____ atrazine
14. _____ DDT
15. _____ malathion
16. _____ toxin
17. _____ insecticide

A. A nerve cell poison that persists in the environment for 2–15 years after application
B. A herbicide known commercially as Roundup, Lasso, or Alar
C. A common organophosphate insecticide
D. An organic compound from one species that has a negative effect on another
E. A class of toxins that disrupts the nerves or muscles of the target organism

---

Self-Quiz

Choice

For questions 1–3, choose from the following answers: [pp.555–556]

a. gibberellins b. ethylene c. abscisic acid d. auxins

_____ 1. Promoting fruit ripening and abscission of leaves, flowers, and fruits is a function ascribed to ________________.

_____ 2. The stimulation of apical dominance is caused by ________________.

_____ 3. ________________ causes stems to lengthen by stimulating cell division and elongation.

_____ 4. ________________ is demonstrated by a germinating seed whose first root always curves down, while the stem always curves up. [p.556]
a. Phototropism
b. Photoperiodism
c. Gravitropism
d. Thigmotropism

_____ 5. Light of ________________ wavelengths is the main stimulus for phototropism. [pp.556–557]
a. blue
b. yellow
c. red
d. green

_____ 6. Plants whose leaves are open during the day but folded at night are exhibiting a ________________. [p.558]
a. growth movement
b. circadian rhythm
c. biological clock
d. both b and c are correct

_____ 7. 2,4-D, a potent dicot weed killer, is a synthetic ________________. [p.555]
a. auxin
b. gibberellin
c. cytokinin
d. phytochrome

_____ 8. All the processes that lead to the death of a plant or any of its organs are called ________________. [p.560]
a. dormancy
b. vernalization
c. abscission
d. senescence

_____ 9. Phytochrome is converted to an active form, ________________, at sunrise and reverts to an inactive form, ________________, at sunset, at night, or in the shade. [p.558]
a. Pr; Pfr
b. Pfr; Pfr
c. Pr; Pr
d. Pfr; Pr

_____ 10. Which of the following classes of compounds is used to control weeds? [p.562]

a. malathion
b. atrazine
c. phytochrome
d. DDT

_____ 11. When a perennial or biennial plant stops growing under conditions suitable for growth, it has entered a state of _______________. [p.560]

a. senescence
b. vernalization
c. dormancy
d. abscission

Chapter Objectives/Review Questions

1. __________ are signaling molecules. [p.550]
2. __________ is the process by which an immature stage in the life cycle resumes growth after a period of arrested development. [p.552]
3. Define *imbibition* and describe its role in germination. [p.552]
4. List the environmental factors that influence germination. [p.552]
5. The primary __________ breaks through the seed coat first. [pp.552–553]
6. The basic patterns of growth and development are heritable, dictated by the plant's __________. [p.552]
7. Compare and contrast the major features of the growth and development of a monocot and a dicot plant. [pp.552–553]
8. Distinguish between the terms *growth* and *development* and identify whether each is measured in quantitative or qualitative terms. [p.554]
9. Explain what is meant by *turgor pressure.* [p.554]
10. Describe the general role of plant hormones. [p.554]
11. __________ affect the lengthening of stems and responses to gravity and light; they also make coleoptiles grow longer. [pp.554–555]
12. __________ promote stem lengthening and help buds and seeds break dormancy to resume growth in the spring. [pp.554–555]
13. Certain synthetic auxins are used as __________, compounds that kill some plant species but not others. [pp.554–555]
14. __________ promote cell division and leaf expansion and retard leaf aging. [pp.554–555]
15. __________ __________ promotes stomatal closure as well as bud and seed dormancy. [pp.554–555]
16. __________ promotes ripening of fruit and abscission of leaves, flowers, and fruits. [pp.554–555]
17. Describe the form of growth inhibition known as apical dominance. [p.555]
18. Define *phototropism, gravitropism,* and *thigmotropism,* and give examples of each. [pp.556–557]
19. Define the role of statoliths in gravitropism. [p.556]
20. Plants exhibit the strongest phototropic response to light of __________ wavelengths; __________ is a yellow pigment molecule that absorbs blue wavelengths. [pp.556–557]
21. Give an example of how mechanical stress can affect plants. [p.557]
22. Plants have internal time-measuring mechanisms called biological __________. [p.558]
23. Describe the role of phytochromes as an alarm button. [p.558]
24. What are circadian rhythms? Give an example. [p.558]
25. Describe the process by which phytochrome is activated and inactivated in a plant. [p.558]
26. __________ is a biological response to a change in the relative length of daylight and darkness in a 24-hour cycle. [p.558]
27. Describe the photoperiodic responses of "long-day," "short-day," and "day-neutral" plants. [p.558]
28. Explain how botanists can influence the photoperiodic response of long-day and short-day plants. [p.559]
29. __________ is the dropping of leaves, flowers, fruits, or other plant parts. [p.560]
30. Describe the events that signal plant senescence. [p.560]
31. List environmental cues that send a plant into dormancy. [p.560]
32. The low-temperature stimulation of flowering is called __________. [p.561]
33. What conditions are instrumental in the dormancy-breaking process? [p.561]

Integrating and Applying Key Concepts

1. An oak tree has grown up in the middle of a forest. A lumber company has just cut down all the surrounding trees except for a narrow strip of woods that includes the oak. How will the oak be likely to respond as it adjusts to its changed environment? To what new stresses will it be exposed? Which hormones will most probably be involved in the adjustment?
2. You have been hired by a company in Costa Rica to raise a rare breed of northern plant for distribution. This plant requires 12 hours of continuous darkness over a three-week period to flower and form seeds. Explain what facilities you will need to make this possible.

33

ANIMAL TISSUES AND ORGAN SYSTEMS

Interactive Exercises

Meerkats, Humans, It's All the Same [pp.566–567]

33.1. EPITHELIAL TISSUE [pp.568–569]

Selected Words: *anatomy* [p.566], *physiology* [p.566], division of labor [p.567], *simple* epithelium [p.568], *stratified* epithelium [p.568], squamous [p.568], *Dendrobates* [p.569], communication [p.569], *secretion* [p.569], excretion [p.564]

Boldfaced, Page-Referenced Terms

[p.566] internal environment ______________________________

____

[p.566] homeostasis ______________________________

____

[p.566] tissue ______________________________

____

[p.566] organ ______________________________

____

[p.567] organ system ______________________________

____

[p.568] epithelium ____

[p.568] tight junctions ____

[p.568] adhering junctions ____

[p.568] gap junctions ____

[p.568] gland cells ____

[p.569] exocrine glands ____

[p.569] endocrine glands ____

Matching

Choose the most appropriate answer for each term.

1. ____ internal environment [p.566]
2. ____ anatomy [p.566]
3. ____ physiology [p.566]
4. ____ homeostasis [p.566]
5. ____ tissue [p.566]
6. ____ organ [p.566]
7. ____ organ system [p.567]
8. ____ division of labor [p.567]

A. Consists of two or more organs that are interacting physically, chemically, or both in a common task
B. How the body functions
C. Consists of interstitial fluid (tissue fluids) and blood that bathe the living cells of any complex animal
D. Consists of different tissues that are organized in specific proportions and patterns
E. Cells, tissues, organs, and organ systems split up the work in ways that contribute to the survival of the animal as a whole
F. How the animal body is structurally put together
G. An interactive group of cells and intercellular substances that take part in one or more particular tasks
H. With respect to the animal body, refers to stable operating conditions in the internal environment

Fill-in-the-Blanks

(9) ______________ [p.568] tissue has a free surface that faces either a body fluid or the outside environment. (10) ______________ ______________ [p.568] has a single layer of cells and functions as a lining for body cavities, ducts, and tubes. (11) ______________ ______________ [p.568] has two or more layers and typically provides protection, as for example, in the skin. (12) ______________ [p.568] junctions are strands of proteins that help stop substances from leaking across a tissue. (13) ______________ [p.568] junctions cement cells together. (14) ______________ [p.568] junctions help communicate by promoting the rapid transfer of ions and small molecules among them. (15) ______________ [p.569] junctions in the epithelium of your stomach help prevent a condition called (16) ______________ [p.569] ulcer. (17) ______________ [p.569] glands secrete mucus, saliva, earwax, milk, oil, digestive enzymes, and other cell products. These products are usually released onto a free (18) ______________ [p.569] surface through ducts or tubes. (19) ______________ [p.569] glands lack ducts; their products are (20) ______________ [p.569], which are secreted directly into the fluid bathing the gland. Typically, the (21) ______________ [p.569] picks up the hormone molecules and distributes them to target cells elsewhere in the body.

33.2. CONNECTIVE TISSUE [pp.570–571]

33.3. MUSCLE TISSUE [p.572]

33.4. NERVOUS TISSUE [p.573]

33.5. FOCUS ON SCIENCE: *Frontiers in Tissue Research* [p.573]

Selected Words: fibroblasts [p.570], tendons [p.570], *plasma* [p.571], *contract* [p.572], *striated* [p.572], *lab-grown epidermis* [p.573], *designer organs* [p.573], type I *diabetes mellitus* [p.573]

Boldfaced, Page-Referenced Terms

[p.570] loose connective tissue ______________________________

____

[p.570] dense, irregular connective tissue ______________________________

____

[p.570] dense, regular connective tissue ______________________________

____

[p.570] cartilage ______________________________

____

[p.571] bone tissue ______________________________

____

[p.571] adipose tissue ______________________________

____

[p.571] blood ____

____

[p.572] skeletal muscle tissue ____

____

[p.572] smooth muscle tissue ____

____

[p.572] cardiac muscle tissue ____

____

[p.573] nervous tissue ____

____

[p.573] neurons ____

____

[p.573] neuroglia ____

____

Choice

For questions 1–10, choose from the following types of connective tissue proper: [p.570]

a. loose b. dense, irregular c. dense, regular

1. ____ Contains many fibers, mostly collagen-containing ones, in no particular orientation, as well as a few fibroblasts
2. ____ Rows of fibroblasts often intervene between the bundles of fibers
3. ____ Has its fibers and cells loosely arranged in a semifluid ground substance
4. ____ Has parallel bundles of many collagen fibers and resists being torn apart
5. ____ Forms protective capsules around organs that do not stretch much
6. ____ Often serves as a support framework for epithelium
7. ____ Found in tendons, which attach skeletal muscle to bones
8. ____ Besides fibroblasts, it contains infection-fighting white blood cells
9. ____ Found in elastic ligaments, which attach bones to each other
10. ____ Is also present in the deeper part of skin

Complete the Table

11. After reading each description, supply the name of the specialized connective tissue.

Specialized Connective Tissue	*Description*
a. [p.571]	Chock-full of large fat cells; stores excess carbohydrates and proteins; richly supplied with blood
b. [pp.570–571]	Intercellular material, solid yet pliable, resists compression; structural models for vertebrate embryo bones; maintains shape of nose, outer ear, and other body parts; cushions joints
c. [p.571]	Derived mainly from connective tissue; has transport functions; circulating within plasma are a great many red blood cells, white blood cells, and platelets
d. [p.571]	Weight-bearing tissue of vertebrate skeletons, which support or protect softer tissues and organs; mineral-hardened with calcium salt–laden collagen fibers and ground substance; interact with skeletal muscles attached to them

Dichotomous Choice

Circle one of two possible answers given between parentheses in each statement.

12. Contractile cells of (skeletal/smooth) muscle tissue taper at both ends. [p.572]
13. The contractile walls of the heart are composed of (striated/cardiac) muscle tissue. [p.572]
14. Walls of the stomach and intestine contain (smooth/skeletal) muscle tissue. [p.572]
15. The only muscle tissue attached to bones is (skeletal/smooth). [p.572]
16. (Smooth/Skeletal) muscle cells are bundled together in parallel. [p.572]
17. "Involuntary" muscle action is associated with (smooth/skeletal) muscle tissue. [p.572]
18. The term *striated* means (bundled/striped). [p.572]
19. (Smooth/Skeletal) muscle tissue has a sheath of tough connective tissue enclosing several bundles of muscle cells. [p.572]
20. The function of smooth muscle tissue is to (pump blood/move internal organs). [p.572]
21. Cell junctions fuse together the plasma membranes of (smooth/cardiac) muscle cells. [p.572]
22. (Muscle/Nervous) tissue exerts the greatest control over the body's responsiveness to changing conditions. [p.573]
23. Excitable cells are the (neuroglia/neurons). [p.573]
24. (Neuroglia/Muscle) cells protect and structurally and metabolically support the neurons. [p.573]
25. When a (neuron/muscle cell) is suitably stimulated, an electrical "message" travels over its plasma membrane that may result in stimulation of other cells of the same type or of other types. [p.573]
26. Different types of (neuroglia/neurons) detect specific stimuli, integrate information, and issue or relay commands for response. [p.573]
27. The lives of people with type I diabetes mellitus might, in the future, be made more normal by means of (a designer organ/a sheet of laboratory-grown epidermis). [p.573]

Labeling and Matching

In the answer blanks below, label each of the following illustrations with one of the following terms: *connective, epithelial, muscle,* or *nervous* tissue. Complete the exercise by writing *all* appropriate letters and numbers from both of the following lists in the parentheses after each label.

28. ________________ (   ) [p.570]
29. ________________ (   ) [p.568]
30. ________________ (   ) [p.572]
31. ________________ (   ) [p.572]
32. ________________ (   ) [p.570]
33. ________________ (   ) [p.571]
34. ________________ (   ) [p.568]
35. ________________ (   ) [p.571]
36. ________________ (   ) [p.573]
37. ________________ (   ) [p.572]
38. ________________ (   ) [p.568]
39. ________________ (   ) [p.571]

A. Adipose
B. Bone
C. Cardiac
D. Dense, regular
E. Loose
F. Simple columnar
G. Simple cuboidal
H. Simple squamous
I. Smooth
J. Skeletal
K. Blood
L. Neurons

1. Absorption
2. Communication by means of electrical signals
3. Energy reserve
4. Contraction for voluntary movements
5. Diffusion
6. Padding
7. Contract to propel substances along internal passageways; not striated
8. Attaches muscle to bone and bone to bone
9. In vertebrates, provides the strongest internal framework of the organism
10. Elasticity
11. Secretion
12. Pumps circulatory fluid; striated
13. Insulation
14. Transport of nutrients and waste products to and from body cells

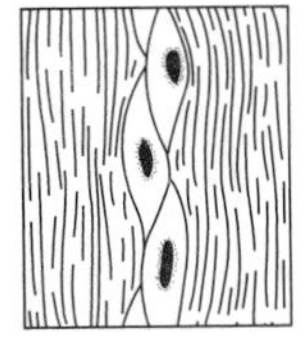
28.

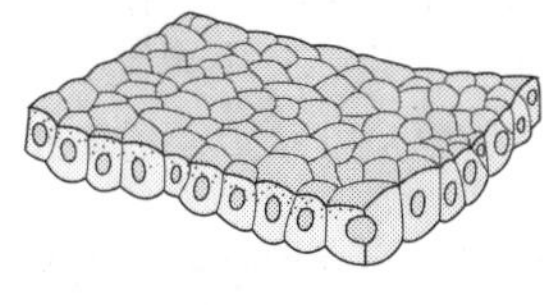
29.

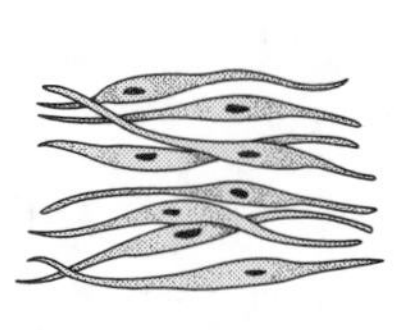
30.

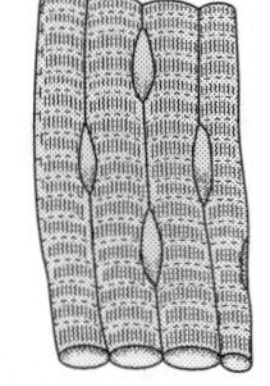
31.

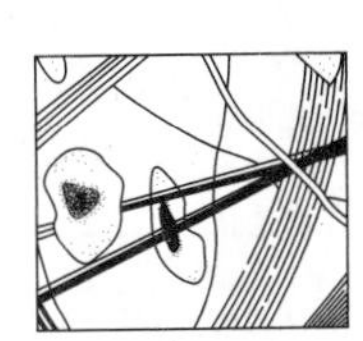
32.

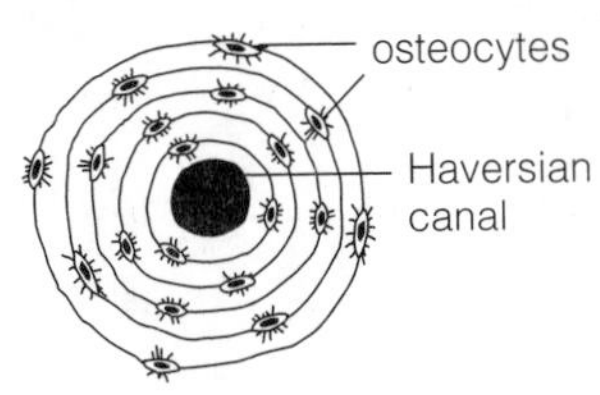

33.

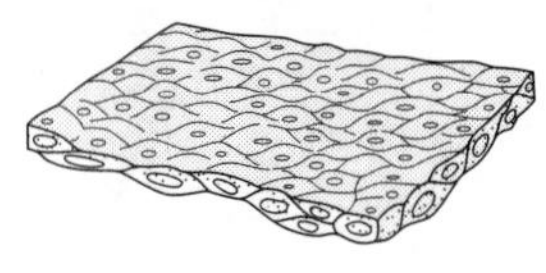
34.

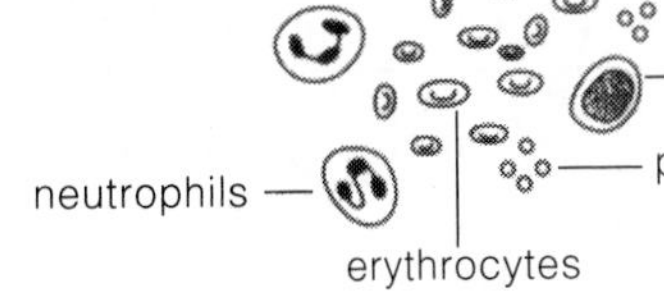

35.

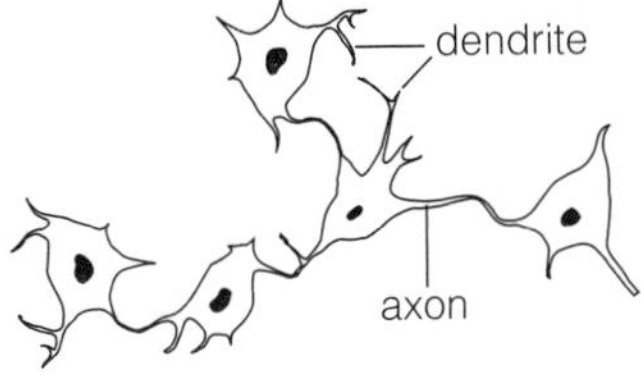

36.

37.

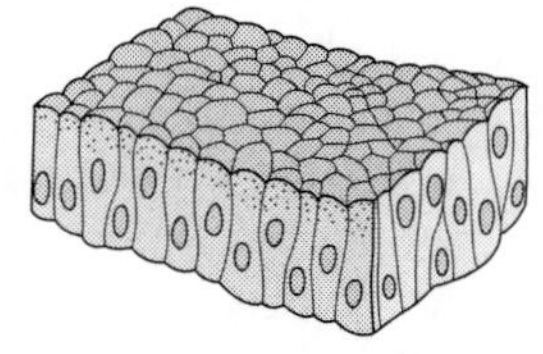
38.

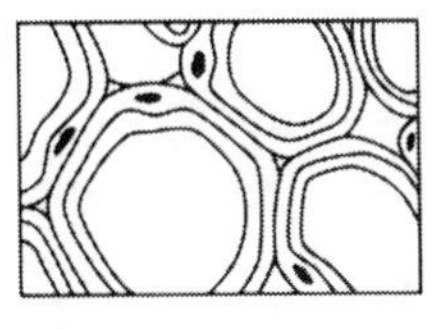
39.

33.6. ORGAN SYSTEMS [pp.574–575]

Selected Words: midsagittal [p.574], dorsal [p.574], ventral [p.574], frontal [p.574], transverse [p.574], *anterior* [p.574], *posterior* [p.574], superior [p.575], inferior [p.575], distal [p.575], proximal [p.575], *germ* cells [p.575], *somatic* [p.575]

Boldfaced, Page-Referenced Terms

[p.575] ectoderm ____

____

[p.575] mesoderm ____

____

[p.575] endoderm ____

____

Fill-in-the-Blanks

The brain is housed in the (1) ______________ [p.574] cavity. The (2) ______________ [p.574] cavity contains the spinal cord and the beginnings of spinal nerves. The heart and lungs are found within the (3) ______________ [p.574] cavity. The stomach, spleen, liver, gallbladder, pancreas, small intestine, most of the large intestine, kidneys, and ureters lie inside the (4) ______________ [p.574] cavity. The (5) ______________ [p.574] cavity contains the urinary bladder, sigmoid colon, rectum, and reproductive organs.

Complete the Table

6. Supply the name of the primary tissue of the embryo that does the job indicated by becoming specialized in particular ways. [all from p.575]

Primary Tissue	*Functions*
a.	Forms internal skeleton and muscle, circulatory, reproductive, and urinary systems
b.	Forms inner lining of gut and linings of major organs formed from the embryonic gut
c.	Forms outer layer of skin and the tissues of the nervous system

Labeling

Identify each numbered part of the accompanying illustration, which reviews the directional terms and planes of symmetry for the human body. [p.569]

7. ____________
8. ____________
9. ____________
10. ____________
11. ____________
12. ____________
13. ____________
14. ____________

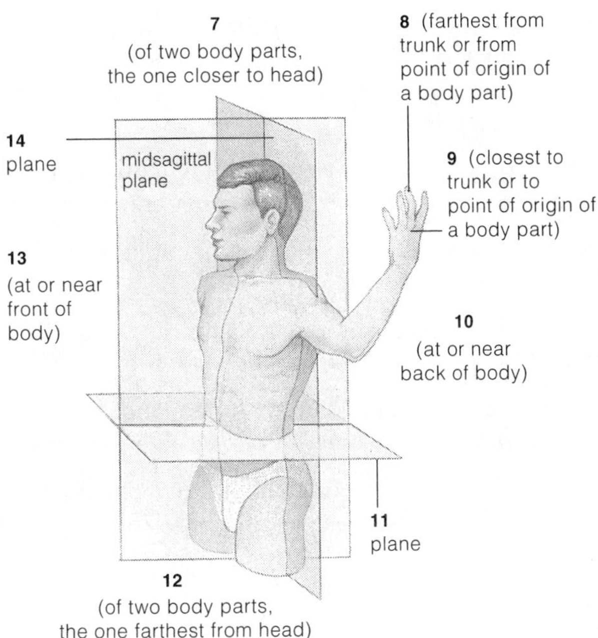

Labeling and Matching

In the answer blank provided, label each organ system described. Complete the exercise by matching and entering the proper letter from the following illustration in the parentheses after each label.

15. ______________ ( ) Rapidly transport many materials to and from cells; help stabilize internal pH and temperature [p.574]

16. ______________ ( ) Rapidly deliver oxygen to the tissue fluid that bathes all living cells; remove carbon dioxide wastes of cells; help regulate pH [p.575]

17. ______________ ( ) Maintain the volume and composition of internal environment; excrete excess fluid and blood-borne wastes [p.575]

18. ______________ ( ) Support and protect body parts; provide muscle attachment sites; produce red blood cells; store calcium and phosphorus [p.574]

19. ______________ ( ) Hormonally control body function; work with nervous system to integrate short-term and long-term activities [p.574]

20. ______________ ( ) *Female:* produce eggs; after fertilization, afford a protected, nutritive environment for the development of a new individual. *Male:* produce and transfer sperm to the female. Hormones of both systems also influence other organ systems [p.575]

21. ______________ ( ) Ingest food and water; mechanically, chemically break down food and absorb small molecules into internal environment; eliminate food residues [p.575]

22. ______________ ( ) Move body and its internal parts; maintain posture; generate heat (by increases in metabolic activity) [p.574]

23. ______________ ( ) Detect both external and internal stimuli; control and coordinate responses to stimuli; integrate all organ system activities [p.574]

24. ______________ ( ) Protect body from injury, dehydration, and some pathogens; control its temperature; excrete some wastes; receive some external stimuli [p.574]

25. ______________ ( ) Collect and return some tissue fluid to the bloodstream; defend the body against infection and tissue damage [p.575]

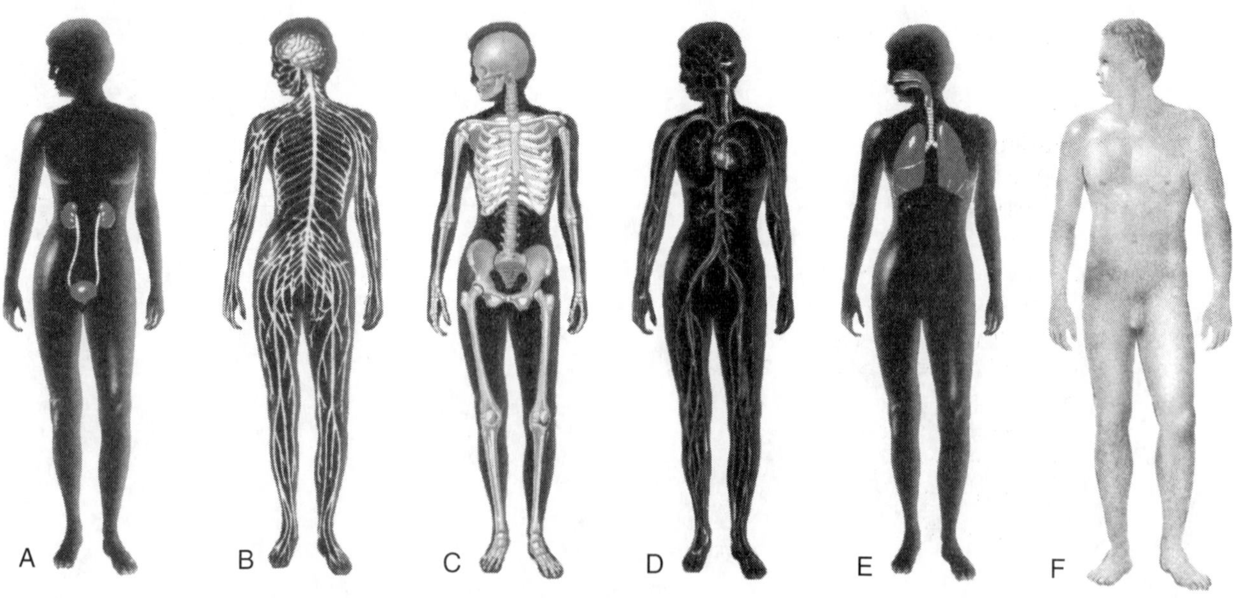

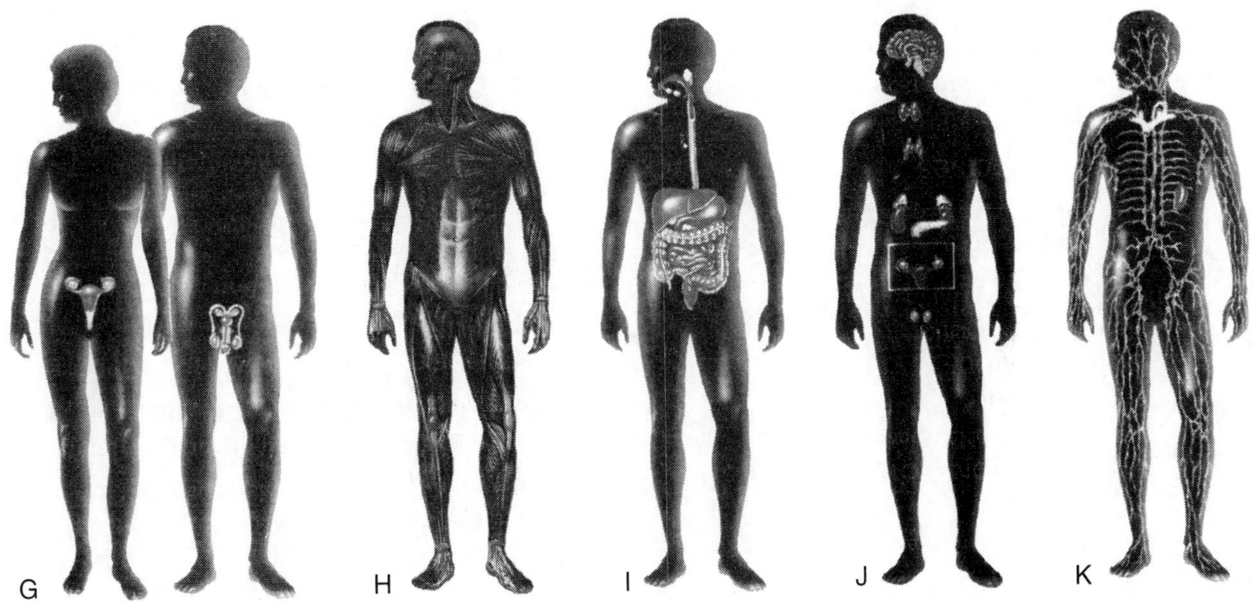

Self-Quiz

____ 1. Which of the following is not included in connective tissues? [pp.570–571]
a. Bone
b. Blood
c. Cartilage
d. Skeletal muscle

____ 2. Gland cells are contained in ______________ tissues. [p.569]
a. muscular
b. epithelial
c. connective
d. nervous

____ 3. Blood is considered to be a(n) ______________ tissue. [p.571]
a. epithelial
b. muscular
c. connective
d. none of these

____ 4. ______________ are abundant in tissues of the heart and skeletal muscle, where they promote diffusion of ions and small molecules from cell to cell. [p.569]
a. Adhesion junctions
b. Filter junctions
c. Gap junctions
d. Tight junctions

____ 5. Muscle that is not striped and is involuntary is ______________. [p.572]
a. cardiac
b. skeletal
c. striated
d. smooth

____ 6. A(n) ______________ is a group of cells and intercellular substances, all interacting in one or more tasks. [p.566]
a. organ
b. organ system
c. tissue
d. cuticle

____ 7. A graduate student in developmental biology accidentally stabbed a fish embryo. Later the embryo developed into a creature that could not move and had no supportive or circulatory systems. Which embryonic tissue had suffered the damage? [p.575]
a. ectoderm
b. endoderm
c. mesoderm
d. protoderm

_____ 8. A tissue whose cells are striated and fused at the ends by cell junctions so that the cells contract as a unit is called _____________ tissue. [p.572]
a. smooth muscle
b. dense fibrous connective
c. supportive connective
d. cardiac muscle

_____ 9. The secretion of tears, milk, sweat, and oil are functions of _____________ tissues. [p.569]
a. epithelial
b. loose connective
c. lymphoid
d. nervous

_____ 10. Memory, decision making, and issuing commands to effectors are functions of _____________ tissue. [p.573]
a. connective
b. epithelial
c. muscle
d. nervous

_____ 11. Which group is arranged correctly from smallest structure to largest, reading left to right? [p.572]
a. muscle cells, muscle bundle, muscle
b. muscle cells, muscle, muscle bundle
c. muscle bundle, muscle cells, muscle
d. none of the above

Matching

Choose the most appropriate answer for each term.

12. _____ circulatory system [p.574]
13. _____ digestive system [p.575]
14. _____ endocrine system [p.574]
15. _____ integumentary system [p.574]
16. _____ muscular system [p.574]
17. _____ nervous system [p.574]
18. _____ reproductive system [p.575]
19. _____ respiratory system [p.575]
20. _____ skeletal system [p.574]
21. _____ urinary system [p.575]

A. Picks up nutrients absorbed from gut and transports them to cells throughout body
B. Helps cells use nutrients by supplying them with oxygen and relieving them of CO_2 wastes
C. Helps maintain the volume and composition of body fluids that bathe the body's cells
D. Provides basic framework for the animal and supports other organs of the body
E. Uses chemical messengers to control and guide body functions
F. Produces younger, temporarily smaller versions of the animal
G. Breaks down larger food molecules into smaller nutrient molecules that can be absorbed by body fluids and transported to body cells
H. Consists of contractile parts that move the body through the environment and propel substances about in the animal
I. Serves as an electrochemical communications system in the animal's body
J. In the meerkat, serves as a heat catcher in the morning and protective insulation at night

Chapter Objectives/Review Questions

1. Explain how the meerkat maintains a rather constant internal environment in spite of changing external conditions. [p.566]
2. Cells are the basic units of life; in a multicellular animal, similar cells are grouped into a(n) __________, and these are organized in specific proportions and patterns that compose a(n) __________. [p.566]
3. Explain how, if each cell can perform all its basic activities, organ systems contribute to cell survival. [pp.566–567,574]
4. __________ has a free surface that faces either a body fluid or the outside environment. [p.568]

5. Distinguish simple epithelium from stratified epithelium. [p.568]
6. Name and describe three kinds of cell junctions that occur in epithelia and other tissues. [p.568]
7. Name and describe the various types of epithelial tissues as well as their location and general functions. [pp.568–569]
8. Define the term *gland.* [p.569]
9. ________ glands usually secrete their products onto a free epithelial surface through ducts or tubes; give examples of their products. [p.569]
10. ________ glands lack ducts; their products are ________, which are secreted directly into the fluid bathing the gland. [p.569]
11. Distinguish among loose connective tissue; dense, irregular connective tissue; and dense, regular connective tissue on the basis of their structures and functions. [p.570]
12. Cartilage, bone, adipose tissue, and blood are known as the specialized connective tissues; describe their structures and various functions. [pp.570–571]
13. Distinguish among skeletal, smooth, and cardiac muscle tissue in terms of location, structure, and function. [p.572]
14. Muscle tissue contains specialized cells that can ________. [p.572]
15. Of all tissues, ________ tissue exerts the greatest control over the body's responsiveness to changing conditions. [p.573]
16. ________ are excitable cells, the communication units of most nervous systems. [p.573]
17. Discuss the implications of lab-grown epidermis, designer organs, and research to put together packages of cells capable of producing specific life-saving substances that are absent in patients who suffer from genetic disorders or chronic diseases. [p.573]
18. List each of the 11 principal organ systems in humans and give the main task of each. [pp.574–575]
19. List the major cavities in the human body and the organs they house. [p.574]
20. Name the directional terms and planes of symmetry used in the description of the human body. [p.575]
21. ________ gives rise to the skin's outer layer and tissues of the nervous system; ________ gives rise to muscles, bones, and most of the circulatory, reproductive, and urinary systems; ________ gives rise to the lining of the digestive tract and to organs derived from it. [p.575]

Interpreting and Applying Key Concepts

Explain why, of all places in the body, marrow is located inside the long bones. Explain why your bones are remodeled after you reach maturity. Why does your body not keep the same mature skeleton throughout life?

34

INTEGRATION AND CONTROL: NERVOUS SYSTEMS

Interactive Exercises

Why Crack the System? [pp.586–587]

34.1. NEURONS — THE COMMUNICATION SPECIALISTS [pp.576–577]

34.2. HOW ARE ACTION POTENTIALS TRIGGERED AND PROPAGATED? [pp.578–579]

Selected Words: *drug* [p.578], *dealer* [p.578], *crack* [p.578], "the shakes" [p.578], *input* zones [p.580], *trigger* zone [p.580], *conducting* zone [p.580], *output* zones [p.580], "at rest" [p.580], *against* the gradient [p.581], *graded* signal [p.582], *local* signal [p.582], *all-or-nothing* event [p.582]

Boldfaced, Page-Referenced Terms

[p.578] nervous system ____

[p.579] sensory neuron ____

[p.579] stimulus ____

[p.579] interneurons ____

[p.579] motor neuron ____

[p.580] dendrites ____

[p.580] axon ____

[p.580] resting membrane potential ____

[p.580] action potential ____

[p.581] sodium–potassium pumps ____

[p.582] positive feedback ____

Matching

Choose the most appropriate description for each of the following terms.

1. ____ dendrite [p.580]
2. ____ sensory neuron [p.579]
3. ____ motor neuron [p.579]
4. ____ axon [p.580]
5. ____ stimulus [p.579]
6. ____ interneurons [p.579]

A. Receive and process sensory input
B. Represent the input zones of a neuron
C. Relay information away from the brain and spinal cord
D. A specific form of energy that is detected by a receptor
E. Represent both the conducting and the output zones of a neuron
F. Relay stimulus information to the brain and spinal cord.

Fill-in-the-Blanks

A(n) (7) ______________ [p.580] has a multinucleated cell body with cytoplasmic extensions that differ in number and length. The input zones for information into the neuron are called (8) ______________ [p.580]. At a nearby patch of plasma membrane called the (9) ______________ [p.580] zone, the input may give rise to signals that travel along a(n) (10) ______________ [p.580], which acts as the neuron's conducting zone. The axon's endings are the neuron's (11) ______________ [p.580] zones.

Matching

Choose the correct answer for each of the following descriptions.

12. ____ A brief reversal in the voltage difference across the plasma membrane [p.580]

13. ____ The steady voltage difference across the plasma membrane, usually about −70 millivolts [p.580]

14. ____ The mechanism by which the ion concentrations across the membrane are maintained [p.581]

15. ____ Two ions that are used in the generation of an action potential across a membrane [pp.580–581]

16. ____ When an event intensifies the result of its own occurrence [p.582]

A. Na^+ and K^+
B. Action potential
C. Resting membrane potential
D. Positive feedback
E. Sodium–potassium pump

Fill-in-the-Blanks

When a neuron is weakly stimulated at its input zone, the (17) ______________ [p.582] balance across the membrane is disturbed. The signals at the input zone are (18) ______________ [p.582], since they vary in magnitude, and (19) ______________ [p.582], since they do not spread far from the site of stimulation. When a signal is (20) ______________ [p.582] or long lasting, graded signals spread from the input zone to an adjoining (21) ______________ [p.582] zone. This is where a certain amount of change in the voltage difference across the plasma membrane triggers a(n) (22) ______________ ______________ [p.582]. The amount of change is the neuron's (23) ______________ [p.582] level.

When the gates open, positively charged (24) ______________ [p.582] ions flow into the neuron, which causes more gates to open and more sodium to enter. This is a case of (25) ______________ [p.582] feedback.

The potential across the membrane will peak once the threshold is reached. All action potentials in a neuron spike to the same level above threshold in what is called a(n) (26) ______________ - ______________ - ______________ [p.582] event. Each spike lasts for about one (27) ______________ [p.582]. About halfway through the spike, the gated (28) ______________ [p.582] channels close and (29) ______________ [p.583] channels open. This restores the (30) ______________ [p.583] difference across the membrane but not the original gradients. The (31) ______________ – ______________ [p.583] pump actively restores the potassium and sodium gradients.

During an action potential, the inward rush of (32) ______________ [p.583] affects the charge distribution across the adjacent membrane patch, causing the gated channels to (33) ______________ [p.583]. This positive feedback is (34) ______________ - ______________ [p.583] and does not diminish in (35) ______________ [p.583]. The action potentials do not spread back to the trigger zone, since the voltage-gated channels there are briefly (36) ______________ [p.583] to stimulation.

34.3. CHEMICAL SYNAPSES [pp.584–585]

34.4. PATHS OF INFORMATION FLOW [pp.586–587]

Selected Words: *pre*synaptic neuron [p.584], *post*synaptic cell [p.584], *excitatory* effect [p.584], *inhibitory* effect [p.584], *neuromodulators* [p.585], *depolarizing* effect [p.585], *hyperpolarizing* effect [p.585], summation [p.585], *divergent* circuit [p.586], *convergent* circuit [p.586], *reverberating* circuit [p.586], *multiple sclerosis* [p.586], *stretch reflex* [p.586]

Boldfaced, Page-Referenced Terms

[p.584] neurotransmitters ____

____

[p.584] chemical synapses ____

____

[p.584] acetylcholine ____

____

[p.585] neuromodulators ____

____

[p.585] synaptic integration ____

____

[p.586] nerves ____

____

[p.586] reflexes ____

____

Matching

Choose the most appropriate description for each of the following

1. ______ acetylcholine [p.584]
2. ______ neuromodulator [p.585]
3. ______ chemical synapse [p.584]
4. ______ presynaptic neuron [p.584]
5. ______ postsynaptic neuron [p.584]
6. ______ serotonin [p.584]
7. ______ GABA [p.584]
8. ______ neurotransmitters [p.584]

A. Has receptors that bind to specific neurotransmitters
B. The most common inhibitory signal in the brain
C. Contain vesicles filled with neurotransmitters that open when an action potential arrives
D. The most common of the vertebrate neurotransmitters
E. A narrow cleft between a neuron's output zone and the input zone of a neighboring cell
F. Neurotransmitter that helps regulate sensory perception, sleep, and body temperature
G. Signaling molecules that diffuse across chemical synapses
H. Reduces or magnifies the effects of a neurotransmitter

Fill-in-the-Blanks

All synaptic signals are (9) ______________ [p.585] potentials. A(n) (10) ______________ [p.585] brings the membrane closer to threshold and has a depolarizing effect. An inhibitory postsynaptic potential (IPSP) drives the membrane away from threshold and either has a(n) (11) ______________ [p.585] effect or maintains the membrane at its resting level. With (12) ______________ ______________ [p.585], competing signals from more than one (13) ______________ [p.585] cell reach the input zone at the same time and are (14) ______________ [p.585]. The signals may be dampened, (15) ______________ [p.585], reinforced, or sent on to other cells of the body.

Signaling depends on the prompt, controlled removal of neurotransmitters from (16) ______________ ______________ [p.585]. Some molecules simply (17) ______________ [p.585] out of the cleft. (18) ______________ [p.585] in the cleft may remove others, as is the case with acetylcholinesterase. (19) ______________ [p.585] proteins in the membrane may actively pump molecules back inside.

Matching

Choose the most appropriate description for each of the following. [p.586]

20. ______ stretch reflex
21. ______ multiple sclerosis
22. ______ nerves
23. ______ reflexes
24. ______ divergent circuits
25. ______ reverberating circuits
26. ______ convergent circuits

A. An autoimmune disorder that causes the inflammation of axons
B. The signal from a neuron fans out to form connections with other blocks of neurons
C. An automatic movement in response to a stimulus
D. Signals from many neurons are relayed to just a few
E. Collections of sensory neurons and motor neurons bundled inside connective tissue
F. When neurons synapse back on themselves to repeat signals
G. The contraction of a muscle after gravity or some other load stretches it

Labeling

Label the indicated parts of the following diagram. [p.587]

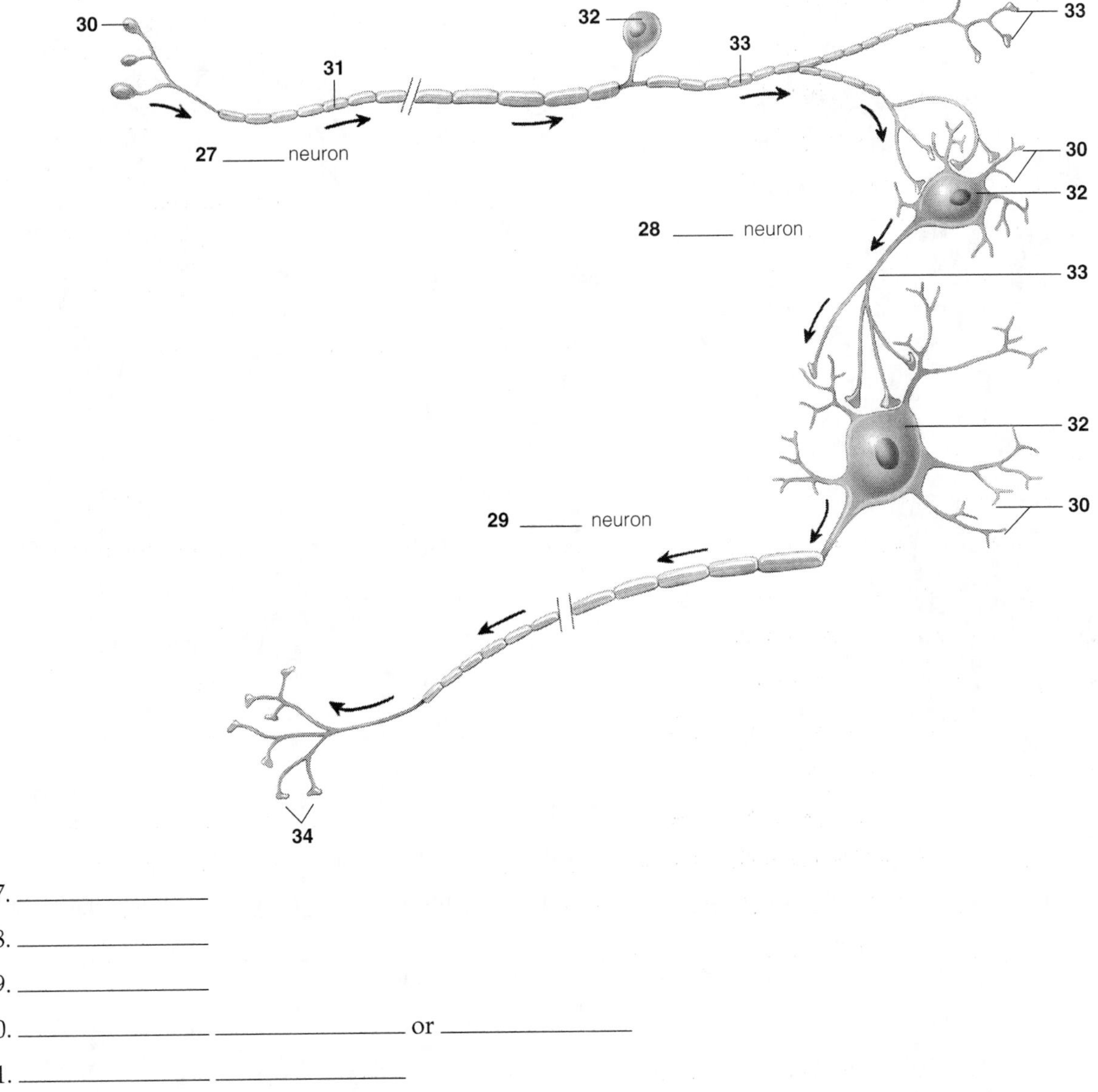

27. ______________

28. ______________

29. ______________

30. ______________ ______________ or ______________

31. ______________ ______________

32. ______________ ______________

33. ______________

34. ______________ ______________

Matching

Match each of the following statements with its correct location in the diagram below. [p.587]

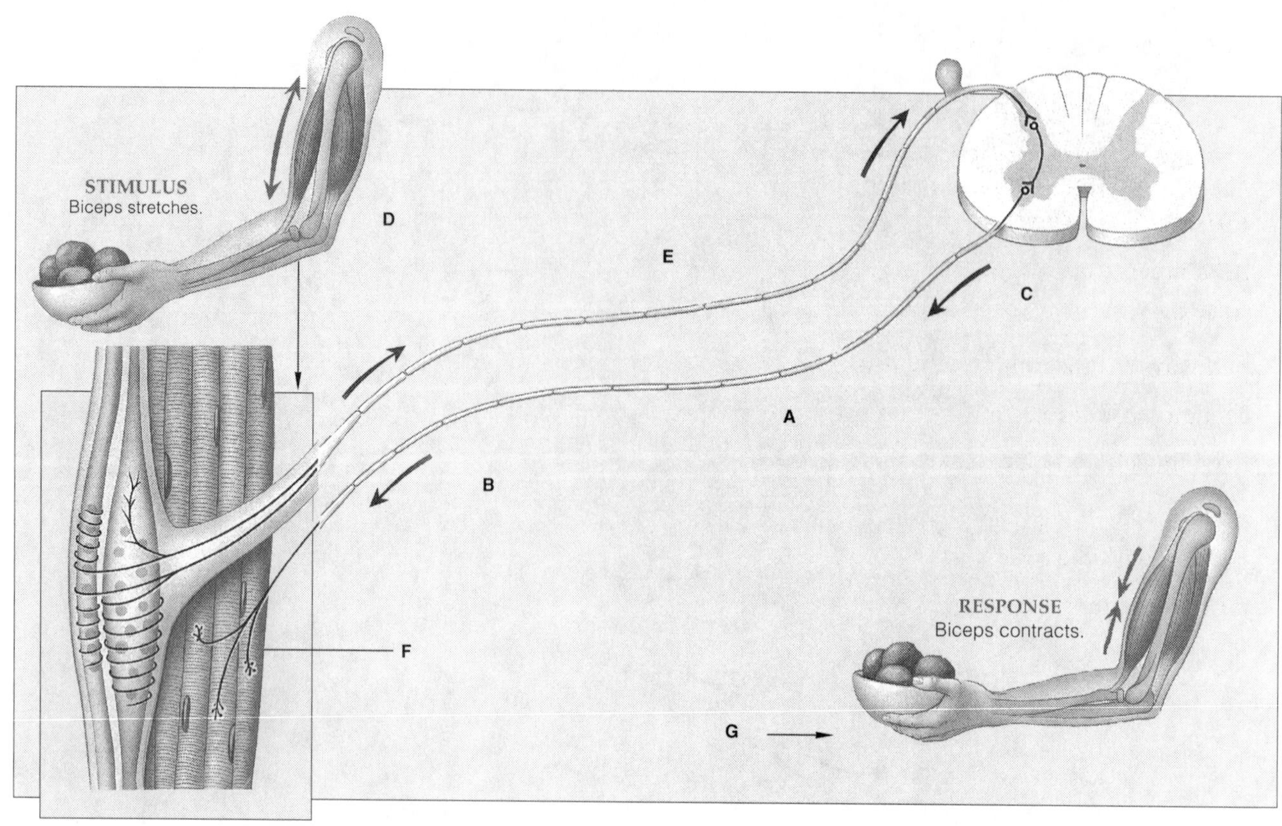

35. ______ Axon endings of the motor neuron synapse with muscle cells
36. ______ Neurotransmitter is released from the sensory neuron and stimulates the motor neuron
37. ______ Neurotransmitter released from the motor neuron stimulates the plasma membrane of muscle cells
38. ______ The action potential is propagated along the axon of the motor neuron
39. ______ A load is placed on the muscle tissue
40. ______ The stretching of the muscle tissue in response to the load stimulates receptors, generating an action potential
41. ______ The muscle is stimulated and contracts

Sequence

Using the diagram from above, place the letters in order according to the correct sequence of events.

42. ____

34.5. INVERTEBRATE NERVOUS SYSTEMS [pp.588–589]

Selected Words: *radial* symmetry [p.588], *bilateral* symmetry [p.588], *ganglion* [p.588]

Boldfaced, Page-Referenced Term

[p.588] nerve net ____

____

Fill-in-the-Blanks

All animals except sponges have some type of (1) ______________ [p.588] system in which nerve cells, such as (2) ______________ [p.588], are oriented in signal-conducting and information-processing pathways. At the minimum, the cells making up the communication lines receive information about changing conditions outside and inside the (3) ______________ [p.588], then elicit suitable responses from muscle and gland cells.

The first animals evolved in the (4) ______________ [p.588]. It is in this environment that we still find animals with the simplest (5) ______________ [p.588] systems. They are sea anemones, jellyfishes, and other cnidarians. These invertebrates display (6) ______________ [p.588] symmetry. Such animals have a(n) (7) ______________ [p.588] net, a loose mesh of nerve cells intimately associated with epithelial tissue. The nerve cells interact with (8) ______________ [p.588] cells and contractile cells along reflex pathways in the same epithelial tissue. In (9) ______________ [p.588] pathways, sensory stimulation triggers simple, stereotyped movements. The nerve net itself extends through the animal's (10) ______________ [p.588], but information flow through it is not highly focused. It simply commands the body wall to slowly contract and expand or move tentacles through the water.

Flatworms are the simplest animals having a(n) (11) ______________ [p.588] nervous system. There are equivalent body parts on the left and right sides of the body's (12) ______________ [p.588] plane. The ladderlike nervous system of the flatworm has (13) ______________ [p.588] cordlike nerves running longitudinally through the body, with many side branches. A(n) (14) ______________ [p.588] is a cluster of nerve cell bodies that is a local integrating center. Inside the head of the flatworm, these ganglia coordinate signals coming from paired (15) ______________ [p.589] organs.

Did (16) ______________ [p.589] nervous systems evolve from nerve nets? Maybe. In nearly all animals more complex than flatworms, we find local nerve nets, or (17) ______________ [p.589], such as the one in your intestinal wall. Chance mutations in ancient planulas may have favored a concentration of (18) ______________ [p.589] cells in the leading end, not the trailing end. This allowed for more rapid and effective responses to varied stimuli. Natural (19) ______________ [p.589] must have favored a concentration of sensory cells at the body's leading end. (20) ______________ [p.589] and bilateral symmetry may have started this way.

34.6. VERTEBRATE NERVOUS SYSTEMS — AN OVERVIEW [pp.590–591]

Selected Words: *afferent* [p.591], *efferent* [p.591], *white* matter [p.591], *gray* matter [p.591]

Boldfaced, Page-Referenced Terms

[p.590] neural tube ______________________________

____

[p.591] central nervous system ______________________________

____

[p.591] peripheral nervous system ______________________________

____

[p.591] brain ______________________________

____

[p.591] spinal cord ______________________________

____

[p.591] somatic subdivision ______________________________

____

[p.591] autonomic subdivision ______________________________

____

Choice

For questions 1–5, choose from the following answers: [p.590]

a. midbrain   b. forebrain   c. hindbrain

1. ____ Coordinates reflex responses
2. ____ Reflex control of respiration, blood circulation
3. ____ Receives and integrates sensory information from nose, eyes, and ears
4. ____ Coordination of sensory input and motor dexterity in complex vertebrates
5. ____ In land vertebrates this is the site of highest integration

Matching

Choose the most appropriate answer for each term.

6. ____ neural tube [p.590]
7. ____ central nervous system [p.591]
8. ____ peripheral nervous system [p.591]
9. ____ tracts [p.591]
10. ____ white matter tracts [p.591]
11. ____ gray matter tracts [p.591]
12. ____ neuroglial cells [p.591]
13. ____ afferent [p.591]
14. ____ efferent [p.591]
15. ____ somatic subdivision [p.591]
16. ____ autonomic subdivision [p.591]

A. Nerves carrying motor output away from the central nervous system to muscles and glands
B. The nerve cord that persists in all vertebrate embryos
C. Protect or structurally and functionally support neurons
D. The spinal cord and brain
E. Nerves carrying sensory input to the central nervous system
F. Consists of unmyelinated axons, dendrites, and nerve cell bodies and neuroglial cells
G. Consists mainly of nerves that thread through the rest of the body and carry signals into and out of the central nervous system
H. The part of the peripheral nervous system that carries signals to and from the internal organs
I. The communication lines inside the brain and spinal cord
J. Contains axons with glistening white myelin sheaths and specializes in rapid signal transmission
K. The part of the peripheral nervous system that carries signals to and from skeletal muscles

34.7. WHAT ARE THE MAJOR EXPRESSWAYS? [pp.592–593]

Selected Words: *spinal* nerves [p.592], *cranial* nerves [p.592], "housekeeping" tasks [p.593], *fight–flight response* [p.593], "rebound effect" [p.593], *meningitis* [p.593]

Boldfaced, Page-Referenced Terms

[p.592] somatic nerves ____________________

____

[p.592] autonomic nerves ____________________

____

[p.592] parasympathetic nerves ____________________

____

[p.593] sympathetic nerves ____________________

____

[p.593] spinal cord ____________________

____

Labeling

Label each numbered part of the accompanying illustration. [p.592]

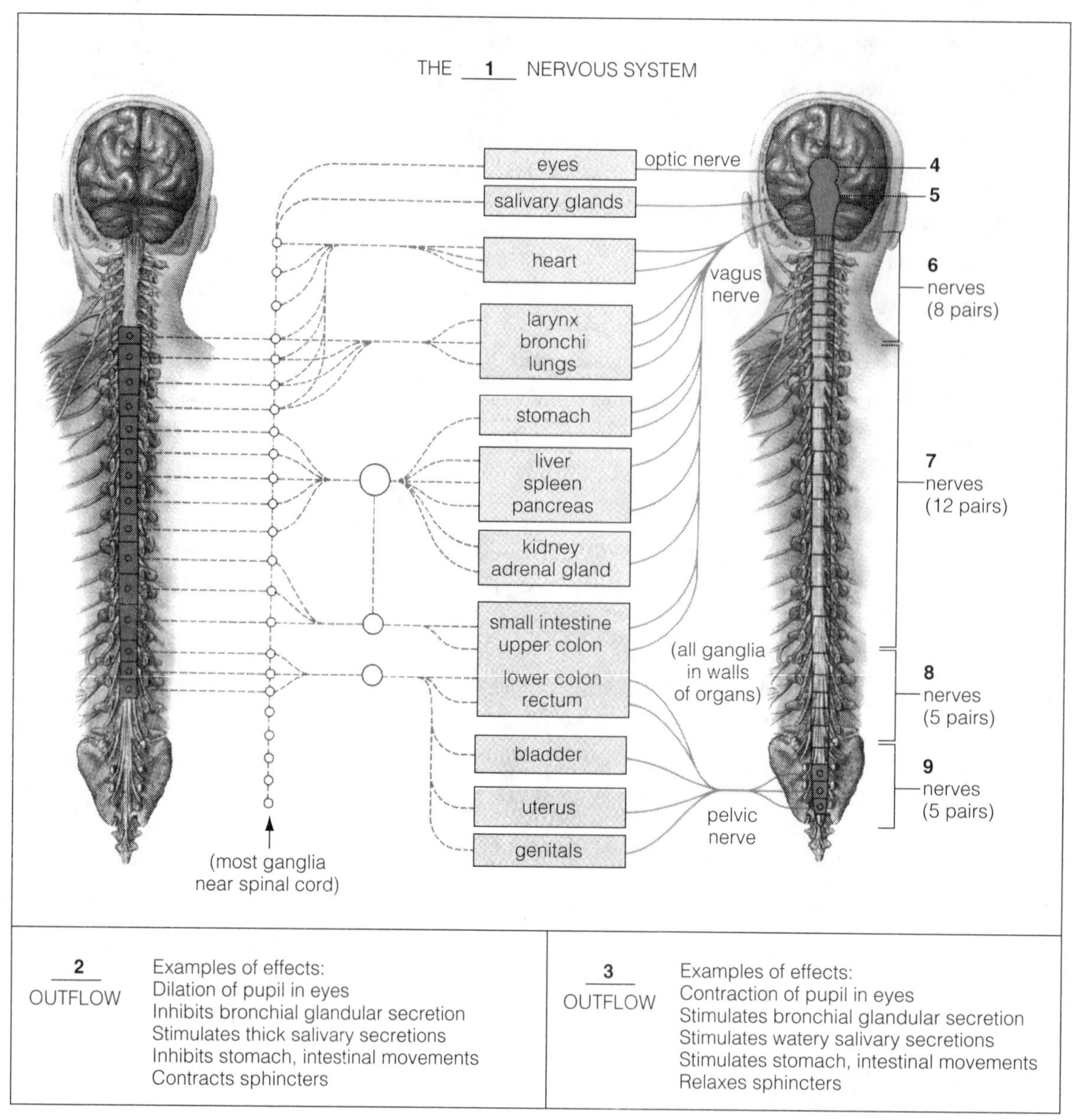

THE 1 NERVOUS SYSTEM
eyes
optic nerve
salivary glands
heart
vagus nerve
larynx bronchi lungs
stomach
liver spleen pancreas
kidney adrenal gland
small intestine upper colon
lower colon rectum
(all ganglia in walls of organs)
bladder
uterus
genitals
pelvic nerve
(most ganglia near spinal cord)
4
5
6 nerves (8 pairs)
7 nerves (12 pairs)
8 nerves (5 pairs)
9 nerves (5 pairs)
2 OUTFLOW
Examples of effects:
Dilation of pupil in eyes
Inhibits bronchial glandular secretion
Stimulates thick salivary secretions
Inhibits stomach, intestinal movements
Contracts sphincters
3 OUTFLOW
Examples of effects:
Contraction of pupil in eyes
Stimulates bronchial glandular secretion
Stimulates watery salivary secretions
Stimulates stomach, intestinal movements
Relaxes sphincters

1. ______________
2. ______________
3. ______________
4. ______________
5. ______________ ______________
6. ______________
7. ______________
8. ______________
9. ______________

Labeling

Identify the numbered parts of the accompanying illustrations. [p.593]

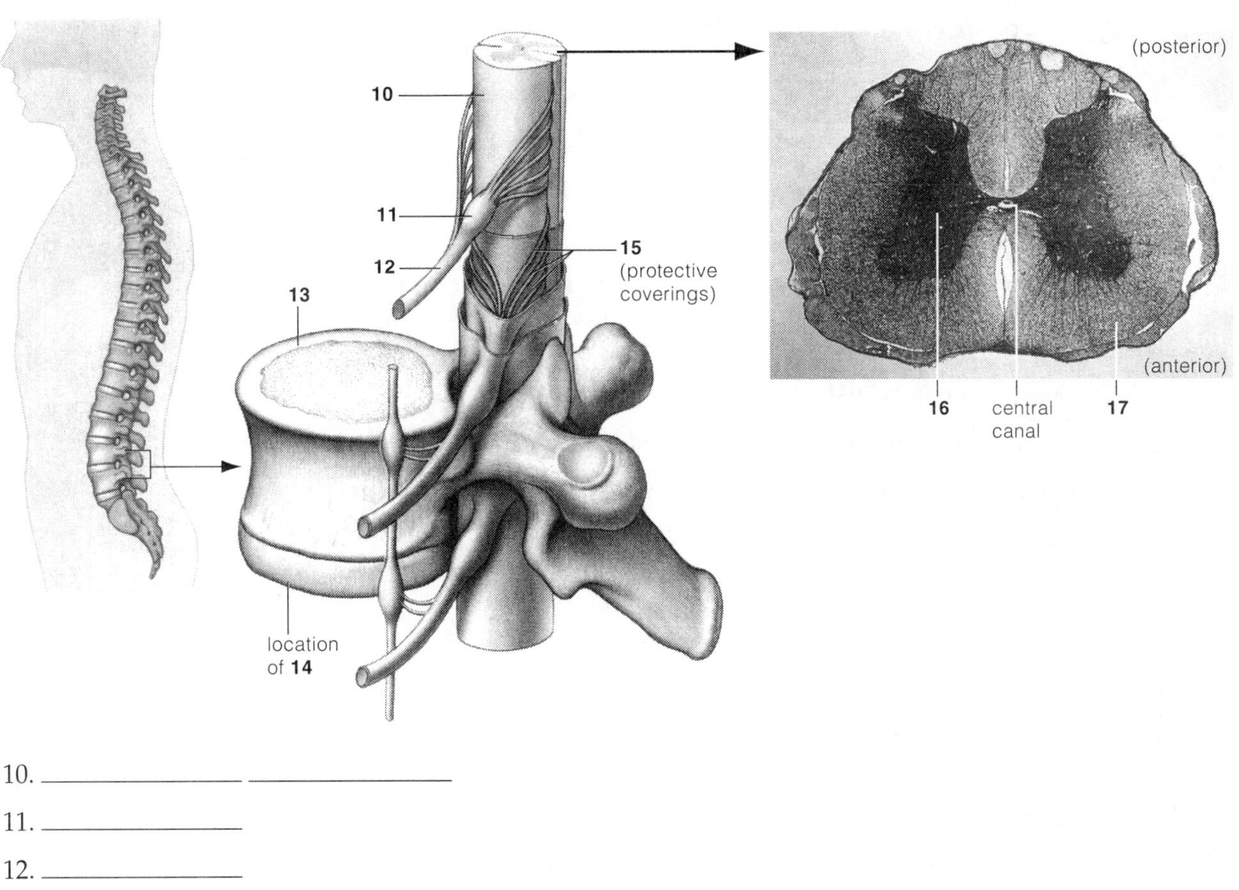

10. ________________ ________________

11. ________________

12. ________________

13. ________________

14. ________________ ________________

15. ________________

16. ________________ ________________

17. ________________ ________________

Choice

For questions 18–34, choose from the following:

a. peripheral — somatic nerves   b. peripheral — autonomic sympathetic nerves
c. peripheral — autonomic parasympathetic nerves   d. spinal cord nerves

18. ____ Dominate when the body is not receiving much outside stimulation [p.592]
19. ____ Carry signals about moving your head, trunk, and limbs [p.592]
20. ____ Dominate in times of sharpened awareness [p.593]
21. ____ Can be attacked by meningitis [p.593]
22. ____ Sensory axons inside these nerves deliver information from receptors in the skin, skeletal muscles, and tendons to the central nervous system [p.592]
23. ____ Tend to slow down the body overall and divert energy to basic "housekeeping" tasks, such as digestion [pp.592–593]
24. ____ The meninges, three tough, tubelike coverings, are part of the structure [p.593]
25. ____ A vital expressway for signals between the peripheral nervous system and the brain [p.593]
26. ____ The fight–flight response [p.593]
27. ____ Threads through a canal formed by bones of the vertebral column [p.593]
28. ____ Commands one's heart to beat faster [p.593]
29. ____ Tend to shelve housekeeping tasks [p.593]
30. ____ Their motor axons deliver commands from the brain and spinal cord to the body's skeletal muscles [p.592]
31. ____ Signals cause the release of epinephrine [p.593]
32. ____ Some of its interneurons exert direct control over certain reflex pathways [p.593]
33. ____ Commands your heart to beat a little slower [p.593]
34. ____ Gray matter that plays an important role in controlling reflexes for limb movement and organ activity [p.593]

34.8. THE VERTEBRATE BRAIN [pp.594–595]

Selected Words: "bridge" [p.594], *tectum* [p.594]

Boldfaced, Page-Referenced Terms

[p.594] brain ____________________

____

[p.594] brain stem ____________________

____

[p.594] medulla oblongata ____________________

____

[p.594] cerebellum ____________________

____

[p.594] pons ____

____

[p.594] cerebrum ____

____

[p.594] thalamus ____

____

[p.594] hypothalamus ____

____

[p.595] reticular formation ____

____

[p.595] blood–brain barrier ____

____

Matching

Choose the most appropriate answer for each term.

1. ____ medulla oblongata [p.594]
2. ____ cerebellum [p.594]
3. ____ pons [p.594]
4. ____ tectum [p.594]
5. ____ cerebrum [p.594]
6. ____ thalamus [p.594]
7. ____ hypothalamus [pp.594–595]
8. ____ reticular formation [p.595]
9. ____ blood–brain barrier [p.595]

A. The roof of the midbrain. In the lower vertebrates, this coordinates sensory input and coordinates motor responses
B. Integrates sensory input from the eyes, ears, and muscle spindles with motor signals from the forebrain; helps control motor dexterity, and more recent expansions may be crucial in language and some other forms of mental dexterity
C. A major network of interneurons; governs many aspects of the nervous system
D. Contains reflex centers for vital tasks, such as respiration and blood circulation; also coordinates motor responses with certain complex reflexes, such as coughing; influences other brain regions, helping you sleep or wake up
E. A pair of outgrowths from the brain stem where olfactory input and responses to it became integrated; these outgrowths expanded greatly, especially during and after the vertebrate invasion of land
F. Bands of many axons extend from both sides of the cerebellum to this area; a major traffic center for information passing between the cerebellum and the higher integrating centers of the forebrain
G. Evolved into the premier center for homeostatic control over the internal environment; became central to behaviors related to internal organ activities, such as thirst, hunger, and sex, and to emotional expression, such as sweating with fear
H. Evolved as coordinating center for sensory input and as a relay station for signals to the cerebrum
I. Limits which solutes enter the cerebrospinal fluid, protecting the brain and spinal cord

Choice

For questions 10–16, indicate to which area of the brain each part belongs. [p.594]

a. forebrain b. hindbrain c. midbrain

10. ____ thalamus
11. ____ hypothalamus
12. ____ tectum
13. ____ medulla oblongata
14. ____ pons
15. ____ cerebellum
16. ____ cerebrum

34.9. THE HUMAN CEREBRUM [pp.596–597]

34.10. FOCUS ON SCIENCE: *Sperry's Split-Brain Experiments* [p.598]

Selected Words: *motor* areas [p.596], *sensory* areas [p.596], *association* areas [p.596], "gut reactions" [p.597]

Boldfaced, Page-Referenced Terms

[p.596] cerebral cortex ____

____

[p.597] limbic system ____

____

Labeling

Identify each numbered part of the accompanying illustration. [p.596]

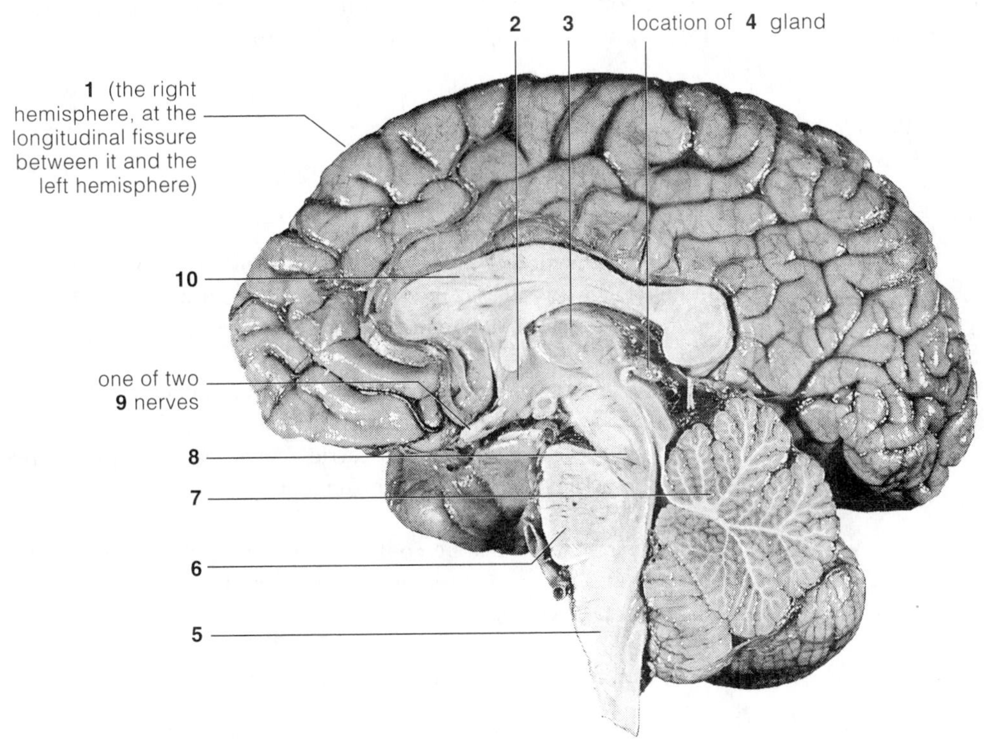

1. ______________
2. ______________
3. ______________
4. ______________
5. ______________ ______________
6. ______________
7. ______________
8. ______________
9. ______________
10. ______________ ______________

Complete the Table

11. Complete the table below by identifying in the left column the area of the brain whose functions are described in the right column. Choose from the following: cerebral cortex, occipital lobe; cerebral cortex, temporal lobe; cerebral cortex, parietal lobe; cerebral cortex, frontal lobe; left cerebral hemisphere; right cerebral hemisphere; cerebral cortex; limbic system; and corpus callosum. [pp.596–597]

Brain Area	*Functions*
a.	Deals more with visual–spatial relationships, music, and other creative enterprises
b.	Primary motor cortex controls coordinated movements of skeletal muscles; thumb, finger, and tongue muscles get much of the area's attention; Broca's area and the frontal eye field are located here
c.	A thin outer layer of gray matter on the left and right cerebral hemispheres
d.	A transverse band of nerve tracts; carries signals back and forth between the hemispheres and coordinates their functioning
e.	The body is spatially mapped out in the primary somatosensory cortex; this area is the main receiving center for sensory input from the skin and joints; also deals with taste perception
f.	Located at the rear of this lobe; primary visual cortex, which receives sensory inputs from the eyes
g.	Deals mainly with speech, analytical skills, and mathematics; dominates the right hemisphere in most people
h.	Located inside the cerebral hemispheres; governs emotions and has roles in memory; distantly related to olfactory lobes and still deals with the sense of smell
i.	Perception of sounds and odors arises in primary cortical areas located here

Short Answer

12. In an effort to relieve the frequent seizures of severe epilepsy, neural surgeon Roger Sperry cut the neural bridge of the corpus callosum in several of these patients. The seizures did subside in frequency and intensity. Summarize Sperry's subsequent findings regarding the function of the corpus callosum. [p.598] ____________________

____

____

____

34.11. HOW ARE MEMORIES TUCKED AWAY? [p.599]

34.12. REFLECTIONS ON THE NOT-QUITE-COMPLETE TEEN BRAIN [pp.600–601]

34.13. FOCUS ON HEALTH: *Drugging the Brain* [pp.602–603]

Selected Words: *short-term* storage [p.599], *long-term* storage [p.599], *facts* [p.599], *skills* [p.599], *amnesia* [p.599], *Parkinson's disease* [p.599], *Alzheimer's disease* [p.599], "gut reaction" [p.600], *MRI* [p.600], *necrotizing* *fasciitis* [p.602], *caffeine* [p.602], *nicotine* [p.602], *cocaine* [p.602], *amphetamines* [p.602], *crank* [p.602], *Ecstasy* [p.602], *club drugs* [p.603], *alcohol* [p.603], "high" [p.603], *analgesics* [p.603], *codeine* [p.603], *heroin* [p.603], *LSD* [p.603], *marijuana* [p.603]

Boldfaced, Page-Referenced Terms

[p.599] memory ______________________________

____

[p.602] drug addiction ______________________________

____

Labeling

Identify each numbered part of the accompanying illustration. [p.599]

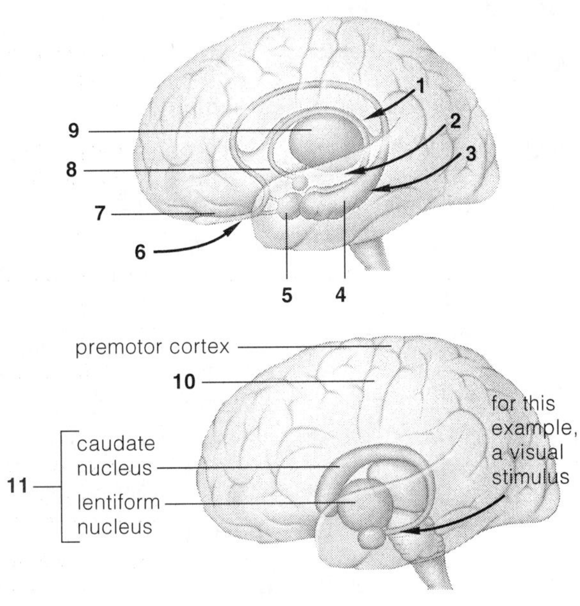

1. ______________
2. ______________
3. ______________
4. ______________
5. ______________
6. ______________
7. ______________ ______________
8. ______________ ______________
9. ______________ and ______________
10. ______________ ______________
11. ______________ ______________

Fill-in-the-Blanks

(12) ______________ [p.599] is the capacity of an individual's brain to store and retrieve information about past sensory experiences. (13) ______________ [p.599] and adaptive modifications of our behavior would be impossible without it. Information is stored in stages. (14) ______________ - ______________ [p.599] storage is a stage of neural excitation that lasts from a few seconds to a few hours; it is limited to a few bits of sensory information. In (15) ______________ - ______________ [p.599] storage, seemingly unlimited amounts of information get tucked away more or less permanently.

Only some (16) ______________ [p.599] input is chosen for transfer to brain structures involved in short-term memory. If information is (17) ______________ [p.599], it is forgotten; otherwise, it is consolidated with banks of information in long-term storage structures.

The human brain processes facts separately from (18) ______________ [p.599]. Explicit bits of information are soon forgotten or filed away in (19) ______________ - ______________[p.599] storage, along with the circumstances in which they were learned. (20) ______________ [p.599] are gained by practicing specific motor activities and are best recalled by actually performing the motor activity involved. Separate memory circuits handle different kinds of (21) ______________ [p.599]. A circuit leading to fact memory starts with inputs at the sensory cortex that flow to the (22) ______________ [p.599] and (23) ______________ [p.599], structures in the limbic system. The (24) ______________ [p.599] is the gatekeeper and connects the sensory cortex with parts of the thalamus and hypothalamus that govern emotional states. The (25) ______________ [p.599] mediates learning and spatial relations. Information flows to the prefrontal cortex, where multiple banks of (26) ______________ [p.599] memories are retrieved and used to stimulate or inhibit other parts of the brain. New input also flows to (27) ______________ [p.599] ganglia, which send it back to the cortex in a feedback loop that reinforces the input until it can be consolidated in (28) ______________ - ______________ [p.599] storage.

(29) ______________ [p.599] memory also begins at the (30) ______________ [p.599] cortex, but this circuit routes sensory input to the (31) ______________ ______________ [p.599], which promotes motor responses. The circuit for motor skills extends to the (32) ______________ [p.599], the brain region that coordinates motor activity.

(33) ______________ [p.599] is a loss of memory, the severity of which depends on whether the hippocampus, amygdala, or both are damaged, as by a severe head blow; this does not affect capacity to learn new (34) ______________ [p.599]. By contrast, basal ganglia are destroyed and learning ability is lost during (35) ______________ [p.599] disease, yet skill memory is retained. With an onset usually during later life, (36) ______________ [p.599] disease is linked to structural changes in the cerebral cortex and (37) ______________ [p.599]. Affected people often can remember long-standing (38) ______________ [p.599] but have difficulty remembering what has just happened to them. In time they become confused, depressed, and unable to complete a train of thought.

Choice

For questions 39–48, choose from the following:

a. stimulants [pp.602–603] b. depressants, hypnotics [p.603] c. analgesics [p.603] d. psychedelics, hallucinogens [p.603]

39. ____ amphetamines
40. ____ caffeine
41. ____ cocaine
42. ____ codeine
43. ____ alcohol
44. ____ heroin
45. ____ lysergic acid diethylamide (LSD)
46. ____ marijuana
47. ____ nicotine
48. ____ MDMA (Ecstasy)

Self-Quiz

____ 1. Which of the following is *not* true of an action potential? [pp.582–583]
   a. It is a short-range message that can vary in size.
   b. It is an all-or-none brief reversal in membrane potential.
   c. It doesn't decay with distance.
   d. It is self-propagating.

____ 2. The conducting zone of a neuron is the ____________. [p.580]
   a. axon
   b. axonal terminals
   c. cell body
   d. dendrite

____ 3. The output zone of a neuron is the ____________. [p.580]
   a. axon
   b. axonal endings
   c. cell body
   d. dendrite

____ 4. An action potential is brought about by ____________. [p.580]
   a. a sudden membrane impermeability
   b. the movement of negatively charged proteins through the neuronal membrane
   c. the movement of lipoproteins to the outer membrane
   d. a local change in membrane permeability caused by a greater-than-threshold stimulus

____ 5. The resting membrane potential ____________. [p.580]
   a. exists as long as a voltage difference sufficient to do work exists across a membrane
   b. occurs because there are more potassium ions outside the neuronal membrane than there are inside
   c. occurs because of the unique distribution of receptor proteins located on the dendrite exterior
   d. is brought about by a local change in membrane permeability caused by a greater-than-threshold stimulus

_____ 6. The phrase *all or none* used in conjunction with discussion about an action potential means that _______________. [p.582]
a. a resting membrane potential has been received by the cell
b. an impulse does not diminish or dissipate as it travels away from the trigger zone
c. the membrane either achieves total equilibrium or remains as far from equilibrium as possible
d. propagation along the neuron is much faster than in other neurons

_____ 7. An action potential passes from neuron to neuron across a synaptic cleft by way of _______________. [p.584]
a. myelin bridges
b. the resting membrane potential
c. neurotransmitter substances
d. neuromodulator substances

_____ 8. _______________ nerves dominate when the body is not receiving much outside stimulation. [p.592]
a. Ganglia
b. Pacemaker
c. Sympathetic
d. Parasympathetic
e. all of the above

_____ 9. What humans comprehend, communicate, remember, and voluntarily act on arises in the _______________. [p.596]
a. medulla oblongata
b. thalamus
c. hypothalamus
d. cerebellum
e. cerebral cortex

_____ 10. The _______________ are the protective coverings of the brain and spinal cord. [p.593]
a. ventricles
b. meninges
c. tectums
d. olfactory bulbs
e. pineal glands

_____ 11. The _______________ persists as a low-level pathway to motor centers of the medulla oblongata and spinal cord; it also can activate centers in the cerebral cortex and thereby govern the activities of the nervous system as a whole. [p.595]
a. medulla oblongata
b. pons
c. thalamus
d. hypothalamus
e. reticular formation

_____ 12. The part of the brain that controls the basic responses necessary to maintain life processes (respiration, blood circulation) is _______________. [p.594]
a. the cerebral cortex
b. the cerebellum
c. the corpus callosum
d. the medulla oblongata

_____ 13. The _______________ integrates sensory input from the eyes, ears, and muscle spindles with motor signals from the forebrain; it also helps control motor dexterity. [p.594]
a. cerebrum
b. pons
c. cerebellum
d. hypothalamus
e. thalamus

_____ 14. The _______________ evolved as a coordinating center for sensory input and a relay station for signals to the cerebrum. [p.595]
a. medulla oblongata
b. pons
c. reticular formation
d. hypothalamus
e. thalamus

Chapter Objectives/Review Questions

1. Define the terms *sensory neuron, stimulus, interneuron,* and *motor neuron.* [p.579]
2. Draw a neuron and label it according to its three general zones, its specific structures, and the specific function(s) of each structure. [p.580]
3. Define *resting membrane potential;* explain what establishes it and how it is used by the cell neuron. [p.580]
4. Define *action potential* and explain how sodium and potassium ions are used to generate an action potential. [pp.580–581]
5. Understand the importance of the sodium–potassium pump in maintaining the resting membrane potential. [p.580]
6. Explain how graded signals differ from action potentials. [p.582]
7. Explain a graph of an action potential. Look at Figure 34.6 in your text and determine which part of the curve represents the following: (a) the point at which the stimulus was applied, (b) the events prior to attainment of the threshold value, (c) the opening of the ion gates and diffusing of the ions, (d) the change from net negative charge inside the neuron to net positive charge and back again to net negative charge, and (e) the active transport of sodium ions out of and potassium ions into the neuron. [pp.582–583]
8. Define *period of insensitivity* and state what causes it. [p.583]
9. Understand the relationship between neurotransmitters and chemical synapses. [p.584]
10. Understand the difference between pre- and postsynaptic cells. [p.584]
11. Distinguish the way excitatory synapses function from the way inhibitory synapses function. [p.584]
12. Understand the process of synaptic integration. [p.585]
13. Understand the three mechanisms by which neurotransmitters are removed from the synaptic cleft. [p.585]
14. Distinguish between divergent, convergent, and reverberating circuits. [p.586]
15. Explain what the stretch reflex is and how it helps an animal survive. [p.587]
16. Explain what a reflex is by drawing and labeling a diagram and telling how it functions. [pp.586–587]
17. Describe a nerve net. [p.588]
18. Fully explain how the shift from radial to bilateral symmetry within invertebrate animals influenced the complexity of nervous systems. [pp.588–589]
19. The nerve cord that persists in all vertebrate embryos is called the ________ ________. [p.590]
20. Define and contrast the vertebrate central and peripheral nervous systems. [p.591]
21. Distinguish between a tract and a nerve. [p.591]
22. Compare the structures of the spinal cord and brain with respect to white matter and gray matter. [p.591]
23. Distinguish between afferent and efferent nerves [p.591]
24. Distinguish between somatic nerves and autonomic nerves. [p.592]
25. Explain how parasympathetic nerve activity balances sympathetic nerve activity. List activities of the sympathetic and parasympathetic nerves in regulating pupil diameter, rate of heartbeat, activities of the gut, and elimination of urine. [pp.592–593]
26. Describe the basic structural and functional organization of the spinal cord. In your answer, distinguish the spinal cord from the vertebral column. [p.593]
27. List the parts of the brain found in the hindbrain, midbrain, and forebrain, and give the basic functions of each. [pp.594–595]
28. The ________ formation is an ancient mesh of interneurons that still persists as a low-level pathway to motor centers of the medulla oblongata and spinal cord. [p.595]
29. In terms of structure and function, explain how the mechanism called the blood–brain barrier protects the brain and spinal cord. [p.595]
30. The cerebral cortex is functionally divided into ________ areas (control of voluntary motor activity), ________ areas (perception of the meaning of sensations), and ________ areas (information integration that precedes conscious action). [p.596]

31. The ________ system, which is located inside the cerebral hemispheres, governs emotions and has roles in memory. [p.597]
32. State what the results of the "split-brain" experiments suggest about the functioning of the cerebral hemispheres. [p.598]
33. Distinguish between short-term and long-term information storage. [p.599]
34. Explain the difference between facts and skills as they relate to memory. [p.599]
35. Define *amnesia, Parkinson's disease,* and *Alzheimer's disease;* list the characteristics of each. [p.599]
36. Cite evidence gathered by researchers showing that the teenage brain is not completely developed, which may account for stereotypic behavior during these years. [pp.600–601]
37. List the major classes of psychoactive drugs and provide an example of each class. [pp.600–601]

Integrating and Applying Key Concepts

1. What do you think might happen to human behavior if inhibitory postsynaptic potentials did not exist and if the threshold stimulus necessary to provoke an EPSP were much higher?
2. Suppose that anger is eventually determined to be caused by excessive amounts of specific transmitter substances in the brains of angry people. Also suppose that an inexpensive antidote to anger that neutralizes these anger-producing transmitter substances is readily available. Can violent murderers now argue that they have been wrongfully punished because they were victimized by their brain's transmitter substances and could not have acted in any other way? Suppose an antidote is prescribed to curb violent tempers in an easily angered person. Suppose also that the person forgets to take the pill and subsequently murders a family member. Can the murderer still claim to have been victimized by transmitter substances?

35

SENSORY RECEPTION

Interactive Exercises

Different Strokes for Different Folks [pp.606–607]

35.1. OVERVIEW OF SENSORY PATHWAYS [pp.608–609]

Selected Words: ultrasounds [p.606], *compound* sensations [p.606], sensory receptors [p.607], *amplitude* [p.608], *frequency* [p.608], baroreceptors [p.608]

Boldfaced, Page-Referenced Terms

[p.606] echolocation ____

____

[p.606] sensory systems ____

____

[p.606] sensation ____

____

[p.606] perception ____

____

[p.607] stimulus ____

____

[p.607] mechanoreceptors ______________________________

____

[p.607] thermoreceptors ______________________________

____

[p.607] pain receptors (nociceptors) ______________________________

____

[p.607] chemoreceptors ______________________________

____

[p.607] osmoreceptors ______________________________

____

[p.607] photoreceptors ______________________________

____

[p.608] action potentials ______________________________

____

[p.609] sensory adaptation ______________________________

____

[p.609] somatic sensations ______________________________

____

[p.609] special senses ______________________________

____

Fill-in-the-Blanks

The specialized peripheral endings of sensory neurons that detect specific kinds of stimuli are (1) ______________ ______________ [p.607]. A sensory system consists of sensory receptors for specific stimuli, (2) ______________ ______________ [p.607] that conduct information from those receptors to the brain, and (3) ______________ ______________ [p.607] where information is evaluated. A(n) (4) ______________ [p.606] is conscious awareness of change in internal or external conditions; this is not to be confused with (5) ______________ [p.606], which is an understanding of what sensation means. A(n) (6) ______________ [p.607] is any form of energy that activates a specific type of sensory receptor. (7) ______________ [p.607] detect the chemical energy of specific substances dissolved in the fluid surrounding them; (8) ______________ [p.607] detect mechanical energy associated with changes in pressure, position, or acceleration; (9) ______________ [p.607] detect the energy of visible and ultraviolet wavelengths of light; and (10) ______________ [p.607] detect radiant energy associated with temperature changes.

Every type of sensation is caused by (11) ______________ ______________ [p.608] arriving from particular nerve pathways and activating specific neurons in the (12) ______________ [p.608]. Besides sensing the kind of stimulus, the brain also interprets variations in (13) ______________ ______________ [p.608]. Interpretation is based on the (14) ______________ [p.608] of action potentials propagated along single axons and the (15) ______________ [p.608] of axons carrying action potentials from a given tissue.

Sometimes the (16) ______________ [p.609] of action potentials decreases or stops, even when a stimulus is being maintained at constant strength; such a decrease is known as (17) ______________ ______________ [p.609].

Some mechanoreceptors only signal a(n) (18) ______________ [p.609] in a stimulus; if a stimulus is constant, but deserves no response, these receptors will quit responding. But there are other types of receptors that adapt slowly or not at all. (19) ______________ [p.609] receptors, which continually inform the brain about the changes in (20) ______________ [p.609] of particular muscles, help maintain balance and posture.

Sensory receptors that are present at more than one body location generally inform the brain about (21) ______________ ______________ [p.609], that is, how different parts of the body feel at any particular time. Special senses are restricted to specific locations, such as inside the eyes or ears.

Matching

Select the best match for each of the following items. [all from pp.607–608]

22. ______ Is associated with vision
23. ______ Is associated with pain
24. ______ Detects odors
25. ______ Detects sounds
26. ______ Detects CO_2 concentration in the blood
27. ______ Detects environmental temperature
28. ______ Detects internal body temperature
29. ______ Detects touch
30. ______ Rods and cones
31. ______ Hair cells in the ear's organ of Corti
32. ______ Pacinian corpuscles in the skin
33. ______ Olfactory receptors in the nose
34. ______ Any stimulus that causes tissue damage
35. ______ Is associated with the movement of fluid in the inner ear

A. chemoreceptors
B. mechanoreceptors
C. nociceptors
D. photoreceptors
E. thermoreceptors

35.2. SOMATIC SENSATIONS [pp.610–611]

35.3. SENSES OF TASTE AND SMELL [p.612]

Selected Words: Meissner's corpuscle [p.610], the bulb of Krause [p.610], Ruffini endings [p.610], Pacinian corpuscle [p.610], *somatic* pain [p.610], *visceral* pain [p.610], bradykinins [p.611], histamine [p.611], prostaglandins [p.611], substance P [p.611], endorphins [p.611], enkephalins [p.611], *hyperalgesia* [p.611], *referred pain* [p.611], *phantom pain* [p.611], *chemical* senses [p.612], olfactory bulbs [p.612], taste buds [p.612]

Boldfaced, Page-Referenced Terms

[p.610] somatosensory cortex ____

____

[p.610] free nerve endings ____

____

[p.610] encapsulated receptors ____

____

[p.610] pain ____

____

[p.612] olfactory receptors ____

____

[p.612] pheromones ____

____

[p.612] taste receptors ____

____

Fill-in-the-Blanks

Signals from receptors in the skin and joints travel to the (1) ______________ ______________ [p.610], which is a strip little more than 2.5 cm wide running from the top of the (2) ______________ [p.610, Fig. 35.4] to just above the (3) ______________ [p.610, Fig. 35.4] on the surface of each (4) ______________ ______________ [p.610]. The largest portion of the somatosensory cortex is A (see accompanying figure), which receives signals coming from the (5) ______________ [p.610]. The second largest region is C, which receives signals coming from the (6) ______________ [p.610].

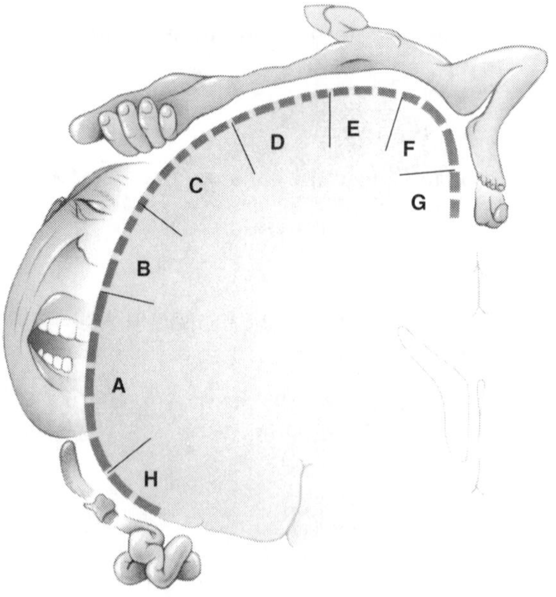

The somatic sensations (awareness of (7) ______________ [p.610], pressure, (8) ______________ [p.610], warmth, and pain) start with receptor endings that are embedded in (9) ______________ [p.610] and other tissues at the body's surfaces, in (10) ______________ [p.610] muscles, and in the walls of internal organs. Skin regions with more (11) ______________ [p.610] are more sensitive to touch, vibration, and pressure. Some (12) ______________ [p.610] nerve endings serve as thermoreceptors, and their firing of action potentials increases with increases in temperature. (13) ______________ [p.610] is the perception of injury to some body region; the perception begins when (14) ______________ [p.610], which include free nerve endings, send signals via the thalamus to the parietal lobe of the brain, where they are interpreted. Superficial (15) ______________ ______________ [p.610] arises at or near the skin surface. The brain sometimes gets confused and may associate perceived pain with a tissue some distance from the damaged area; this phenomenon is called (16) ______________ ______________ [p.611]. Usually the nerve pathways to both the injured and the mistaken area pass through the same segment of spinal cord. (17) ______________ [p.610] in skeletal muscle, joints, tendons, ligaments, and (18) ______________ [p.610] are responsible for awareness of the body's position in space and of limb movements.

Fill-in-the-Blanks

(19) ______________ [p.612] receptors detect molecules that become dissolved in fluid next to some body surface. Receptors are the modified dendrites of (20) ______________ [p.612] neurons. Animals smell substances by means of (21) ______________ [p.612] receptors, such as the ones in your (22) ______________ [p.612]; humans have about (23) ______________ [p.612] million of these in the nose. Sensory nerve pathways lead from the nasal cavity to the region of the brain where odors are identified and associated with their sources — the (24) ______________ [p.612] bulb and nerve tract. (25) ______________ [p.612] are signaling molecules secreted by one individual that influence the behavior of another. These molecules also target (26) ______________ [p.612] receptors. In the case of taste, these receptors, when located on animal tongues, are often part of sensory organs called (27) ______________ ______________ [p.612], which are enclosed by circular papillae.

35.4. SENSE OF BALANCE [p.613]

35.5. SENSE OF HEARING [pp.614–615]

Selected Words: *equilibrium* [p.613], utricle [p.613], saccule [p.613], "semicircular" canals [p.613], *static* equilibrium [p.613], cristae [p.613], *dynamic* equilibrium [p.613], cupola [p.613], vestibular nerve [p.613], *vertigo* [p.613], *amplitude* [p.614], intensity [p.614], *frequency* [p.614], pitch [p.614], pinna [p.614], auditory canal [p.614], eardrum [p.614], hammer, anvil, stirrup [p.614], oval window [p.614], *basilar* membrane [p.615], organ of Corti [p.615], *tectorial* membrane [p.615], auditory nerve [p.615]

Boldfaced, Page-Referenced Terms

[p.613] vestibular apparatus ____

[p.613] otoliths ("ear stones") ____

[p.613] hair cells ____

[p.614] hearing ____

[p.614] inner ear ____

[p.614] middle ear ____

[p.614] external ears ____

[p.615] cochlea ____

[p.615] acoustical receptors ____

Fill-in-the-Blanks

Hair cells are (1) ______________ [p.613] that detect vibrations and are important in our sense of (2) ______________ [p.613]. The (3) ______________ [p.614] (perceived loudness) of sound depends on the height of the sound wave. The (4) ______________ [p.614] (perceived pitch) of sound depends on how fast the wave changes occur. The faster the vibrations, the (5) [choose one] ☐ higher ☐ lower [p.614] the sound. The hammer, anvil, and stirrup are located in the (6) [choose one] ☐ inner ☐ middle [p.614] ear. The (7) ______________ [p.614] is a coiled tube that resembles a snail shell and contains the (8) ______________ ______________ ______________ [pp.614–615]—the organ that changes vibrations into electrochemical impulses.

Labeling

Identify each indicated part of the accompanying illustrations.

9. ______________ ______________ ______________ [p.614]

10. ______________ [p.614]

11. ______________ ______________ [p.614]

12. ______________ ______________ [p.614]

13. ______________ ______________ [pp.614–615]

14. ______________ ______________ [p.615]

15. ______________ ______________ [p.615]

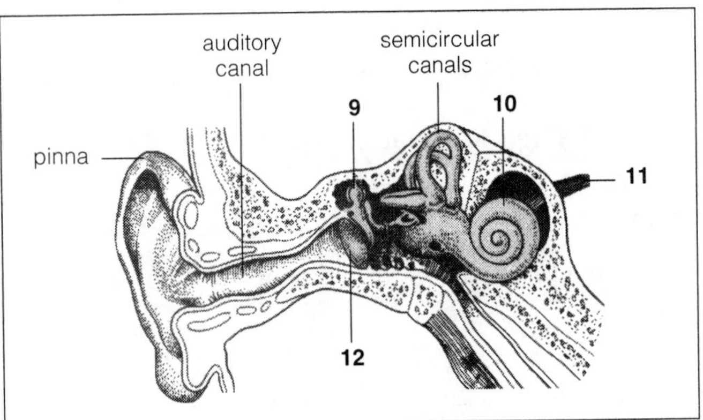

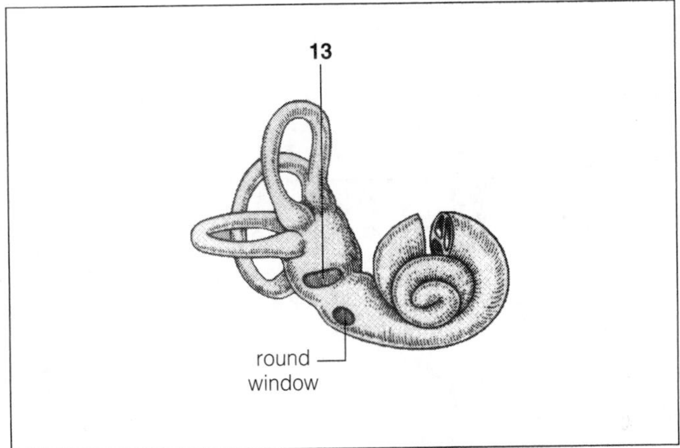

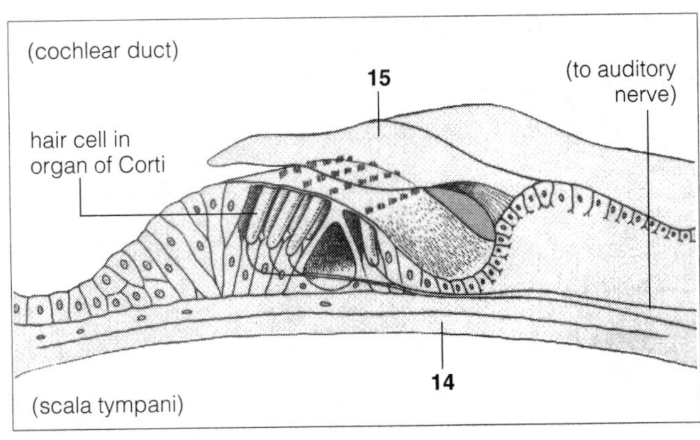

35.6. SENSE OF VISION [pp.616–617]

35.7. STRUCTURE AND FUNCTION OF VERTEBRATE EYES [pp.618–619]

35.8. CASE STUDY—FROM SIGNALING TO VISUAL PERCEPTION [pp.620–621]

35.9. FOCUS ON HEALTH: *Disorders of the Human Eye* [pp.622–623]

Selected Words: *rhabdomeric* photoreceptors [p.616], *ciliary* photoreceptors [p.616], ocellus (plural, ocelli) [p.616], ommatidia [p.617], sclera [p.618], choroid [p.618], ciliary body [p.618], iris [p.617], pupil [p.617], aqueous humor [p.618], vitreous body [p.618], optic nerve [p.620], rhodopsin [p.620], fovea [p.621], visual cortex [p.621], *color-blind* [p.622], *red–green color blindness* [p.622], *astigmatism* [p.622], *nearsightedness* [p.622], *farsightedness* [p.622], *histoplasmosis* [p.622], *herpes simplex* [p.622], *trachoma* [p.622], *cataracts* [p.623], *glaucoma* [p.623], *retinal detachment* [p.623], *corneal transplant* [p.623], *radial keratotomy* [p.623], *laser coagulation* [p.623]

Boldfaced, Page-Referenced Terms

[p.616] vision ____________________

____

[p.616] eyes ____________________

____

[p.616] visual field ____________________

____

[p.616] lens ____________________

____

[p.616] cornea ____________________

____

[p.617] compound eye ____________________

____

[p.617] camera eyes ____________________

____

[p.618] retina ____________________

____

[p.619] visual accommodation ____________________

____

[p.620] rod cell ____________________

____

[p.620] cone cells ____________________

____

Fill-in-the-Blanks

Light is a stream of (1) ________________ [p.620]— discrete energy packets. (2) ________________ [p.616] is a process in which photons are absorbed by pigment molecules in receptors and photon energy is transformed into the electrochemical energy of a nerve signal. (3) ________________ [p.616] requires precise light focusing onto a layer of photoreceptive cells that are dense enough to sample details of the light stimulus, followed by image formation in the brain. (4) ________________ [p.616] are simple clusters of photosensitive cells, usually arranged in a cuplike depression in the epidermis. (5) ________________ [p.616] are well-developed photoreceptor organs that allow at least some degree of image formation. The (6) ________________ [p.616] is the transparent cover of the lens area, and the (7) ________________ [p.617] consists of tissue containing densely packed photoreceptors. Compound eyes contain several thousand photosensitive units known as (8) ________________ [p.617]. In the vertebrate eye, lens adjustments assure that the (9) ________________ ________________ [p.619] for a specific group of light rays lands on the retina. (10) ________________ ________________ [p.619] refers to the lens adjustments that bring about precise focusing onto the retina. Rod cells contain the visual pigment (11) ________________ [p.620] and can respond even in dim light. (12) ________________ [p.621] cells are concerned with daytime vision and, usually, color perception. A(n) (13) ________________ [p.621] is a funnel-shaped pit on the retina that provides the greatest visual acuity.

Labeling

Identify each indicated part of the accompanying illustration. [all from p.618]

14. ________________ ________________
15. ________________
16. ________________
17. ________________
18. ________________ ________________
19. ________________ ________________
20. ________________
21. ________________
22. ________________ ________________
23. ________________ ________________
24. ________________

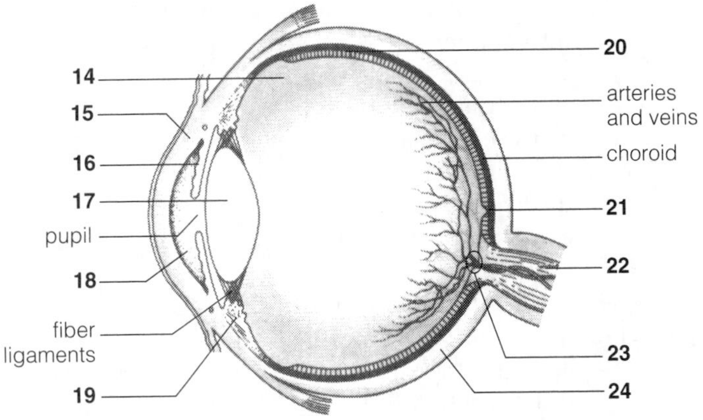

Fill-in-the-Blanks

(25) ________________ [p.622] occurs when irregularities of the cornea prevent the light from converging on the same focal point. When light is focused on a point in front of the retina, (26) ________________ [p.622] occurs. People with severe (26) sometimes opt for a surgical procedure called (27) ________________ ________________ [p.623]. (28) ________________ [p.623] are a clouding of the lens that is related to aging. Another age-associated problem is (29) ________________ [p.623], caused by increased pressure in the eye.

Self-Quiz

____ 1. According to the mosaic theory, ____________. [p.617]
   a. the basement membrane's pigment molecules prevent the scattering of light
   b. light falling on the inner area of an "on-center" field activates firing of the cells
   c. hair cells in the semicircular canals cooperate to detect rotational acceleration
   d. each ommatidium detects information about only one small region of the visual field; many ommatidia contribute "bits" to the total image
   e. all of the above

____ 2. The principal place in the human ear where sound waves are amplified is ____________. [p.614]
   a. the pinna
   b. the ear canal
   c. the middle ear
   d. the organ of Corti
   e. none of the above

____ 3. The place where vibrations are translated into patterns of nerve impulses is ____________. [p.615]
   a. the pinna
   b. the ear canal
   c. the middle ear
   d. the organ of Corti
   e. none of the above

For questions 4–8, choose from the following answers:

   a. fovea [p.621]
   b. cornea [p.618]
   c. iris [pp.618–619]
   d. retina [p.620]
   e. sclera [p.618]

____ 4. The white protective fibrous tissue of the eye is the ____________.

____ 5. Rods and cones are located in the ____________.

____ 6. The highest concentration of cones is in the ____________.

____ 7. The adjustable ring of contractile and connective tissues that controls the amount of light entering the eye is the ____________.

____ 8. The outer transparent protective covering of part of the eyeball is the ____________.

____ 9. Visual accommodation involves the ability to ____________. [p.619]
   a. change the sensitivity of the rods and cones by means of transmitters
   b. change the width of the lens by relaxing or contracting certain muscles
   c. change the curvature of the cornea
   d. adapt to large changes in light intensity
   e. all of the above

____ 10. Nearsightedness is caused by ____________. [p.622]
   a. eye structure that focuses an image in front of the retina
   b. uneven curvature of the lens
   c. eye structure that focuses an image posterior to the retina
   d. uneven curvature of the cornea
   e. none of the above

Chapter Objectives/Review Questions

1. Define and distinguish among chemoreceptors, mechanoreceptors, photoreceptors, and thermoreceptors. Name at least one example of each type that appears in an animal. [p.608]
2. Distinguish the types of stimuli detected by tactile and stretch receptors from those detected by hearing and equilibrium receptors. [pp.608,610,613–614]
3. Explain how a taste bud works and distinguish the types of stimuli it detects from those detected by touch or stretch receptors. [pp.610,612]
4. Explain how the three semicircular canals of the human ear detect changes of position and acceleration in a variety of directions. [p.613]
5. Follow a sound wave from pinna to organ of Corti; mention the name of each structure it passes and state where the sound wave is amplified and where the pattern of pressure waves is translated into electrochemical impulses. [pp.614–615]
6. State how low- and high-amplitude sounds affect the organ of Corti. [pp.614–615]
7. State how low- and high-pitch sounds affect the organ of Corti. [pp.614–615]
8. Explain what a visual system is and list four of the five aspects of a visual stimulus that are detected by different components of a visual system. [p.616]
9. Contrast the structure of compound eyes with the structures of invertebrate eyespots and the human eye. [pp.616–618]
10. Define *nearsightedness* and *farsightedness* and relate each to eyeball structure. [pp.622–623]
11. Describe how the human eye perceives color and how it perceives black-and-white. [pp.620–621]
12. Explain the general principles that affect how light is detected by photoreceptors and changed into electrochemical messages. [pp.620–621]
13. Indicate the causes of the following disorders of the human eye: (a) red–green color blindness, (b) astigmatism, (c) cataracts, (d) glaucoma, and (e) retinal detachment. [pp.622–623]

Integrating and Applying Key Concepts

How might human behavior be changed if human eyes were compound eyes composed of ommatidia and if humans perceived only vibrations — as fish do — rather than sounds?

36

INTEGRATION: ENDOCRINE CONTROL

Interactive Exercises

Hormone Jamboree [pp.626–627]

36.1. THE ENDOCRINE SYSTEM [pp.628–629]

Selected Words: *signaling molecules* [p.628], target cells [p.628], vomeronasal organ [p.628], pituitary gland [p.628], adrenal glands [p.628], pancreatic islets [p.628], thyroid gland [p.628], parathyroid glands [p.628], pineal gland [p.628], thymus [p.628], gonads [p.628], hypothalamus [p.628]

Boldfaced, Page-Referenced Terms

[p.628] animal hormones ______________________________

[p.628] neurotransmitters ______________________________

[p.628] local signaling molecules ______________________________

[p.628] pheromones ______________________________

[p.629] endocrine system ______________________________

Matching

Choose the most appropriate answer for each term. [p.628]

1. ____ hormones
2. ____ neurotransmitters
3. ____ vomeronasal organ
4. ____ target cells
5. ____ local signaling molecules
6. ____ pheromones

A. Signaling molecules released from axon endings of neurons that act swiftly on target cells
B. A pheromone detector discovered in humans
C. Released by many types of body cells; alter conditions within localized regions of tissues
D. Nearly odorless secretions of particular exocrine gland; common signaling molecules that act on cells of other animals of the same species and help integrate social behavior
E. Secretions from endocrine glands, endocrine cells, and some neurons that the bloodstream distributes to nonadjacent target cells
F. Cells that have receptors for any given type of signaling molecule

Complete the Table

7. Complete the following table by identifying the numbered components of the endocrine system shown in the illustration on the facing page, together with the hormones produced by each. [p.629]

Gland Name	*Number*	*Hormones Produced*
a. Hypothalamus		
b. Pituitary, anterior lobe		
c. Pituitary, posterior lobe		
d. Adrenal glands (cortex)		
e. Adrenal glands (medulla)		
f. Ovaries (two)		
g. Testes (two)		
h. Pineal		
i. Thyroid		
j. Parathyroids (four)		
k. Thymus		
l. Pancreatic islets		

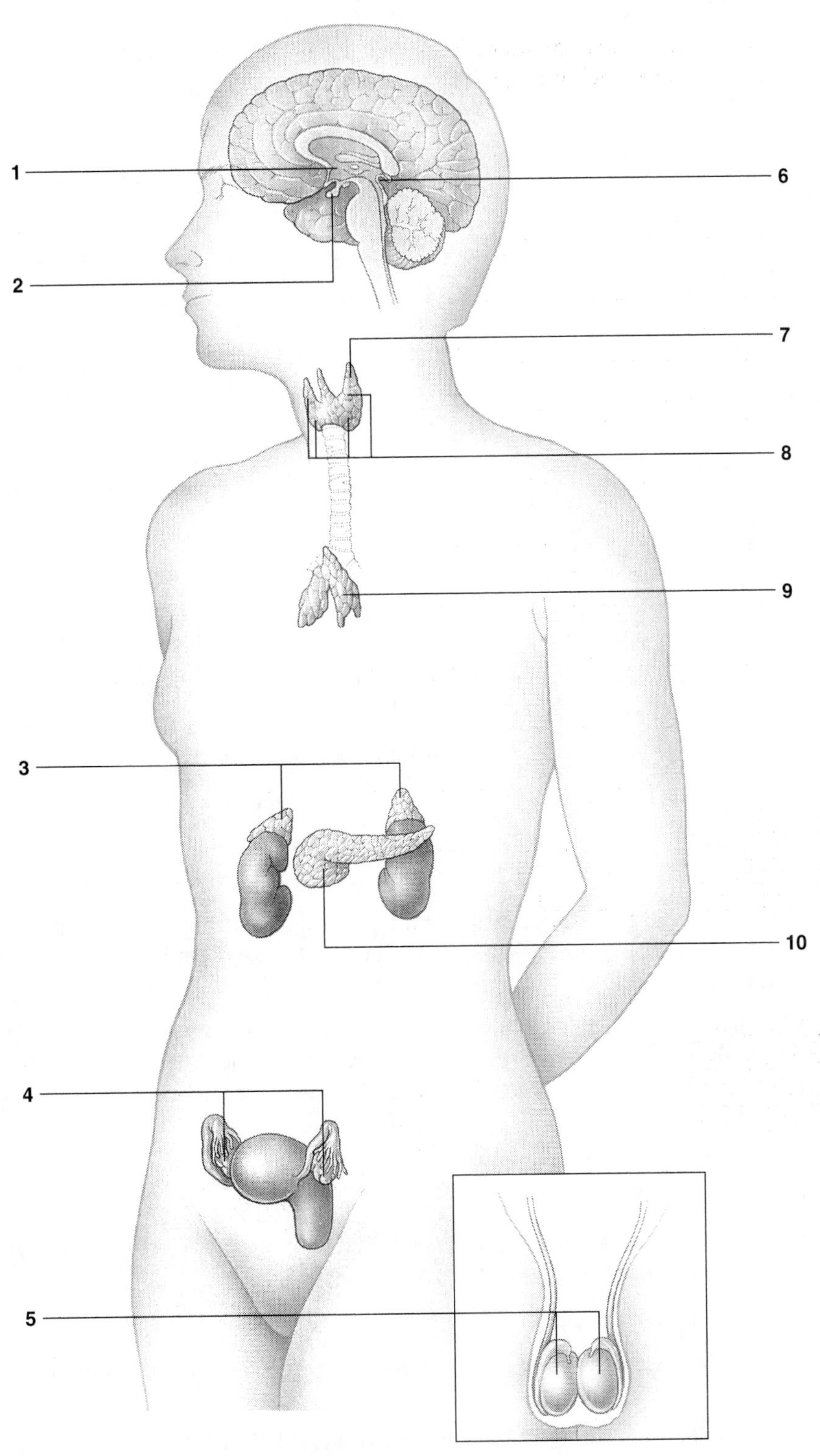
1
2
3
4
5
6
7
8
9
10

36.2. SIGNALING MECHANISMS [pp.630–631]

Selected Words: *androgen insensitivity syndrome* [p.630], cAMP (cyclic adenosine monophosphate) [p.631]

Boldfaced, Page-Referenced Terms

[p.630] steroid hormones ______________________________

____

[p.631] peptide hormones ______________________________

____

[p.631] second messenger ______________________________

____

Choice

For questions 1–10, choose from the following:

a. steroid hormones b. peptide hormones

1. ____ Lipid-soluble molecules derived from cholesterol; can diffuse directly across the lipid bilayer of a target cell's plasma membrane [p.630]
2. ____ Various peptides, polypeptides, and glycoproteins [p.631]
3. ____ One example involves testosterone, defective receptors, and a condition called androgen-insensitivity syndrome [p.630]
4. ____ Hormones that often require assistance from second messengers [p.631]
5. ____ Hormones that bind to receptors at the plasma membrane of a cell; the receptor then activates specific membrane-bound enzyme systems, which in turn initiate reactions leading to the cellular response [p.631]
6. ____ Lipid-soluble molecules that move through the target cell's plasma membrane to the nucleus, where they bind to some type of protein receptor. The hormone–receptor complex moves into the nucleus and interacts with specific DNA regions to stimulate or inhibit transcription of mRNA [p.630]
7. ____ Water-soluble signaling molecules that may incorporate anywhere from 3 to 180 amino acids [p.631]
8. ____ Involves molecules such as cyclic AMP that activate many enzymes in cytoplasm, which, in turn, causes alteration in some cell activity [p.631]
9. ____ Glucagon is an example [p.631]
10. ____ Cyclic AMP relays a signal into the cell's interior to activate protein kinase A [p.631]

36.3. THE HYPOTHALAMUS AND PITUITARY GLAND [pp.632–633]

Selected Words: *posterior* lobe [p.632], *anterior* lobe [p.632], *second* capillary bed [p.632]

Boldfaced, Page-Referenced Terms

[p.632] hypothalamus ______________________________

____

[p.632] pituitary gland ____

____

[p.633] releasers ____

____

[p.633] inhibitors ____

____

Choice/Match

Label each hormone listed below with an "A" if it is secreted by the anterior lobe of the pituitary, a "P" if it is released from the posterior pituitary, or an "I" if it is released from intermediate tissue. Complete the exercise by entering the letter of the corresponding action in the parentheses after each label. [p.632]

1. _____ ( ) ACTH
2. _____ ( ) ADH
3. _____ ( ) FSH
4. _____ ( ) STH (GH)
5. _____ ( ) LH
6. _____ ( ) MSH
7. _____ ( ) OCT
8. _____ ( ) PRL
9. _____ ( ) TSH

A. Stimulates egg and sperm formation in ovaries and testes
B. Targets pigmented cells in skin and other surface coverings; induces color changes in response to external stimuli and affects some behaviors
C. Stimulates and sustains milk production in mammary glands
D. Stimulates progesterone secretion, ovulation, and corpus luteum formation in females; promotes testosterone secretion and sperm release in males
E. Induces uterine contractions and milk movement into secretory ducts of the mammary glands
F. Stimulates release of thyroid hormones from the thyroid gland
G. Acts on the kidneys to conserve water required for control of extracellular fluid volume
H. Stimulates release of adrenal steroid hormones from the adrenal cortex
I. Promotes growth in young; induces protein synthesis and cell division; roles in adult glucose and protein metabolism

Dichotomous Choice

Circle one of two possible answers given between parentheses in each statement.

10. The (hypothalamus/pituitary gland) region of the brain monitors internal organs and activities related to their functioning, such as eating and sexual behavior; it also secretes some hormones. [p.632]
11. The (posterior/anterior) lobe of the pituitary stores and secretes two hormones, ADH and OCT, that are produced by the hypothalamus. [p.632]
12. The (posterior/anterior) lobe of the pituitary produces and secretes its own hormones, which govern the release of hormones from other endocrine glands. [p.632]
13. Humans lack the (posterior/intermediate) lobe of the pituitary gland, one that is possessed by many other vertebrates. [p.632]
14. Most hypothalamic hormones acting in the anterior pituitary lobe are (releasers/inhibitors) and cause target cells there to secrete hormones of their own. [p.633]
15. Some hypothalamic hormones slow secretion from their targets in the anterior pituitary; these are classed as (releasers/inhibitors). [p.632]

36.4. FOCUS ON HEALTH: *Abnormal Pituitary Output* [p.634]

Selected Words: *gigantism* [p.634], *pituitary dwarfism* [p.634], *acromegaly* [p.634], *diabetes insipidus* [p.634]

Complete the Table

1. Complete the following table summarizing examples of abnormal pituitary output. [p.634]

Condition	*Hormone/Abnormality*	*Characteristics*
a.	Excessive somatotropin produced during childhood	Affected adults are proportionally similar to a normal person but larger
b.	Insufficient somatotropin produced during childhood	Affected adults are proportionally similar to a normal person but much smaller
c.	Diminished ADH secretion by a damaged posterior pituitary lobe	Large volumes of dilute urine are secreted, causing life-threatening dehydration
d.	Excessive somatotropin output during adulthood, when long bones can no longer lengthen	Abnormal thickening of bone, cartilage, and other connective tissues in the hands, feet, and jaws

36.5. SOURCES AND EFFECTS OF OTHER HORMONES [p.635]

Complete the Table

1. Complete the following table by matching the gland/organ and the hormone(s) produced by it to the descriptions of hormone action. Refer to Table 36.3, p.635 in the text.

Gland/Organ

A. adrenal cortex
B. adrenal medulla
C. thyroid
D. parathyroids
E. testes
F. ovaries
G. pancreas (alpha cells)
H. pancreas (beta cells)
I. pancreas (delta cells)
J. thymus
K. pineal

Hormones

a. thyroxine and triiodothyronine
b. glucagon
c. PTH
d. androgens
e. somatostatin
f. thymosins
g. glucocorticoids
h. estrogens (progesterone also important)
i. epinephrine
j. melatonin
k. insulin
l. progesterone
m. mineralocorticoids
n. calcitonin
o. norepinephrine

Gland/Organ	*Hormone*	*Hormone Action*
a.		Elevates calcium levels in blood
b.		Influences carbohydrate metabolism of insulin-secreting cells
c.		Required in egg maturation and release, as well as in preparation of uterine lining for pregnancy and its maintenance in pregnancy; influences growth, development, and genital development; maintains sexual traits
d.		Promote protein breakdown and conversion to glucose in most cells
e.		Lowers blood sugar level in muscle and adipose tissue
f.		In general, required in sperm formation, genital development, and maintenance of sexual traits; influences growth and development
g.		In most cells, regulates metabolism; plays roles in growth and development
h.		In the gonads, influences daily biorhythms; influences gonad development and reproductive cycles
i.		In liver, muscle, and adipose tissues; raises blood sugar level, fatty acids; increases heart rate and force of contraction
j.		Targets lymphocytes; has roles in immunity
k.		Raises blood sugar level
l.		Prepares and maintains uterine lining for pregnancy; stimulates breast development
m.		In kidneys, promotes sodium reabsorption and control of salt–water balance
n.		In smooth muscle cells of blood vessels, promotes constriction or dilation of blood vessels
o.		In bone, lowers calcium levels in blood

36.6. FEEDBACK CONTROL OF HORMONAL SECRETIONS [pp.636–637]

Selected Words: *inhibit* [p.636], *stimulate* [p.636], *fight–flight response* [p.636], *simple goiter* [p.637], *hypothyroidism* [p.637], *toxic goiter* [p.637], *hyperthyroidism* [p.637], *primary* reproductive organs [p.637]

Boldfaced, Page-Referenced Terms

[p.636] negative feedback ____

____

[p.636] positive feedback ____

____

[p.636] adrenal cortex ____

____

[p.636] adrenal medulla ____

____

[p.636] thyroid gland ____

____

[p.637] gonads ____

____

Matching

Choose the most appropriate answer for each term relating to the adrenal glands. [p.636]

1. _____ adrenal medulla
2. _____ ACTH
3. _____ nervous system
4. _____ glucocorticoids
5. _____ CRH
6. _____ cortisol
7. _____ fight–flight response
8. _____ cortisol-like drugs
9. _____ adrenal cortex
10. _____ feedback mechanisms

A. Results from actions of epinephrine and norepinephrine in times of excitement or stress
B. Outer portion of each adrenal gland; some of its cells secrete hormones such as glucocorticoids
C. Negative ultimately inhibits hormone secretion, positive stimulates further hormone secretion
D. Initiates a stress response during chronic stress, injury, or illness; cortisol helps to suppress inflammation
E. Inner portion of the adrenal gland; neurons located here release epinephrine and norepinephrine
F. Stimulates the adrenal cortex to secrete cortisol; this helps raise the level of glucose by preventing muscle cells from taking up more blood glucose
G. Used to counter asthma and other chronic inflammatory disorders
H. Help maintain blood glucose concentration and help suppress inflammatory responses
I. Secreted by the hypothalamus in response to falling glucose blood levels; stimulates the anterior pituitary to secrete ACTH
J. Secreted by the adrenal cortex; blocks the uptake and use of blood glucose by muscle cells; also stimulates liver cells to form glucose from amino acids

Fill-in-the-Blanks

Thyroxine and triiodothyronine are the main hormones secreted by the human (11) ______________ [p.636] gland. They are critical for normal development of many tissues, and they control overall (12) ______________ [p.636] rates in humans and other warm-blooded animals. The synthesis of thyroid hormones requires (13) ______________ [p.637], which is obtained from food. In the absence of iodide, blood levels of thyroid hormones (14) ______________ [p.637]. The anterior pituitary responds by secreting (15) ______________ [p.637]. When thyroid hormones cannot be synthesized, the feedback signal continues — and so does TSH secretion. As TSH secretion continues, the thyroid gland is overstimulated, causing (16) ________________ [p.637], an enlargement of the thyroid gland. (17) ________________ [p.637] results from insufficient blood-level concentrations of thyroid hormones. Affected adults are often (18) ________________ [p.637], sluggish, dry-skinned, intolerant of cold, and sometimes confused and depressed. (19) ________________ [p.637] results from excess concentrations of thyroid hormones. Affected adults show an increased heart rate, heat intolerance, elevated blood pressure, profuse sweating, and weight loss even when caloric intake increases. Affected individuals typically are nervous and agitated and have trouble sleeping.

The primary reproductive organs are known as (20) ________________ [p.637]. These organs produce and secrete sex (21) ________________ [p.637] essential to reproduction. These organs are known as the (22) ________________ [p.637] in human males and the (23) ________________ [p.637] in females. Testes secrete (24) ________________ [p.637]; ovaries secrete estrogens and progesterone. All these hormones influence (25) ________________ [p.637] sexual traits. Both types of organs also produce (26) ________________ [p.637], or sex cells.

36.7. DIRECT RESPONSES TO CHEMICAL CHANGES [pp.638–639]

Selected Words: *rickets* [p.638], nerve growth factor (NGF) [p.638], *exocrine* cells [p.638], *endocrine* cells [p.638], *alpha* cells [p.638], *glucagon* [p.638], *beta* cells [p.638], *insulin* [p.638], *delta* cells [p.638], *diabetes mellitus* [p.639], *type 1 diabetes* [p.639], *type 2 diabetes* [p.639]

Boldfaced, Page-Referenced Terms

[p.638] parathyroid glands ____

____

[p.638] pancreatic islet ____

____

Fill-in-the-Blanks

Humans have four (1) ________________ [p.638] glands positioned next to the posterior (or back) of the human thyroid; they secrete (2) ________________ [p.638] in response to a low (3) ________________ [p.638] level in the blood. This hormone induces living bone cells to secrete enzymes that digest bone tissue, thereby releasing (4) ________________ [p.638] and other minerals to interstitial fluid and then to the blood. PTH enhances calcium (5) ________________ [p.638] from the filtrate flowing from the nephrons of the kidneys. It also induces some kidney cells to secrete enzymes that act on blood-borne precursors of the active form of vitamin (6) ________________ [p.638], a hormone. This hormone stimulates (7) ________________ [p.638] cells to increase calcium absorption from the gut lumen. In a child with vitamin D deficiency, too little calcium and phosphorus are absorbed, and so rapidly growing bones develop improperly. The resulting bone disorder is called (8) ________________ [p.638], characterized by bowed legs, a malformed pelvis, and in many cases a malformed skull and rib cage.

Complete the Table

9. Complete the following table summarizing the function of the pancreatic islets.

Pancreatic Islet Cells	*Hormone Secreted*	*Hormone Action*
[p.638] a. Alpha cells		
[p.638] b. Beta cells		
[p.638] c. Delta cells		

Dichotomous Choice

Circle one of two possible answers given between parentheses in each statement.

10. Insulin deficiency can lead to diabetes mellitus, a disorder in which the glucose level (rises/decreases) in the blood, then in the urine. [p.638]
11. In a person with diabetes mellitus, urination becomes (reduced/excessive), and the body's water–solute balance becomes disrupted; people become abnormally dehydrated and thirsty. [p.639]
12. Lacking a steady glucose supply, body cells of people with diabetes mellitus begin breaking down their own fats and proteins for (energy/water). [p.639]
13. Weight loss occurs and (amino acids/ketones) accumulate in blood and urine; this promotes excessive water loss with a life-threatening disruption of brain function. [p.639]
14. After a meal, blood glucose rises; pancreatic beta cells secrete (glucagon/insulin); targets use glucose or store it as glycogen. [p.639]
15. Blood glucose levels decrease between meals. (Glucagon/Insulin) is secreted by stimulated pancreas alpha cells; targets convert glycogen back to glucose, which then enters the blood. [p.639]
16. In (type 1 diabetes/type 2 diabetes) the body mistakenly mounts an autoimmune response against its own insulin-secreting beta cells and destroys them. [p.639]
17. Juvenile-onset diabetes is also known as (type 1 diabetes/type 2 diabetes); these patients survive by means of insulin injections. [p.639]
18. In (type 1 diabetes/type 2 diabetes), insulin levels are close to or above normal, but target cells fail to respond to insulin. [p.639]
19. (Type 1 diabetes/Type 2 diabetes) is usually manifested during middle age and is less dramatically dangerous than the other type; beta cells produce less insulin as a person ages. [p.639]

36.8. HORMONES AND THE EXTERNAL ENVIRONMENT [pp.640–641]

Selected Words: melatonin [p.640], jet lag [p.640], *seasonal affective disorder [SAD]* [p.640], "winter blues" [p.640], *Xenopus laevis* [p.640]

Boldfaced, Page-Referenced Terms

[p.640] pineal gland ____________________

[p.640] biological clock ____________________

[p.640] puberty ____________________

[p.641] molting ____________________

[p.641] ecdysone ____________________

Choice

For questions 1–10, choose from the following:

a. melatonin b. ecdysone

1. ____ The hormone that controls molting [p.641]
2. ____ Hormone secreted by the pineal gland [p.640]
3. ____ High blood levels of this hormone in winter (long nights) suppress sexual activity in hamsters. [p.640]
4. ____ Winter blues [p.640]
5. ____ Chemical interactions that cause an old cuticle to detach from the epidermis and muscles [p.641]
6. ____ Triggers puberty [p.640]
7. ____ Suppresses growth of a bird's gonads in fall and winter [p.640]
8. ____ Jet lag [p.640]
9. ____ Waking up at sunrise [p.640]
10. ____ Produced and stored in molting glands by insects and crustaceans [p.641]

Short Answer

11. What gland and its secretions are suspected of being involved in recently observed developmental deformities in frogs? Why? [pp.640–641] ____________________

Self-Quiz

_____ 1. The _______________ region of the forebrain monitors internal organs, influences certain forms of behavior, and secretes some hormones. [p.632]
a. hypothalamus
b. pancreas
c. thyroid
d. pituitary
e. thalamus

_____ 2. Neurons of the _______________ produce ADH and oxytocin, which are stored within axon endings of the _______________. [p.632]
a. anterior pituitary, posterior pituitary
b. adrenal cortex, adrenal medulla
c. posterior pituitary, hypothalamus
d. posterior pituitary, thyroid
e. hypothalamus, posterior pituitary

_____ 3. If you were lost in the desert and had no fresh water to drink, the level of _______________ in your blood would increase as a means of conserving water. [p.632]
a. insulin
b. corticotropin
c. oxytocin
d. antidiuretic hormone
e. salt

Choice

For questions 4–6, choose from the following answers:

a. estrogen b. PTH c. FSH d. somatotropin e. prolactin

_____ 4. _______________ stimulates bone cells to release calcium and phosphate and the kidneys to conserve it. [p.638]

_____ 5. _______________ stimulates and sustains milk production in mammary glands. [p.632]

_____ 6. _______________ is the hormone associated with pituitary dwarfism, gigantism, and acromegaly. [p.634]

For questions 7–9, choose from the following answers:

a. adrenal medulla b. adrenal cortex c. thyroid d. anterior pituitary e. posterior pituitary

_____ 7. The _______________ produces glucocorticoids that help increase the level of glucose in blood. [p.636]

_____ 8. The gland that is most closely associated with emergency situations is the _______________. [p.636]

_____ 9. The overall metabolic rates of warm-blooded animals, including humans, depend on hormones secreted by the _______________ gland. [pp.636–637]

For question 10, choose from the following answers:

a. parathyroid hormone b. aldosterone c. calcitonin d. mineralocorticoids e. none of the above

_____ 10. If all sources of calcium were eliminated from your diet, your body would secrete more _______________ in an effort to release calcium stored in your body and send it to the tissues that require it. [p.638]

Matching

Choose the most appropriate answer for each term.

11. ____ ADH [p.632]
12. ____ ACTH and TSH [p.636]
13. ____ FSH and LH [pp.632–633]
14. ____ GnRH [p.633]
15. ____ STH [pp.632,634]
16. ____ glucocorticoids [p.636]
17. ____ cortisol [p.636]
18. ____ epinephrine and norepinephrine [p.636]
19. ____ thyrosine and triiodothyronine [p.636]
20. ____ estrogens and progesterone [pp.629,635]
21. ____ PTH [p.638]
22. ____ glucagon [p.638]
23. ____ testosterone [pp.629,635]
24. ____ insulin [p.638]
25. ____ pheromones [p.628]
26. ____ somatostatin [p.633]
27. ____ melatonin [p.640]
28. ____ ecdysone [p.641]

A. In times of excitement or stress, these adrenal medulla hormones help adjust blood circulation together with fat and carbohydrate metabolism
B. Hormones secreted by the ovaries; influence secondary sexual traits
C. Hormone secreted by beta pancreatic cells; lowers blood glucose level
D. Abnormal amount of this anterior pituitary lobe hormone has different effects on human growth during childhood and adulthood
E. Anterior pituitary lobe hormones that orchestrate secretions from the adrenal gland and thyroid gland, respectively
F. Adrenal cortex hormones that help increase the level of glucose in the blood
G. Hormone secreted by alpha pancreatic cells; raises the blood glucose level
H. The hormone that largely controls molting in insects and crustaceans
I. Major thyroid hormones having widespread effects such as controlling the overall metabolic rates of warm-blooded animals
J. A glucocorticoid that comes into play when the body is under stress, as when the blood glucose level declines below a set point
K. Hypothalamic hormone that brings about secretion of FSH and LH, which are gonadotropins
L. Hormone secreted by the testes; influences secondary sexual traits
M. Hormone secreted by the pineal gland; influences the growth and development of gonads
N. Antidiuretic hormone produced by the posterior pituitary lobe; promotes water reabsorption when the body must conserve water
O. Hormone secreted by delta pancreatic cells; helps control digestion and absorption of nutrients; can also block secretion of insulin and glucagon
P. Nearly odorless hormonelike secretions of certain exocrine glands; they diffuse through water or air to cellular targets outside the animal body
Q. Anterior pituitary lobe hormones that act through the gonads to influence gamete formation and secretion of the sex hormones required in sexual reproduction
R. Hormone secreted by the parathyroid glands in response to low blood calcium levels

Chapter Objectives/Review Questions

1. Hormones, neurotransmitters, local signaling molecules, and pheromones are all known as ________ molecules that carry out integration. [p.628]
2. ________ are the secretory products of endocrine glands, endocrine cells, and some neurons. [p.628]
3. Define the terms *neurotransmitters, local signaling molecules,* and *pheromones.* [p.628]
4. Collectively, the body's sources of hormones came to be called the ________ system. [p.629]
5. Locate and name the components of the human endocrine system on a diagram such as Figure 36.2b in the text. [p.629]
6. Target cells have ________ receptors that hormones and other signaling molecules interact with, producing diverse effects on physiological processes. [p.630]
7. Contrast the proposed mechanisms of hormonal action on target cell activities by (a) steroid hormones and (b) peptide hormones that are proteins or are derived from proteins. [pp.630–631]
8. The ________ and the pituitary gland interact closely as a major neural–endocrine control center. [p.632]
9. Explain how, even though the anterior and posterior lobes of the pituitary are compounded as one gland, the tissues of each part differ in character. [pp.632–633]
10. Identify the hormones released from the posterior lobe of the pituitary and state their target tissues. [p.632]
11. Identify the hormones produced by the anterior lobe of the pituitary and tell which target tissues or organs each acts on. [pp.632–633]
12. Most hypothalamic hormones acting in the anterior lobe are ________; they cause target cells to secrete hormones of their own. But others are ________, which slow secretion from their targets. [p.633]
13. Pituitary dwarfism, gigantism, and acromegaly are all associated with abnormal secretion of ________ by the pituitary gland. [p.634]
14. One cause of ________ ________ is damage to the pituitary's posterior lobe and the diminished secretion or lack of ADH. [p.634]
15. Outline the major human hormone sources and their secretions, main targets, and primary actions, as shown in Table 36.3 in the text. [p.635]
16. With ________ feedback, an increase or decrease in the concentration of a secreted hormone triggers events that inhibit further secretion. [p.636]
17. With ________ feedback, an increase in the concentration of a secreted hormone triggers events that stimulate further secretion. [p.636]
18. The adrenal ________ secretes glucocorticoids. [p.636]
19. The ________ system initiates the stress response. [p.636]
20. Describe the role of cortisol in a stress response. [p.636]
21. The adrenal ________ contains neurons that secrete epinephrine and norepinephrine. [p.636]
22. List the features of the fight–flight response. [p.636]
23. Over time, a sustained response is made to a low blood level of TSH produced by the thyroid gland; this causes an enlargement of the gland and is known as a form of ________. [p.637]
24. Describe the characteristics of hypothyroidism and hyperthyroidism. [p.637]
25. The ________ are primary reproductive organs that produce and secrete hormones with essential roles in reproduction. [p.637]
26. Name the glands that secrete PTH and state the function of this hormone. [p.638]
27. Describe an ailment called rickets and give its cause. [p.638]
28. Give two examples that illustrate the effects of local signaling molecules. [p.638]
29. Name the hormones secreted by alpha, beta, and delta pancreatic cells, respectively, and describe the effect of each. [p.638]
30. Describe the symptoms of diabetes mellitus, and distinguish between type 1 and type 2 diabetes. [pp.638–639]
31. The pineal gland secretes the hormone ________; give two examples of the action of this hormone. [p.640]

32. Explain the cause of "jet lag" and "winter blues." [p.640]
33. Discuss possible causes of frog deformities, a relatively recent observation in nature. [pp.640–641]
34. The invertebrate hormone __________ is related to the control of the phenomenon known as __________, which occurs among crustaceans and insects. [p.641]

Integrating and Applying Key Concepts

Suppose you suddenly quadruple your already high daily consumption of calcium. State which body organs would be affected, and tell how they would be affected. Name two hormones whose levels would most probably be affected, and tell whether your body's production of them would increase or decrease. Suppose you continue this high rate of calcium consumption for 10 years. Predict which organs would be subject to the most stress as a result.

37

PROTECTION, SUPPORT, AND MOVEMENT

Interactive Exercises

Of Men, Women, and Polar Huskies [pp.644–645]

37.1. EVOLUTION OF VERTEBRATE SKIN [pp.646–647]

37.2. FOCUS ON HEALTH: *Sunlight and Skin* [p.648]

Selected Words: integere [p.645], "split ends" [p.647], *hirsutism* [p.647], *cold sores* [p.648]

Boldfaced, Page-Referenced Terms

[p.645] integument ____

[p.646] vertebrate skin ____

[p.646] epidermis ____

[p.646] dermis ____

[p.646] keratinocytes ____

[p.646] melanocytes ____

[p.647] hair ____

[p.648] vitamin D ____

[p.648] folate ____

[p.648] Langerhans cells ____

[p.648] Granstein cells ____

Fill-in-the-Blanks

Vertebrate skin consists of an outer (1) ______________ [p.646] and an underlying (2) ______________ [p.646] made of dense (3) ______________ [p.646] tissue. There are two evolutionary trends manifested by the vertebrates that moved onto dry land: (4) ______________ [p.646] and the emergence of recessed (5) ______________ [p.646] with ducts to the skin's surface. First in (6) ______________ [p.646] and then in reptiles, birds, and mammals, (7) ______________ [p.646] and (8) ______________ [p.646] developed in the outer epidermal layers. Keratinocytes are cells that produce (9) ______________ [p.646], a water-resistant protein. Melanocytes produce a brownish-black pigment called (10) ______________ [p.646], which protects the body from harmful (11) ______________ [p.646] radiation from the sun.

Labeling

Label the numbered parts of the following illustration. [p.646]

12. ________________ ________________

13. ________________ ________________

14. ________________ ________________

15. ________________ ________________

16. ________________ ________________

17. ________________ ________________

18. ________________

19. ________________

20. ________________

21. ________________

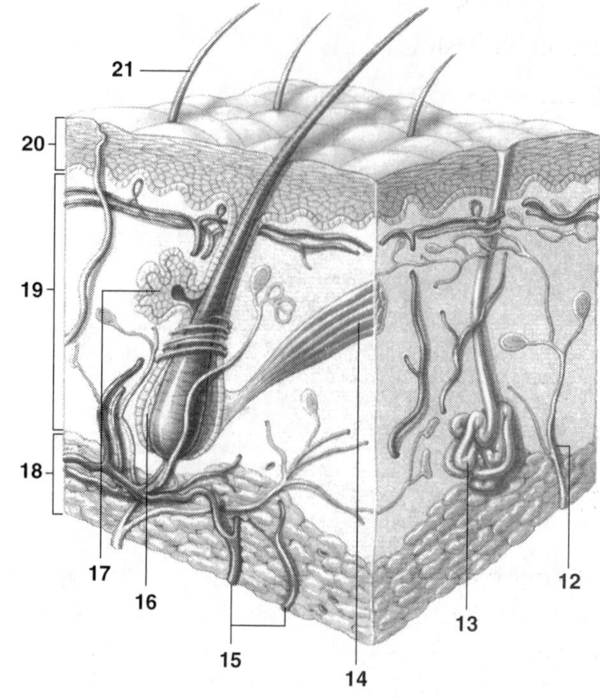

Matching

Match each of the following terms with its definition.

22. ____ epidermis [p.646]
23. ____ hypodermis [p.646]
24. ____ dermis [p.646]
25. ____ melanin [p.646]
26. ____ keratin [p.646]
27. ____ hair [p.647]
28. ____ hirsutism [p.647]

A. The layer that consists of dense connective tissue and supportive collagen fibers
B. A site of continuous mitotic cell division
C. This layer has adipose tissue that helps in insulation
D. Excessive hairiness produced by excessive amounts of testosterone
E. The protein associated with skin coloration
F. A fibrous protein found in scales, horns, hooves, beaks, and claws
G. A flexible structure made of keratin that is rooted within the skin

Matching

Choose the most appropriate answer for each of the following. [p.648]

29. ____ Langerhans cells
30. ____ Granstein cells
31. ____ a tan
32. ____ cold sores
33. ____ folate
34. ____ vitamin D

A. Phagocytes in the skin that are used to mobilize the immune system
B. A condition caused by the increased secretion of melanin
C. A B vitamin that is broken down by ultraviolet radiation
D. A group of cells that slow the immune response of the skin
E. An infection caused by the *Herpes simplex* virus
F. A vitamin manufactured in the skin that is used to help the body absorb calcium from food

37.3. TYPES OF SKELETONS [p.649]

37.4. EVOLUTION OF THE VERTEBRATE SKELETON [pp.650–651]

37.5. A CLOSER LOOK AT BONES AND JOINTS [pp.652–653]

Selected Words: *external* body parts *[p.649], internal* body parts [p.649], *hydraulic* [p.649], *appendicular* [p.650], *axial* [p.650], *herniated* [p.651], *compact* bone tissue [p.652], *spongy* bone tissue [p.652], *fibrous* joints [p.652], *cartilaginous* joints [p.652], *synovial* joints [p.652], *strain* [p.653], *sprain* [p.653], "arthritis" [p.653], *osteoarthritis* [p.653], *rheumatoid arthritis* [p.653], *osteoporosis* [p.653]

Boldfaced, Page-Referenced Terms

[p.649] hydrostatic skeleton ____________________

____

[p.649] exoskeleton ____________________

____

[p.649] endoskeleton ____________________

____

[p.650] vertebrae ____________________

____

[p.651] intervertebral disks ____________________

____

[p.652] bones ____________________

____

[p.652] osteocytes ____________________

____

[p.652] red marrow ____________________

____

[p.652] yellow marrow ____________________

____

[p.652] osteoblasts ____________________

____

[p.652] bone remodeling ____________________

____

[p.652] osteoclasts ____________________

____

[p.652] joints ____________________

____

[p.652] ligaments ____________________

____

Choice

For questions 1–6, choose from the following answers: [p.649]

a. hydrostatic skeleton b. endoskeleton
c. exoskeleton d. all of the above

1. ____ Uses external body parts to receive the applied force of muscle contractions
2. ____ Earthworms and soft-bodied invertebrates use this form of skeleton
3. ____ The type of structural element against which the force of muscle contraction is applied
4. ____ Uses internal body parts to receive the applied force of muscle contractions
5. ____ In this case muscles work against an internal body fluid
6. ____ The skeleton type that is typical of the arthropods

Fill-in-the-Blanks

As vertebrates progress through time, they move from the (7) ____________ [p.650] skeleton of fish to the bony (8) ____________ [p.650] that evolved among the land vertebrates. Some of these features are conserved in the human skeleton, with its (9) ____________ (number) [p.650] bones. The (10) ____________ [p.650] portion of the human skeleton consists of the pectoral girdle, the (11) ____________ [p.650] girdle, and paired arms, hands, feet, and legs. The axial skeleton consists of the jaws and (12) ____________ [p.650] bones, 12 pairs of (13) ____________ [p.650], the breastbone, and 26 (14) ____________ [p.650], which are the bony segments of the backbone. Between each pair of vertebrae are the (15) ____________ ____________ [p.651], which are cartilaginous shock absorbers. A severe or rapid shock can result in a(n) (16) ____________ [p.651] disk, causing chronic pain.

Complete the Table

17. Using Figure 37.11 from page 651 of the text, complete the following table by providing the missing information. [p.651]

Name	*Description or Function*
a.	Supports the weight of the vertebral column
b. Rib cage	
c.	The strongest weight-bearing bone of the body; key role in locomotion
d.	Name for bones found both in the hands and feet
e. Intervertebral disks	
f. Cranial bones	
g.	The bone of the leg used for muscle attachment; not load-bearing
h.	The name of the collarbone
i.	The names of the two bones of the forearm
j. Vertebrae	
k.	The name of the breastbone
l. Carpals and metacarpals	

Labeling

Identify each indicated part of the accompanying illustration. [p.652]

18. ____________________

____

19. ____________________

____

20. ____________________

____

21. ____________________

____

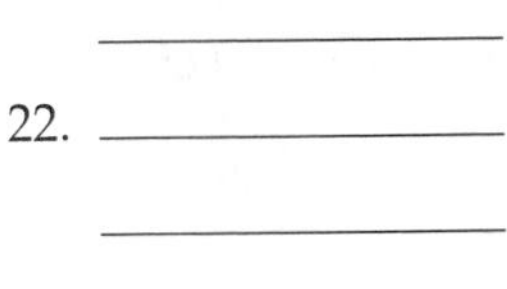

22. ____________________

____

____

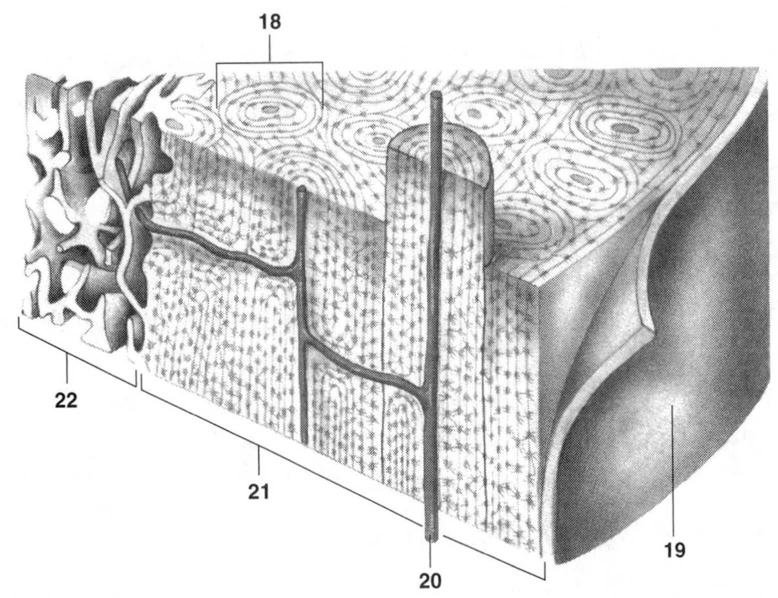

Matching

Choose the most appropriate definition for each of the following.

23. ____ yellow marrow [p.652]
24. ____ osteoclasts [p.652]
25. ____ osteocytes [p.652]
26. ____ ligaments [p.652]
27. ____ joints [p.652]
28. ____ red marrow [p.652]
29. ____ osteoblasts [p.652]
30. ____ bones [p.652]

A. Bone-forming cells
B. Complex organs that function in support, protection, blood cell formation, and mineral storage
C. The cells that break down bone using digestive enzymes
D. Long straps of connective tissue
E. Mature bone cells
F. A fatty region of the bone; may be used for blood cell formation
G. The location within the bone of red blood cell formation
H. Examples are fibrous, cartilaginous, and synovial

Complete the Table

31. For each of the following problems of the skeletal system, give the name of the condition based on the symptoms provided. [p.653]

Name of Condition	*Symptoms*
a.	The inflammation of synovial joints as a result of an autoimmune response
b.	Bending a joint too far
c.	The tearing of ligaments or tendons supporting a bone
d.	A decrease in bone mass, usually as a result of aging
e.	The breakdown of the cartilage at freely moving joints, usually as a result of aging

37.6. SKELETAL–MUSCULAR SYSTEMS [pp.654–655]

37.7. HOW DOES SKELETAL MUSCLE CONTRACT? [pp.656–657]

37.8. WHAT CONTROLS CONTRACTION? [pp.658–659]

37.9. ENERGY FOR CONTRACTION [p.659]

37.10. PROPERTIES OF WHOLE MUSCLES [p.660]

37.11. FOCUS ON HEALTH: *A Bad Case of Tetanic Contraction* [p.661]

Selected Words: *skeletal muscle* [p.654], *excitable* cells [p.658], *isotonically* contracting [p.660], *isometrically* contracting [p.660], *lengthening* contraction [p.660], "charley horses" [p.660], *muscular dystrophies* [p.660], *aerobic exercise* [p.661], *strength training* [p.661], *botulism* [p.661], *tetanus* [p.661]

Boldfaced, Page-Referenced Terms

[p.654] skeletal muscles ____________________

____

[p.654] tendons

[p.656] sarcomeres

[p.656] myofibrils

[p.656] myosin

[p.656] actin

[p.657] titin

[p.657] sliding-filament model

[p.657] cross-bridge

[p.658] action potential

[p.658] sarcoplasmic reticulum

[p.659] creatine phosphate

[p.659] oxygen debt

[p.660] muscle tension

[p.660] motor unit

[p.660] muscle twitch

[p.660] tetanus

[p.660] muscle fatigue

[p.661] exercise

Fill-in-the-Blanks

The functional partner of bone is (1) ________________ [p.654] muscle. Each skeletal muscle contains bundles of (2) ________________ [p.654] cells. Muscle tissues (3) ________________ [p.654] in response to adequate stimulation; they lengthen in response to (4) ________________ [p.654] and other loads. Dense connective tissue, called (5) ________________ [p.654], bundles many muscle cells together. Each tendon attaches a muscle to a(n) (6) ________________ [p.654]. They form a(n) (7) ________________ [p.654] system, in which a rigid rod is attached to a fixed point and can move about it. Muscles connect to bones near a(n) (8) ________________ [p.654]. When they contract, they transmit (9) ________________ [p.654] to bones to make them move. Skeletal muscles also interact with one another. Some work in pairs to (10) ________________ [p.654] movement; others work in (11) ________________ [p.654]. The human body consists of over (12) ________________ [p.655] skeletal muscles.

Complete the Table

13. Using Figure 37.18 from p. 655 of the text, complete the following table by providing the missing information. [p.655]

Name	*Function or Description*
a. Deltoid	
b.	Flexes the foot toward the shin
c. Gluteus maximus	
d. Biceps femoris	
e.	Draws the arm forward and in toward the body
f.	Bends the lower leg at the knee when walking
g. Trapezius	
h. Sartorius	
i.	Depresses the thoracic cavity
j. Triceps brachii	
k. Quadriceps femoris	

Matching

Choose the most appropriate description for each of the following.

14. ____ myosin [p.656]
15. ____ actin [p.656]
16. ____ titin [p.657]
17. ____ ATP [p.657]
18. ____ sliding filament model [p.657]
19. ____ sarcomeres [p.656]
20. ____ myofibrils [p.656]
21. ____ Z band [p.656]

A. The basic unit of muscle contraction
B. A cytoskeletal element that flanks the sarcomere and anchors its components
C. Threadlike, cross-banded cell structures that are arranged in parallel
D. Thin filaments found within the sarcomere
E. The sliding of the actin filaments over the myosin using ATP
F. A motor protein that has a club-shaped head
G. The energy source for contraction of the sarcomere
H. This protein keeps the myosin filaments centered during contraction

Sequence

Place the following statements in sequential order, with "1" representing the first event, "2" the second event, and so forth. [p.657]

22. ____ Myosin heads bind to nearby actin filaments.
23. ____ ATP binds to myosin head, causing myosin to release its grip on actin.
24. ____ All myosin heads tilt toward the center of the sarcomere.
25. ____ ATP hydrolysis returns myosin heads to their original orientation.
26. ____ Increase in calcium triggers changes that expose binding site for myosin on actin filaments.
27. ____ The actin binding site is blocked until local concentration of calcium increases.

Fill-in-the-Blanks

Like other cells, a muscle cell shows a difference in (28) ____________ [p.658] charge across its plasma membrane. However, only in (29) ____________ [p.658] cells does that difference reverse abruptly and briefly. A(n) (30) ____________ ____________ [p.658] is the name for the abrupt reversal of charge across a plasma membrane.

When action potentials arise in a muscle cell, they spread rapidly from the point of stimulation, then along (31) ____________ [p.658] extensions of the plasma membrane. These tubes have anchor points for the (32) ____________ [p.658] filaments of the myofibril's (33) ____________ [p.658]. A system of membranous chambers, called the (34) ____________ ____________ [p.658], wraps around the myofibrils. This system takes up, stores, and releases (35) ____________ [p.658] ions.

The arrival of the action potential causes a(n) (36) ____________ [p.658] flow of calcium ions. These ions diffuse into the (37) ____________ [p.658], reaching the actin filaments. The calcium ions (38) ____________ [p.658] the binding sites, so that (39) ____________ [p.658] can proceed.

In a muscle at rest, the interaction of (40) ________________ ________________ [p.658] and troponins blocks the binding site. However, when calcium ions arrive, the (41) ________________ [p.658] changes shape, giving it a different molecular grip with the tropomyosin. The tropomyosin is now free to move and expose the binding site.

Labeling

Label the sections of the following diagram. [p.659]

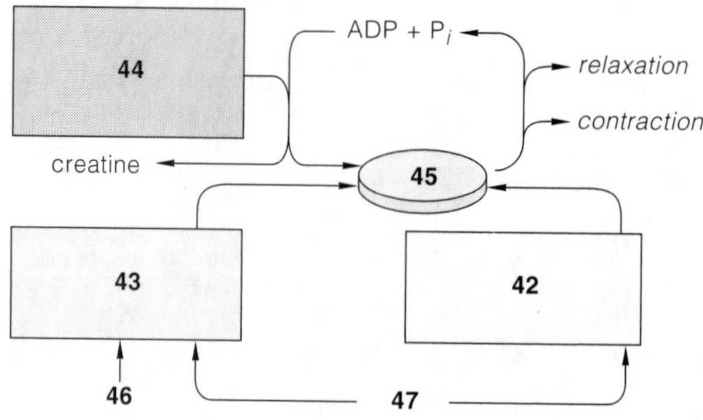

42. ________________ ________________

43. ________________ ________________

44. ________________ ________________ ________________ ________________

45. ________________

46. ________________

47. ________________ ________________ ________________ ________________ ________________

Matching

Choose the most appropriate description for each of the following.

48. ____ charley horses [p.660]
49. ____ muscle fatigue [p.660]
50. ____ muscular dystrophies [p.660]
51. ____ tetanus (disease) [p.661]
52. ____ botulism [p.661]
53. ____ tetanus [p.660]
54. ____ motor unit [p.660]
55. ____ muscle twitch [p.660]

A. A disease in which overstimulated muscles stiffen and contract, sometimes called lockjaw
B. A disease caused by bacteria which affects the neurons that synapse with muscle cells
C. The result of a sustained state of contraction due to high-frequency stimulation
D. Sustained contraction that results in the repeated stimulation of a motor unit
E. Large, painful contractions of muscles, probably the result of muscle fatigue
F. A genetic disorder in which the muscles progressively weaken and degenerate
G. The 50-millisecond interval in which the tension in a motor unit increases, peaks, and then declines
H. A motor neuron and all of the muscle cells that form junctions with its endings

Short Answer

56. Explain the effects of aerobic exercise and strength training on muscles [p.661]. ____________________

____

____

____

57. Explain the difference between isotonically contracting, isometrically contracting, and lengthening contracting muscles. [p.660] ____________________

____

____

____

____

Self-Quiz

For questions 1–7, choose from the following answers:

a. bone [p.652]
b. vertebrae [p.650]
c. epidermis [p.646]
d. dermis [p.646]
e. hypodermis [p.646]

____ 1. Fat cells in adipose tissue are most likely to be located in this

____ 2. Keratinized, flattened cells are most likely to be located in this

____ 3. Melanin in melanocytes is most likely to be here

____ 4. Smooth muscles attached to hairs are probably here

____ 5. Attachment site for paired muscles; encloses spinal cord

____ 6. The receiving ends of sensory receptors are most likely here

____ 7. Serves as a "bank" for withdrawing and depositing calcium and phosphate ions

For questions 8–11, choose from the following answers:

a. ligaments [p.652]
b. osteoblasts [p.652]
c. osteoclasts [p.652]
d. red marrow [p.652]
e. tendons [p.654]

_____ 8. Secretes bone-dissolving enzymes

_____ 9. Major site of blood cell formation

_____ 10. Remove Ca^{2+} and PO_4^{3-} ions from blood and build bone

_____ 11. Attach muscles to bone

For questions 12–16, choose from the following answers:

a. an action potential [p.658]
b. cross-bridge formation [p.657]
c. the sliding-filament model [p.657]
d. muscle tension [p.660]
e. tetanus [p.660]

_____ 12. The mechanical force that causes muscle cells to shorten if it is not exceeded by opposing forces

_____ 13. A wave of electrical disturbance that moves along a neuron or muscle cell in response to a threshold stimulus

_____ 14. A large contraction caused by repeated stimulation of motor units that are not allowed to relax

_____ 15. Assisted by calcium ions and ATP

_____ 16. Explains how myosin filaments move to the centers of sarcomeres and back

For questions 17–21, choose from the following answers:

a. actin [p.656]
b. myofibril [p.656]
c. myosin [p.656]
d. sarcomere [p.656]
e. sarcoplasmic reticulum [p.658]

_____ 17. Contains many repetitive units for muscle contraction

_____ 18. The basic unit of muscle contraction

_____ 19. Thin filaments that depend on calcium ions to clear their binding sites so that they can attach to parts of thick filaments

_____ 20. Stores calcium ions and releases them in response to an action potential

_____ 21. A motor protein

Chapter Objectives/Review Questions

1. Define *integument.* [p.645]
2. Name the two evolutionary trends in vertebrate skin. [p.646]
3. Describe the layered structure of human skin, and identify the items located in each layer. [p.646]
4. Distinguish between melanocytes and keratinocytes. [p.646]
5. Describe the role of keratin in the formation of human hair. [p.647]
6. Describe several ways in which sunlight affects human skin. [p.648]
7. Describe the relationship between vitamin D and folate and human skin. [p.648]
8. What is the importance of Langerhans and Granstein cells? [p.648]
9. Define and give examples of the three types of skeletons. [p.649]
10. Give examples of how the vertebrate skeleton enables life on land. [p.650]
11. Distinguish between the axial and the appendicular skeleton. [p.650]
12. Explain the importance of intervertebral disks for upright locomotion. [p.651]
13. Recognize the major bones of the human body and their roles. [p.651]
14. Explain the various roles of osteoblasts, osteoclasts, osteocytes, red marrow, and yellow marrow. [p.652]
15. Give the five functions of bones. [p.652]
16. Distinguish between joints, tendons, and ligaments. [pp.652,654]
17. Understand the relationship between strains, sprains, osteoarthritis, rheumatoid arthritis, and osteoporosis, on the one hand, and the human skeleton, on the other. [p.653]

18. Recognize the roles of calcitonin and PTH in maintaining bone calcium levels. [p.653]
19. Understand the interaction of skeletal muscle and bones. [p.654]
20. Refer to Figure 37.18 in your text and indicate (a) a muscle used in sit-ups, (b) another used in dorsally flexing and inverting the foot, and (c) another used in flexing the elbow joint. [p.655]
21. Describe the fine structure of a muscle fiber; use terms such as *myofibril, sarcomere, motor unit, actin,* and *myosin.* [pp.656–657]
22. Understand the sliding-filament model for muscle contraction. [p.657]
23. Understand the importance of calcium in muscle contraction. [p.658]
24. Identify the major sources of energy for muscle contraction. [p.659]
25. Distinguish between isotonically contracting, isometrically contracting, and lengthening contracting muscles. [p.660]
26. Recognize how the terms *tetanus, muscle fatigue, muscle tension,* and *muscle twitch* relate to the properties of whole muscles. [p.660]
27. Identify how the diseases muscular dystrophy, tetanus, and botulism interact with the muscle system. [pp.660–661]
28. Distinguish between aerobic and strength-training exercise. [p.661]
29. Distinguish twitch contractions from tetanic contractions. [p.662]

Integrating and Applying Key Concepts

1. If humans had an exoskeleton rather than an endoskeleton, would they move differently from the way they do now? Name any advantages or disadvantages that having an exoskeleton instead of an endoskeleton would present in human locomotion.
2. Why are pregnant mothers required to increase their calcium intake? Where would the fetus get its calcium supply from naturally? Why should mothers continue increased calcium consumption following pregnancy?

38

CIRCULATION

Interactive Exercises

Heartworks [pp.664–665]

38.1. EVOLUTION OF CIRCULATORY SYSTEMS [pp.666–667]

Selected Words: electrocardiogram (ECG) [p.664], "internal environment" [p.666], *closed* circulatory system [p.666], *open* circulatory system [p.666], blood flow *velocity* [p.666], blood flow *rate* [p.666], gills [p.667], lungs [p.667], one circuit [p.667], double circuit [p.667]

Boldfaced, Page-Referenced Terms

[p.665] circulatory system ____________________

____

[p.666] interstitial fluid ____________________

____

[p.666] blood ____________________

____

[p.666] heart ____

____

[p.666] capillary beds ____

____

[p.666] capillaries ____

____

[p.667] pulmonary circuit ____

____

[p.667] systemic circuit ____

____

[p.667] lymphatic system ____

____

Fill-in-the-Blanks

Cells survive by taking in from their surroundings what they need, substances called (1) ______________ [p.665], and giving back to their surroundings materials that they don't need: (2) ______________ [p.665]. In humans and many other animals, substances move rapidly to and from living cells by way of a(n) (3) ______________ [p.666] circulatory system. (4) ______________ [p.668], a fluid connective tissue within the (5) ______________ [p.666] and blood vessels, is the transport medium. Most of the cells of animals are bathed in a(n) (6) ______________ ______________ [p.666]; blood is constantly delivering nutrients and removing wastes from that fluid. The (7) ______________ [p.666] generates the pressure that keeps blood flowing. Blood flows (8) [choose one] □ rapidly □ slowly [p.666] through large-diameter vessels to and from the heart; but where the exchange of nutrients and wastes occurs, in the (9) ______________ [p.666] beds, the blood is divided up into vast numbers of smaller diameter vessels with tremendous surface area, which enables the exchange to occur by diffusion. An elaborate network of drainage vessels attracts excess interstitial fluid and reclaimable (10) ______________ [p.667] and returns them to the circulatory system. This network is part of the (11) ______________ [p.679] system, which also helps clean disease agents out of the blood.

Nutrients are absorbed into the blood from the (12) ______________ [p.665] and (13) ______________ [p.665] systems. Carbon dioxide is given to the (14) ______________ [p.657] system for elimination, and excess water, solutes, and wastes are eliminated by the (15) ______________ [p.665] system.

Labeling

Label the numbered parts in the following illustrations. [all from p.666]

16. ________________

17. ________________ ________________

18. ________________

Short Answer

Name and describe the kind of circulatory system in:

19. Creature A. ____
____
____
____

20. Creature B. ____
____
____
____

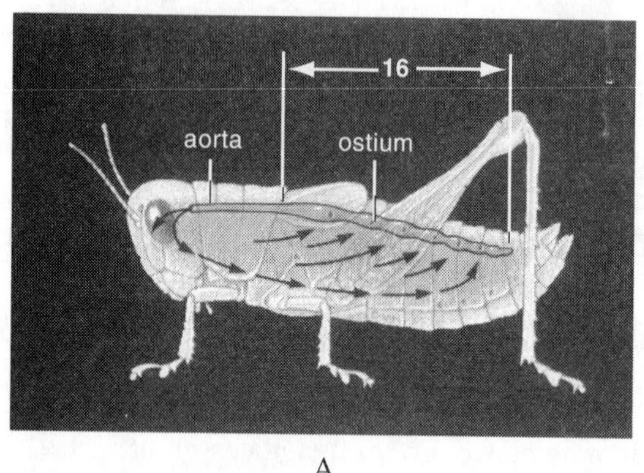

A

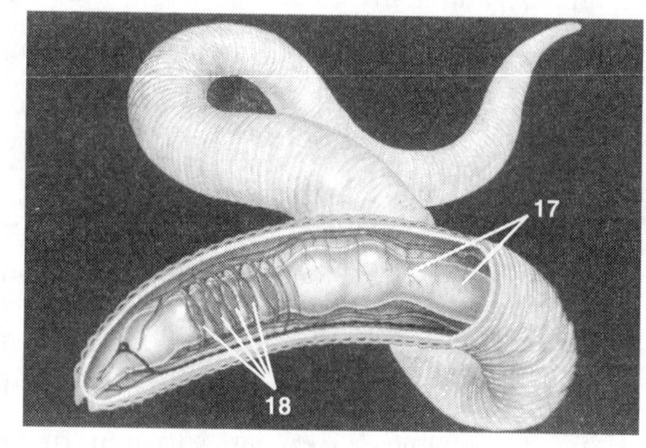

B

38.2. CHARACTERISTICS OF BLOOD [pp.668–669]

38.3. FOCUS ON HEALTH: *Blood Disorders* [p.670]

38.4. BLOOD TRANSFUSION AND TYPING [pp.670–671]

Selected Words: plasma proteins [p.668], erythrocytes [p.668], leukocytes [p.669], NK cells [p.669], neutrophils [p.669], eosinophils [p.669], basophils [p.669], monocytes [p.669], lymphocytes [p.669], macrophages [p.669], megakaryocytes [p.669], *hemorrhagic, chronic, hemolytic, iron deficiency,* and *B_{12} deficiency anemias* [p.670], *sickle-cell* anemia [p.670], *thalassemias* [p.670], *polycythemias* [p.670], *blood doping* [p.670], *infectious mononucleosis* [p.670], *leukemias* [p.670], *"self"* markers [p.670], *erythroblastosis fetalis* [p.671], RhoGam [p.671].

Boldfaced, Page-Referenced Terms

[p.668] plasma ______

[p.668] red blood cells ______

[p.668] white blood cells ______

[p.668] platelets ______

[p.669] stem cells ______

[p.669] cell count ______

[p.670] anemias ______

[p.670] blood transfusions ______

[p.670] agglutination ______

[p.671] ABO blood typing ______

[p.671] Rh blood typing ______

Complete the Table

1. Complete the following table, which describes the components of blood. [all from p.668]

Components	*Relative Amounts*	*Functions*
Plasma portion (50–60% of total volume):		
Water	91–92% of plasma volume	Solvent
a. (albumin, globulins, fibrinogen, etc.)	7–8%	Defense, clotting, lipid transport, roles in extracellular fluid volume, and so forth
Ions, sugars, lipids, amino acids, hormones, vitamins, dissolved gases	1–2%	Roles in extracellular fluid volume, pH, and so on
Cellular portion (40%–50% of total volume):		
b.	4,800,000–5,400,000	O_2, CO_2 transport per microliter
White blood cells:		
c.	3,000–6,750	Phagocytosis
d.	1,000–2,700	Immunity
Monocytes (macrophages)	150–720	Phagocytosis
Eosinophils	100–360	Defense against parasites
Basophils	25–90	Roles in inflammatory response, fat removal
e.	250,000–300,000	Roles in clotting

Fill-in-the-Blanks

Blood is a highly specialized fluid (2) ______________ [p.668] tissue that helps stabilize internal (3) ______________ [p.668] and equalize internal temperature throughout an animal's body. Oxygen binds with the (4) ______________ [p.668] atom in a hemoglobin molecule. The red blood (5) ______________ ______________ [p.669] in males is about 5.4 million cells per microliter of blood; in females, it is 4.8 million per microliter. The plasma portion constitutes approximately (6) ______________ to ______________ [p.668] percent of the total blood volume. Erythrocytes are produced in the (7) ______________ ______________ [p.669]. (8) ______________ ______________ [p.669] are immature cells that are not yet fully differentiated. (9) ______________ [p.669] and monocytes are highly mobile and phagocytic; they chemically detect, ingest, and destroy bacteria, foreign matter, and dead cells. (10) ______________ [p.669] (thrombocytes) are membrane-bound cell fragments that aid in forming blood clots by releasing substances that initiate the process.

In humans, red blood cells lack their (11) ________________ [p.669], but they contain enough resources to sustain them for about (12) ________________ [p.669] months. Platelets also have no (13) ________________ [p.669], but they last a maximum of (14) ________________ [p.669] days in the human bloodstream.

If you are blood type (15) ________________ [p.671], you have no antibodies against A or B markers in your plasma.

(16) ________________ [pp.670–671] is a response in which antibodies act against "foreign" cells bearing specific markers and cause them to clump together.

Labeling and Matching

Identify the numbered cell types in the following illustration. Complete the exercise by matching and entering the letter of the appropriate function in the parentheses after the given cell types. A letter may be used more than once. [Blanks are from p.669; parentheses are from p.668.]

17. ________________ ( )
18. ________________ ________________ ( )
19. ________________ ( )
20. ________________ ( )
21. ________________ ( )
22. ________________ ( )
23. ________________
24. ________________ ( )

A. Phagocytosis
B. Plays a role in the inflammatory response
C. Plays a role in clotting
D. Immunity
E. O_2, CO_2 transport
F. Immature, unspecialized blood cells

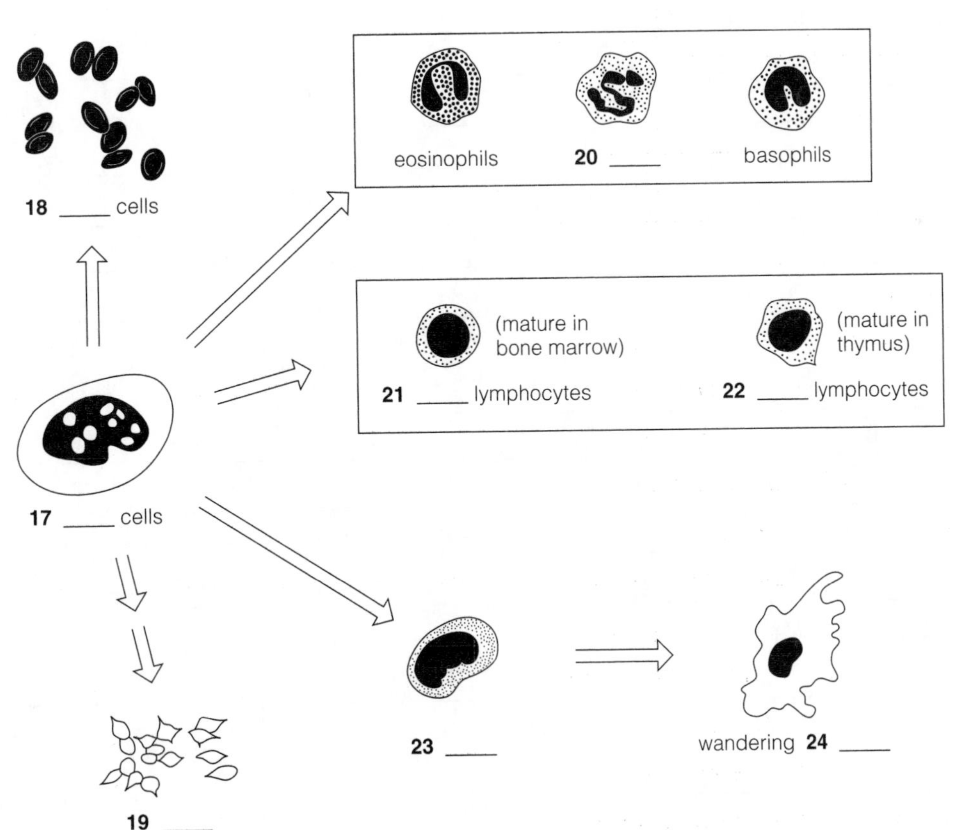

Matching (one letter is used twice) [all from p.670 except no. 27]

25. _____ B_{12} deficiency
26. _____ chronic anemias
27. _____ erythroblastosis fetalis [p.671]
28. _____ hemolytic anemias
29. _____ hemorrhagic anemias
30. _____ infectious mononucleosis
31. _____ leukemias
32. _____ polycythemias
33. _____ sickle-cell anemia
34. _____ thalassemias

A. A potential hazard for strict vegetarians and alcoholics
B. A category of cancers that suppress or impair white blood cell formation in bone marrow
C. Disorder caused by sluggish blood flow due to far too many red blood cells; "blood doping" and some bone marrow cancers
D. Caused by mixing Rh^+ and Rh^- blood types; RhoGam can prevent this
E. Abnormal forms of hemoglobin caused by a gene mutation
F. Results from sudden blood loss, as from a severe wound
G. Disorders caused by specific infectious bacteria and parasites as they replicate inside red blood cells and then lyse them
H. An Epstein–Barr virus causes this highly contagious disease, which results from too many monocytes and lymphocytes
I. These disorders result from ongoing but slight blood loss; hemorrhoids, a bleeding ulcer, or monthly blood loss by premenopausal women could be the cause

38.5. HUMAN CARDIOVASCULAR SYSTEM [pp.672–673]

38.6. THE HEART IS A LONELY PUMPER [pp.674–675]

Selected Words: "cardiovascular" [p.672], *pulmonary* circuit [p.672], *systemic* circuit [p.672], pericardium [p.674], myocardium [p.674], *coronary* circulation [p.674], endothelium [p.674], atrium, atria [p.674], ventricle [p.674], *one-way* valves [p.674], systole [p.674], diastole [p.674], SA node [p.675], AV node [p.675]

Boldfaced, Page-Referenced Terms

[p.672] arteries ____

____

[p.672] arterioles ____

____

[p.672] venules ____

____

[p.672] veins ____

____

[p.672] aorta ____

____

[p.674] cardiac cycle ____

____

[p.675] cardiac conduction system ____

____

[p.676] cardiac pacemaker ____

____

Fill-in-the-Blanks

The heart is a pumping station for two major blood transport routes: the (1) ________________ [p.672] circulation to and from the lungs and the (2) ________________ [p.672] circulation to and from the rest of the body. In the pulmonary circuit, the heart pumps (3) ________________ [p.672] -poor blood to the lungs through the pulmonary arteries; then the (4) ________________ [p.672] -enriched blood flows back to the (5) ________________ [p.672] through the pulmonary veins.

In an amphibian, reptile, bird, or mammal heart, blood first enters into chambers called (6) ________________ [p.674]; blood is pumped from the heart out of the thicker walled chambers called (7) ________________ [p.674]. Each contraction period is called (8) ________________ [p.674]; each relaxation period is (9) ________________ [p.674]. During a cardiac cycle, contraction of the (10) ________________ [p.675] is the driving force for blood circulation; (11) ________________ [p.674] contraction helps fill the ventricles.

Fill-in-the-Blanks

Look at Figure 38.11 in your text and, using a red pen, redden all tubes in the illustration below (except the pulmonary artery) that are indicated by the words *aorta* and *artery*. Memorize their names, then fill in all the following answer blanks. Also redden the parts of the heart that contain oxygen-rich blood. [all from p.675]

12. ________________ VEINS (from brain, neck, head tissues)

13. ________________ ________________ ________________ (from neck, shoulder, arms — veins of upper body)

14. ________________ VEINS (from lungs to heart)

15. ________________ ________________ VEIN (from liver to inferior vena cava)

16. ________________ VEIN (from kidneys back to heart)

17. ________________ ________________ ________________ (receives blood from all veins below the diaphragm)

18. ________________ VEINS (carry blood from pelvic organs and lower abdominal wall)

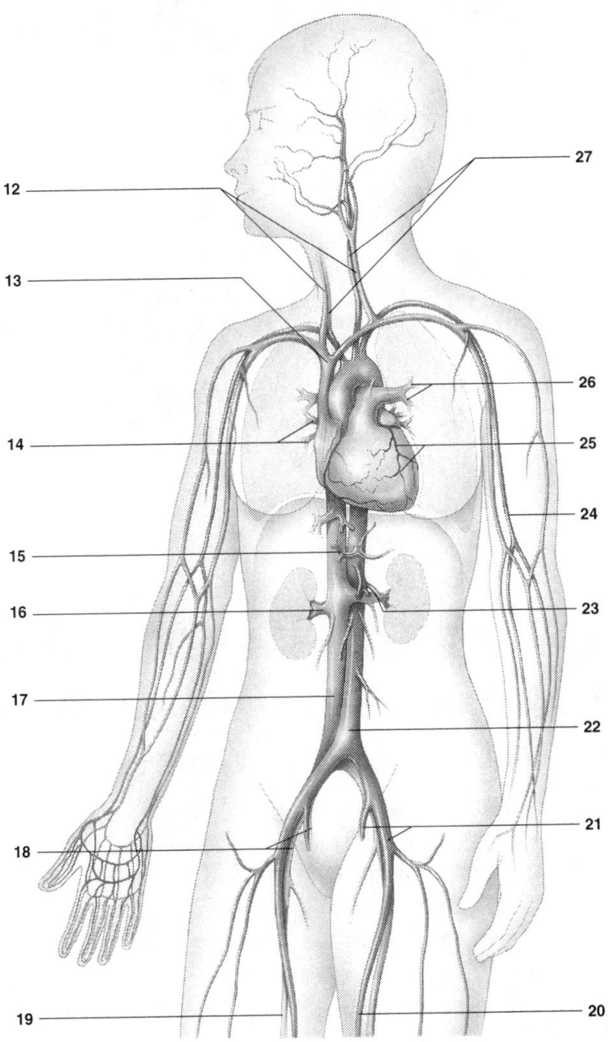

27. ________________ ARTERIES

26. ________________ ARTERIES (to lungs)

25. ________________ ARTERIES (to cardiac muscle)

24. ________________ ARTERY (to arm, hand)

23. ________________ ARTERY (to kidney)

22. ________________ AORTA (to digestive tract, pelvic organs)

21. ________________ ARTERIES (to pelvic organs, lower abdominal wall)

20. ________________ ARTERY (to thigh and inner knees)

19. ________________ VEIN (from thigh and inner knee)

Labeling

Identify each indicated part of the accompanying illustrations. [all from p.674]

28. ______________

29. ______________ ______________ ______________

30. ______________ ______________ ______________

31. ______________ ______________

32. ______________ ______________ ______________

33. ______________ ______________ ______________

34. ______________ ______________ ______________

35. ______________ ______________ ______________

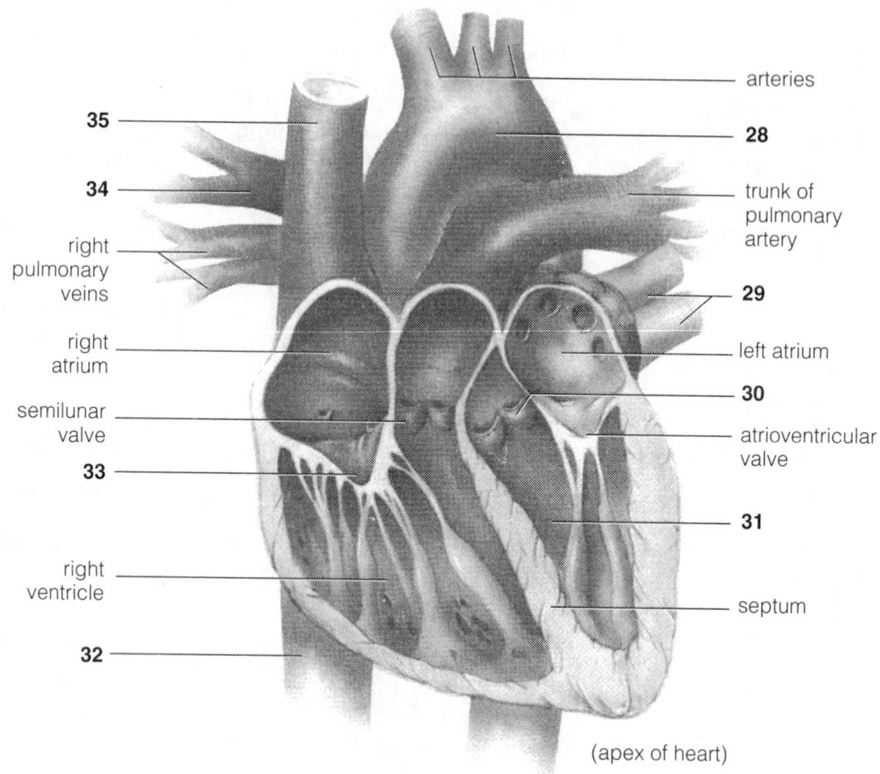

38.7. BLOOD PRESSURE IN THE CARDIOVASCULAR SYSTEM [pp.676–677]

38.8. FROM CAPILLARY BEDS BACK TO THE HEART [pp.678–679]

38.9. FOCUS ON HEALTH: *Cardiovascular Disorders* [pp.680–681]

Selected Words: *systolic* and *diastolic* pressure [p.676], *pulse* pressure [p.676], mean arterial pressure [p.677], epinephrine [p.677], angiotensin [p.677], carotid arteries [p.677], medulla oblongata [p.677], *edema* [p.679], *elephantiasis* [p.679], *hypertension* [p.680], *atherosclerosis* [p.680], *heart attack* [p.680], *strokes* [p.680], *arteriosclerosis* [p.680], *atherosclerotic plaque* [p.681], *thrombus* [p.681], *embolus* [p.613], *angina pectoris* [p.613], *stress electrocardiograms* [p.681], *angiography* [p.681], *coronary bypass surgery* [p.681], *laser angioplasty* [p.681], *balloon angioplasty* [p.681], *arrhythmias* [p.681], *bradycardia* [p.681], *tachycardia* [p.681], *atrial fibrillation* [p.681], *ventricular fibrillation* [p.681]

Boldfaced, Page-Referenced Terms

[p.676] blood pressure ____

____

[p.677] vasodilation ____

____

[p.677] vasoconstriction ____

____

[p.677] baroreceptor reflex ____

____

[p.679] ultrafiltration ____

____

[p.679] reabsorption ____

____

[p.680] LDLs, low-density lipoproteins ____

____

[p.681] HDLs, high-density lipoproteins ____

____

Fill-in-the-Blanks

Blood pressure is normally high in the (1) ______________ [p.676] immediately after leaving the heart, but then it drops as the fluid passes along the circuit through different kinds of blood vessels. (2) ______________ [p.676] are pressure reservoirs that keep blood flowing smoothly away from the heart while the (3) ______________ [p.676] are relaxing. Energy in the form of (4) ______________ [p.676] is lost as it overcomes (5) ______________ [p.676] to the flow of blood. Arterial walls are thick, muscular, and (6) ______________ [p.676] and have large diameters. Arteries present [choose one] (7) ☐ much ☐ little [p.676] resistance to blood flow, so pressure [choose one] (8) ☐ drops a lot ☐ does not drop much [p.676] in the arterial portion of the systemic and pulmonary circuits.

The greatest drop in pressure occurs at [choose one] (9) ☐ capillaries ☐ arterioles ☐ veins [p.677]. With this slowdown, blood flow can now be allotted in different amounts to different regions of the body in response to signals from the (10) ______________ [p.677] system and endocrine system or even changes in local chemical conditions. (11) ______________ [p.677] are control points where adjustments can be made in the volume of blood flow to be delivered to different capillary beds. They offer great resistance to flow, so there is a major drop in (12) ______________ [p.677] in these tubes.

When a person is resting, blood pressure is influenced most by reflex centers in the (13) ______________ [p.677]. When the resting level of blood pressure increases, reflex centers command the heart to [choose one] (14) ☐ beat more slowly ☐ beat faster [p.677] and command smooth muscle cells in arteriole walls to [choose one] (15) ☐ contract less forcefully ☐ relax [p.677], which results in [choose one] (16) ☐ vasodilation ☐ vasoconstriction [p.677].

A(n) (17) ______________ [p.678] is a blood vessel with such a small diameter that red blood cells must flow through it single file; its wall consists of no more than a single layer of (18) ______________ [p.678] cells resting on a basement membrane. In each (19) ______________ ______________ [p.678], small molecules move between the bloodstream and the (20) ______________ [p.678] fluid. Capillary beds are (21) ______________ [p.678] zones, where substances are exchanged between blood and interstitial fluid. Capillaries have the thinnest walls across which (22) ______________ ______________ [p.679] drives fluid, forcing molecules of water and solutes to move in the same direction. (23) ______________ [p.679] serve as large-diameter, low-resistance transport tubes back to the heart. Leading into (23) are (24) ______________ [p.679]. (25) ______________ [p.679] serve as temporary reservoirs for blood volume. (26) ______________ [p.679] are narrower in diameter; some solutes diffuse across their somewhat thinner walls, and some control over capillary pressure is also exerted at these vessels. Veins contain (27) ______________ [p.679] that prevent blood from flowing backward; they contain [choose one] (28) □ 20–30 □ 35–45 □ 50–60 [p.679] percent of the total blood volume.

Labeling

Identify each indicated part of the accompanying illustrations. [all from p.676]

29. ______________ 32. ______________

30. ______________ 33. ______________ ______________, ______________ ______________

31. ______________ 34. ______________

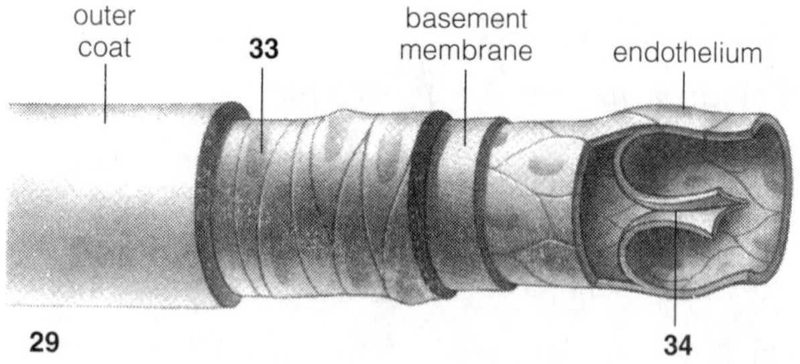

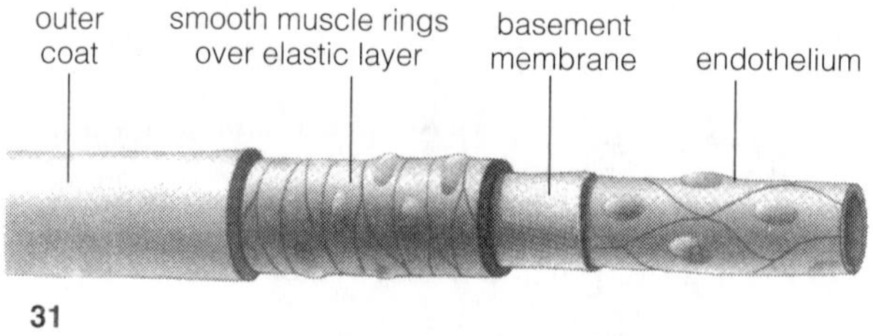

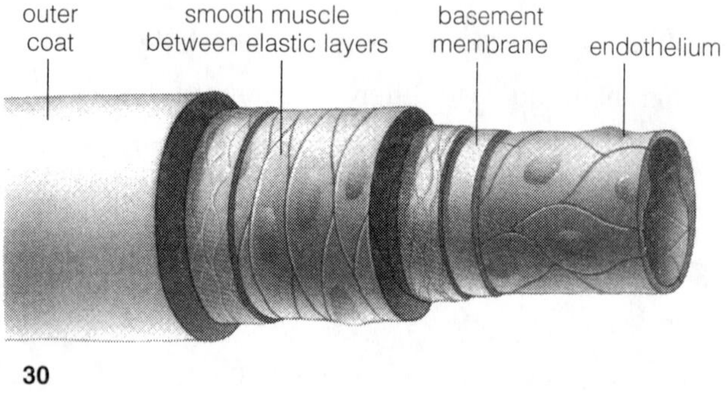

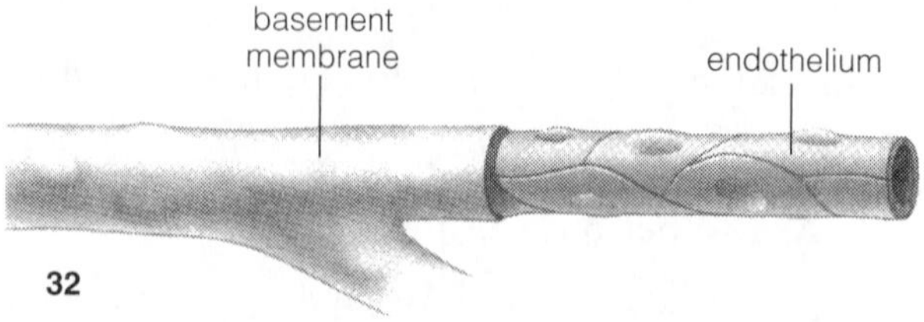

True/False

If the statement is true, write a "T" in the blank. If false, make it true by changing the underlined word. [p.676]

______________ 35. The pulse rate is the difference between the systolic and the diastolic pressure readings.

______________ 36. Because the total volume of blood remains constant in the human body, blood pressure must also remain constant throughout the circuit.

Fill-in-the-Blanks

Sustained high blood pressure is called (37) ______________ [p.680]; it is often referred to as a(n) (38) ______________ ______________ [p.680] because affected people often show no symptoms. Another cause of cardiovascular disease is (39) ______________ [p.680], a progressive thickening of the arterial walls. This is often due to a high concentration of (40) ______________ [p.681], which is deposited in and around endothelial cells. The addition of calcium deposits and a fibrous net forms a(n) (41) ______________ ______________ [p.681], which can partially block the arterial lumen and initiate the formation of clots by (42) ______________ [p.681]. A stationary clot is called a thrombus, while a travelling clot is a(n) (43) ______________ [p.681].

38.10. HEMOSTASIS [p.682]

38.11. LYMPHATIC SYSTEM [pp.682–683]

Selected Words: lymph capillaries [p.682], lymph vessels [p.682], "valves" [p.682], *lymphoid* organs and tissues [p.682], tonsils [p.682], *red* pulp [p.683], *white* pulp [p.683]

Boldfaced, Page-Referenced Terms

[p.682] hemostasis ____

____

[p.682] lymph vascular system ____

____

[p.682] lymph nodes ____

____

[p.683] spleen ____

____

[p.683] thymus gland ____

____

Fill-in-the-Blanks

Bleeding is stopped by several mechanisms that are referred to as (1) _______________ [p.682]; the mechanisms include blood vessel spasm, (2) _______________ _______________ _______________ [p.682], and blood (3) _______________ [p.682]. Rod-shaped proteins called fibrinogens react with (4) _______________ [p.682] fibers in damaged vessel walls. Reactions cause these proteins to assemble into long, (5) _______________ [p.682] fibers, which trap blood cells and platelets and begin the formation of a clot.

Choice

Choose from these possibilities for questions 6–10:

a. bone marrow b. lymph nodes c. lymph vascular system d. spleen e. thymus gland

6. _____ A huge reservoir of red blood cells and a filter of pathogens and used-up blood cells from the blood; contains red pulp [p.683]
7. _____ Delivers water and plasma proteins from capillary beds to the blood circulation; delivers fats from the small intestine to the blood; and delivers pathogens, foreign cells, and material and cellular debris to the organized disposal centers [pp.682–683]
8. _____ Immature T lymphocytes become mature here; hormones are produced here [p.683]
9. _____ Contain white blood cells that destroy invading bacteria and viruses as they are filtered from the lymph [p.682]
10. _____ Lymphocytes, which are a kind of white blood cell, are produced here before they go on to other parts of the lymphatic system [pp.682–683]

Identification/Fill-in-the-Blanks

Refer to the accompanying illustration, then supply the missing terms indicated by each answer blank (or blanks).

11. ________________ [p.683]

12. ________________ gland [p.683]

13. ________________ duct [p.683]

14. ________________ [p.683]

15. ________________ ________________ [p.683]

16. organized arrays of ________________ [p.683]

17. ________________ ________________ [p.683]

18. ________________ [p.682] vessels reclaim fluid lost from the bloodstream, purify the blood of microorganisms,

19. and transport ________________ [p.682] from the

20. ________________ ________________ [p.682] to the bloodstream.

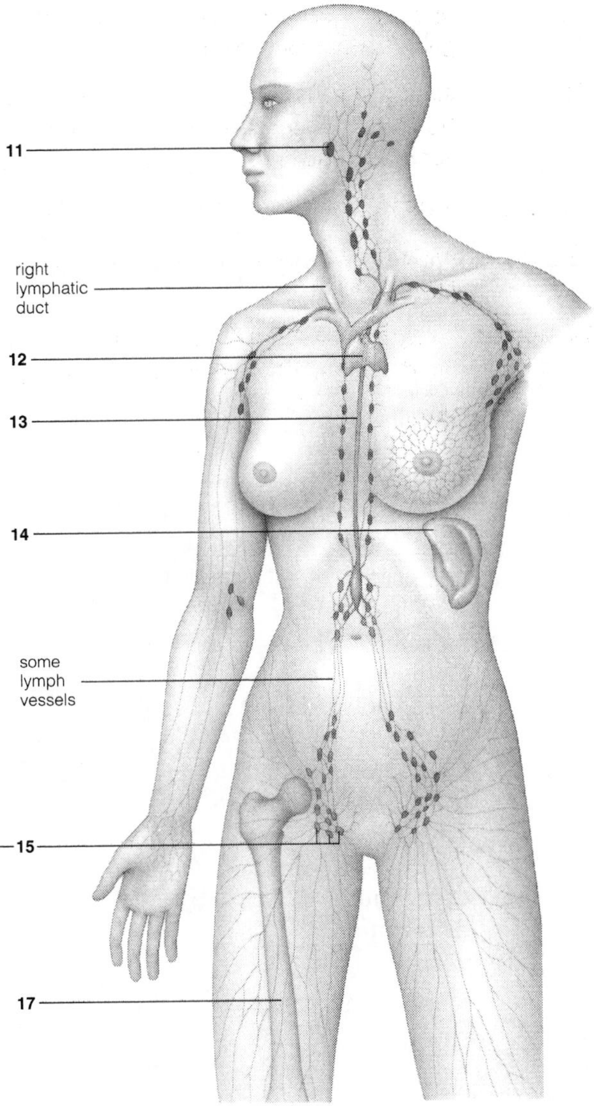

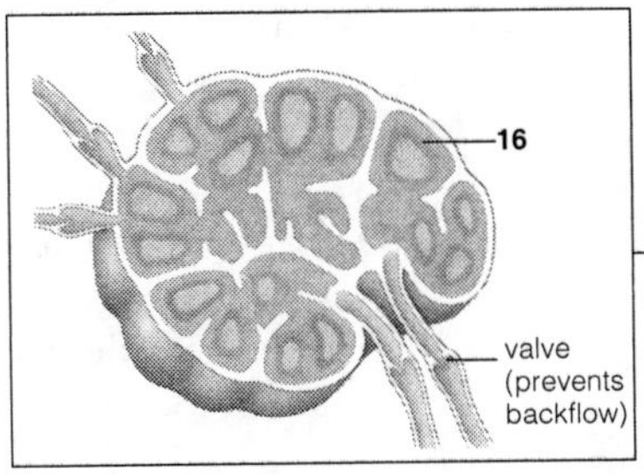

Self-Quiz

Multiple Choice

____ 1. Most of the oxygen in human blood is transported by ________________. [pp.668–669]
a. plasma
b. serum
c. platelets
d. hemoglobin
e. leukocytes

____ 2. Of all the different kinds of white blood cells, two classes of ____________ are the ones that respond to specific invaders and confer immunity to a variety of disorders. [p.669]
a. basophils
b. eosinophils
c. monocytes
d. neutrophils
e. lymphocytes

____ 3. Open circulatory systems generally lack ____________. [p.666]
a. a heart
b. arterioles
c. capillaries
d. veins
e. arteries

____ 4. Red blood cells originate in the ____________. [p.669]
a. liver
b. spleen
c. yellow bone marrow
d. thymus gland
e. red bone marrow

____ 5. Hemoglobin contains ____________. [p.669]
a. copper
b. magnesium
c. sodium
d. calcium
e. iron

____ 6. The pacemaker of the human heart is the ____________. [p.675]
a. sinoatrial node
b. semilunar valve
c. inferior vena cava
d. superior vena cava
e. atrioventricular node

____ 7. During systole, ________________. [p.674]
a. oxygen-rich blood is pumped to the lungs
b. the heart muscle tissues contract
c. the atrioventricular valves suddenly open
d. oxygen-poor blood from all parts of the human body, except the lungs, flows toward the right atrium
e. none of the above

____ 8. ____________ are reservoirs of blood pressure in which resistance to flow is low. [p.676]
a. Arteries
b. Arterioles
c. Capillaries
d. Venules
e. Veins

____ 9. Begin with a red blood cell located in the superior vena cava and travel with it in proper sequence as it goes through the following structures. Which will be *last* in sequence? [p.674]
a. aorta
b. left atrium
c. pulmonary artery
d. right atrium
e. right ventricle

____ 10. The lymphatic system is the principal avenue in the human body for transporting ____________. [p.624]
a. fats
b. wastes
c. carbon dioxide
d. amino acids
e. interstitial fluids

Matching

11. ____ agglutination [pp.670–671]
12. ____ angiogram, angiography [p.681]
13. ____ atherosclerosis [pp.680–681]
14. ____ carotid arteries [p.677]
15. ____ coronary arteries [p.673]
16. ____ edema [p.679]
17. ____ embolism, embolus [p.681]
18. ____ erythroblastosis fetalis [p.671]
19. ____ hypertension [p.680]
20. ____ inferior vena cava [p.673]
21. ____ jugular veins [p.673]
22. ____ renal arteries [p.673]
23. ____ stroke [p.680]
24. ____ tachycardia [p.681]
25. ____ thrombosis, thrombus [p.681]

A. Receive(s) blood from the brain, tissues of the head and neck
B. The heart's own blood supplier(s)
C. Excessively fast rate of heart beat; occurs during heavy exercising
D. Damage to brain caused by damaged blood vessels
E. High blood pressure
F. A blood clot that is on the move from one place to another
G. Diagnostic technique that uses opaque dyes and X rays to discover blocked blood vessels
H. Deliver(s) blood to the head, neck, brain
I. The clumping of red blood cells or of antibodies with antigens
J. Delivers blood to kidneys, where its composition and volume are adjusted
K. A blood clot that is lodged in a blood vessel and is blocking it
L. Receive(s) blood from all veins below the diaphragm
M. Progressive thickening of the arterial wall and narrowing of the arterial lumen (space)
N. Blood cell destruction caused by mismatched Rh blood types (Rh^- mother and Rh^+ fetus)
O. Accumulation of excess fluid in interstitial spaces; extreme in elephantiasis

---

Chapter Objectives/Review Questions

1. Distinguish between open and closed circulatory systems. [pp.666–667]
2. Describe the composition of human blood, using percentages of volume. [p.668]
3. Distinguish the five types of leukocytes from each other in terms of structure and function. [pp.668–669]
4. State where erythrocytes, leukocytes, and platelets are produced. [p.669]
5. Describe how blood is typed for the ABO blood group and the Rh factor. [pp.670–671]
6. Trace the path of blood in the human body. Begin with the aorta and name all major components of the circulatory system through which the blood passes before it returns to the aorta. [pp.673–674]
7. Explain what causes a heart to beat. Then describe how the rate of heartbeat can be slowed down or speeded up. [p.675]
8. Describe how the structures of arteries, capillaries, and veins differ. [p.676]
9. Explain what causes high pressure and low pressure in the human circulatory system. Then show where major drops in blood pressure occur in humans. [p.677]
10. List the factors that cause blood to leave the heart and those that cooperate to return blood to the heart. [pp.676–680]
11. Explain how veins and venules can act as reservoirs of blood volume. [pp.678–679]
12. Describe how hypertension develops, how it is detected, and whether it can be corrected. [p.680]
13. Distinguish a stroke from a coronary artery blockage, or occlusion. [pp.680–681]
14. State the significance of high- and low-density lipoproteins with regard to cardiovascular disorders. [p.681]
15. List in sequence the events that occur in the formation of a blood clot. [p.682]
16. Describe the composition and function of the lymphatic system. [pp.682–683]

Integrating and Applying Key Concepts

You observe that some people appear as though fluid had accumulated in their lower legs and feet; their lower extremities resemble those of elephants. You inquire about what is wrong and are told that the condition is caused by the bite of a mosquito that is active at night. Construct a testable hypothesis that would explain (1) why the fluid was not being returned to the torso, as normal, and (2) what the mosquito did to its victims.

39

IMMUNITY

Interactive Exercises

Russian Roulette, Immunological Style [pp.686–687]

39.1. THREE LINES OF DEFENSE [p.688]

39.2. COMPLEMENT PROTEINS [p.689]

39.3. INFLAMMATION [pp.690–691]

Selected Words: smallpox [p.686], "immune" [p.686], Jenner [p.686], Pasteur [p.687], Koch [p.687], anthrax [p.687], *Bacillus anthracis* [p.687], *athlete's foot* [p.688], *Lactobacillus* [p.688], *nonspecific* and *specific* responses [p.688], complement "coat" [p.689], *chemotaxins* [p.691], *interleukins* [p.691], *lactoferrin* [p.691], *endogenous pyrogen* [p.691], set point [p.691], *interleukin-1* [p.691]

Boldfaced, Page-Referenced Terms

[p.687] vaccination ____

____

[p.688] pathogens ____

____

[p.688] lysozyme ____

____

[p.689] complement system ____

____

[p.689] lysis ____

____

[p.692] neutrophils ____

____

[p.690] eosinophils ____

____

[p.690] basophils ____

____

[p.690] mast cells ____

____

[p.690] macrophages ____

____

[p.690] acute inflammation ____

____

[p.690] histamine ____

____

[p.691] fever ____

____

Fill-in-the-Blanks

The best way to deal with damaging foreign agents is to prevent their entry. Several barriers prevent pathogens from crossing the boundaries of your body. Intact skin and (1) ______________ [p.688] membranes are effective barriers. (2) ______________ [p.688] is an enzyme that destroys the cell walls of many bacteria. (3) ______________ [p.688] fluid destroys many food-borne pathogens in the gut. Normal (4) ______________ [p.688] residents of the skin, gut, and vagina outcompete pathogens for resources and help keep their numbers under control.

Even simple aquatic invertebrates defend themselves with (5) ______________ [p.688] cells and antimicrobial substances, including lysozymes, that cause bacteria to burst open. In most animals, when a sharp object cuts through the skin and foreign microbes enter, some plasma proteins come to the rescue and seal the wound with a(n) (6) ______________ [p.688] mechanism. (7) ______________ [p.688] white blood cells engulf bacteria soon thereafter.

Vertebrates have about 20 plasma proteins, collectively referred to as the (8) ______________ ______________ [p.689], that are activated one after another in a cascade of reactions to help destroy invading microorganisms. When the complement system is activated, circulating basophils and mast cells in tissues release (9) ______________ [p.690], which increases the permeability of (10) ______________ [p.690] and makes them "leaky," so that fluid seeps out and causes the inflamed area to become swollen and warm. In addition to fluid, both (11) ______________ [p.691] and (12) ______________ [p.691] exit the capillaries and begin to engulf the invaders. Chemicals called (13) ______________ [p.691] attract more phagocytes, while (14) ______________ ______________ [p.691] triggers a fever.

Complete the Table

Complete the following table by providing the specific functions carried out by each of the four different kinds of white blood cells listed. [recall p.669]

Type of Cells	*Functions*
Basophils	15.
Eosinophils	16.
Neutrophils	17.
Macrophages	18.

39.4. OVERVIEW OF THE IMMUNE SYSTEM [pp.692–693]

39.5. HOW LYMPHOCYTES FORM AND DO BATTLE [pp.694–695]

39.6. ANTIBODY-MEDIATED RESPONSE [pp.696–697]

39.7. CELL-MEDIATED RESPONSE [pp. 698–699]

Selected Words: self/nonself recognition [p.692], *specificity* [p.692], *diversity* [p.692], *memory* [p.692], self markers [p.692], nonself markers [p.692], effectors [p.692], *memory* cells [p.692], dendritic cells [p.692], *cell-mediated* responses [p.692], *antibody-mediated* responses [p.692], "touch killing" [p.693], *primary* and *secondary* immune responses [p.693], *clonal selection* [p.694], clone [p.694], plasma cell [p.696], *IgM, IgD, IgG, IgA, IgE* [pp.696–697], perforins [p.699]

Boldfaced, Page-Referenced Terms

[p.692] immune system ____

____

[p.692] antigen ____

____

[p.692] MHC markers ____

____

[p.692] antigen-presenting cell ____

____

[p.692] antigen–MHC complexes ____

____

[p.692] helper T cells ____

____

[p.692] cytotoxic T cells ____

____

[p.692] natural killer (NK) cells ____

____

[p.692] B cells ____

____

[p.692] antibodies ____

____

[p.693] supressor T cells ____

____

[p.694] TCR ____

____

[p.696] immunoglobulins, Igs ____

____

[p.699] apoptosis ____

____

Fill-in-the-Blanks

If the (1) ______________ ______________ [p.692] and inflammation fail to repel microbial invaders, then the body calls on its (2) ______________ ______________ [p.692], which identifies *specific* targets to kill and *remembers* the identities of its targets. Your own unique (3) ______________ ______________ [p.692] patterns identify your cells as "self" cells. Any other surface pattern is, by definition, (4) ______________ [p.692] and doesn't belong in your body.

The principal actors of the immune system are (5) ______________ [p.692], descended from stem cells (consult Fig. 39.7) in the bone marrow, which have two different strategies for dealing with their different kinds of enemies. (6) ______________ ______________ [pp.692,694] clones mediate the antibody response and act principally against the extracellular enemies that are pathogens in blood or on the cell surfaces of body tissues. (7) ______________ ______________ [pp.692,694] clones descend from lymphocytes that matured in the (8) ______________ [p.694] gland, where they acquired specific markers on their cell surfaces; they defend principally against intracellular pathogens such as (9) ______________ [pp.698–699] and against any cells that are perceived as abnormal or foreign, such as (10) ______________ [p.698] cells and cells of organ transplants.

Each kind of cell, virus, or substance bears unique molecular configurations (patterns) that give it a unique (11) ______________ [p.692]. A(n) (12) ______________ [p.692] is any molecular configuration that causes the formation of lymphocyte armies. Any cell that processes and displays (12) together with a suitable MHC molecule is known as a(n) (13) ______________-______________ [p.692] cell that can activate lymphocytes to undergo rapid cell divisions. Lymphocyte subpopulations that fight and destroy enemies are known as (14) ______________ [p.692] cells; among these are (15) ______________ ______________ [p.692] cells, which promote the immune response, and (16) ______________ ______________ [p.692] cells, which eliminate infected body cells and tumor cells by "touch killing." Together with other "killers," they execute the (17) ______________-______________ [pp.692,698] immune responses. By contrast, B lymphocytes produce Y-shaped antigen-binding receptor molecules called (18) ______________ [p.692]. (6) and (18) carry out the (19) ______________-______________ [pp.692–696] immune responses.

Other lymphocyte subpopulations, (20) ______________ [p.692] cells, enter a resting phase; but they "remember" the specific agent that was conquered and will undertake a larger, more rapid response if it shows up again.

Choice

Match each of the white blood cell types with the appropriate function.

a. effector cytotoxic T cells b. effector helper T cells c. macrophages
d. memory cells e. effector B cells

21. ____ Lymphocytes that directly destroy by "touch-killing" tumor cells, cells of organ transplants, or body cells already infected by certain viruses [pp.698,699]
22. ____ A portion of B and T cell populations, set aside as the result of a first encounter, now circulate freely and respond rapidly to any later attacks by the same type of invader [p.692]
23. ____ Lymphocytes and their progeny that produce antibodies [pp.696,697]
24. ____ Lymphocytes that promote immune responses; stimulate the rapid division of B cells and cytotoxic T cells [p.692]
25. ____ "Big eater" white blood cells that develop from monocytes, engulf anything perceived as foreign, and alert helper T cells to the presence of specific foreign agents [recall p.690]

Ordering

Put the following steps of the antibody response in the correct order. [pp.696–697]

26. ____ Step 1
27. ____ Step 2
28. ____ Step 3
29. ____ Step 4
30. ____ Step 5
31. ____ Step 6
32. ____ Step 7
33. ____ Step 8

A. Antibody molecules secreted by effector cells
B. Helper cells bind to the B cell's antigen
C. Naive B cell bristles with many identical antibody molecules
D. Helper cell secretes interleukins
E. Antigen binds to antibodies on naive B cell and is taken into cell
F. B cell becomes an antigen-presenting cell
G. Effector and memory B cells mature
H. Interleukin binds to B cells, triggering mitosis

Fill-in-the-Blanks

After a first-time contact with an antigen, the (34) ____________ ____________ ____________ [p.695] is more rapid because patrolling battalions of (35) ____________ ____________ [p.695] are in the bloodstream on the lookout for enemies they have conquered before. When they meet up with recognizable (36) ____________ [p.695], they divide at once, producing large clones of B or T cells within two to three days.

Complete the Table [pp.696–697]

Immunoglobulin Type	*Function*
37.	Enter mucus-coated surfaces of the respiratory, digestive, and reproductive tracts, where they neutralize pathogens
38.	Trigger inflammation when parasitic worms attack the body; play a role in allergic responses
39.	Activate complement proteins; neutralize many toxins; long lasting; can cross placenta and protect developing fetus; also present in colostrum from mammary glands
40.	First to be secreted during immune responses; after binding to antigen, trigger complement cascade; also tag invaders and bind them in clumps for later phagocytosis

Fill-in-the-Blanks

In addition to effector B cells and cytotoxic T cells (executioner lymphocytes that mature in the thymus), (41) ____________ ____________ [p.697] cells mature in other lymphoid tissues and search out any cell that either is coated with complement proteins or antibodies or bears any foreign molecular pattern. When cytotoxic T cells find foreign cells, they secrete (42) ____________ [pp.696–697] and other toxic substances to touch-kill and induce cell death by apoptosis.

39.8. FOCUS ON SCIENCE: *Cancer and Immunotherapy* [p.699]

39.9. DEFENSES ENHANCED, MISDIRECTED, OR COMPROMISED [pp.700–702]

39.10. FOCUS ON HEALTH: *AIDS — The Immune System Compromised* [pp.702–703]

Selected Words: *carcinomas* [p.699], *sarcomas* [p.699], *leukemia* [p.699], *immunotherapy* [p.699], C. Milstein and G. Kohler [p.699], *monoclonal antibodies* [p.699], "plantibodies" [p.699], *LAK* cells [p.699], *therapeutic vaccines* [p.699], melanoma [p.699], *active* immunization [p.700], booster injection [p.700], *passive* immunization [p.700], *asthma* [p.700], *hay fever* [p.700], *anaphylactic shock* [p.700], antihistamine [p.700], *rheumatoid arthritis* [p.701], Graves' disorder [p.701], *multiple sclerosis* [p.701], *primary* immune deficiency [p.701], *SCID* (severe combined immunodeficiency) [p.701], *secondary* immune deficiencies [p.701], *AIDS* (acquired immunodeficiency syndrome) [p.701], *Pneumocystis carinii* [p.702], Kaposi's sarcoma [p.702], AZT and ddI [p.703], protease inhibitors [p.703]

Boldfaced, Page-Referenced Terms

[p.700] immunization ____

____

[p.700] vaccine ____

____

[p.700] allergens ____

____

[p.700] allergy ____

____

[p.701] autoimmune response ____

____

Fill-in-the-Blanks

Various strategems have been developed that enhance immunological defenses against tumors as well as against certain pathogens; collectively they are referred to as (1) ______________ [p.699].

The term (2) ______________ [p.699] refers to cells that have lost control over cell division. Milstein and Kohler developed a means of producing large amounts of (3) ______________ ______________ [p.699], which are produced by clones of proliferating (4) ______________ [p.699] cells: mouse B cells fused with tumor cells that divide nonstop. A different therapy involves researchers extracting (5) ______________ [p.699] from tumors, activating them by exposing them to a(n) (6) ______________ [p.699], and injecting these *LAK* cells back into the patient, where they actively kill tumor cells.

Complete the Table

Indicate with a checkmark (√) the recommended age(s) of vaccination. [all from p.702]

Age Vaccination Is Administered

	(a)	*(b)*	*(c)*	*(d)*	*(e)*	*(f)*	*(g)*	*(h)*	*(i)*	*(j)*
Disease	*At birth*	*At 2 mos.*	*1–4 mos.*	*4 mos.*	*6 mos.*	*6–18 mos.*	*12–15 mos.*	*12–18 mos.*	*4–6 yrs.*	*11–12 yrs.*
7. Diphtheria										
8. *Hemophilus influenzae*										
9. Hepatitis B										
10. Measles										
11. Mumps										
12. Pneumococcus										
13. Polio										
14. Rubella										
15. Tetanus										
16. Varicella										
17. Whooping cough (pertussis)										

Fill-in-the-Blanks

(18) ______________ [p.700] is an altered secondary response to a normally harmless substance that may actually cause injury to tissues. In rare instances, the immune reactions are life threatening and cause (19) ______________ ______________ [p.700]. (20) ______________ ______________ [p.701] is a disorder in which the body mobilizes its forces against certain of its own tissues; (21) ______________ ______________ [p.701] is an example of this kind of disorder, in which T cells trigger inflammation of myelin sheaths. (22) ______________ ______________ [p.701] is a similar kind of disorder, in which skeletal joints are chronically inflamed. In humans, autoimmunity is far more frequent in (23) ______________ [p.701].

AIDS is a constellation of disorders that follow infection by the (24) ______________ ______________ ______________ [p.702]. In the United States, transmission has occurred most often among intravenous drug abusers who share needles and among (25) ______________ ______________ [p.702]. HIV is a(n)

(26) ______________ [p.702]; its genetic material is RNA rather than DNA, and it has several copies of an enzyme, (27) ______________ ______________ [p.702], that uses the viral RNA as a template for making DNA, which is then inserted into a host chromosome.

HIV is transmitted when (28) ______________ ______________ [p.702] of an infected person enter another person's tissues. The virus cripples the immune system by attacking (29) ______________ ______________ [p.702] cells. Drugs such as (30) ______________ [p.703] and (31) ______________ [p.703] target reverse transcriptase, while (32) ______________ ______________ [p.703] block conversion of precursors to active viral proteins. Although research is ongoing, development of (33) ______________ [p.703] is a major challenge.

Self-Quiz

Multiple Choice

____ 1. All the body's phagocytes are derived from stem cells in the ______________. [p.690]
a. spleen
b. liver
c. thymus
d. bone marrow
e. thyroid

____ 2. The plasma proteins that are activated when they contact a bacterial cell are collectively known as the ______________ system. [p.689]
a. shield
b. complement
c. IgG
d. MHC
e. HIV

____ 3. ______________ are divided into two groups: T cells and B cells. [p.694]
a. Macrophages
b. Lymphocytes
c. Platelets
d. Complement cells
e. Cancer cells

____ 4. ______________ produce and secrete antibodies that set up bacterial invaders for subsequent destruction by macrophages. [p.696]
a. B cells
b. Phagocytes
c. T cells
d. Bacteriophages
e. Thymus cells

____ 5. Antibodies are shaped like the letter ______________. [p.696]
a. Y
b. W
c. Z
d. H
e. E

____ 6. The markers for every cell in the human body are referred to by the letters ______________. [p.692]
a. HIV
b. MBC
c. RNA
d. DNA
e. MHC

____ 7. Effector B cells ______________. [pp.666–697]
a. fight against extracellular pathogens and toxins circulating in tissues
b. develop from antigen-presenting cells
c. manufacture and secrete antibodies
d. do not divide and form clones
e. all of the above

_____ 8. Clones of B or T cells are __________ . [p.692]
a. being produced continually
b. sometimes known as memory cells if they keep circulating in the bloodstream
c. only produced when their surface proteins recognize other specific proteins previously encountered
d. produced and mature in the bone marrow
e. both (b) and (c)

_____ 9. Whenever the body is re-exposed to a specific sensitizing agent, IgE antibodies cause __________ . [pp.697,704]
a. prostaglandins and histamine to be produced
b. clonal cells to be produced
c. histamine to be released
d. the immune response to be suppressed
e. none of the above

_____ 10. The clonal selection hypothesis explains __________ . [p.694]
a. how self cells are distinguished from nonself cells
b. how B cells differ from T cells
c. how so many different kinds of antigen-specific receptors can be produced by lymphocytes
d. how memory cells are set aside from effector cells
e. how antigens differ from antibodies

Matching

Choose the most appropriate description for each term.

11. _____ allergy [p.700]
12. _____ antibody [p.696]
13. _____ antigen [p.692]
14. _____ macrophage [p.690]
15. _____ clone [p.694]
16. _____ complement [p.689]
17. _____ histamine [p.690]
18. _____ MHC marker [p.692]
19. _____ effector B cell [p.696]
20. _____ T cell [p.698]

A. Begins its development in bone marrow but matures in the thymus gland
B. Cells that have directly or indirectly descended from the same parent cell
C. A potent chemical that causes blood vessels to dilate and let protein pass through the vessel walls
D. Y-shaped immunoglobulin
E. A nonself marker
F. The progeny of turned-on B cells
G. A group of about 20 proteins that participate in the inflammatory response
H. An altered secondary immune response to a substance that is normally harmless to other people
I. The basis for self-recognition at the cell surface
J. Principal perpetrator of phagocytosis

Chapter Objectives/Review Questions

1. Describe typical external barriers that organisms present to invading organisms. [p.688]
2. List and discuss four nonspecific defense responses that serve to exclude microbes from the body. [pp.688–690]
3. List the three general types of cells that form the basis of the vertebrate immune system. [p.688]
4. Explain how the complement system is related to an inflammatory response. [pp.689–691]
5. Describe the sequence of events that occur during inflammatory responses. [pp.690–691]

6. Explain why the immune system of mammals usually does not attack so-called self tissues. Understand how vertebrates (especially mammals) recognize and discriminate between self and nonself tissues. [p.692]
7. Distinguish the roles of T cells from those of B cells. [pp.694–695]
8. Explain what is meant by *primary immune response,* in contrast to *secondary immune response.* [p.693]
9. Distinguish between the antibody-mediated and the cell-mediated response patterns. [pp.696–699]
10. Explain what monoclonal antibodies are and tell how they are currently being used in passive immunization and cancer treatment. [p.699]
11. Describe how recognition proteins and antibodies are made. State how they are used in immunity. [pp.696–697]
12. Describe the clonal selection hypothesis and tell what it helps to explain. [p.694]
13. Describe two ways in which people can be immunized against specific diseases. [p.700]
14. Distinguish allergy from autoimmune disorder. [pp.700–701]
15. Describe some examples of immune failure and identify as specifically as you can which weapons in the immunity arsenal failed in each case. [pp.701–703]
16. Describe how AIDS specifically interferes with the human immune system. [pp.702–703]

Integrating and Applying Key Concepts

Suppose you wanted to get rid of 47 warts that you have on your hands by treating them with monoclonal antibodies. Outline the steps you would have to take.

40

RESPIRATION

Interactive Exercises

Of Lungs and Leatherbacks [pp.706–707]

40.1. THE NATURE OF RESPIRATION [p.708]

40.2. INVERTEBRATE RESPIRATION [p.709]

40.3. VERTEBRATE RESPIRATION [pp.712–713]

Selected Words: "partial" *pressure* [p.708], *internal* respiratory surface [p.709], *external* gills [p.710], *internal* gills [p.710]

Boldfaced, Page-Referenced Terms

[p.707] respiration ____

____

[p.707] respiratory systems ____

____

[p.708] pressure gradients ____

____

[p.708] partial pressure ______________________________

[p.708] respiratory surface ______________________________

[p.708] Fick's law ______________________________

[p.708] hemoglobin ______________________________

[p.708] myoglobin ______________________________

[p.709] integumentary exchange ______________________________

[p.709] gills ______________________________

[p.709] tracheal respiration ______________________________

[p.710] countercurrent flow ______________________________

[p.710] lungs ______________________________

[p.711] vocal cords ______________________________

[p.711] glottis ______________________________

Fill-in-the-Blanks

Each animal has a body plan adapted to the (1) ____________ [p.707] levels of a particular habitat. The body plan allows for (2) ____________ [p.707], the movement of oxygen into the (3) ____________ [p.707] environment and the movement of (4) ____________ ____________ [p.707] out of it. The high energy demands of animals are based on the ATP output of (5) ____________ [p.707] respiration, a metabolic pathway that releases energy from (6) ____________ [p.707] compounds and produces carbon dioxide.

(7) ____________ [p.707] systems function in the exchange of gases between the body and the environment. Together with other organ systems they contribute to (8) ____________ [p.707].

Labeling

Identify the numbered parts of the following diagram, which illustrates the interconnection of the respiratory system with other body systems. [p.707]

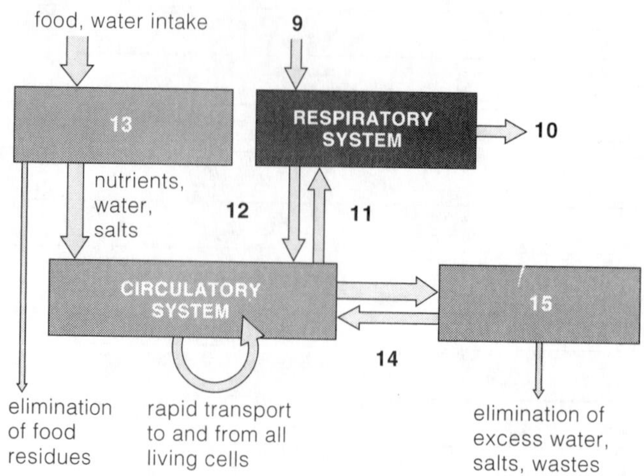

9. ________________

10. ________________ ________________

11. ________________ ________________

12. ________________

13. ________________ ________________

14. ________________ and ________________

15. ________________ ________________

Matching

Choose the most appropriate description of each of the following terms. [p.708]

16. ____ hemoglobin

17. ____ Fick's law

18. ____ ventilation

19. ____ surface-to-volume ratio

20. ____ myoglobin

21. ____ partial pressure

22. ____ pressure gradients

23. ____ respiratory surface

A. The name given to the adaptations that increase gas exchange above the level of diffusion

B. States that the rate of diffusion will be greater with an increase in the partial pressure gradient and surface area

C. Animal body plans are governed by this principle

D. The main transport pigment of humans

E. The term for *concentration gradients* when dealing with gases

F. The contribution of an individual gas to the total atmospheric pressure

G. Gases enter and leave an organism by first crossing this structure

H. The transport pigment that resists fatigue in muscle and skeletal cells

Choice

For questions 24–28, choose from the following examples of invertebrate respiration. [p.709]

a. tracheal respiration   b. integumentary exchange   c. gills

24. ____ The diffusion of a gas directly across the body covering
25. ____ A form of respiratory system commonly found in aquatic organisms
26. ____ An internal system of tubes that acts as a respiratory surface
27. ____ Commonly found in insects, millipedes, and centipedes
28. ____ Thin-walled respiratory surfaces that exchange gases between a body fluid and its surroundings

Choice

For questions 29–33, match the characteristics of the respiratory system to the following groups of vertebrates. [pp.710–711]

a. fish   b. amphibians   c. birds

29. ____ Utilize a countercurrent flow to effectively exchange gases
30. ____ Utilize a flow-through respiratory system
31. ____ These organisms force air into their lungs by contracting muscles in their body wall
32. ____ The lungs first developed in this group in response to an oxygen-poor environment
33. ____ These organisms use their skin to supplement gas exchange

40.4. HUMAN RESPIRATORY SYSTEM [pp.712–713]

40.5. CYCLIC REVERSALS IN AIR PRESSURE GRADIENTS [pp.714–715]

Selected Words: *pulmo* [p.713], *laryngitis* [p.713], *pleurisy* [p.713], *"bronchial tree"* [p.713], *respiratory* bronchioles [p.713], *inhalation* [p.714], *exhalation* [p.714], *atmospheric* pressure [p.714], *intrapulmonary* pressure [p.714], *intrapleural* pressure [p.714]

Boldfaced, Page-Referenced Terms

[p.712] alveoli ____________________

____

[p.713] pharynx ____________________

____

[p.713] larynx ____________________

____

[p.713] epiglottis ____________________

____

[p.713] trachea ____________________

____

[p.713] bronchus ______________________________

____

[p.713] diaphragm ______________________________

____

[p.713] bronchioles ______________________________

____

[p.714] respiratory cycle ______________________________

____

[p.715] vital capacity ______________________________

____

[p.715] tidal volume ______________________________

____

Complete the Table

1. For each of the following items, first identify the structure to which the description belongs. Second, refer to the numbered diagram below and identify the location of the structure. Place this number within the parentheses. [pp.712–713]

Structure Name (Location)		*Description*
a.	( )	The muscles that assist the rib cage in breathing
b.	( )	The vocal cords are located here
c.	( )	Also called the windpipe
d.	( )	Air sacs arranged into clusters
e.	( )	Stops food and liquids from entering the respiratory system
f.	( )	A supplemental airway that may be used in times of increased demand on the respiratory system
g.	( )	Connects the nasal cavity and mouth with the larynx
h.	( )	The muscular partition between the thoracic and abdominal cavities; assists in breathing
i.	( )	Serves to filter, warm, and moisten incoming air
j.	( )	A double-layer membrane that separates the lungs from other organs
k.	( )	The tiny air sacs where gas exchange occurs
l.	( )	A series of branching tubes that end at the air sacs
m.	( )	Lobed, elastic organ of breathing

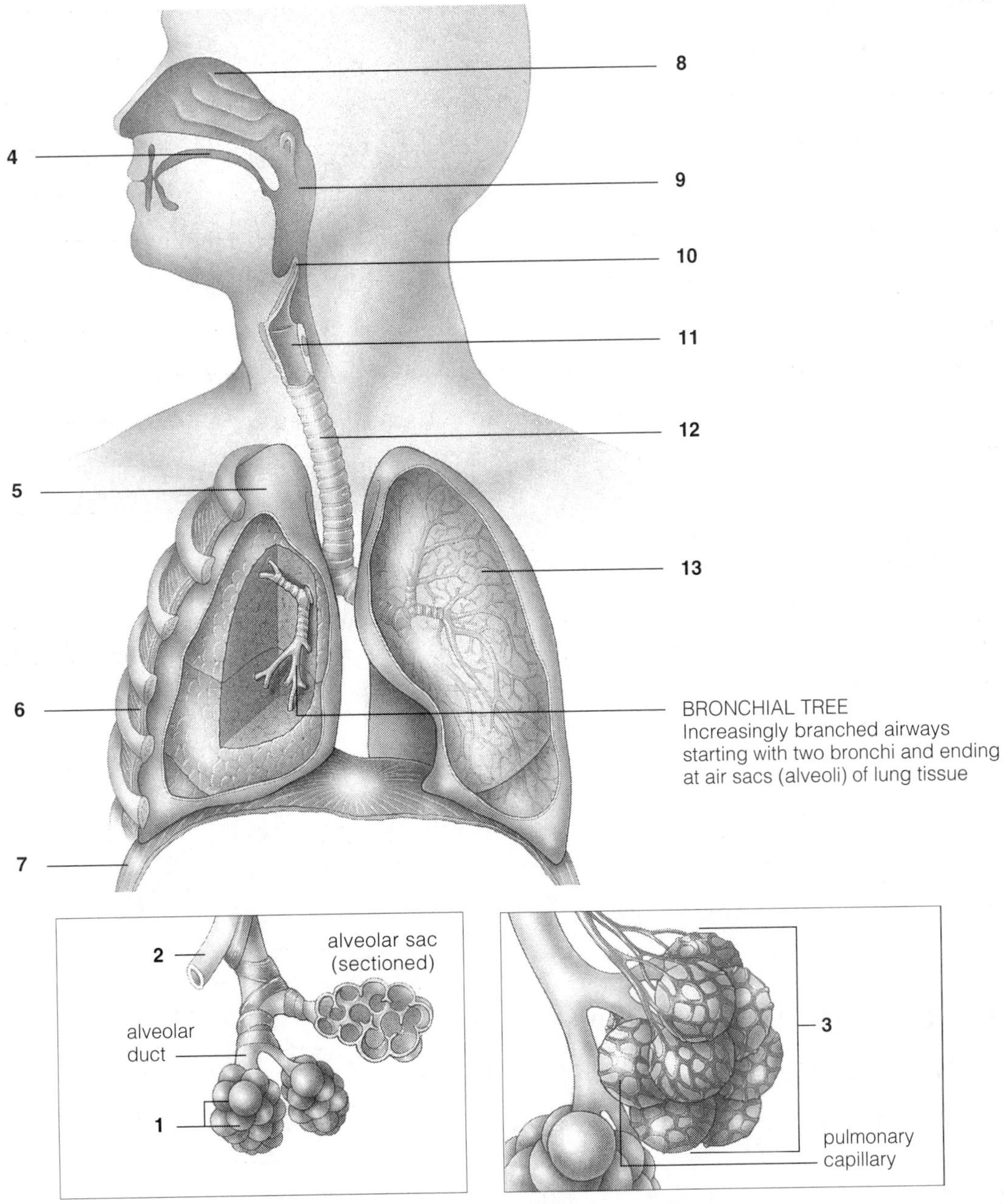
8
4
9
10
11
12
5
13
6
BRONCHIAL TREE
Increasingly branched airways starting with two bronchi and ending at air sacs (alveoli) of lung tissue
7
2
alveolar sac (sectioned)
alveolar duct
1
3
pulmonary capillary

Fill-in-the-Blanks

Each respiratory cycle consists of two actions: (2) ______________ [p.714] and (3) ______________ [p.714]. Inhalation is always a(n) (4) ______________ [p.714] process. During inhalation, the (5) ______________ [p.714] contracts, leading to a(n) (6) ______________ [p.714] in thoracic cavity volume. The movement of air during breathing is based on pressure gradients. The combined weight of all of the gases in the atmosphere is called (7) ______________ [p.714] pressure, while the pressure at the alveoli is called the (8) ______________ [p.714] pressure. As you inhale, the diaphragm flattens and moves (9) ______________ [p.714], while the rib cage is lifted upward and (10) ______________ [p.714]. At this time, the air pressure in the alveolar sacs is (11) ______________ [p.715] than atmospheric pressure. Fresh air flows (12) ______________ [p.715] the airways. The second part of the respiratory cycle, exhalation, is (13) ______________ [p.715] when you are breathing quietly. The muscles relax; the lungs passively recoil. This (14) ______________ [p.715] lung volume, compressing air inside the alveolar sacs. At this time, pressure in the sacs is (15) ______________ [p.715] than atmospheric pressure. Air follows the gradient, flowing (16) ______________ [p.715] from the lungs. Exhalation can become active during intense (17) ______________ [p.715]. The abdominal walls (18) ______________ [p.715] during active exhalation, thereby (19) ______________ [p.715] abdominal pressure.

Short Answer

20. Distinguish between tidal volume and vital capacity. ______________

____

____

____

40.6. GAS EXCHANGE AND TRANSPORT [pp.716–717]

40.7. FOCUS ON HEALTH: *When the Lungs Break Down* [pp.718–719]

40.8. HIGH CLIMBERS AND DEEP DIVERS [pp.720–721]

Selected Words: *buffer* [p.717], *rhythm* [p.717], *magnitude* [p.717], *apnea* [p.717], *sudden infant death syndrome* (SIDS) [p.717], *bronchitis* [p.718], *emphysema* [p.718], *secondhand smoke* [p.718], "smoker's cough" [p.718], *pot* [p.718], *hypoxia* [p.720], *carbon monoxide poisoning* [p.720], *nitrogen narcosis* [p.720], "raptures of the deep" [p.720], "the bends" [p.721]

Boldfaced, Page-Referenced Terms

[p.716] heme groups ______________

____

[p.716] oxyhemoglobin (HbO_2) ______________

____

[p.716] carbamino hemoglobin (HbO_2) ______________

____

[p.716] carbonic anhydrase ______

______

[p.720] acclimatization ______

______

[p.720] erythropoietin ______

______

Choice

Questions 1–6 address the transport of oxygen and carbon dioxide. For each statement, choose which gas the statement is associated with. [p.716]

a. oxygen b. carbon dioxide

1. ____ When bound to hemoglobin, forms carbamino hemoglobin
2. ____ Heme groups with iron are designed to bind with this gas
3. ____ This is released where the blood is warmer and pH is lower
4. ____ The enzyme carbonic anhydrase is involved in the transport of this gas
5. ____ When bound to hemoglobin, forms oxyhemoglobin
6. ____ In water, this gas forms carbonic acid

Fill-in-the-Blanks

Gas exchange is most efficient when the rate of air flow matches the rate of (7) ______ [p.717] flow. The nervous system acts to balance them by controlling the (8) ______ [p.717] of breathing and its (9) ______ [p.717], or rate and depth. A respiratory system inside the (10) ______ ______ [p.717] controls rhythmic patterns of breathing. Higher in the brain stem, an area called the (11) ______ [p.717] acts to smooth out the rhythm. Control of the magnitude of breathing is done by (12) ______ [p.717] that respond to H^+ increases in the cerebrospinal fluid. The brain responds by signaling the (13) ______ [p.717] and other muscles to alter activity.

Matching

Match the following disorders of the respiratory system with their correct description.

14. ____ sudden infant death syndrome [p.717]
15. ____ apnea [p.717]
16. ____ bronchitis [p.718]
17. ____ emphysema [p.718]
18. ____ hypoxia [p.720]
19. ____ nitrogen narcosis [p.720]
20. ____ carbon monoxide poisoning [p.720]

A. Cellular oxygen deficiencies
B. Small amounts can tie up as much as 50% of the hemoglobin and cause death
C. The breakdown of the thin walls of the alveoli
D. An irritation of the respiratory lining, resulting in an excessive secretion of mucus
E. Also called the bends, or decompression sickness
F. A brief interruption in the respiratory cycle
G. May be caused by an irregular heartbeat or result from exposure to cigarette smoke during pregnancy

Self-Quiz

Matching

1. ______ bronchioles [pp.712–713]
2. ______ bronchitis [p.718]
3. ______ carbonic anhydrase [p.716]
4. ______ emphysema [p.718]
5. ______ Fick's law [p.708]
6. ______ glottis [p.711]
7. ______ hypoxia [p.720]
8. ______ intercostal muscles [p.712]
9. ______ larynx [pp.712–713]
10. ______ oxyhemoglobin [p.716]
11. ______ pharynx [pp.712–713]
12. ______ pleurisy [p.713]
13. ______ tidal volume [p.715]
14. ______ ventilation [p.708]
15. ______ vital capacity [p.715]

A. Membrane that encloses human lung becomes inflamed and swollen; painful breathing generally results
B. HbO_2
C. The amount of air inhaled and exhaled during normal breathing of a human at rest; generally about 500 ml
D. Throat passageway that connects to *both* the respiratory tract below *and* the digestive tract
E. One gas will diffuse faster than another if its surface area is more extensive and its partial pressure is greater than that of the other gas
F. Inflammation of the two principal passageways that lead air into the human lungs
G. Contract when air is leaving the lungs; relax when lungs are filling with air
H. The opening into the "voicebox"
I. Finer and finer branchings that lead to alveoli
J. Maximum volume of air that can move out of your lungs after a single, maximal inhalation
K. An enzyme that increases the rate of production of H_2CO_3 from CO_2 and H_2O
L. Lungs have become distended and inelastic so that walking, running, and even exhaling are difficult
M. Where sound is produced by vocal cords
N. Movements that keep air or water moving across a respiratory surface
O. Too little oxygen is being distributed in the body's tissues

Multiple Choice

______ 16. The transport pigment that is abundant in muscle and skeletal cells of the human body is ______________. [p.708]
a. myoglobin
b. hemoglobin
c. carbonic acid
d. carbonic anhydrase

______ 17. ______________ is the most abundant gas in Earth's atmosphere. [p.708]
a. Water vapor
b. Oxygen
c. Carbon dioxide
d. Hydrogen
e. Nitrogen

______ 18. With respect to respiratory systems, countercurrent flow is a mechanism that explains how ______________. [p.710]
a. oxygen uptake by blood capillaries in the lamellae of fish gills occurs
b. ventilation occurs
c. intrapleural pressure is established
d. sounds originating in the vocal cords of the larynx are formed
e. all of the above

______ 19. A flow-through respiratory system is found in ______________. [p.711]
a. amphibians
b. reptiles
c. birds
d. mammals
e. humans

_____ 20. During inhalation, _______________ [p.714]

a. the pressure in the thoracic cavity (intrapleural pressure) is less than the pressure within the lungs (intrapulmonary pressure)
b. the pressure in the chest cavity (intrapleural pressure) is greater than the pressure within the lungs (intrapulmonary pressure)
c. the diaphragm moves upward and becomes more curved
d. the thoracic cavity volume decreases
e. all of the above

_____ 21. Oxygen moves from alveoli to the bloodstream _______________. [p.716]

a. by diffusion when the concentration of oxygen is greater in alveoli than in the blood
b. by means of active transport
c. with the assistance of carbaminohemoglobin
d. principally due to the activity of carbonic anhydrase in the red blood cells
e. by all of the above

_____ 22. Immediately before reaching the alveoli, air passes through the _______________. [p.713]

a. bronchioles
b. glottis
c. larynx
d. pharynx
e. trachea

_____ 23. Oxyhemoglobin _______________. [p.716]

a. releases oxygen more readily in metabolically active tissues
b. tends to release oxygen in places where the temperature is lower
c. tends to hold onto oxygen when the pH of the blood drops
d. tends to give up oxygen in regions where partial pressure of oxygen exceeds that in the lungs
e. all of the above

_____ 24. In which of the following ways is carbon dioxide *not* transported? [pp.716–717]

a. bound to hemoglobin to form carbamino hemoglobin
b. dissolved directly in the blood
c. transported as biocarbonate ions
d. bound to myoglobin for transport to alveoli

_____ 25. Which of the following respiratory ailments is common among smokers? [p.718]

a. emphysema
b. bronchitis
c. lung cancer
d. smoker's cough
e. all of the above

Chapter Objectives/Review Questions

1. Understand how the human respiratory system is related to the circulatory system, to cellular respiration, and to the nervous system. [p.707]
2. Understand the physical properties of gases and the limitations of a respiratory surface. [p.708]
3. Recognize the difference between myoglobin and hemoglobin in human respiration. [p.708]
4. List the types of invertebrate respiratory surfaces that participate in gas exchange and give an example of each. [p.709]
5. Define *countercurrent flow* and explain how it works. State where such a mechanism is found. [p.710]
6. Describe the major developments in the evolution of the paired lungs in vertebrates. [p.710]
7. Recognize the difference between the respiratory system of birds and those of amphibians and mammals. [pp.710–711]
8. List all the principal parts of the human respiratory system, and explain how each structure contributes to transporting oxygen from the external world to the bloodstream. [pp.712–713]
9. Describe the respiratory cycle and the processes of inhalation and exhalation. [pp.714–715]
10. Explain the difference between vital capacity and tidal volume. [p.715]

11. Trace oxygen transport from the air to the tissues of the body. [p.716]
12. Explain the factors that influence the release of oxygen to the tissues. [p.716]
13. Trace the transport of carbon dioxide from the tissues of the body to the lungs. [p.716]
14. List the three methods by which carbon dioxide may be transported. [p.716]
15. Describe the interaction of the nervous system with the respiratory system. [p.717]
16. Explain how bronchitis, emphysema, and lung cancer are all related to smoking. [p.718]
17. List some of the challenges to respiration at high altitude and during diving. [pp.720–721]
18. Explain why carbon monoxide is a dangerous gas. [p.720]

Integrating and Applying Key Concepts

Consider the amphibians — animals that generally have aquatic larval forms (tadpoles) and terrestrial adults. Outline the respiratory changes that you think might occur as an aquatic tadpole metamorphoses into a land-going juvenile.

41

DIGESTION AND HUMAN NUTRITION

Interactive Exercises

Lose It—And It Finds Its Way Back [pp.724–725]

41.1. THE NATURE OF DIGESTIVE SYSTEMS [pp.726–727]

41.2. VISUAL OVERVIEW OF THE HUMAN DIGESTIVE SYSTEM [p.728]

41.3. INTO THE MOUTH, DOWN THE TUBE [p.729]

Selected Words: *anorexia nervosa* [p.724], *bulimia* [p.724], *food-gathering* region [p.726], *food-processing* region [p.726], *crop* [p.726], gizzard [p.726], "chewing cud" [p.727], *lumen* [p.728], *caries* [p.729], *gingivitis* [p.729], *periodontal disease* [p.729], epiglottis [p.729]

Boldfaced, Page-Referenced Terms

[p.725] nutrition

[p.725] digestive system

[p.726] incomplete digestive system

[p.726] complete digestive system

[p.726] mechanical processing and motility

[p.726] secretion

[p.726] digestion

[p.726] absorption

[p.726] elimination

[p.727] ruminants

[p.728] gut

[p.729] tooth

[p.729] tongue

[p.729] saliva

[p.729] pharynx

[p.729] esophagus

[p.729] sphincter

Fill-in-the-Blanks

(1) ______________ [p.725] is a large concept that encompasses the processes by which food is ingested, digested, absorbed, and later converted into the body's own (2) ______________ [p.725], lipids, proteins, and nucleic acids. A digestive system is some form of body cavity or tube in which food is reduced first to (3) ______________ [p.725] and then to small (4) ______________ [p.725]. Unabsorbed residues are then (5) ______________ [p.725] into the internal environment. The (6) ______________ [p.725] system distributes nutrients to cells throughout the body. The (7) ______________ [p.725] system supplies oxygen to the cells so that they can oxidize the carbon atoms of food molecules, thereby changing them into the waste product (8) ______________ ______________ [p.725], which is eliminated by the same system. And if excess water, salts, and wastes accumulate in the blood, the (9) ______________ [p.725] system and skin will maintain the volume and composition of blood and other body fluids. A(n) (10) ______________ [p.726] digestive system has only one opening and two-way traffic. Flatworms have a highly branched gut cavity that serves both digestive and (11) ______________ [p.726] functions. A(n) (12) ______________ [p.726] digestive system has a tube or cavity with regional specializations and a(n) (13) ______________ [p.726] at each end. (14) ______________ [p.726] that break up food involve the muscular contraction of the gut wall, and (15) ______________ [p.726] is the release into the lumen of enzyme fluids and other substances required to carry out digestive functions. The pronghorn antelope feeds on plant material that breaks down slowly. Antelopes are (16) ______________ [p.727]: hoofed mammals that have multiple stomach chambers; nutrients are released slowly when the animal rests. In comparison with human molars, the antelope molar has a much higher (17) ______________ [p.727]; its teeth wear down rapidly because their plant diet is mixed with abrasive bits of dirt.

The human digestive system is a tube, 21–30 feet long in an adult, that has regions specialized for different aspects of digestion and absorption; they are, in order, the mouth, pharynx, esophagus, (18) ______________ [p.728], (19) ______________ ______________ [p.728], large intestine, rectum, and (20) ______________ [p.728]. Various (21) ______________ [p.728] structures secrete enzymes and other substances that are also essential to the breakdown and absorption of nutrients; these include the salivary glands, liver, gallbladder, and (22) ______________ [p.728].

Saliva contains an enzyme, (23) ______________ ______________ [p.729], that breaks down starch. Contractions force the larynx against a cartilaginous flap called the (24) ______________ [p.729], which closes off the trachea. The (25) ______________ [p.729] is a muscular tube that propels food to the stomach.

Labeling

Identify each numbered structure in the accompanying illustration. [all from p.732]

26. ______________ ______________

27. ______________

28. ______________

29. ______________

30. ______________

31. ______________ ______________

32. ______________ ______________

33. ______________

34. ______________

35. ______________

36. ______________

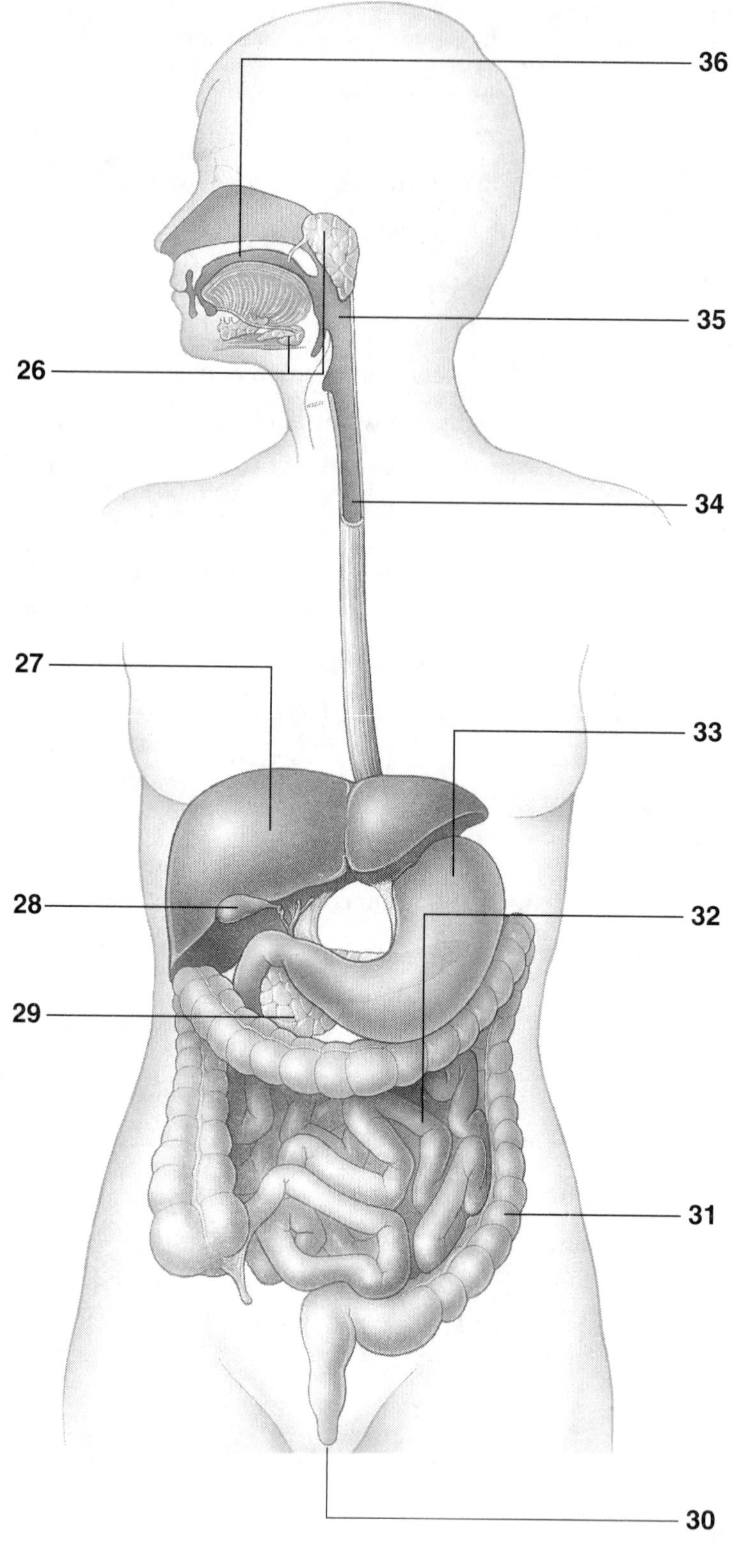

41.4. DIGESTION IN THE STOMACH AND SMALL INTESTINE [pp.730–731]

41.5. ABSORPTION IN THE SMALL INTESTINE [pp.732–733]

41.6. DISPOSITION OF ABSORBED ORGANIC COMPOUNDS [p.734]

41.7. THE LARGE INTESTINE [p.735]

Selected Words: *heartburn* [p.730], *peptic ulcer* [p.730], *Helicobacter pylori* [p.730], peristalsis [p.730], duodenum, jejunum, and ileum [p.730], trypsin and chymotrypsin [p.731], "emulsion" [p.731], gastrin [p.731], secretin [p.731], CCK (cholecystokinin) [p.731], GIP (glucose insulinotropic peptide) [p.731], serosa [p.731], mucosa [p.731], chylomicrons [p.733], urea [p.734], cecum [p.735], *constipation* [p.735], *appendicitis* [p.735], *colon cancer* [p.735]

Boldfaced, Page-Referenced Terms

[p.730] stomach ____________________

____

[p.730] gastric fluid ____________________

____

[p.730] chyme ____________________

____

[p.730] pancreas ____________________

____

[p.730] liver ____________________

____

[p.730] gallbladder ____________________

____

[p.731] bile ____________________

____

[p.731] emulsification ____________________

____

[p.732] villi (singular, villus) ____________________

____

[p.732] microvilli (singular, microvillus) ____________________

____

[p.733] segmentation ____________________

____

[p.733] micelle formation ____________________

____

[p.735] colon ____________________

____

[p.735] bulk ____

____

[p.735] appendix ____

____

Complete the Table

1. Complete the following table by naming the organs described.

Organ	*Main Functions*
a. [recall p.729]	Mechanically breaks down food and mixes it with saliva
b. [recall p.729]	Moisten food; start polysaccharide breakdown; buffer acidic foods in mouth
c. [p.730]	Stores, mixes, and dissolves food; kills many microorganisms; starts protein breakdown; empties in a controlled way
d. [p.732]	Digests and absorbs most nutrients
e. [p.730]	Produces enzymes that break down all major food molecules; produces buffers against hydrochloric acid from stomach
f. [p.731]	Secretes bile for fat emulsification; secretes bicarbonate, which buffers hydrochloric acid from stomach
g. [p.731]	Stores and concentrates bile from liver
h. [p.735]	Stores and concentrates undigested matter by absorbing water and salts
i. [p.735]	Controls elimination of undigested and unabsorbed residues

Fill-in-the-Blanks

Carbohydrates include sugars and (2) ______________ [recall pp.729–730], the name commonly given to digestible plant polysaccharides. Rice, cereal, pasta, bread, and white potatoes are composed of many polysaccharide molecules that are too large to be absorbed into the internal environment. If these foods are chewed thoroughly, (3) ______________ ______________ [p.730] in the mouth digests them to the (4) ______________ [p.730] (double sugar) level. Since little carbohydrate digestion occurs in the (5) ______________ [p.730], if you gulped down your food, starch digestion would again begin in the (6) ______________ ______________ [p.730], where (7) ______________ [p.730] produced by the pancreas would do what should have been done in the mouth. Digestion of disaccharides to monosaccharides (simple sugars) also occurs in the (8) ______________ ______________ [p.730]. The enzymes responsible are (9) ______________ [p.730], with names such as *sucrase, lactase,* and *maltase.*

Proteins are digested to protein fragments, beginning in the (10) ______________ [p.730] by (11) ______________ [p.730] secreted by the lining of the stomach. Protein fragments are subsequently digested to smaller protein fragments in the (12) ______________ ______________ [p.730] by enzymes known as trypsin and chymotrypsin produced by the (13) ______________ [p.730]. Eventually the smaller protein fragments are digested to (14) ______________ ______________ [p.730] by means of carboxy-peptidase produced by the pancreas and by aminopeptidase produced by glands in the intestinal lining.

Another name for fat is *triglycerides.* (15) ______________ [p.730], produced by the pancreas but acting in the (16) ______________ ______________ [p.730], breaks down one triglyceride molecule into three (17) ______________ ______________ [pp.730–731] molecules and one glycerol molecule. (18) ______________ [p.731], which was made by the liver and stored in the (19) ______________ [p.731] and which does not contain digestive enzymes, emulsifies the fat droplet (converts it into small droplets coated with bile salts), thereby increasing the surface area of the substrate on which (20) ______________ [p.731] can act. (21) ______________ ______________ [p.734] are made in the pancreas but convert DNA and RNA into nucleotides in the small intestine. Nutrients are mostly digested and absorbed in the (22) ______________ ______________ [p.732]. Any alternating progression of contracting and relaxing muscle movements along the length of a tube is known as (23) ______________ [p.734].

True/False

If the statement in true, write a "T" in the blank. If false, make it true by changing the underlined word.

______________ 24. Amylase digests starch, lipase digests lipids, and <u>peptidases</u> break peptide bonds in protein fragments, yielding amino acids. [p.730]

______________ 25. The appendix has no known <u>digestive</u> functions. [p.735]

______________ 26. Water and <u>sodium ions</u> are absorbed into the bloodstream from the lumen of the large intestine. [p.735]

______________ 27. Fatty acids and monoglycerides recombine into fats inside epithelial cells lining the <u>colon</u>. [p.733]

Short Answer

To answer the following questions, consult Fig. 41.12 and pages 734 and 736–737 in the text.

28. What is the pool of amino acids used for in the human body? ______________

____

____

____

29. Which breakdown products result from carbohydrate and fat digestion? ______________

____

____

____

30. Monosaccharides, free fatty acids, and monoglycerides all have three uses; identify them. ______________

____

____

____

41.8. HUMAN NUTRITIONAL REQUIREMENTS [pp.736–737]

41.9. VITAMINS AND MINERALS [pp.738–739]

41.10. FOCUS ON SCIENCE: *Weighty Questions, Tantalizing Answers* [pp.740–741]

Selected Words: *Mediterranean diet* [p.736], *olive oil* [p.736], antioxidant [p.736], *high-glycemic* index [p.737], *complete* proteins [p.737], *incomplete* proteins [p.737], ketosis [p.737], fat-soluble vitamins [p.739], *body mass index* (BMI) [p.740], *caloric intake* [p.740], *energy output* [p.740], *identical twins* [p.741], "set point" [p.741]

Boldfaced, Page-Referenced Terms

[p.736] food pyramids ______________________________

____

[p.736] obesity ______________________________

____

[p.737] essential fatty acids ______________________________

____

[p.737] essential amino acids ______________________________

____

[p.738] vitamins ______________________________

____

[p.738] minerals ______________________________

____

[p.740] kilocalorie ______________________________

____

[p.741] *ob* gene ______________________________

____

[p.741] leptin ______________________________

____

[p.741] ghrelin ______________________________

____

Complete the Table

1. Complete the following table by determining how many kilocalories the people described should take in daily, given the stated exercise level, in order to *maintain* their weight [p.740].

Height	*Age*	*Sex*	*Level of Physical Activity*	*Present Weight (lbs.)*	*Number of Kilocalories/Day*
5′6″	25	Female	Moderately active	138	a.
5′10″	18	Male	Very active	145	b.
5′8″	53	Female	Not very active	143	c.

Fill-in-the-Blanks

If your body mass index (BMI) is 27 or higher, your risk for developing type 2 diabetes, heart disease, hypertension, breast cancer, colon cancer, gout, gallstones, and/or osteoarthritis increases dramatically. You can calculate your BMI as follows: Multiply your weight in pounds times (2) ______________ [p.740], and divide the result by your height in inches squared. (3) ______________ ______________ [p.736] are the body's main sources of energy; they should make up (4) ______________ to ______________ [p.736] percent of the human daily caloric intake. (5) ______________ [p.737] and cholesterol are components of animal cell membranes.

Fat deposits are used primarily as (6) ______________ ______________ [p.737], but they also cushion many organs and provide insulation. Fats should constitute (7) ______________ to ______________ [p.736] percent of the human diet. One teaspoon a day of polyunsaturated oil supplies all (8) ______________ ______________ ______________ [p.737] that the body cannot synthesize. (9) ______________ [p.737] are digested to 20 common amino acids, of which eight are (10) ______________ [p.741], cannot be synthesized, and must be supplied by the diet. Animal proteins are (11) ______________ [p.737] and contain large amounts of all essential amino acids. However, most plant proteins are considered (12) ______________ [p.737], since they lack one or more essential amino acids. (13) ______________ [p.738] are organic substances needed in small amounts in order to build enzymes or to help them catalyze metabolic reactions. (14) ______________ [p.738] are inorganic substances needed for a variety of uses.

Review Problems

You are a 19-year-old male, very sedentary (TV, sleep, and computers), 6 feet tall, medium frame, and you weigh 195 lbs.

15. Use Figure 41.16 in the text to calculate the number of calories required to sustain your desired weight. Are you underweight, overweight, or just right? Explain your answer. [p.740] ______________

____

____

____

16. How many calories are you allowed to ingest every day? Explain. [p.740] ______________

____

____

____

Complete the Table

Use the new, improved food pyramid to construct a one-day diet that would eventually allow you to reach that weight if you ate a similar diet every day. Place your choices in the following table as a diet for the person described above. [p.736]

How many servings from each group below are you allowed to have daily? [see legend for Figure 41.14 in the text]	*What, specifically, could you choose to eat? [What are your preferences?]*
17a. complex carbohydrates	17b.
18a. fruits	18b.
19a. vegetables	19b.
20a. dairy group	20b.
21a. assorted proteins	21b.
22a. The "sin" group at the top	22b.

Fill-in-the-Blanks [all from p.736]

Use the (23) ________________ ________________ diagram at the right to devise a well-balanced diet for yourself. Group 1, the trapezoidal base, represents the group of complex (24) ________________, which includes rice, pasta, cereal, and (25) ________________. From this group, (26) (choose 1) ☐ 0, ☐ 2–3, ☐ 2–4, ☐ 3–5, ☐ 6–11 servings from this group every day are needed to supply energy and fiber. Group 2 represents the (27) ________________ group. Use the choices in (26) to indicate the number of servings that are needed from this group each day: (28) ________________. Group 3 is the (29) ________________ group, from which (30) ________________ servings are needed each day. Choices include mangoes, oranges, (31) ________________, cantaloupe, pineapple, or 1 cup of fresh (32) ________________. Group 4 includes foods that are a source of nitrogen: nuts, poultry, fish, legumes, and (33) ________________. From this group, the body's (34) ________________ and nucleic acids are constructed. (35) ________________ servings are required every day because the human body cannot synthesize eight of the 20 essential (36) ________________ ________________ that are used to construct proteins and so must get them in its food supplies. The foods in Group 5, the (37) ________________, yogurt, and cheese group, supply calcium and vitamins A, D, B_2, and B_{12}. You need (38) ________________ servings every day. The foods in Group 6 provide extra calories but few vitamins and minerals; (39) ________________ servings are needed every day.

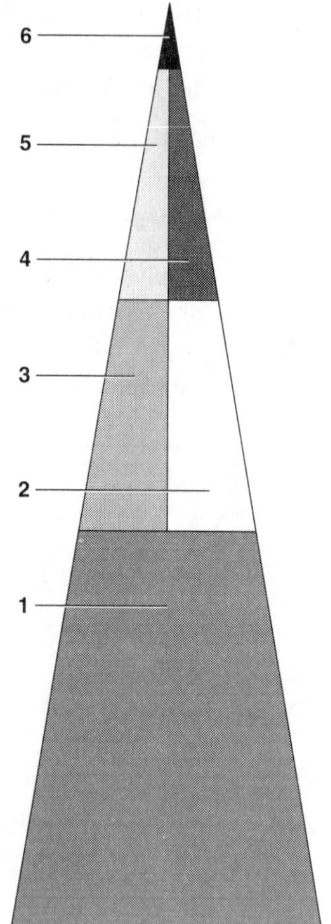

Self-Quiz

Multiple Choice

_____ 1. The process that moves nutrients into the blood or lymph is _______________ . [p.726]
a. ingestion
b. absorption
c. assimilation
d. digestion
e. none of the above

_____ 2. The enzymatic digestion of proteins begins in the _______________ . [p.730]
a. mouth
b. stomach
c. liver
d. pancreas
e. small intestine

_____ 3. The enzymatic digestion of starches begins in the _______________ . [p.730]
a. mouth
b. stomach
c. liver
d. pancreas
e. small intestine

_____ 4. The greatest amount of absorption of digested nutrients occurs in the _______________ . [p.732]
a. stomach
b. pancreas
c. liver
d. colon
e. small intestine

_____ 5. Glucose moves through the membranes of the small intestine mainly by _______________ . [p.733]
a. peristalsis
b. osmosis
c. diffusion
d. active transport
e. bulk flow

_____ 6. Which of the following is *not* found in bile? [p.731]
a. lecithin
b. salts
c. digestive enzymes
d. cholesterol
e. pigments

_____ 7. Soft drinks, sport drinks, and fruit drinks account for what percentage of the average American's calories? [p.737]
a. 1
b. 5
c. 10
d. 20
e. 50

_____ 8. Of the following, _______________ is low in both lysine and tryptophan. [Fig. 41.15]
a. soy beans
b. eggs
c. corn
d. hazelnuts
e. wheat

_____ 9. How many Americans are estimated to suffer from an eating disorder? [p.725]
a. 1 million
b. 2 million
c. 7 million
d. 8 million
e. 10 million

_____ 10. The element needed by humans for blood clotting, nerve impulse transmission, and bone and tooth formation is _______________ . [p.739]
a. magnesium
b. iron
c. calcium
d. iodine
e. zinc

Matching

Match the most appropriate lettered item with each numbered item below.

11. ____ anorexia nervosa [p.724]
12. ____ vitamins C and E [p.738]
13. ____ bulimia [p.724]
14. ____ complex carbohydrates [p.740]
15. ____ essential amino acids [p.737]
16. ____ essential fatty acids [p.737]
17. ____ mineral [p.739]
18. ____ rickets [p.738]
19. ____ scurvy [p.738]
20. ____ vitamin [p.738]

A. Linoleic acid is one example
B. Phenylalanine, lysine, and methionine are three of eight
C. Combine with free radicals; counteract their destructive effects on DNA and cell membranes
D. Obsessive dieting + skewed perception of body weight
E. Vitamin C deficiency
F. Vitamin D deficiency in young children
G. Organic substance that helps enzymes to do their jobs; required in small amounts for good health
H. Feasting followed by vomiting or taking laxatives
I. Inorganic substance required for good health
J. Long chains of simple sugars; in pasta and white potatoes

Chapter Objectives/Review Questions

1. Distinguish between incomplete and complete digestive systems and tell which is characterized by (a) specialized regions, (b) two-way traffic, and (c) discontinuous feeding. [pp.726–727]
2. Define and distinguish among motility, secretion, digestion, and absorption. [p.726]
3. List all parts (in order) of the human digestive system through which food actually passes. Then list the auxiliary organs that contribute one or more substances to the digestive process. [p.728]
4. Explain how, during digestion, food is mechanically broken down. Then explain how it is chemically broken down. [p.729]
5. Tell which foods undergo digestion in each of the following parts of the human digestive system and state what the food is broken down into: mouth, stomach, small intestine, large intestine. [p.730]
6. List the enzyme(s) that act in (a) the oral cavity, (b) the stomach, and (c) the small intestine. Then tell where each enzyme was originally made. [p.730]
7. Describe how the digestion and absorption of fats differ from the digestion and absorption of carbohydrates and proteins. [pp.729–733]
8. Explain how the human body manages to meet the energy and nutritional needs of the various body parts even though the person may be feasting sometimes and fasting at other times. [p.734]
9. Describe the cross-sectional structure of the small intestine, and explain how its structure is related to its function. [pp.731–733]
10. List the items that leave the digestive system and enter the circulatory system during the process of absorption. [pp.732–733]
11. State which processes occur in the colon (large intestine). [p.735]
12. Reproduce from memory the food pyramid diagram. Identify each of the six components, give the numerical range of servings permitted from each group, and list some of the choices available. [p.736]
13. Compare the contributions of carbohydrates, proteins, and fats to human nutrition with the contributions of vitamins and minerals. [pp.736–739]
14. Construct an ideal diet for yourself for one 24-hour period. Calculate the number of calories necessary to maintain your weight [see p.740] and then use the food pyramid [p.736] to choose exactly what to eat and how much.
15. Summarize current ideas for promoting health by eating properly. [pp.736–740]
16. Summarize the daily nutritional requirements of a 25-year-old man who weighs 135 pounds, works at a desk job, and exercises very little. State what he needs in energy, carbohydrates, proteins, and lipids, and name at least six vitamins and six minerals that he needs to include in his diet every day. [pp.736–740]

17. Distinguish vitamins from minerals. [pp.738–739]
18. Name five minerals that are important in human nutrition, and state the specific role of each. [p.739]

Integrating and Applying Key Concepts

Suppose you could not eat solid food for two weeks and had only water to drink. List in correct sequential order the measures your body would take to try to preserve your life. Mention the command signals that are given as, one after another, critical points are reached, and tell which parts of the body are the first and the last to make up for the deficit.

42

THE INTERNAL ENVIRONMENT

Interactive Exercises

Tale of the Desert Rat [pp.744–745]

42.1. URINARY SYSTEM OF MAMMALS [pp.746–747]

Selected Words: "metabolic water" [p.744], *glomerular* capillaries [p.751], *peritubular* capillaries [p.751]

Boldfaced, Page-Referenced Terms

[p.745] interstitial fluid ____

____

[p.745] blood ____

____

[p.745] extracellular fluid ____

____

[p.745] urinary excretion ____

____

[p.746] urea ____

____

[p.747] urinary system ____

____

[p.747] kidneys ____

____

[p.747] urine ____

____

[p.747] ureter ____

____

[p.747] urinary bladder ____

____

[p.747] urethra ____

____

[p.747] nephrons ____

____

[p.747] Bowman's capsule ____

____

[p.747] renal corpuscle ____

____

[p.747] proximal tubule ____

____

[p.747] loop of Henle ____

____

[p.747] distal tubule ____

____

[p.747] collecting duct ____

____

Fill-in-the-Blanks

In most animals, the fluid that fills the spaces between cells and tissues is called (1) ______________ [p.745] fluid. The fluid connective tissues is called (2) ______________ [p.745], which moves substances by way of the circulatory system. Combined, interstitial fluid and blood are called (3) ______________ [p.745] fluids. In animals a well-developed (4) ______________ [p.745] system helps keep the (5) ______________ [p.745] and composition of this fluid within tolerable ranges. Other major organ systems interact with the urinary system in the performance of this (6) ______________ [p.745] task.

Choice

For questions 7–14, indicate if the process is associated with the gain or loss of solutes or water from the body. Some questions may have more than one answer. [p.746]

a. gain of solutes b. loss of solutes c. gain of water d. loss of water

7. ____ Respiration
8. ____ Metabolism
9. ____ Urinary excretion
10. ____ Sweating
11. ____ Absorption from the gut
12. ____ Cellular secretions
13. ____ Elimination in feces
14. ____ Evaporation from the lungs and skin

Fill-in-the-Blanks

The (15) ______________ [p.746] of mammals includes various wastes formed by the breakdown of (16) ______________ [p.746] compounds. (17) ______________ [p.746], a major waste, forms in the (18) ______________ [p.746] when two ammonia molecules join with (19) ______________ ______________ [p.746]. Also in the urine is (20) ______________ ______________ [p.746], formed from nucleic acid breakdown, (21) ______________ [p.746] breakdown products, drugs, and food additives.

Labeling

Identify each indicated part of the accompanying illustrations.

22. ________________ [p.746]
23. ________________ [p.746]
24. ________________ ________________ [p.746]
25. ________________ [p.746]
26. ________________ ________________ [p.747]
27. ________________ ________________ [p.747]
28. ________________ [p.747]

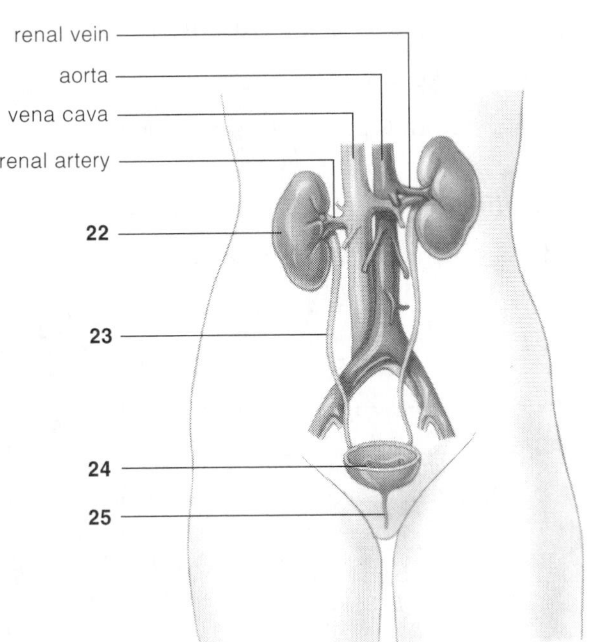

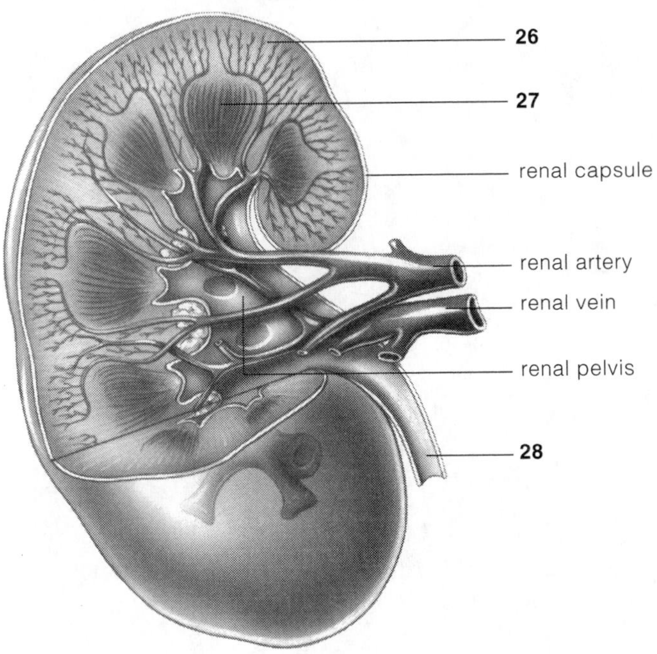

Fill-in-the-Blanks

Mammalian (29) ________________ [p.747] filter water, mineral ions, organic wastes, and other substances from the blood. They return all but about (30) ________________ [p.747] percent of the filtered material to the blood. The small portion of unclaimed water is called (31) ________________ [p.747]. This fluid flows from each kidney into a(n) (32) ________________ [p.747], which empties into a muscular sac called the (33) ________________ ________________ [p.747]. Urine is stored here before flowing into the (34) ________________ [p.747], which opens at the body's surface. Flow from the bladder is a(n) (35) ________________ [p.747] action. (36) ________________ [p.747] muscle surrounds the urethra and is under (37) ________________ [p.747] control.

Each human kidney contains over a million (38) ________________ [p.747], where water and solutes are filtered from the blood. A nephron starts at the (39) ________________ ________________ [p.747], which together with the glomular capillaries is called the (40) ________________ ________________ [p.747]. Next is a tubular region closest to the capsule, called the (41) ________________ ________________ [p.747]. Following this is a hairpin structure called the (42) ________________ ________________ ________________ [p.747] and a distal tubule. A(n) (43) ________________ ________________ [p.747] is a part of the duct system leading to the kidney's central cavity and into the (44) ________________ [p.747].

Labeling

Identify each indicated part of the accompanying illustration. [p.747]

45. ________________ ________________
46. ________________ ________________
47. ________________ ________________
48. ________________ ________________
49. ________________ ________________
    ____
50. ________________ ________________
51. ________________ ________________

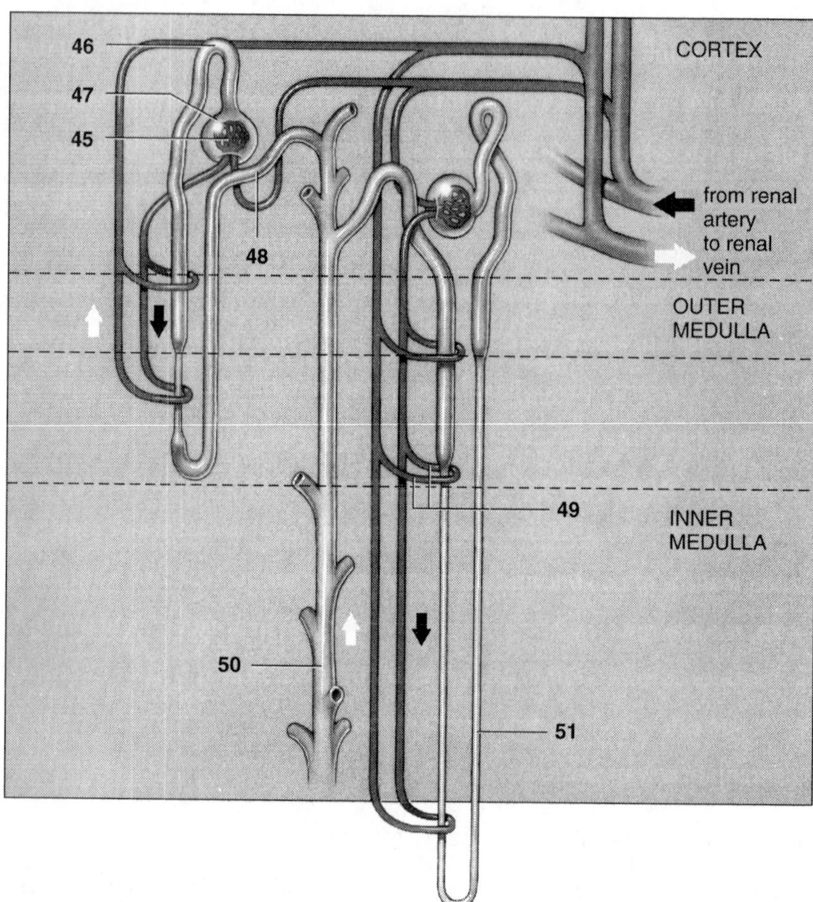

42.2. URINE FORMATION [pp.748–749]

42.3. FOCUS ON HEALTH: *When Kidneys Break Down* [p.750]

42.4. THE BODY'S ACID–BASE BALANCE [p.750]

42.5. ON FISH, FROGS, AND KANGAROO RATS [p.751]

Selected Words: *ADH* [p.748], aldosterone [p.748] *uremic toxicity* [p.750], *glomerulonephritis* [p.750], *kidney stones* [p.750], *kidney dialysis machine* [p.750], "dialysis" [p.750], *hemodialysis* [p.750], *peritoneal dialysis* [p.750], *metabolic acidosis* [p.750], *bicarbonate–carbon dioxide* buffer system [p.750]

Boldfaced, Page-Referenced Terms

[p.748] filtration ______________________________

____

[p.748] tubular reabsorption ______________________________

____

[p.748] tubular secretion ______________________________

____

[p.749] ADH ______________________________

____

[p.749] aldosterone ______________________________

____

[p.749] angiotensin II ______________________________

____

[p.749] thirst center ______________________________

____

[p.750] renal failure ______________________________

____

[p.750] acid–base balance ______________________________

____

Choice

For questions 1–5, choose from the following answers: [p.748]

a. tubular secretion b. tubular reabsorption c. filtration

1. _____ Driven by blood pressure
2. _____ Excess ions enter the nephron
3. _____ Water and solutes are reclaimed
4. _____ Involves the activity of the Bowman's capsule and glomerular capillaries
5. _____ Releases drugs, toxicants, and other metabolites such as urea

Matching

Choose the most appropriate description for each of the following. [p.749]

6. ____ angiotensin II
7. ____ thirst center
8. ____ ADH
9. ____ aldosterone
10. ____ renin

A. Splits a plasma protein in response to decreased extracellular fluid volume
B. A hormone produced by renin that influences aldosterone activity
C. Promotes reabsorption of sodium
D. Hormone released by the pituitary to control water content of the urine
E. A section of the hypothalamus that induces water-seeking behavior

Complete the Table

11. For each of the following (a–g), provide the name of the described condition of the urinary system. [p.750]

Name of Condition	*Description*
a.	Abnormal retention of metabolic wastes
b.	Accumulation of uric acid, calcium salts, and other wastes in the renal pelvis
c.	Caused by damage to or failure of the nephrons of both kidneys
d.	When antibody–antigen complexes get caught in the glomeruli, causing inflammation and tissue damage
e.	Use of a kidney dialysis machine to clean the blood of the body
f.	Kidneys are unable to secrete enough H^+ ions to meet the demands of metabolism
g.	Use of a patient's abdominal cavity to clean metabolic wastes from the body

42.6. HOW ARE CORE TEMPERATURES MAINTAINED? [pp.752–753]

42.7. TEMPERATURE REGULATION IN MAMMALS [pp.754–755]

Selected Words: *core* [p.752], *behavioral* temperature regulation [p.752], *evaporate* [p.754], "panting" [p.754], *hyperthermia* [p.754], *brown* adipose tissue [p.755], *hypothermia* [p.755], *frostbite* [p.755].

Boldfaced, Page-Referenced Terms

[p.752] core temperature ______________________________

____

[p.752] radiation ______________________________

____

[p.752] conduction ______________________________

____

[p.752] convection ____

[p.752] evaporation ____

[p.752] ectotherms ____

[p.753] endotherms ____

[p.753] heterotherms ____

[p.754] peripheral vasodilation ____

[p.754] evaporative heat loss ____

[p.755] peripheral vasoconstriction ____

[p.755] pilomotor response ____

[p.755] shivering response ____

[p.755] nonshivering heat production ____

Matching

Choose the most appropriate answer to match with the following terms.

1. ____ conduction [p.752]
2. ____ convection [p.752]
3. ____ ectotherm [p.752]
4. ____ endotherm [p.753]
5. ____ evaporation [p.752]
6. ____ heterotherm [p.753]
7. ____ radiation [p.752]

A. Body temperature determined more by heat exchange with the environment than by metabolic heat
B. Heat transfer by means of heat-bearing currents of air or water away from or toward a body
C. Body temperature determined largely by metabolic activity and by precise controls over heat produced and lost
D. Direct transfer of heat energy between two objects in direct contact with each other
E. Emission of energy in the form of infrared or other wavelengths that are converted to heat by the absorbing body
F. Body temperature fluctuating at some times and heat balance controlled at other times
G. In changing from the liquid state to the gaseous state, the energy required is supplied by the heat content of the liquid

Choice

For questions 8–17, choose from the following answers:

a. fever [pp.754–755] b. response to heat stress [p.754] c. response to cold stress [p.755] d. both b and c

8. ____ An inflammatory response to a tissue injury
9. ____ If the response is not successful, frostbite or hypothermia may occur
10. ____ May involve the nonshivering response of brown adipose tissue
11. ____ Involves peripheral vasodilation
12. ____ May be controlled by the use of anti-inflammatory drugs
13. ____ May involve rhythmic tremors of skeletal muscles to generate heat
14. ____ Involves the peripheral vasoconstriction of capillaries
15. ____ May involve behavioral responses such as panting
16. ____ If uncontrolled can result in hyperthermia
17. ____ Is partially controlled by the hypothalamus

Self-Quiz

____ 1. An entire subunit of a kidney that purifies blood and restores solute and water balance is called a ________. [p.747]
a. glomerulus
b. loop of Henle
c. nephron
d. ureter
e. none of the above

____ 2. The last portion of the excretory system passed by urine before it is eliminated from the body is the ________. [p.747]
a. renal pelvis
b. bladder
c. ureter
d. collecting ducts
e. urethra

____ 3. Filtration of the blood in the kidney takes place in the ________. [p.748]
a. loop of Henle
b. proximal tubule
c. distal tubule
d. Bowman's capsule
e. all of the above

____ 4. ________ primarily controls the concentration of water in urine. [p.749]
a. Aldosterone
b. Antidiuretic hormone
c. Angiotensin II
d. Glucagon
e. Renin

____ 5. ________ primarily controls the concentration of sodium in urine. [p.749]
a. Insulin
b. Glucagon
c. Antidiuretic hormone
d. Aldosterone
e. Epinephrine

____ 6. Hormonal control over the reabsorption or tubular secretion of sodium primarily affects ________. [pp.748–749]
a. Bowman's capsules
b. distal tubules and collecting ducts
c. proximal tubules
d. the urinary bladder
e. loops of Henle

_____ 7. During reabsorption, sodium ions cross the proximal tubule walls into the interstitial fluid principally by means of _______________. [p.749]
a. osmosis
b. countercurrent multiplication
c. bulk flow
d. active transport
e. all of the above

_____ 8. In humans, the thirst center is located in the _______________. [p.749]
a. adrenal cortex
b. thymus
c. heart
d. adrenal medulla
e. hypothalamus

_____ 9. Which of the following represents the process by which there is heat transfer between an animal and another object of differing temperature that is in direct contact with it? [p.752]
a. evaporation
b. conduction
c. radiation
d. convection

_____ 10. Which of the following terms represents a group of mammals whose high metabolic rates keep them active under a wide variety of temperature ranges? [pp.752–753]
a. endotherms
b. ectotherms
c. heterotherms
d. none of the above

_____ 11. Which of the following is a response to heat stress? [p.754]
a. panting
b. peripheral vasodilation
c. evaporative heat loss
d. all of the above

Chapter Objectives/Review Questions

1. Distinguish between blood, interstitial fluid, and extracellular fluid. [p.745]
2. List the factors that contribute to the gain and loss of water in mammals. [p.746]
3. List the factors that contribute to the gain and loss of solutes in mammals. [p.746]
4. Distinguish between urea and urine. [pp.746–747]
5. Be able to identify the major components of the urinary system from a diagram. [pp.746–747]
6. Trace the flow of fluid through a human nephron, listing the major structures involved. [p.747]
7. Locate the processes of filtration, reabsorption, and tubular secretion along a nephron, and tell what makes each process happen. [pp.748–749]
8. Give the functions of aldosterone, ADH, and angiotensin II in regulating the activity of the urinary system. [p.749]
9. Understand the interaction of the hypothalamus with the urinary system. [p.749]
10. List two kidney disorders and explain what can be done if kidneys become too diseased to work properly. [p.750]
11. Describe the role of the kidney in maintaining the pH of the extracellular fluids between 7.37 and 7.43. [p.750]
12. Distinguish between evaporation, radiation, conduction, and convection as methods of heat transfer. [p.752]
13. Explain how endotherms, ectotherms, and heterotherms regulate body temperature and give an example of each. [pp.752–753]
14. Explain the mechanisms by which a mammal may respond to heat stress. [p.754]
15. Explain the mechanisms by which a mammal may respond to cold stress. [p.755]

Crossword Puzzle — Urinary Systems

1 2 3 4 5 6 7 8 9 10 11 12 13 14 15 16 17 18 19 20 21 22 23 24 25 26 27

Across

1. Connects urinary bladder to exterior [p.747]
4. Enhances sodium reabsorption and is produced by the adrenal cortex [p.749]
6. Extracellular ________ bathes the body's cells [p.745]
7. ________ acid is the least toxic nitrogenous waste and is constructed from nucleic acid breakdown [p.746]
8. Bowman's ________ encloses the glomerulus [p.747]
10. An enzyme secreted by kidney cells that detaches part of a protein circulating in the blood so that the new protein can be made into a hormone that acts on the adrenal cortex [p.749]
11. A cluster of capillaries enclosed by a Bowman's capsule that filters blood [pp.747–748]
13. Connects a kidney to the urinary bladder [p.747]
14. An organ that adjusts the volume and composition of blood and helps maintain the composition of extracellular fluid [p.747]
16. A small tube [p.747]
17. Relating to kidney function [p.750]
19. of Henle [p.747]
20. Nitrogenous waste formed from an amino group [p.746]
21. Nitrogenous waste formed in the liver and relatively harmless [p.746]
22. Water and solutes move out of the nephron tubule, then into adjacent capillaries [p.748]
24. The urinary ________ includes three organs and three larger waste-containing tubes [p.747]
25. More than a million of these units are packed inside each fist-sized kidney [p.747]
26. The elimination of fluid wastes [p.746]
27. The ________ tubule is the part of the nephron that is farthest from the Bowman's capsule [p.747]

Down

1. Product of kidneys [p.747]
2. ________ fluid bathes the body's cells [p.745]
3. Loop of ________ [p.747]
4. ________ glands are perched on top of the kidneys [p.746]
5. Moves excess H^+ and a few other substances by active transport from the capillaries into the cells of the nephron wall and then into the urine [p.748]
6. Water and small-molecule solutes are forced from the blood into the Bowman's capsule [p.748]
9. Urinary ________ stores urine [pp.746–747]
12. The middle region of the kidney [p.747]
13. ________ excretion is a process that dumps excess mineral ions and metabolic wastes [p.748]
15. The ________ tubule is nearest to its Bowman's capsule [p.747]
18. The outer region of kidney, adrenal gland, or brain [p.747]
23. The "water conservation" hormone produced by the posterior pituitary [p.749]

Integrating and Applying Key Concepts

The hemodialysis machine used in hospitals is expensive and time consuming. So far, artificial kidneys that would allow people who have nonfunctional kidneys to purify their blood by themselves, without having to go to a hospital or clinic, have not been developed. Which aspects of the hemodialysis procedure do you think have presented the greatest problems in development of a method of home self-care? If you had an unlimited budget and were appointed head of a team to develop such a procedure and its instrumentation, what strategy would you pursue?

43

PRINCIPLES OF REPRODUCTION AND DEVELOPMENT

Interactive Exercises

From Frog to Frog and Other Mysteries [pp.758–759]

43.1. THE BEGINNING: REPRODUCTIVE MODES [pp.760–761]

43.2. STAGES OF DEVELOPMENT — AN OVERVIEW [pp.762–763]

Selected Words: *reproductive timing* [p.760], *oviparous* [p.761], hermaphrodite [p.761], *ovoviviparous* [p.761], *viviparous* [p.761], internal fertilization [p.761]

Boldfaced, Page-Referenced Terms

[p.758] zygotes ______________________________

____

[p.760] sexual reproduction ______________________________

____

[p.760] asexual reproduction ______________________________

____

[p.761] yolk ____

____

[p.762] embryos ____

____

[p.762] gamete formation ____

____

[p.762] fertilization ____

____

[p.762] cleavage ____

____

[p.762] blastomeres ____

____

[p.762] gastrulation ____

____

[p.762] ectoderm ____

____

[p.762] endoderm ____

____

[p.762] mesoderm ____

____

[p.762] organ formation ____

____

[p.762] growth and tissue specialization ____

____

Fill-in-the-Blanks

New sponges budding from parent sponges and a flatworm dividing into two flatworms represent examples of (1) ______________ [p.760] reproduction. This type of reproduction is useful when gene-encoded traits are strongly adapted to a limited set of (2) ______________ [p.760] conditions. Separation into male and female sexes requires special reproductive structures, control mechanisms, and behaviors; this cost is offset by a selective advantage: (3) ______________ [p.760] in traits among the offspring. Males and females of the same species must synchronize their (4) ______________ ______________ [p.760] to ensure that their gametes mature and find each other at the correct time for fertilization to occur. (5) ______________ [p.761] animals are nourished by maternal tissues, not only yolk, until the time of birth.

Copperheads are (6) ______________ [p.761]: Fertilization is internal; the fertilized eggs develop inside the mother's body without additional nourishment, and the young are born live. Birds are

(7) ______________ [p.761]: Eggs with large yolk reserves are released from and develop outside the mother's body.

(8) ______________ ______________ [p.762] is considered the first stage of animal development. Rich stores of substances such as yolk become assembled in localized regions of the (9) ______________ [p.761] cytoplasm. When sperm and egg unite and their DNA mingles and is reorganized, the process is referred to as (10) ______________ [p.762]. At the end of fertilization, a(n) (11) ______________ [p.762] is formed. (12) ______________ [p.762] includes the repeated mitotic divisions of a zygote that segregate the egg cytoplasm into a cluster of cells known as (13) ______________ [p.762]; the entire cluster is known as a blastula. (14) ______________ [p.762] is the process that arranges cells into three primary tissue layers. Ectoderm will eventually give rise to skin epidermis and the (15) ______________ [p.762] system. Endoderm forms the inner lining of the (16) ______________ [p.762] and associated digestive glands. Mesoderm forms the circulatory system, the (17) ______________ [p.762], the muscles, and connective tissues. Each stage of (18) ______________ [p.762] development builds on structures that were formed during the stage preceding it. (19) ______________ [p.763] cannot proceed properly unless each stage is successfully completed before the next begins.

True/False

If the statement is true, write a "T" in the blank. If false, explain why.

____ 20. Yolk is a substance rich in carbohydrates that nourishes embryonic stages. [p.761] ______________

____ 21. Most animals reproduce sexually. [p.760] ______________

____ 22. Sexual reproduction is less advantageous in predictable environments; asexual reproduction is more advantageous in predictable environments. [p.760] ______________

____ 23. The eggs of leopard frogs (*Rana pipiens*) are fertilized externally in the water. [p.758] ______________

____ 24. Gastrulation precedes organ formation. [p.762] ______________

Sequence

Arrange the following events in correct chronological sequence. Write the letter of the first step next to 25, the letter of the second step next to 26, and so on. [all from p.762]

25. ____
26. ____
27. ____
28. ____
29. ____
30. ____

A. Gastrulation
B. Fertilization
C. Cleavage
D. Growth, tissue specialization
E. Organ formation
F. Gamete formation

Complete the Table

31. Complete the following table by entering the name of the germ layer (ectoderm, mesoderm, or endoderm) that forms the tissues and organs listed. [all from p.762]

Tissues/Organs	*Germ Layer* (=*primary* tissue layer)
Muscle, circulatory organs	a.
Nervous tissues	b.
Inner lining of the gut	c.
Circulatory organs (blood vessels, heart)	d.
Outer layer of the integument	e.
Reproductive and excretory organs	f.
Organs derived from the gut	g.
Most of the skeleton	h.
Connective tissues of the gut and integument	i.

43.3. EARLY MARCHING ORDERS [pp.764–765]

43.4. HOW DO SPECIALIZED TISSUES AND ORGANS FORM? [pp.766–767]

Selected Words: "maternal messages" [p.764], cleavage furrow [p.764], *vegetal* pole [p.765], *animal* pole [p.765], *radial* cleavage [p.765], morula [p.765], *incomplete* cleavage [p.765], *rotational* cleavage [p.765], inner cell mass [p.765], *identical twins* [p.765], *fraternal twins* [p.765], crystallins [p.766]

Boldfaced, Page-Referenced Terms

[p.764] oocyte ______________________________

[p.764] sperm ______________________________

[p.764] gray crescent ______________________________

[p.765] cytoplasmic localization ______________________________

[p.765] blastula ______________________________

[p.765] blastocyst ______________________________

[p.766] neural tube ______________________________

[p.766] cell differentiation ______________________________

[p.767] morphogenesis ______________________________

True/False

If the statement is true, write a "T" in the blank. If statement 1, 2, or 3 is false, explain why; if 4 or 5 is false because of the underlined term, write the correct term in the blank.

_____ 1. During gastrulation, maternal controls over gene activity are activated and begin the process of differentiation in each cell's nucleus. [pp.766–767] ______________________________

_____ 2. In the zygote, different cells have different genes. [p.766] ______________________________

_____ 3. In a developing chick embryo, the heart begins to beat at some time between 30 and 36 hours after fertilization. [p.766] ______________________________

_____ 4. <u>Sperm penetration</u> into the cytoplasm of the egg brings about specific structural changes and chemical reactions. [p.764] ______________________________

_____ 5. Body parts become folded, tubes become hollowed out, and eyelids, lips, noses, and ears all become slit or perforated by <u>apoptosis.</u> [p.767] ______________________________

Fill-in-the-Blanks

In amphibian eggs, sperm penetration on one side of an egg causes pigment granules on the opposite side of the egg to flow toward the point of (6) ______________ ______________ [p.764]. A lightly pigmented area near the frog egg's midsection called the (7) ______________ ______________ [p.764] results. It is a visible marker of the site where the (8) ______________ ______________ [p.764] will be established and where gastrulation will begin.

The third stage of animal development, (9) ______________ [recall pp.762,764], is characterized by the subdividing and compartmentalizing of the zygote; no growth occurs at this stage, and usually a hollow ball of cells, the (10) ______________ [p.765], is formed. Much of the information that determines how structures will be spatially organized in the embryo begins with the distribution of (11) ______________ ______________ [pp.764–765] in the oocyte. These consist of mRNAs, regionally distributed enzymes and other proteins, yolk, and other factors. As cleavage membranes divide up the cytoplasm, various cytoplasmic determinants become localized in different daughter cells; this process, called (12) ______________ ______________ [p.765], helps seal the developmental fate of the descendants of those cells.

The fourth stage, (13) ______________ [recall pp.762,766], is concerned with the formation of ectoderm, mesoderm, and endoderm, the (14) ______________ [p.766] tissue layers of the embryo; at the end of this stage, the (15) ______________ [p.766] is formed, which resembles an early embryo. Cleavage of a frog's zygote is complete, because there is so little yolk that cleavage membranes can subdivide the entire cytoplasmic mass; in the chick, however, there is so much yolk that cleavage membranes cannot subdivide the entire mass. Cleavage is therefore said to be incomplete, and the chick grows from an embryonic primitive streak floating on the surface of the (16) ______________ [p.765] mass into a chick embryo complete with wing and leg buds and beating heart during the first (17) ______________ [p.766] days.

Through (18) ______________ ______________ [p.766], a single fertilized egg gives rise to an assortment of different types of specialized cells; these differentiated cells have the same number and same kinds of (19) ______________ [p.766] because they are all descended by mitosis from the same zygote. However, from gastrulation onward, certain genes are (20) ______________ [p.766] in some cells but not in others.

In every vertebrate, an imaginary straight line connecting the head to the tail end defines where a(n) (21) ______________ ______________ [p.766], the forerunner of the brain and spinal cord, will form. All organs of the adult begin formation by two principal processes. (22) ______________ ______________ [p.766], in which a cell selectively activates specific genes (and not others) and synthesizes some proteins not found in other cell types, lays the groundwork for (23) ______________ [p.767], which is a program of orderly changes in an embryo's size, shape, and proportions that results in tissues becoming specialized to function in a specific way and form the early stages of organs. (24) ______________ [p.767] (from root words that mean "giving rise to shape") creates the specialized tissues and early organs characteristic of that species.

During (25) ________ ________ ________ [p.767], cells send out pseudopods and use them to move themselves along prescribed routes; forerunners of neurons interconnect in this way as a nervous system is forming. (26) ________ [p.767] cues tell the cells when to stop migration. As (27) ________ [p.767] lengthen and rings of (28) ________ [p.767] in cells constrict, sheets of cells expand and fold inward and outward, as in neural tube formation. Sometimes, as in the formation of the human hand, specific cells in the early form of the organ die on schedule according to cues in their genetic programs; this form of programmed cell death, called (29) ________ [p.767], rids the developing embryo of cells not needed in the next developmental phase. Sometimes entire organs (such as testes in human males) change position in the developing organism; but the inward or outward folding of (30) ________ ________ [p.767] is seen more often.

43.5. PATTERN FORMATION [pp.768–769]

43.6. FOCUS ON SCIENCE: *To Know a Fly* [pp.770–771]

43.7. WHY DO ANIMALS AGE? [p.772]

43.8. CONNECTIONS: *Death in the Open* [p.773]

Selected Words: dorsal lip [p.768], cytoplasmic localization [p.768], *physical, architectural,* and *phyletic* constraints [p.769], superficial cleavage pattern [p.770], *maternal effect* genes [p.771], *gap* genes [p.771], *pair-rule* genes [p.771], *segment polarity* genes [p.771], telomerase [p.772], *Werner's syndrome* [p.772]

Boldfaced, Page-Referenced Terms

[p.768] embryonic induction ________________

[p.768] pattern formation ________________

[p.768] morphogens ________________

[p.768] theory of pattern formation ________________

[p.768] master genes ________________

[p.768] homeotic genes ________________

[p.771] fate map ________________

[p.772] telomeres ________________

Fill-in-the-Blanks

(1) ______________ ______________ [p.768] consists of several processes that transform a gastrula (with its three germ layers) into an embryo with established developmental axes and organ rudiments (undeveloped lumps of tissue) in place and recognizable. As the embryo develops, one group of cells may produce a substance (say, a growth factor) that diffuses to another group of cells and turns on protein synthesis in those cells. Such interaction among embryonic cells is called (2) ______________ ______________ [p.768].

Spemann demonstrated that the process known as (3) ______________ ______________ [p.768] occurs in salamander embryos, where one body part differentiates because of signals it receives from an adjacent body part. (4) ______________ [p.768] are slowly degradable proteins that form concentration gradients as they diffuse from an inducing tissue into adjoining tissues. These chemicals then activate (5) ______________ [p.768] genes. Mutations in these genes, which cause them to malfunction, can lead to (6) ______________ [p.769]. Many of the studies of these genes have used (7) ______________ [pp.770–772] as their subject. Normal cells cap the ends of chromosomes with repetitive DNA sequences called (8) ______________ [p.772]; a bit of each is lost during each nuclear division, and when only a nub remains, cells stop dividing and eventually die.

Matching

In *Drosophila,* specific types of genes act as master organizers that establish the location, shape, and size of various body parts in the animal during pattern formation. Match the correct gene with its function. [all from p.771]

9. ____ gap genes
10. ____ homeotic genes
11. ____ maternal effect genes
12. ____ pair-rule genes
13. ____ segment polarity genes

A. Divide the embryo into segment-size units
B. Produces substances that accumulate in bands each of which corresponds to two body segments
C. Map out broad regions of the body; different concentrations of these switch on pair-rule genes
D. Directly and collectively govern the developmental fate of each body segment
E. Transcribe specific mRNAs and indirectly cause specific regulatory proteins to be translated; these products become localized in different parts of the egg cytoplasm and are activated in the zygote, where they activate or suppress gap genes.

True/False

If the statement is true, write a "T" in the blank. If the underlined term makes the statement false, write the correct term in the blank.

______________ 14. The aging and death of a cell may be coded in large part in its DNA; <u>external</u> signals activate these DNA messages and tell the cell that it is time to die. [p.772]

______________ 15. A process of predictable cellular deterioration is built into the life cycle of all organisms which consist of <u>differentiated cells that show considerable specialization.</u> [p.772]

Self-Quiz

_____ 1. Animals such as birds lay eggs with large amounts of yolk; embryonic development takes place within the egg covering outside the mother's body. Birds are said to be _______________ . [p.761]
   a. ovoviviparous
   b. viviparous
   c. oviparous
   d. parthenogenetic
   e. none of the above

_____ 2. The process of cleavage most commonly produces a(n) _______________ . [pp.762,765]
   a. zygote
   b. blastula
   c. gastrula
   d. third germ layer
   e. organ

_____ 3. Incomplete cleavage of cells at the yolk periphery is characteristic of the cleavage pattern of _______________ . [p.765]
   a. frogs
   b. sea urchins
   c. chickens
   d. humans
   e. none of the above

_____ 4. The formation of three germ (embryonic) tissue layers occurs during _______________ . [p.762]
   a. gastrulation
   b. cleavage
   c. pattern formation
   d. morphogenesis
   e. neural plate formation

_____ 5. The differentiation of a body part in response to signals from an adjacent body part is _______________ . [p.768]
   a. contact inhibition
   b. ooplasmic localization
   c. embryonic induction
   d. pattern formation
   e. none of the above

_____ 6. A homeotic mutation _______________ . [pp.768–769]
   a. may cause a leg to develop on the head where an antenna should grow
   b. may affect pattern formation
   c. affects morphogenesis
   d. may alter the path of development
   e. all of the above

_____ 7. Shortly after fertilization, the zygote is subdivided into a multicelled embryo during a process known as _______________ . [p.762]
   a. meiosis
   b. parthenogenesis
   c. embryonic induction
   d. cleavage
   e. invagination

_____ 8. Muscles differentiate from _______________ tissue. [p.762]
   a. ectoderm
   b. mesoderm
   c. endoderm
   d. parthenogenetic
   e. yolky

_____ 9. The gray crescent is _______________ . [p.764]
   a. formed where the sperm penetrates the egg
   b. part of only one blastomere after the first cleavage
   c. the yolky region of the egg
   d. where the first mitotic division begins
   e. formed opposite from where the sperm enters the egg

_____ 10. The nervous system differentiates from _______________ tissue. [p.762]
   a. ectoderm
   b. mesoderm
   c. endoderm
   d. yolky
   e. homeotic

Chapter Objectives/Review Questions

1. Explain how a spherical zygote becomes a multicellular adult with arms and legs. [pp.767–769]
2. Describe how asexual reproduction differs from sexual reproduction. Know the advantages and problems associated with having separate sexes. [pp.760–761]
3. Define the terms *oviparous, viviparous,* and *ovoviviparous.* For each of the three developmental strategies, cite an example of an animal that goes through it. [p.761]
4. Explain why evolutionary trends in many groups of organisms tend toward developing more complex, sexual strategies rather than toward retaining simpler, asexual strategies. [pp.760–761]
5. Explain how the amount of yolk in an ovum can influence an animal's cleavage pattern. [p.765]
6. Name each of the three embryonic tissue layers and the organs formed from it. [p.762]
7. Describe early embryonic development and distinguish among the following: gamete formation, fertilization, cleavage, gastrulation, organ formation, and tissue specialization. [p.762]
8. Compare the early stages of frog and chick development (see Figs. 43.4, 43.5, and 43.9 in the text) with respect to egg size and type of cleavage pattern (incomplete or complete). [pp.762–766]
9. Explain what causes polarity to occur during oocyte maturation in the mother, and how polarity influences later development. [pp.764–765]
10. Define the term *gastrulation* and tell what process begins at this stage that did not take place during cleavage. [p.766]
11. Define the term *differentiation* and give two examples of cells in a multicellular organism that have undergone differentiation. [pp.766–767]

Integrating and Applying Key Concepts

If embryonic induction did not occur in a human embryo, how would the eye region appear? What would happen to the forebrain and epidermis? If controlled cell death did not take place in a human embryo, how would its hands appear? Its face?

44

HUMAN REPRODUCTION AND DEVELOPMENT

Interactive Exercises

Sex and the Mammalian Heritage [pp.776–777]

44.1. REPRODUCTIVE SYSTEM OF HUMAN MALES [pp.778–779]

44.2. MALE REPRODUCTIVE FUNCTION [pp.780–781]

Selected Words: gametes [p.777], gonads [p.777], puberty [p.778], scrotum [p.778], epididymis [p.778], vasa deferentia (singular, vas deferens) [p.778], ejaculatory ducts [p.778], urethra [p.778], penis [p.778], seminal vesicles [p.778], prostate gland [p.779], bulbourethral glands [p.779], *prostate cancer* [p.779], *testicular cancer* [p.779], PSA (prostate-specific antigen) test [p.779]

Boldfaced, Page-Referenced Terms

[p.777] testes (singular, testis) ______________________________

____

[p.777] ovaries (singular, ovary) ______________________________

____

[p.777] secondary sexual traits ______________________________

____

[p.778] seminiferous tubules ______________________________

____

[p.778] semen ______________________________

____

[p.780] Sertoli cells ______________________________

____

[p.780] Leydig cells ______________________________

____

[p.780] testosterone ______________________________

____

[p.780] LH, luteinizing hormone ______________________________

____

[p.780] FSH, follicle-stimulating hormone ______________________________

____

[p.781] GnRH ______________________________

____

Fill-in-the-Blanks

The numbered items in the illustrations that follow represent missing information; fill in the corresponding numbered blanks to complete the narrative below. Some illustrated structures are numbered more than once to aid identification.

Within each testis and after repeated (1) ______________ [p.780] divisions of undifferentiated diploid cells just inside the (2) ______________ [p.780] tubule walls, (3) ______________ [p.780] occurs to form haploid, eventually mature (4) ______________ [p.780]. Males produce sperm continuously from puberty onward. Sperm leaving a testis enter a long, coiled duct, the (5) ______________ [p.778]; the sperm are stored in the last portion of this organ. When a male is sexually aroused, muscle contractions quickly propel the sperm through a thick-walled tube, the (6) ______________ ______________ [p.778], then to ejaculatory ducts, and finally to the (7) ______________ [p.778], which opens at the tip of the penis. During the trip to the urethra, glandular secretions become mixed with the sperm to form semen.

(8) ______________ ______________ [p.778] secrete fructose to nourish the sperm and prostaglandins in order to induce contractions in the female reproductive tract. (9) ______________ ______________ [p.779] secretions help neutralize vaginal acids. (10) ______________ [p.779] glands secrete mucus to lubricate the penis, aid vaginal penetration, and improve sperm motility.

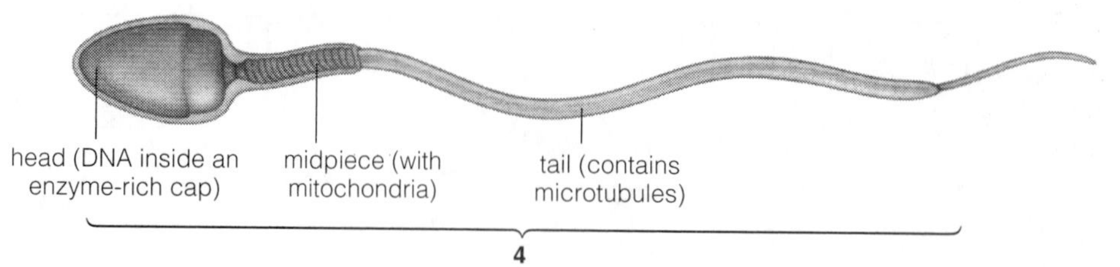

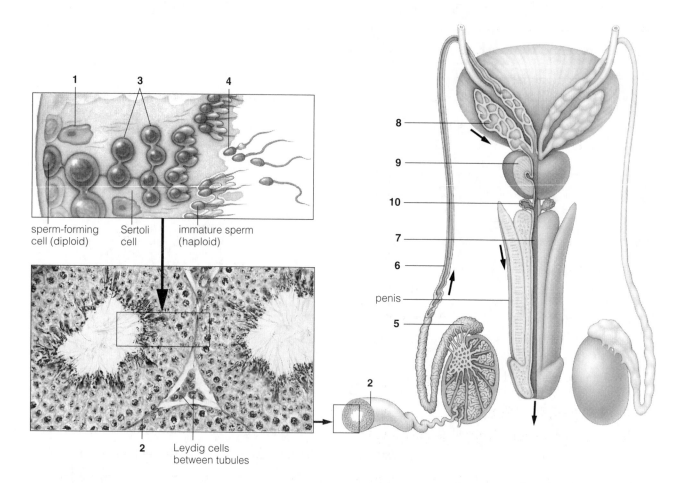

1
3
4
sperm-forming cell (diploid)
Sertoli cell
immature sperm (haploid)
2
Leydig cells between tubules
8
9
10
7
6
penis
5
2

Dichotomous Choice

Circle one of two possible answers given between parentheses in each statement.

11. Testosterone is secreted by (Leydig/hypothalamus) cells. [p.780]
12. (Testosterone/FSH) governs the growth, form, and functions of the male reproductive tract. [p.780]
13. Sexual behavior, aggressive behavior, and secondary sexual traits are associated with (LH/testosterone). [p.780]

14. LH and FSH are secreted by the (anterior/posterior) lobe of the pituitary gland. [p.780]
15. The (testes/hypothalamus) govern(s) sperm production by controlling interactions among testosterone, LH, and FSH. [p.781]
16. When blood levels of testosterone (increase/decrease), the hypothalamus stimulates the pituitary to release LH and FSH, which travel the bloodstream to the testes. [p.781]
17. Within the testes, (LH/FSH) acts on Leydig cells; they secrete testosterone, which enters the sperm-forming tubes. [p.781]
18. FSH enters the sperm-forming tubes and diffuses into (Sertoli/Leydig) cells to improve testosterone uptake. [p.781, Fig. 44.5]
19. When blood testosterone levels (increase/decrease) past a set point, negative feedback loops to the hypothalamus slow down testosterone secretion. [p.781]

Labeling [all from p.781]

20. ________________
21. ________________ ________________
22. ________________ ________________
23. ________________ ________________

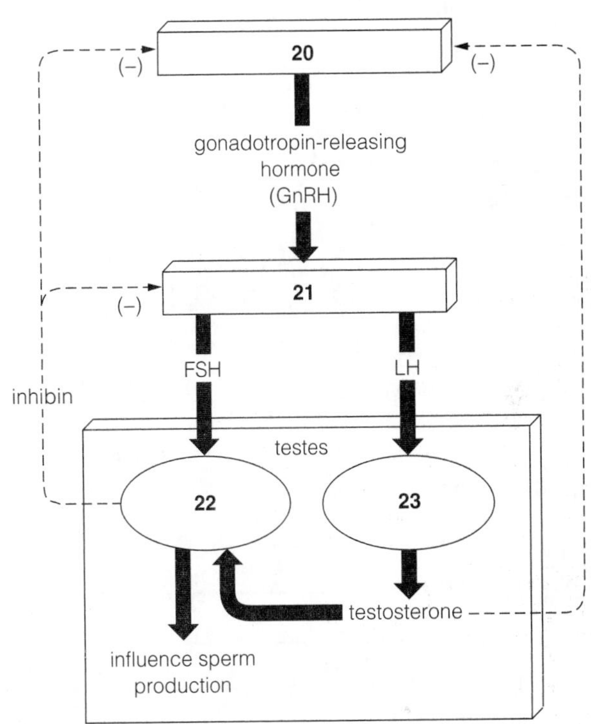

44.3. REPRODUCTIVE SYSTEM OF HUMAN FEMALES [pp.782–783]

44.4. FEMALE REPRODUCTIVE FUNCTION [pp.784–785]

44.5. VISUAL SUMMARY OF THE MENSTRUAL CYCLE [p.786]

Selected Words: oviducts [p.782], myometrium [p.782], cervix [p.782], vagina [p.782], vulva [p.782], labia majora [p.782], labia minora [p.782], clitoris [p.782], *estrous* cycle [p.782], menstruation [p.782], *follicular* phase [p.782], luteal phase [p.782], menopause [p.783], *endometriosis* [p.783], granulosa cells [p.784]

Boldfaced, Page-Referenced Terms

[p.782] oocytes ____

____

[p.782] uterus ____

____

[p.782] endometrium ____

____

[p.782] menstrual cycle ____

____

[p.782] ovulation ____

____

[p.783] estrogens ____

____

[p.783] progesterone ____

____

[p.784] follicle ____

____

[p.784] zona pellucida ____

____

[p.784] secondary oocyte ____

____

[p.784] polar bodies ____

____

[p.785] corpus luteum ____

____

[p.785] blastocyst ____

____

Fill-in-the-Blanks

The numbered items in the illustrations that follow represent missing information; fill in the corresponding numbered blanks to complete the narrative below. Some illustrated structures are numbered more than once to aid identification. [pp.782–783]

An immature egg (oocyte) is released from one (1) ______________ of a pair. From each ovary, a(n) (2) ______________ forms a channel for transport of the immature egg to the (3) ______________, a hollow, pear-shaped organ where the embryo grows and develops. The lower, narrowed part of the uterus is the (4) ______________. The uterus has a thick layer of smooth muscle, the (5) ______________, lined inside with connective tissue, glands, and blood vessels; this lining is called the (6) ______________. The

(7) ______________, a muscular tube, extends from the cervix to the body surface; this tube receives sperm and functions as part of the birth canal. At the body surface are external genitals (vulva) that include organs for sexual stimulation. Outermost is a pair of fat-padded skin folds, the (8) ______________ ______________. Those folds enclose a smaller pair of skin folds, the (9) ______________ ______________. The smaller folds partly enclose the (10) ______________, an organ sensitive to stimulation. The location of the opening to the (11) ______________ is about midway between the clitoris and the vaginal opening.

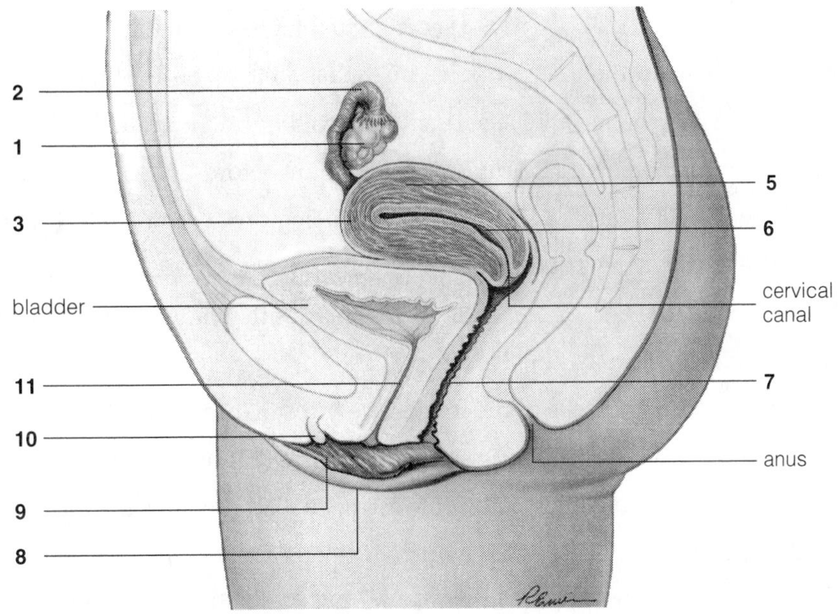

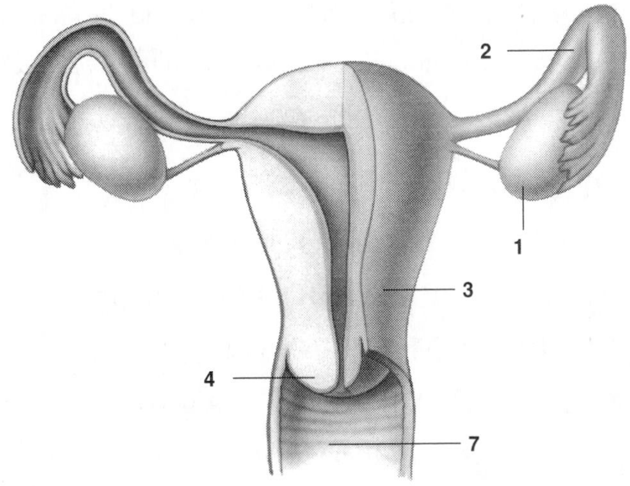

Fill-in-the-Blanks

(12) ______________ [p.784] occurs in the ovaries, so that a normal female infant has about 2 million primary oocytes, with the division process halted in the first stage of meiosis. By age seven, only about (13) ______________ [p.784] remain. A primary oocyte surrounded by a nourishing layer of granulosa cells is called a(n) (14) ______________ [p.784]. Between the oocyte and the granulosa cells, a protective layer, the (15) ______________ ______________, forms.

When a female enters puberty, the (16) ______________ [p.784] secretes a hormone (GnRH) that makes the (17) ______________ ______________ [p.784] secrete the follicle-stimulating hormone (FSH) and luteinizing hormone (LH). These hormones are carried by the blood to all parts of the body, but each month, one or more follicles respond by growing and secreting the steroid hormones called (18) ______________ [p.784]. The primary oocyte completes the (12) division 8–10 hours before (19) ______________ [p.784] occurs in response to a surge of (20) ______________ [p.784] being produced by the (17) on day 12, 13, or 14 of a 28-day cycle. The first five days of the cycle are occupied by (21) ______________ [p.786], the deterioration and expulsion of (22) ______________ [p.785] tissues that line the uterus. During the next week, (18) stimulate new (22) tissues to be constructed.

Following ovulation, the granulosa cells of the follicle are transformed into a(n) (23) ______________ ______________ [pp.785–786] by the midcycle surge of LH. The (23) secretes its key hormone, (24) ______________ [pp.785–786], together with some estrogen. (24) prepares the reproductive tract for the arrival of the (25) ______________ [p.785], which develops after an egg is fertilized. (24) also maintains the (26) ______________ [p.785] during pregnancy. If a(n) (25) does not burrow into the (26), the (23) self-destructs after approximately 12 days by secreting prostaglandins that disrupt its own functioning. After this, (24) and estrogen levels in the blood decline rapidly, and the endometrial tissues die and disintegrate. The sloughing tissue together with released blood form the (27) ______________ ______________ [p.785].

44.6. PREGNANCY HAPPENS [p.787]

44.7. FORMATION OF THE EARLY EMBRYO [pp.788–789]

44.8. EMERGENCE OF THE VERTEBRATE BODY PLAN [p.790]

44.9. WHY IS THE PLACENTA SO IMPORTANT? [p.791]

44.10. EMERGENCE OF DISTINCTLY HUMAN FEATURES [pp.792–793]

44.11. FOCUS ON HEALTH: *Mother as Provider, Protector, Potential Threat* [pp.794–795]

Selected Words: *erection* [p.787], *ejaculation* [p.787], spongy tissue [p.787], glans penis [p.787], *orgasm* [p.787], *embryonic* period [p.788], *fetal* period [p.788], *first, second, third* trimesters [p.788], *amniotic* cavity [p.789], *pregnancy tests* [p.789], "primitive streak" [p.790], embryonic inductions [p.790], neural tube [p.790], notochord [p.790], pharyngeal arches [p.790], chorionic villi [p.791], umbilical cord [p.791], folate (folic acid) [p.794], *teratogens* [p.794], *rubella* [p.795], *thalidomide* [p.795], *anti-acne drugs* [p.795], *fetal alcohol syndrome* (FAS) [p.795], *secondhand smoke* [p.795]

Boldfaced, Page-Referenced Terms

[p.787] coitus ______________________________

[p.787] ovum (plural, ova) ______________________________

[p.788] fetus ______________________________

[p.788] implantation ______________________________

[p.789] amnion ______________________________

[p.789] yolk sac ______________________________

[p.789] chorion ______________________________

[p.789] allantois ______________________________

[p.789] HCG, human chorionic gonadotropin ______________________________

[p.790] gastrulation ______________________________

[p.790] somites ______________________________

[p.791] placenta ______________________________

Fill-in-the-Blanks

Fertilization generally takes place in the (1) ______________ [pp.787–788, Fig. 44.10]; six or seven days after conception, (2) ______________ [p.788] begins as the blastocyst sinks into the endometrium. Extensions from the chorion fuse with the endometrium of the uterus to form a(n) (3) ______________ [p.789], the organ of interchange between mother and fetus. By the end of the (4) ______________ [pp.788,792] week, all major organs have formed; the offspring is now referred to as a(n) (5) ______________ [pp.788,792].

Labeling

Identify each numbered part of the accompanying illustrations.

6. ________________ ________________ [p.789]
7. ________________ ________________ [p.789]
8. ________________ [p.792]
9. ________________ ________________ [p.792]
10. ________________ [p.792]
11. ________________ [p.799]
12. ________________ [p.792]

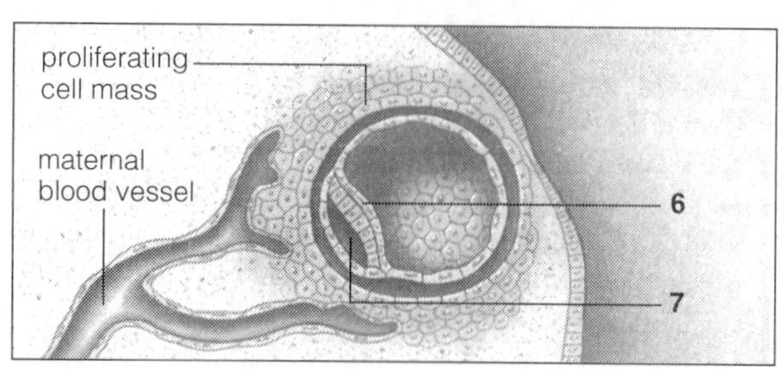

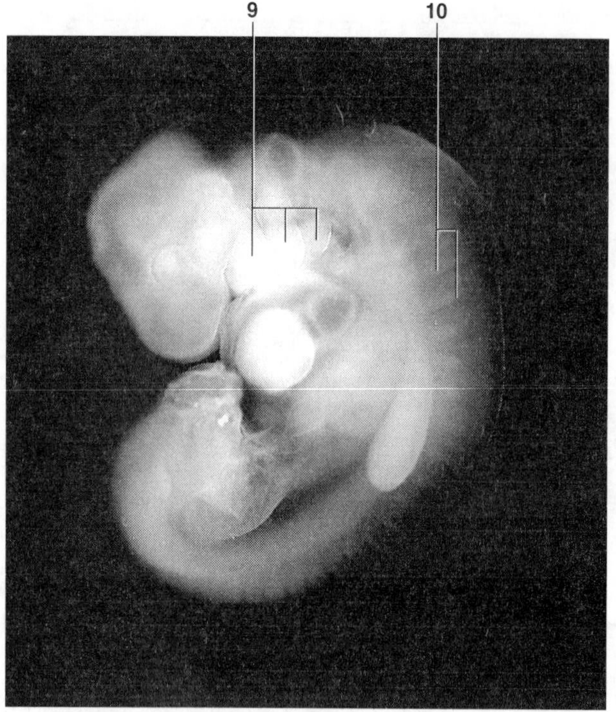

A human embryo at (**8**) weeks after conception.

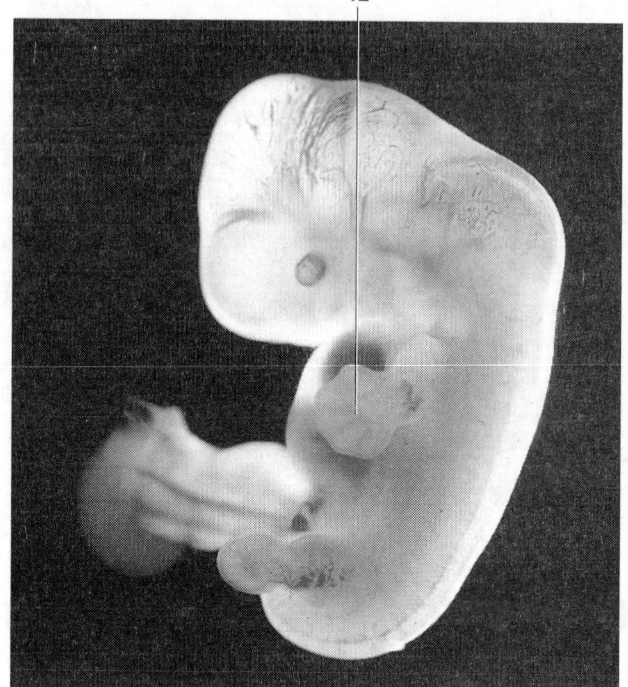

A human embryo at (**11**) weeks after conception.

Matching

Match each of the four extra-embryonic membranes with its function. [all from p.789]

13. _____ allantois
14. _____ amnion
15. _____ chorion
16. _____ yolk sac

A. Outermost membrane; will become part of the spongy, blood-engorged placenta
B. Directly encloses the embryo and cradles it in a buoyant protective fluid
C. In humans, some of this membrane becomes a site for blood cell formation, some will become the forerunner of gametes
D. In humans, this membrane serves in early blood formation and formation of the urinary bladder

Fill-in-the-Blanks

At-home human pregnancy tests use a treated "dip-stick" that changes color when (17) ______________ ______________ ______________ [p.789] secreted by the blastocyst is present in the mother's urine. (17) also stimulates the (18) ______________ ______________ [p.789] to continue secreting progesterone and estrogen. By the time a woman has missed her first menstrual period, generally during the third week after conception, cleavage is completed and (19) ______________ [p.790] is well under way as the three primary tissue layers (ectoderm, mesoderm, and endoderm) are forming. On day 15, a faint band, the (20) ______________ ______________ [p.790], appears around a depression along the axis of the embryonic disk; it marks the beginning of (19) in some vertebrate embryos. During days 18–23, morphogenetic processes cause neural folds to arch upward and merge to form the (21) ______________ ______________ [p.790]. (22) ______________ [p.790] appear that will later form most of the axial skeleton, skeletal muscles, and much of the dermal layer of the skin. By days 24–25 (during the fourth week), (23) ______________ ______________ [p.790] are visible that will contribute in forming the face, neck, mouth, larynx, pharynx, and nasal cavities.

During the third week after conception, the (24) ______________ [p.791] was also forming from endometrial and extra-embryonic membrane tissues; a(n) (25) ______________ ______________ [p.791] connects the developing embryo to it. Oxygen and (26) ______________ [p.791] exit from maternal blood vessels into the embryo. Coming from the embryo are carbon dioxide and other (27) ______________ [p.791] that will quickly be disposed of by the mother's lungs and kidneys.

By the end of the fourth week, (28) ______________ [p.792] form from limb buds and fingers and toes form from embryonic paddles. During the months that follow, recognizable organs develop from the lumps and bumps that preceded them. Growth of the (29) ______________ [p.792] surpasses that of any other region of the body. During the second trimester, the fetus is moving most of its facial (30) ______________ [p.792] and is covered with hair. The eyes are sealed shut. During the last trimester, the fetus gains weight and develops its lungs, immune system, and temperature regulation. By the ninth month, the rate of surviving birth has increased to (31) ______________ [p.793] percent.

Proper nutrition of the mother is important to the well-being of the fetus. For example, a maternal diet deficient in the B vitamin (32) ______________ [p.794] often leads to neural tube defects in the fetus. Maternal infection can also cause fetal problems. A woman who contracts (33) ______________ [p.795] during the first trimester may give birth to a child with various organ malformations. The consumption of (34) ______________ [p.795] can result in fetuses with reduced brain size, facial deformities, and other problems. Maternal use of cocaine affects the fetus's (35) ______________ ______________ [p.795], as does the antibiotic (36) ______________ [p.795].

44.12. FROM BIRTH ONWARD [pp.796–797]

44.13. FOCUS ON BIOETHICS: ***Control of Human Fertility*** [p.798]

44.14. BIRTH CONTROL OPTIONS [pp.798–799]

44.15. FOCUS ON HEALTH: ***Sexually Transmitted Diseases*** [pp.800–801]

44.16. FOCUS ON BIOETHICS: ***To Seek or End Pregnancy*** [p.802]

Selected Words: "afterbirth" [p.796], *afterbaby blues* [p.796], *breast cancer* [p.797], prenatal stage [p.797], postnatal stage [p.797], *abstinence* [p.798], *rhythm method* [p.798], *withdrawal* [p.798], *douching* [p.798], *vasectomy* [p.798], *tubal ligation* [p.798], *spermicidal foam* [p.799], *spermicidal jelly* [p.799], *IUDs* [p.799], *diaphragm* [p.799], *condoms* [p.799], *birth control pill* [p.799], *Depo-Provera* injection [p.799], Norplant [p.799], *morning-after pill* [p.799], RU-486 [p.799], Type II *Herpes simplex* [p.800], *AIDS* [p.800], *gonorrhea* [p.800], *syphilis* [p.801], chancre [p.801], *pelvic inflammatory disease* (PID) [p.801], *Chlamydia trachomatis* [p.801], *NGU* (chlamydial nongonococcal urethritis) [p.801], *genital warts* [p.801]

Boldfaced, Page-Referenced Terms

[p.796] labor ____

____

[p.796] relaxin ____

____

[p.796] lactation ____

____

[p.797] prolactin ____

____

[p.797] oxytocin ____

____

[p.796] CRH (corticotropin-releasing hormone) ____

____

[p.800] sexually transmitted diseases (STDs) ____

____

[p.802] *in vitro* fertilization ____

____

[p.802] abortion ____

____

Fill-in-the-Blanks

On Earth each day, (1) ________________ [p.798] babies are born every hour. Each year in the United States, we still have about (2) ________________ [p.798] teenage pregnancies and (3) ________________ [p.798] abortions. The most effective method of preventing conception is complete (4) ________________ [p.798]. (5) ________________ [p.799] are up to 95 percent effective when used with spermicide and help prevent venereal disease. A(n) (6) ________________ [p.799] is a flexible, dome-shaped disk, used with a spermicidal foam or jelly, that is placed over the cervix. In the United States, the most widely used contraceptive is the Pill—an oral contraceptive of synthetic (7) ________________ [p.799] and (8) ________________ [p.799], which suppress the release of FSH and LH from the (9) ________________ ________________ [recall p.786] and thereby prevent the cyclic maturation and release of eggs. Two forms of surgical sterilization are vasectomy and (10) ________________ ________________ [p.798].

Matching

Match each of the following with *all* applicable diseases. Choose from: [pp.800–801]

A. AIDS B. chlamydial infection C. genital herpes D. gonorrhea
E. pelvic inflammatory disease F. syphilis

11. _____ Can damage the brain and spinal cord in ways leading to various forms of insanity and paralysis
12. _____ Has no cure
13. _____ Can cause violent cramps, fever, vomiting, and sterility due to scarring and blocking of the oviducts
14. _____ Caused by a motile, corkscrew-shaped bacterium, *Treponema pallidum*
15. _____ Infected women typically have miscarriages, stillbirths, or sickly infants
16. _____ Caused by a bacterium with pili, *Neisseria gonorrhoeae*
17. _____ Caused by direct contact with the viral agent; about 25 million people in the United States are infected by it
18. _____ Produces a chancre (localized ulcer) 1–8 weeks following infection
19. _____ Chronic infections by this can lead to cervical cancer
20. _____ Can lead to lesions in the eyes that cause blindness in babies born to mothers infected with it
21. _____ Acyclovir decreases the healing time and may also decrease the pain and viral shedding from the blisters
22. _____ Eight weeks following infection, a flattened, painless chancre harbors many motile bacteria
23. _____ May be cured by antibiotics but can be infected again
24. _____ Can be treated with tetracycline and sulfonamides
25. _____ Generally preventable by correct condom usage

Self-Quiz

For questions 1–5, choose from the following answers: [pp.800–801]

a. AIDS
b. chlamydial infection
c. genital herpes
d. gonorrhea
e. syphilis

____ 1. ______________ is a disease caused by a spherical bacterium (*Neisseria*) with pili; it is curable by prompt diagnosis and treatment.

____ 2. ______________ is a disease caused by a spiral bacterium (*Treponema*) that produces a localized ulcer (a chancre).

____ 3. ______________ is an incurable disease caused by a retrovirus (an RNA-based virus).

____ 4. ______________ is a disease caused by an intracellular parasite that lives in the genital and urinary tracts; it also causes NGU.

____ 5. ______________ is an extremely contagious viral infection (DNA-based) that causes sores on the facial area and reproductive tract; it is also incurable.

For questions 6–8, choose from the following answers:

a. blastocyst
b. allantois
c. yolk sac
d. oviduct
e. cervix

____ 6. The ______________ lies between the uterus and the vagina. [p.782]

____ 7. The ______________ is a pathway from the ovary to the uterus. [p.782]

____ 8. The ______________ results from the process known as cleavage. [p.785]

For questions 9–12, choose from the following answers:

a. Leydig cells
b. seminiferous tubules
c. vas deferens
d. epididymis
e. prostate

____ 9. The ______________ delivers sperm to the ejaculatory duct. [p.778]

____ 10. Testosterone is produced by the ______________. [p.780]

____ 11. Meiosis occurs in the ______________. [p.778]

____ 12. Sperm mature and become motile in the ______________. [p.778]

Chapter Objectives/Review Questions

1. Distinguish between primary and secondary sexual traits and between gonads and accessory reproductive organs. [p.777]
2. Follow the path of a mature sperm from the seminiferous tubules to the urethral exit. List every structure encountered along the path and state its contribution to the nurture of the sperm. [p.780]
3. Describe how a man examines himself for testicular cancer. [p.779]
4. Diagram the structure of a sperm, label its components, and state the function of each. [pp.780–781]
5. List in order the stages that make up spermatogenesis. [pp.780–781]
6. Name the four hormones that directly or indirectly control male reproductive function. Diagram the negative feedback mechanisms that link the hypothalamus, anterior pituitary, and testes in controlling gonadal function. [pp.780–781]

7. Compare the function of the Leydig cells of the testis with those of the ovarian follicle and the corpus luteum. [pp.780,784–785]
8. Distinguish the follicular phase of the menstrual cycle from the luteal phase and explain how the two cycles are synchronized by hormones from the anterior pituitary, hypothalamus, and ovaries. [pp.782–783]
9. Trace the path of a sperm from the urethral exit to the place where fertilization normally occurs. Mention in correct sequence all major structures of the female reproductive tract that are passed along the way, and state the principal function of each structure. [p.787]
10. State which hormonal event brings about ovulation and which other hormonal events bring about the onset and finish of menstruation. [pp.784–786]
11. List the physiological factors that bring about erection of the penis during sexual stimulation and those that bring about ejaculation. [p.787]
12. List the similar events that occur in both male and female orgasm. [p.787]
13. Describe the events that occur during the first month of human development. State how much time cleavage and gastrulation require, when organogenesis begins, and what is involved in implantation and placenta formation. [pp.788–792]
14. Tell when it is that the embryo begins to be referred to as a fetus and when is the earliest that a fetus born prematurely can survive. [pp.792–793]
15. Explain why the mother must be particularly careful of her diet, health habits, and lifestyle during the first trimester after fertilization (especially during the first six weeks). [pp.794–795]
16. Describe two different types of sterilization. [p.798]
17. Identify the factors that encourage and those that discourage methods of human birth control. [pp.798–799]
18. State which birth control methods help prevent venereal disease. [pp.798–799]
19. Identify the three most effective birth control methods used in the United States and the four least effective birth control methods. [pp.798–799]
20. For each STD described in the Focus section, state the causative organism together with the symptoms of the disease. [pp.800–801]
21. State the physiological circumstances that would prompt a couple to try *in vitro* fertilization. [p.802]

Integrating and Applying Key Concepts

What rewards do you think a society should give potential parents who have at most two children during their lifetimes? In cases of rape and/or incest that causes a pregnancy, should society be able to dictate the fate of the victimized female? Should the man responsible be punished? In the absence of rewards or punishments, how can a society encourage women not to have abortions and yet ensure that the human birth rate does not continue to increase?

45

POPULATION ECOLOGY

Interactive Exercises

Tales of Nightmare Numbers [pp.806–807]

45.1. CHARACTERISTICS OF POPULATIONS [p.808]

45.2. FOCUS ON SCIENCE: *Elusive Heads to Count* [p.819]

45.3. POPULATION SIZE AND EXPONENTIAL GROWTH [pp.810–811]

Selected Words: *pre-reproductive, reproductive,* and *postreproductive* ages [p.808], *habitat* [p.808], *crude* density [p.808], *per capita* [p.810]

Boldfaced, Page-Referenced Terms

[p.807] ecology

[p.808] demographics

[p.808] population size

[p.808] age structure

[p.808] reproductive base

[p.808] population density

[p.808] population distribution

[p.809] capture–recapture method

[p.810] immigration

[p.810] emigration

[p.810] migration

[p.810] zero population growth

[p.810] per capita

[p.810] net reproduction per individual per unit time, or *r*

[p.811] exponential growth

[p.811] doubling time

[p.811] biotic potential

Matching

Choose the most appropriate answer for each term. [p.808]

1. ____ demographics
2. ____ population size
3. ____ population density
4. ____ habitat
5. ____ population distribution
6. ____ age structure
7. ____ reproductive base
8. ____ crude density
9. ____ prereproductive, reproductive, and postreproductive
10. ____ clumped dispersion
11. ____ nearly uniform dispersion
12. ____ random dispersion

A. Includes prereproductive and reproductive age categories
B. The general pattern in which the individuals of a population are dispersed throughout a specified area
C. When individuals of a population are more evenly spaced than they would be by chance alone
D. The number of individuals in some specified area or volume of a habitat
E. Occurs only when individuals of a population neither attract nor avoid one another, when conditions are fairly uniform through the habitat, and when resources are available all the time
F. The number of individuals in each of several-to-many age categories
G. The measured number of individuals in a specified area
H. The number of individuals that contribute to a population's gene pool
I. The type of place where a species normally lives
J. Categories of a population's age structure
K. The vital statistics of a population
L. Individuals of a population form aggregations at specific habitat sites; most common dispersion pattern

Short Answer

13. A zoologist wishes to estimate the population size of a rare salamander. Initially, 25 salamanders are caught and marked with an orange, waterproof dye. After six months the scientist returns and captures five marked salamanders out of a total catch of 50. What is the population size of the salamanders? [p.809] ____________________

____

14. List three variables that may have caused error in the above estimate of population size. [p.809] ____________________

____

____

____

Fill-in-the-Blanks

Population size increases as a result of births and (15) ______________ [p.810], that is, the arrival of new residents from other populations of the species. Population size decreases as a result of deaths and (16) ______________ [p.810], whereby individuals permanently move out of the population. Population size also changes on a predictable basis as a result of daily or seasonal events called (17) ______________ [p.810]. Since migration is a recurring round-trip between two areas, however, these transient effects need not initially be considered. If we assume that immigration is balancing emigration over time, the effects of both on population size may be ignored. This allows the definition of (18) ______________ ______________ [p.810] growth as an interval during which the number of births is balanced out by the number of deaths. The population size undergoes no overall increase or decrease.

Births, deaths, and other variables that might affect population size can be measured in terms of (19) ______________ ______________ [p.810] rates or rates per individual. Consider 2,000 mice living in a large cornfield. If the female mice collectively give birth to 1,000 mice per month, the birth rate would be (20) ______________ [p.810] per mouse per month. If 200 of the 2,000 die during the month, the death rate would be (21) ______________ [p.810] per mouse per month.

If it is assumed that the birth rate and death rate remain constant, both can be combined into a single variable — the net (22) ______________ [p.810] per individual per unit (23) ______________ [p.810], or r for short. For the mouse population in the example just given, the value of r is (24) ______________ [p.810] per mouse per month.

If the monthly increases in the number of individuals in a population are plotted against time, one finds a graph line in the shape of a(n) (25) ______________ [p.811]. Then one knows that (26) ______________ [p.811] growth is being tracked. The length of time it takes for a population to double in size is its (27) ______________ [p.811] time. When a population displays its (28) ______________ ______________ [p.811] in terms of growth, this is the maximum rate of increase per individual under ideal conditions.

Problems

For questions 29–33, consider the equation $G = rN$, where G = population growth rate per unit time, r = net population growth rate per individual per unit time, and N = number of individuals in the population. [p.810]

29. Assume that r remains constant at 0.2.
    a. As the value of G increases, what happens to the value of N? ______________

    ____

    b. If the value of G decreases, what happens to the value of N? ______________

    ____

    c. If the net reproduction per individual stays the same and the population grows faster, ______________

    ____

    then what must happen to the number of individuals in the population? ______________

30. If a society decides it is necessary to lower its value of N through reproductive means because supportive resources are dwindling, it must lower either its net reproduction per individual per unit or its [pp.810–811] ______________

    ______________.

31. The equation $G = rN$ expresses a direct relationship between G and $r \times N$. If G remains constant and N increases, what must the value of r do? (In this situation, r varies inversely with N.) [pp.810–811] ______________

    ____

32. Look at line (a) on the graph at the right. After seven hours have elapsed, approximately how many individuals are in the population? [pp.810–811] ______

______

33. Look at line (b) in the same graph.
   a. After 24 hours have elapsed, approximately how many individuals are in the population? [pp.816–817] ______

   ______

   b. After 28 hours have elapsed, approximately how many individuals are in the population? [pp.816–817] ______

   ______

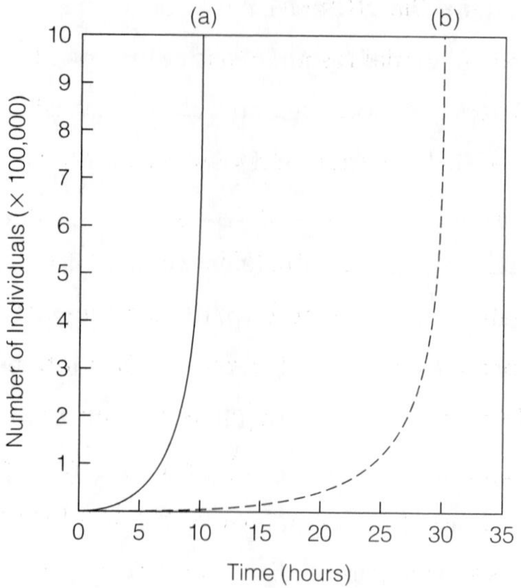

45.4. LIMITS ON THE GROWTH OF POPULATIONS [pp.812–813]

45.5. LIFE HISTORY PATTERNS [pp.814–815]

45.6. NATURAL SELECTION AND THE GUPPIES OF TRINIDAD [pp.816–817]

Selected Words: *sustainable* supply of resources [p.812], *bubonic plague* [p.813], *pneumonic plague* [p.813], "survivorship" schedules [p.814], *Type I* curves [p.815], *Type II* curves [p.815], *Type III* curves [p.815]

Boldfaced, Page-Referenced Terms

[p.812] limiting factor ______

______

[p.812] carrying capacity ______

______

[p.812] logistic growth ______

______

[p.812] density-dependent control ______

______

[p.813] density-independent factors ______

______

[p.814] life history pattern ______

______

[p.814] cohort ______

______

[p.815] survivorship curve ______

______

Fill-in-the-Blanks

Any essential resource that is in short supply is a(n) (1) ______________ [p.812] on population growth. (2) ______________ ______________ [p.812] refers to the maximum number of individuals of a population that can be sustained indefinitely by the environment. S-shaped growth curves are characteristic of (3) ______________ [p.812] population growth. The plot of (3) growth levels off once the (4) ______________ ______________ [p.812] is reached. In the equation $G = r_{max}N [(K - N)/K]$, as the value of N approaches the value of K and K and r_{max} remain constant, the value of G (5) ______________ [p.812]. As the value of r_{max} increases and G and N remain constant, the value of K (the carrying capacity) (6) ______________ [p.812]. In an overcrowded population, predators, parasites, and disease agents serve as (7) ______________ - ______________ [p.812] controls. When an event such as a freak summer snowstorm in the Colorado Rockies causes more deaths or fewer births in a butterfly population (with no regard to crowding or dispersion patterns), the controls are said to be (8) ______________ - ______________ [p.813] factors. Food availability is a density- (9) ______________ [p.813] factor that works to cut back population size when it approaches the environment's (10) ______________ ______________ [p.813]. Environmental disruptions such as forest fires and floods are density- (11) ______________ [p.813] factors that may push a population above or below its tolerance range for a given variable.

Each species exhibits a(n) (12) ______________ ______________ [p.814] pattern, which is a set of adaptations that influence survival, fertility, and age at first reproduction. Life and health (13) ______________ [p.814] companies use information summarized from life tables, which show trends in mortality and life expectancy. A(n) (14) ______________ [p.814] is a group of individuals that is tracked from the time of birth until the last one dies. A(n) (15) ______________ [p.814] table lists the completed data on a population's age-specific death schedule. Such tables can be converted to cheerier (16) "______________" [p.814] schedules, which give the number of individuals that reach some specified age (x).

(17) ______________ [p.815] curves are the graph lines that emerge when ecologists plot the age-specific survival of a cohort in a particular habitat. Type (18) ______________ [p.815] survivorship curves characterize populations having low survivorship early in life. Organisms whose life cycles reflect high survivorship until fairly late in life, with a later large increase in deaths, have a type (19) ______________ [p.815] survivorship curve. Organisms such as lizards, small mammals, and some songbirds are just as likely to be killed or die of disease at any age; they are characterized by the type (20) ______________ [p.815] survivorship curve. Provided people have access to good health care, human populations typically exhibit a type (21) ______________ [p.815] survivorship curve.

Choice

For questions 22–27, choose from the following answers:

a. example of a density-dependent control [pp.812–813]
b. example of a density-independent control [p.813]

22. ____ Drought, floods, earthquakes
23. ____ Bubonic plague and pneumonic plague
24. ____ Food availability
25. ____ Adrenal enlargement in wild rabbits in response to crowding
26. ____ Heavy applications of pesticides in your backyard
27. ____ Freak snowstorms in the Rocky Mountains

45.7. HUMAN POPULATION GROWTH [pp.818–819]
45.8. CONTROL THROUGH FAMILY PLANNING [pp.820–821]
45.9. POPULATION GROWTH AND ECONOMIC DEVELOPMENT [pp.822–823]
45.10. SOCIAL IMPACT OF NO GROWTH [p.823]

Selected Words: *cholera* [p.819], *baby-boomers* [p.820], *preindustrial stage* [p.822], *transitional* stage [p.822], *industrial* stage [p.822], *postindustrial* stage [p.822], *postpone* [p.823]

Boldfaced, Page-Referenced Terms

[p.820] total fertility rate ______________________________

____

[p.822] demographic transition model ______________________________

____

Graph Construction

1. Construct a graph of the following data in the space provided on the next page. [p.819]

Year	*Estimated World Population*
1650	500,000,000+
1850	1,000,000,000+
1930	2,000,000,000+
1975	4,000,000,000+
1986	5,000,000,000+
1993	5,500,000,000+
1995	5,700,000,000+
1997	5,800,000,000+
2050 (projected)	9,000,000,000+

a. Estimate the year that the world contained 3 billion humans. ____________. (Consult your graph.)

b. Estimate the year that Earth will house 8 billion humans. ____________. (Consult your graph.)

c. Do you expect Earth to house 8 billion humans within your lifetime? ____________. (Consult your graph.)

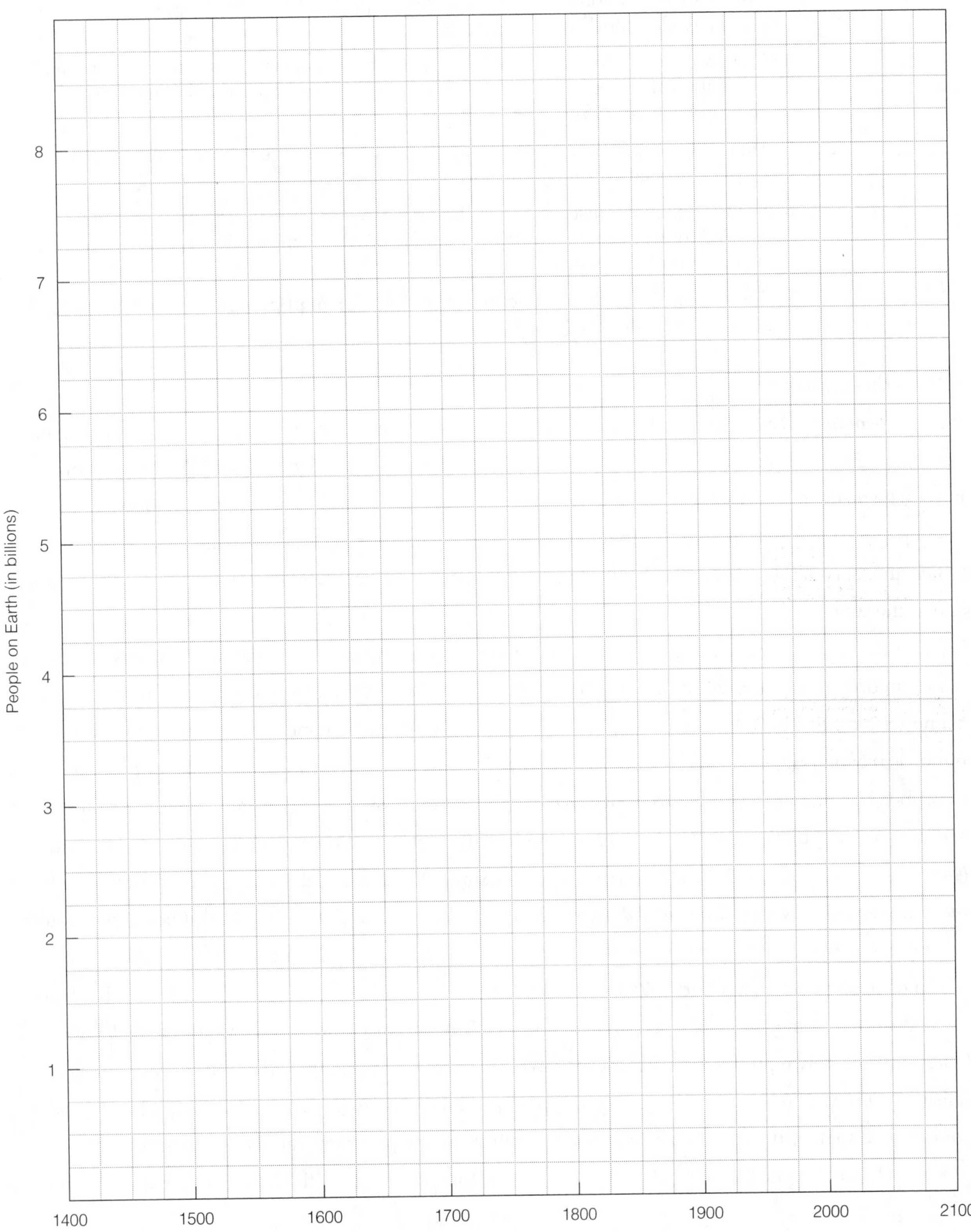

True/False

If the statement is true, write a "T" in the blank. If the statement is false, make it correct by changing the underlined word(s) and writing the correct word(s) in the answer blank.

_______________ 2. Currently, the human population has surpassed <u>8 billion</u>. [p.818]

_______________ 3. Even if we could double our present food supply, death from starvation could still claim <u>20 million to 40 million people</u> a year. [p.818]

_______________ 4. Compared with the geographic spread of other organisms, it has taken the human population an exceptionally <u>long</u> period of time to expand into diverse new environments. [p.818]

_______________ 5. Managing food supplies through agriculture has had the effect of increasing the <u>carrying capacity</u> for the human population. [p.818]

_______________ 6. By bringing many disease agents under control and by tapping into concentrated existing forms of energy, humans <u>used</u> factors that had previously limited their population growth. [p.819]

_______________ 7. The stupendously accelerated growth of the human population continues and <u>can</u> be sustained indefinitely. [p.819]

Fill-in-the-Blanks

If the current annual average population growth rate of 1.47 percent is maintained for the human population, there may be more than (8) _______________ [p.820] billion people by the year 2050. Large efforts will have to be made to increase the amount of (9) _______________ _______________ [p.820] to sustain that many people. Such gross manipulation of resources is likely to intensify (10) _______________ [p.820], which almost certainly will adversely affect our water supplies, the atmosphere, and productivity on land and in the seas.

(11) _______________ _______________ [p.820] programs educate individuals about choosing how many children they will have and when. Carefully developed and administered programs may bring about a long-term decline in (12) _______________ [p.820] rates. A more useful measure of global population trends is the total (13) _______________ [p.820] rate. This is defined as the average number of children born to women during their reproductive years, as estimated on the basis of current age-specific birth rates. To arrive at zero population growth, the average "replacement rate" would have to be slightly higher than (14) _______________ [p.820] children per couple, for some (15) _______________ [p.820] children die before reaching reproductive age. Currently the average rate is (16) _______________ [p.820] children per woman, an impressive decline from 1950, when the rate was (17) _______________ [p.820].

In the United States, a cohort of 78 million (18) _______________ - _______________ [p.820] has been tracked since it formed in 1946, when soldiers returned home from World War II. Although the United States has a narrow reproductive base, more than (19) _______________ [p.820] of the world population falls in the broad (20) _______________ [p.820] prereproductive base. (21) _______________ [p.820] is the country said to have the world's most extensive family planning program. Family planning programs on a global scale are designed to help (22) _______________ [p.821] the size of the human population. Even if the human population reaches a level of zero growth, however, it will continue to grow for 60 years, because its (23) ______________ [p.821] base already consists of a staggering number of individuals.

Choice

For questions 24–29, refer to Figure 45.14 in the text and choose from the following age structure descriptions: [p.821]

a. rapid growth b. slow growth c. zero growth d. negative growth

24. _____ United States
25. _____ China
26. _____ Canada
27. _____ Mexico
28. _____ Australia
29. _____ India

Sequence

Arrange the following stages of the demographic transition model in correct chronological sequence. Write the letter of the first step next to 30, the letter of the second step next to 31, and so on. [p.822]

30. _____
31. _____
32. _____
33. _____

A. *Industrial stage:* population growth slows and industrialization is in full swing
B. *Preindustrial stage:* harsh living conditions, high birth rates and low death rates, slow population growth
C. *Postindustrial stage:* zero population growth is reached; birth rate falls below death rate, and population size slowly decreases
D. *Transitional stage:* industrialization begins, food production rises, and health care improves; death rates drop, birth rates remain high to give rapid population growth

Fill-in-the-Blanks

Today, the United States, Canada, Australia, Japan, nations of the former Soviet Union, and most countries of western Europe are in the (34) ______________ [p.822] stage of the demographic transition model. Their growth rate is slowly (35) ______________ [p.822]. In Germany, Bulgaria, Hungary, and some other countries, death rates exceed birth rates, and the populations are getting (36) ______________ [p.822]. Mexico and other less-developed countries are in the (37) ______________ [p.822] stage. In many countries, population growth outpaces economic growth, death rates will increase, and they may be stuck in the (38) ______________ [p.822] stage. Claiming that (39) ______________ [p.823] growth affects economic health, many governments restrict (40) ______________ [p.823].

India has (41) ______________ [p.823] percent of the human population. By comparison, the United States has (42) ______________ [p.823] percent. The highly industrialized United States produces (43) ______________ [p.823] percent of all the world's goods and services. It uses (44) ______________ [p.823] percent of the world's processed minerals and available nonrenewable sources of energy. Its people consume (45) ______________ [p.823] times more goods and services than the average person in India. The United States generates at least (46) ______________ [p.823] percent of the global pollution and trash. By contrast, India produces about (47) ______________ [p.823] percent of all goods and services and uses (48) ______________ [p.823] percent of the available minerals and nonrenewable energy resources. India generates only about (49) ______________ [p.823] percent of the pollution and trash. Tyler Miller, Jr. estimated that it would take (50) ______________ [p.823] billion impoverished individuals in India to have as much environmental impact as (51) ______________ [p.823] million Americans.

For all species on our planet, the biological implications of extremely rapid (52) ______________ [p.823] are staggering. Yet so are the (53) ______________ [p.823] implications of what will happen when and if the human population declines to the point of zero population growth—and stays there. If the population ever does reach and maintain zero growth over time, a larger proportion of individuals will end up in the (54) ______________ [p.823] age brackets. To care for older, nonproductive citizens, fewer productive ones must carry more and more of the (55) ______________ [p.823] burden. Through our special abilities to undergo rapid cultural evolution, we have (56) ______________ [p.823] the action of most of the factors that limit growth. But no amount of (57) ______________ [p.823] intervention can repeal the laws ultimately governing population growth, as imposed by the (58) ______________ ______________ [p.823] of the environment.

Self-Quiz

____ 1. The number of individuals that contribute to a population's gene pool is ______________. [p.808]
a. the population density
b. the population growth
c. the population birth rate
d. the population size

____ 2. The number of individuals in a given area or volume of a habitat is ______________. [p.808]
a. the population density
b. the population growth
c. the population birth rate
d. the population size

____ 3. A population that is growing exponentially in the absence of limiting factors can be illustrated accurately by a(n) ______________. [pp.810–811]
a. S-shaped curve
b. J-shaped curve
c. curve that terminates in a plateau phase
d. tolerance curve

____ 4. How are the individuals in a population most often dispersed? [p.808]
a. clumped
b. very uniform
c. nearly uniform
d. random

____ 5. Assuming immigration is balancing emigration over time, ______________ may be defined as an interval in which the number of births is balanced by the number of deaths. [p.810]
a. the lack of a limiting factor
b. exponential growth
c. saturation
d. zero population growth

____ 6. Assuming the birth and death rate remain constant, both can be combined into a single variable, *r*, or ______________. [p.810]
a. the per capita rate
b. the minus migration factor
c. exponential growth
d. the net reproduction per individual per unit time

_____ 7. _______________ is a way to express the growth rate of a given population. [p.811]

a. Doubling time
b. Population density
c. Population size
d. Carrying capacity

_____ 8. The maximum rate of increase per individual under ideal conditions is called the _______________. [p.811]

a. biotic potential
b. carrying capacity
c. doubling time
d. population size

_____ 9. The maximum number of individuals of a population (or species) that a given environment can sustain indefinitely defines _______________. [p.812]

a. the carrying capacity of the environment
b. exponential growth
c. the doubling time of a population
d. density-independent factors

_____ 10. In natural communities, some feedback mechanisms operate whenever populations change in size; they are _______________. [p.812]

a. density-dependent factors
b. density-independent factors
c. always intrinsic to the individuals of the community
d. always extrinsic to the individuals of the community

_____ 11. _______________ are any essential resources that are in short supply for a population. [p.812]

a. Density-independent factors
b. Extrinsic factors
c. Limiting factors
d. Intrinsic factors

_____ 12. Which of the following is *not* characteristic of logistic growth? [p.812]

a. S-shaped curve
b. leveling off of growth as carrying capacity is reached
c. unrestricted growth
d. slow growth of a low-density population followed by rapid growth

_____ 13. The population growth rate (*G*) is equal to the _______________ the net population growth rate per individual (*r*) and number of individuals (*N*). [p.810]

a. sum of
b. product of
c. doubling of
d. difference between

_____ 14. $G = r\max N(K - N/K)$ represents _______________. [p.812]

a. exponential growth
b. population density
c. population size
d. logistic growth

_____ 15. The beginning of industrialization, a rise in food production, improvement of health care, rising birth rates, and declining death rates describe the _______________ stage of the demographic transition model. [p.822]

a. preindustrial
b. transitional
c. industrial
d. postindustrial

_____ 16. A group of individuals that is typically tracked from the time of birth until the last member dies is a _______________. [p.814]

a. cohort
b. species
c. type I curve group
d. density-independent group

_____ 17. The survivorship curve typical of industrialized human populations is type _______________. [p.815]

a. I
b. II
c. III
d. none of the above

Chapter Objectives/Review Questions

1. Define the term *ecology*. [p.807]
2. Define the following terms: *demographics, habitat, population size, population density, population distribution, age structure,* and *reproductive base.* [p.808]
3. List and describe the three patterns of dispersion illustrated by populations in a habitat. [p.808]
4. Given data on a mark–recapture experiment, how would you estimate the population size of the species? [p.809]
5. Distinguish immigration from emigration; define the term *migration*. [p.810]
6. Define *zero population growth* and describe how achieving it would affect population size. [p.810]
7. In the equation $G = rN$, as long as r holds constant, any population will show ________ growth. [p.810]
8. Given a set of values, calculate the population growth rate (G). [p.810]
9. Explain the relationship between exponential growth and doubling time. [p.811]
10. ________ ________ is the maximum rate of increase per individual under ideal conditions. [p.817]
11. The ________ rate of increase in population growth depends on the age at which each generation starts to reproduce, how often each individual reproduces, and how many offspring are produced. [p.811]
12. List several examples of limiting factors and explain how they influence population curves. [p.812]
13. Explain what is meant by *carrying capacity*. [p.812]
14. Explain the meaning of the logistic growth equation and calculate values of G by using the equation. Explain the meaning of r max and K. [p.812]
15. Compare logistic and exponential growth. [pp.811–812]
16. Define the term *density-dependent growth* of populations; cite one example. [pp.812–813]
17. Define the term *density-independent factors* and give two examples; indicate how such factors affect populations. [p.813]
18. Most species have a ________ ________ pattern in that their individuals exhibit particular morphological, physiological, and behavioral traits that are adaptive to different conditions at different times in the life cycle. [p.814]
19. Life insurance companies and ecologists track a ________, a group of individuals, from the time of their birth until the last one dies. [p.814]
20. Explain how the construction of life tables and survivorship curves can be useful to humans in managing the distribution of scarce resources. [p.814]
21. Explain the significance and use of life tables; list and interpret the three survivorship curves. [p.815]
22. Guppy populations targeted by killifish tend to be larger, less streamlined, and more brightly colored; guppy populations targeted by pike-cichlids tend to be smaller, more streamlined, and duller in color patterning. Other life history pattern differences exist between the two groups. After consideration of the research results obtained by Reznick and Endler, provide an explanation for these differences. [pp.816–817]
23. List three possible reasons why growth of the human population is out of control. [p.818]
24. Most governments are trying to lower birth rates by ________ ________ programs. [p.820]
25. Define the term *total fertility rate*. [p.820]
26. What is the significance of the cohort of 78 million baby-boomers? [p.820]
27. Be able to analyze age structure diagrams to determine patterns of growth. [p.821]
28. List and describe the four stages of the demographic transition model. [p.822]
29. Generally compare the implications of the amount of resource consumption by India and by the United States. [p.823]
30. In the final analysis, no amount of ________ intervention can repeal the ultimate laws governing population growth, as imposed by the ________ ________ of the environment. [p.823]

Integrating and Applying Key Concepts

Assume that the world has reached zero population growth. The year is 2110, and there are 10.5 billion individuals of *Homo pollutans* on Earth. You have seen stories on the community television screen about how people used to live 120 years ago. List the ways that life has changed, and comment on the events that no longer happen because of the enormous human population.

46

SOCIAL INTERACTIONS

Interactive Exercises

Deck the Nest with Sprigs of Green Stuff [pp.826–827]

46.1. BEHAVIOR'S HERITABLE BASIS [pp.828–829]

46.2. LEARNED BEHAVIOR [p.830]

Selected Words: intermediate response [p.828]

Boldfaced, Page-Referenced Terms

[p.827] animal behavior ____________________

____

[p.829] song system ____________________

____

[p.829] instinctive behavior ____________________

____

[p.829] sign stimuli ______________________________

____

[p.829] fixed action pattern ______________________________

____

[p.830] learned behavior ______________________________

____

[p.830] imprinting ______________________________

____

Matching

Choose the most appropriate answer for each term.

1. ____ intermediate response [p.828]
2. ____ song system [p.829]
3. ____ instinctive behavior [p.829]
4. ____ sign stimuli [p.829]
5. ____ fixed action pattern [p.829]
6. ____ learned behavior [p.830]
7. ____ imprinting [p.830]
8. ____ animal behavior [p.827]

A. Animals process and integrate information gained from experience, then use it to vary or change responses to stimuli
B. Well-defined environmental cues that trigger suitable responses
C. Term applied to genetically based behavioral reactions of hybrid offspring
D. Time-dependent form of learning; triggered by exposure to sign stimuli and usually occurring during sensitive periods for young animals
E. A behavior performed without having been learned through actual environmental experience
F. Consists of several brain structures that will govern activity of a vocal organ's muscles
G. A program of coordinated muscle activity that runs to completion independent of feedback from the environment
H. Observable, coordinated responses to stimuli

Dichotomous Choice

Circle one of two possible answers given between parentheses in each statement.

9. For garter snake populations living along the California coast, the food of choice is (the banana slug/tadpoles and small fishes). [p.828]
10. In Stevan Arnold's experiments, newborn garter snakes that were offspring of coastal parents usually (ate/ignored) a chunk of slug as the first meal. [p.828]
11. Newborn garter snake offspring of (coastal/inland) parents ignored cotton swabs drenched in essence of slug and only rarely ate the slug meat. [p.828]
12. The differences in the behavioral eating responses of coastal and inland garter snakes (were/were not) learned. [p.828]
13. Hybrid garter snakes with coastal and inland parents exhibited a feeding response that indicated a(n) (environmental/genetic) basis for this behavior. [p.828]
14. In zebra finches and some other songbirds, singing behavior is an outcome of seasonal differences in the secretion of melatonin, a hormone secreted by the (gonads/pineal gland). [p.828]
15. In songbirds in spring, melatonin secretion is suppressed and gonads are released from hormonal suppression; they (increase/decrease) in size and step up their secretions of estrogen and testosterone. [p.829]

16. Even before a male bird hatches, a high (estrogen/testosterone) level stimulates development of a masculinized brain. [p.829]
17. (Hormones/Genes) underlie animal behavior–coordinated responses to stimuli. [p.829]

Complete the Table

18. Complete the following table with examples of instinctive and learned behavior.

Category	*Examples*
[p.829] a. Instinctive behavior	
[p.830] b. Learned behavior	

Matching

Choose the most appropriate answer for each category of learned behavior. [p.836]

19. ____ insight learning
20. ____ imprinting
21. ____ classical conditioning
22. ____ operant conditioning
23. ____ habituation
24. ____ spatial or latent learning

A. Birds living in cities learn not to flee from humans or cars, which pose no threat to them.
B. Chimpanzees abruptly stack several boxes and use a stick to reach suspended bananas out of reach.
C. In response to a bell, dogs salivate even in the absence of food.
D. Bluejays store information about dozens or hundreds of places where they have stashed food.
E. Baby geese formed an attachment to Konrad Lorenz if separated from the mother shortly after hatching.
F. A toad learns to avoid stinging or bad-tasting insects after its first attempt to eat them.

46.3. THE ADAPTIVE VALUE OF BEHAVIOR [p.831]

Boldfaced, Page-Referenced Terms

[p.831] natural selection ____________________

[p.831] reproductive success ____________________

[p.831] adaptive behavior ____________________

[p.831] social behavior ____________________

[p.831] selfish behavior ____________________

[p.831] altruistic behavior ____

____

[p.831] territory ____

____

Matching

Choose the most appropriate answer for each of the following. [p.831]

1. ____ territory
2. ____ natural selection
3. ____ adaptive behavior
4. ____ reproductive success
5. ____ social behavior
6. ____ selfish behavior
7. ____ altruistic behavior

A. Any behavior that promotes propagation of an individual's genes
B. The number of surviving offspring that the individual produces
C. Increases the success of the individual over the group
D. Self-sacrificing behavior
E. Interdependent interactions among individuals of the species
F. An area that one or more individuals defend against competitors
G. A result of differences in reproductive success among individuals that vary for heritable traits

46.4. COMMUNICATION SIGNALS [pp.832–833]

Selected Words: *intraspecific* interactions [p.832], *signaling* pheromones [p.832], *priming* pheromones [p.832], *round* dance [p.833], *waggle* dance [p.833], *straight* run [p.833], "recruited" [p.833], *ritualized* [p.833]

Boldfaced, Page-Referenced Terms

[p.832] communication signals ____

____

[p.832] signaler ____

____

[p.832] signal receivers ____

____

[p.832] pheromones ____

____

[p.832] composite signal ____

____

[p.832] communication display ____

____

[p.832] threat display ____

____

[p.833] courtship displays ______________________________

____

[p.833] tactile displays ______________________________

____

[p.833] illegitimate receiver ______________________________

____

[p.833] illegitimate signalers ______________________________

____

Choice

For questions 1–12, choose from the following:

a. chemical signal, signaling pheromones   b. communication display, threat display
c. chemical signal, priming pheromones   d. communication display, courtship   e. acoustical signal
f. communication display, tactile signal   g. composite signal   h. illegitimate signaler
i. communication display, social signal   j. illegitimate receiver

1. ____ Assassin bugs hook dead termite bodies on their dorsal surfaces and acquire termite scent; this deception allows assassin bugs to hunt termite victims more easily [p.833]
2. ____ Male songbirds sing to secure territory and attract a female [p.832]
3. ____ The play bow of dogs and wolves [p.832]
4. ____ Bombykol molecules released by female silk moths serve as sex attractants [p.832]
5. ____ A male bird might emit calls while bowing low, as if to peck the ground for food [p.833]
6. ____ Termites act defensively when detecting scents from invading ants whose scent signals are meant to elicit cooperation from other ants [p.833]
7. ____ A dominant male baboon's "yawn" that exposes large canines [pp.832–833]
8. ____ A volatile odor in the urine of certain male mice triggers and enhances estrus in female mice [p.832]
9. ____ After finding a source of pollen or nectar, a foraging honeybee returns to its colony (a hive) and performs a complex dance [p.833]
10. ____ Tungara frogs issue nighttime calls to females and rival males; the call is a "whine" followed by a "chuck" [p.832]
11. ____ A male firefly has a light-generating organ that emits a bright, flashing signal [p.833]
12. ____ Ears laid back against the head of a zebra convey hostility, but ears pointing up convey its absence; a zebra with laid-back ears isn't too riled up when its mouth is open just a bit, but when the mouth is gaping, watch out [p.832]

46.5. MATES, PARENTS, AND INDIVIDUAL REPRODUCTIVE SUCCESS [pp.834–835]

Selected Words: "nuptial gift" [p.834]

Boldfaced, Page-Referenced Terms

[p.834] sexual selection ____

____

[p.834] lek ____

____

Complete the Table

1. Complete the following table to supply the common names of the animals that fit the text examples of sexual selection. [pp.840–841]

Animals	*Descriptions of Sexual Selection*
a.	Extended parental care improves the likelihood that the current generation of offspring will survive; this behavior comes at a reproductive cost to the adults.
b.	Females select the males that offer them superior material goods; females permit mating only after they have eaten the "nuptial gift" for about five minutes.
c.	Females of a species cluster in defendable groups at a time when they are sexually receptive; males compete for access to the clusters; combative males are favored.
d.	Males congregate in a lek or communal display ground; each male stakes out a few square meters as his territory; females are attracted to the lek to observe male displays and usually select and mate with only one male.

46.6. COSTS AND BENEFITS OF LIVING IN SOCIAL GROUPS [pp.836–837]

46.7. THE EVOLUTION OF ALTRUISM [pp.838–839]

46.8. FOCUS ON SCIENCE: *Why Sacrifice Yourself?* [p.840]

Selected Words: "safe" sites [p.836], caring for one's *relatives* [p.839], *indirect* genetic contribution [p.839], "self-sacrifice" genes [p.839], *DNA fingerprinting* [p.840]

Boldfaced, Page-Referenced Terms

[p.836] selfish herd ____

____

[p.836] dominance hierarchies ____

____

[p.839] inclusive fitness ____

____

[p.841] adoption ____

____

Matching

Choose the most appropriate answer for each term.

1. ____ disadvantages to sociality [p.837]
2. ____ dominance hierarchy [pp.836–837]
3. ____ cooperative predator avoidance [p.836]
4. ____ the selfish herd [p.835]

A. Competition for resources, rapid depletion of food resources, cannibalism, and greater vulnerability to disease
B. A simple society brought together by reproductive self-interest; larger, more powerful male bluegills tend to claim the central locations
C. Adult musk oxen form a circle around their young while they face outward, and the "ring of horns" successfully deters the wolves; writhing, regurgitating reaction of Australian sawfly caterpillars to a disturbance
D. Some individuals of a baboon troop adopt a subordinate status with respect to the other members

Fill-in-the-Blanks

A(n) (5) ________________ [p.838] animal that gives way to a dominant one is acting in its own self-interest. Such (6) ________________-________________ [p.838] animals are found among many vertebrate groups. Altruistic behavior is most extreme in certain (7) ________________ [p.845], honeybee, and ant societies. When a bee plunges its stinger into an invader of the hive, it commits suicide.

Altruistic individuals of a social group do not contribute their (8) ________________ [p.839] to the next generation yet seem to perpetuate the genetic basis for their altruistic behavior over evolutionary time. According to William Hamilton's theory of (9) ________________ ________________ [p.839], those genes associated with caring for relatives—not one's direct descendants—tend to be favored in certain situations. This form of altruism can be thought of as an extension of (10) ________________ [p.839]. For example, if an uncle helps his niece survive long enough to reproduce, he has made a(n) (11) ________________ [p.839] genetic contribution to the next generation, as measured in terms of the genes that he and his relatives share. Altruism costs him; he may lose his own opportunities to (12) ________________ [p.839]. Similarly, nonbreeding workers in insect societies indirectly promote their "self-sacrifice" genes through (13) ________________ [p.839] behavior directed toward relatives. Thus, when a guard bee drives her stinger into a raccoon, she inevitably dies—but her siblings in the hive will perpetuate some of her (14) ________________ [p.839].

Sterility and extreme self-sacrifice are rare among social groups of (15) ________________ [p.840]. Unlike any other known vertebrate, the highly social naked mole rat individuals live out their lives as (16) ________________ [p.840] helpers in their social group. In each mole rat clan, there is a single reproducing female, who mates with one to three males. All other members of the clan care for the (17) ________________ [p.840] and "king" (or kings) and their offspring. A self-sacrificing naked mole rat helps to (18) ________________ [p.840] a very high proportion of the genes that it carries. Studies using a technique called DNA fingerprinting suggest that the (19) ______________ [p.840] of the helpers and the helped among mole rats may be as much as 90 percent identical.

46.9. AN EVOLUTIONARY VIEW OF HUMAN SOCIAL BEHAVIOR [p.841]

Selected Words: "adaptive" [p.841], "morally right" [p.841], "morality" [p.841], "desirability" [p.841], "maladaptive error" [p.841] *redirected* adaptive behaviors [p.841]

Boldfaced, Page-Referenced Term

[p.849] adoption ____

____

Dichotomous Choice

Circle one of two possible answers given between parentheses in each statement. [all from p.841]

1. Many people seem to believe that attempts to identify the adaptive value of a particular (animal/human) trait is an attempt to define its moral or social advantage.
2. *Adaptive* refers to (a trait with moral value/a trait valuable in gene transmission).
3. Cardinals feeding goldfish or Emperor penguins fighting to adopt orphans represent examples of (adopting by mistake/redirecting adaptive behaviors).
4. John Alcock suggests that husbands and wives who have lost an only child or who fail to produce children themselves should be especially prone to adopt (strangers/relatives).
5. The human adoption process (can/cannot) be considered adaptive when indirect selection favors adults who direct parenting assistance to relatives.
6. Joan Silk showed that in some traditional societies, people (will not/will) adopt related children far more often than nonrelated ones.
7. In large, industrialized societies in which agencies and other means of adoption exist, individuals become parents of (related/nonrelated) children.
8. It (is/is not) possible to test evolutionary hypotheses about the adaptive value of human behaviors.
9. *Adaptive* behavior and *socially desirable* behavior (are not/are) separate issues.
10. Strong parenting mechanisms evolved in the past, and it may be that their redirection toward a (relative/nonrelative) says more about human evolutionary history than it does about the transmission of one's genes.

Self-Quiz

____ 1. The observable, coordinated responses that animals make to stimuli are what we call animal ______________. [p.827]
a. imprinting
b. instincts
c. behavior
d. learning

____ 2. In ______________, a particular behavior is performed without having been learned by actual experience in the environment. [p.829]
a. natural selection
b. altruistic behavior
c. sexual selection
d. instinctive behavior

_____ 3. Newly hatched goslings follow any large moving object to which they are exposed shortly after hatching; this is an example of _______________. [p.830]
   a. homing behavior
   b. imprinting
   c. piloting
   d. migration

_____ 4. A young toad flips its sticky-tipped tongue and captures a bumblebee that stings its tongue; in the future, the toad leaves bumblebees alone. This is _______________. [p.830]
   a. instinctive behavior
   b. a fixed reaction pattern
   c. altruistic
   d. learned behavior

_____ 5. Pavlov's dog experiments represent an example of _______________. [p.830]
   a. classical conditioning
   b. latent learning
   c. selfish behavior
   d. habituation

_____ 6. _______________ provides an example of an illegitimate signaler. [p.833]
   a. A soldier termite killing an ant on cue
   b. An assassin bug with acquired termite odor
   c. A termite pheromone alarm signal
   d. The "yawn" of a dominant male baboon

_____ 7. The claiming of the more protected central locations of the bluegill colony by the largest, most powerful males suggests _______________. [p.836]
   a. cooperative predator avoidance
   b. the selfish herd
   c. a huge parent cost
   d. self-sacrificing behavior

_____ 8. A chemical odor in the urine of male mice triggers and enhances estrus in female mice. The source of stimulus for this response is a _______________. [p.832]
   a. generic mouse pheromone
   b. signaling pheromone
   c. priming pheromone
   d. cue from male mice

_____ 9. Female insects often attract mates by releasing sex pheromones. This is an example of a(n) _______________ signal. [p.832]
   a. chemical
   b. visual
   c. acoustical
   d. tactile

_____ 10. Male songbirds sing to stake out territories, attract females, and discourage males. This is an example of a(n) _______________ signal. [p.832]
   a. chemical
   b. visual
   c. acoustical
   d. tactile

_____ 11. When musk oxen form a "ring of horns" against predators, it is _______________. [p.836]
   a. selfish herd
   b. cooperative predator avoidance
   c. self-sacrificing behavior
   d. dominance hierarchy

_____ 12. Caring for nondescendant relatives favors the genes associated with helpful behavior and is classified as _______________. [p.837]
   a. dominance hierarchy
   b. indirect selection
   c. altruism
   d. both b and c

_____ 13. "_______________" means only that a given trait has proved valuable in transmitting an individual's genes. [p.843]
   a. Dominance hierarchy
   b. Indirect selection
   c. Adaptive
   d. Altruism

_____ 14. _______________ also favors adults who direct parenting behavior toward relatives and so indirectly perpetuate their shared genes. [p.839]
   a. Indirect selection
   b. Moral selection
   c. Redirected selection
   d. Perpetuated selection

Chapter Objectives/Review Questions

1. Define the term *animal behavior.* [p.827]
2. What explains the fact that coastal and inland garter snakes of the same species have different food preferences? [p.828]
3. Describe the intermediate response obtained in Arnold's experiment with coastal and inland garter snakes. [p.828]
4. Describe the origin and formation of a song system. [pp.828–829]
5. Define the term *sign stimuli.* [p.829]
6. Describe and cite an example of a fixed action pattern. [p.829]
7. Distinguish learned behavior from instinctive behavior. [pp.829–830]
8. Define each of the following categories of learned behavior and give one example of each: imprinting, classical conditioning, operant conditioning, habituation, spatial or latent learning, and insight learning. [p.830]
9. What is meant by *reproductive success?* [p.831]
10. __________ behavior is any behavior that promotes the propagation of an individual's genes and tends to occur with increased frequency in successive generations. [p.831]
11. __________ behavior refers to the cooperative, interdependent relationships among individuals of the species. [p.831]
12. Distinguish between selfish behavior and altruistic behavior. [p.831]
13. A(n) __________ is an area that one or more individuals defend against competitors. [p.831]
14. Examples of __________ signals are chemical, visual, acoustical, and tactile. [p.832]
15. Define the roles of signalers and signal receivers. [p.832]
16. Distinguish between signaling and priming pheromones and cite an example of each. [p.832]
17. A(n) __________ signal is illustrated by a zebra with laid-back ears and a gaping mouth. [p.832]
18. Describe one example of a threat display. [pp.832–833]
19. Ritualization is often developed to an amazing degree in __________ displays between potential mates. [p.833]
20. An example of a(n) __________ signal is the physical contact of bees in a hive maintaining physical contact during the dance of a returning foraging bee. [p.833]
21. When soldier termites detect ant scents meant for other ants and kill ants on cue, the termites are said to be __________ of a signal meant for individuals of a different species. [p.833]
22. Assassin bugs covered with termite scent are able to use deception to hunt termite victims more easily and as such are acting as __________ signalers. [p.833]
23. Natural selection tends to favor communication signals that promote __________ success. [p.833]
24. Competition among members of one sex for access to and selection of mates is the result of a microevolutionary process called __________. [p.834]
25. Discuss mate selection processes on the part of female hangingflies and the female sage grouse. [pp.834–835]
26. List the costs and benefits of parenting in the example of adult Caspian terns. [p.835]
27. Explain the "cost–benefit approach" that evolutionary biologists use to find answers to questions about social life. [p.836]
28. Studies of Australian sawfly caterpillars indicate __________ predator avoidance. [p.836]
29. Define the term *selfish herd;* cite an example. [p.836]
30. List the disadvantages to sociality. [p.837]
31. Hamilton's theory of __________ concerns caring for nondescendant relatives and how this favors genes associated with helpful behavior. [p.839]
32. Describe the possible benefits of self-sacrificing behavior with regard to indirect selection [p.839]
33. Explain how DNA fingerprinting was used to establish that self-sacrificing mole rats help to perpetuate a very high proportion of the genes (alleles) that they carry—even though they are not the reproducing mole rats. [p.840]
34. Define what is meant by an adaptive behavior. [p.841]

Integrating and Applying Key Concepts

1. Think about communication signals that humans use and list them. Do you believe a dominance hierarchy exists in human society? Think of examples.
2. Apply the concept of self-sacrifice to parenting. Based on the information in this chapter, what are the benefits of a parent protecting its child?

47

COMMUNITY INTERACTIONS

Interactive Exercises

No Pigeon Is an Island [pp.844–845]

47.1. WHICH FACTORS SHAPE COMMUNITY STRUCTURE? [p.846]

47.2. MUTUALISM [p.847]

Selected Words: *partition* the fruit supply [p.845], *fundamental* niche [p.846], *realized* niche [p.846], *obligatory* mutualism [p.847]

Boldfaced, Page-Referenced Terms

[p.845] community structure ______________________________

____

[p.846] habitat ______________________________

____

[p.846] community ______________________________

____

[p.846] niche ______________________________

____

[p.846] commensalism ______________________________

____

[p.846] mutualism ______________________________

____

[p.846] interspecific competition ______________________________

____

[p.846] predation ______________________________

____

[p.846] parasitism ______________________________

____

[p.846] symbiosis ______________________________

____

Fill-in-the-Blanks

The type of place where you will normally find an organism is its (1) ______________ [p.846]. The populations of all species in a habitat associate with one another as a(n) (2) ______________ [p.846]. The (3) ______________ [p.846] of an organism is defined by its "profession" in a community: the sum of activities and relationships in which it engages to secure and use the resources necessary for its survival and reproduction. The (4) ______________ [p.846] niche is the one that could prevail in the absence of competition and other factors that might constrain an organism's acquisition and use of resources. The (5) ______________ [p.846] niche is one that is more constrained and shifts in large and small ways over time, as individuals of the species respond to a mosaic of changes. Tree-roosting birds are (6) ______________ [p.846] with trees; the trees get nothing but are not harmed. In (7) ______________ [p.846], benefits flow both ways between the interacting species; in (8) ______________ [p.846] competition, disadvantages flow both ways. (9) ______________ [p.846] and (10) ______________ [p.846] are interactions that directly benefit one species and directly hurt the other. Commensalism, mutualism, and parasitism are all forms of (11) ______________ [p.846], which means "living together."

Complete the Table

12. Complete the following table to describe how each of the organisms listed is intimately dependent on the other for survival and reproduction in an obligatory mutualistic symbiotic interaction. [p.847]

Organism	*Dependency*
a. Yucca moth	
b. Yucca plant	

47.3. COMPETITIVE INTERACTIONS [pp.848–849]

47.4. PREDATOR–PREY INTERACTIONS [pp.850–851]

47.5. CONNECTIONS: *An Evolutionary Arms Race* [pp.852–853]

Selected Words: *intraspecific* competition [p.848], *interspecific* competition [p.848], *exploitative* competition [p.848], *interference* competition [p.848], *three-level* interaction [p.851], *model* for deception [p.852], *mimic* [p.852]

Boldfaced, Page-Referenced Terms

[p.849] competitive exclusion ____________________

[p.849] resource partitioning ____________________

[p.850] predators ____________________

[p.850] prey ____________________

[p.850] coevolution ____________________

[p.850] carrying capacity ____________________

[p.852] camouflage ____________________

[p.852] warning coloration ____________________

[p.852] mimicry ____________________

Fill-in-the-Blanks

(1) ______________ [p.848] competition means individuals of the same species compete with one another. (2) ______________ [p.848] competition occurs between populations of different species. (3) ______________ [p.848] competition is usually the least intense of the two types. According to the concept of (4) ______________ ______________ [p.849], two species that require identical resources cannot coexist indefinitely.

In (5) ______________-______________ species [p.849], competing species coexist by subdividing a category of similar resources. A good example is the nine species of fruit-eating (6) ______________ [p.849] in the same New Guinea forest. They all require the same resource: fruit. The pigeons overlap only slightly in their use of the resource, because each species specializes in fruits of a particular (7) ______________ [p.849]. Another example is three species of annual plants, each adapted to exploiting a different portion of the habitat. Each has a(n) (8) ______________ [p.849] system that grows to a different depth in the soil from the others.

(9) ______________ [p.850] are animals that feed on other living organisms, called their (10) ______________ [p.850], but do not take up residence on or in them. Their (11) ______________ [p.850] usually die from the interaction. Without predators, prey species increase in number. The maximum population size that a given environment can support is its (12) ______________ ______________ [p.850].

Many adaptations of predators (or parasites) and their victims arose through (13) ______________ [p.850], the joint evolution of two (or more) species that exert selection pressure on each other as a result of close ecological interaction. Organisms that exhibit (14) ______________ [p.852] have adaptations in form, patterning, color, and behavior that help them blend in with their surroundings and escape detection. Toxic types of prey often have (15) ______________ ______________ [p.852], or conspicuous patterns and colors that predators learn to recognize as "avoid me" signals. (16) ______________ [p.852] is a term applied to prey organisms that bear close resemblance to dangerous, unpalatable, or hard-to-catch species. (17) ______________-of-______________ [p.852] *defenses* refers to last-ditch tricks used by prey species for their survival when they are cornered or under attack. Predators may counter prey defenses with their own marvelous (18) ______________ [p.853].

Matching

Match each of the following with the most appropriate answer. The same letter may be used more than once. Use only one letter per blank.

19. ____ Cornered earwigs, skunks, and stink beetles producing awful odors [p.852]
20. ____ A young, inexperienced bird will spear a yellow-banded wasp or an orange-patterned monarch butterfly — once [p.852]
21. ____ Different species of New Guinea pigeons coexisting in the same environment by eating fruits of different sizes [p.849]
22. ____ Canadian lynx and snowshoe hare [p.851]
23. ____ Grasshopper mice plunging the noxious chemical–spraying tail end of their beetle prey into the ground to feast on the head end [p.853]
24. ____ Resemblance of *Lithops,* a desert plant, to a small rock [p.852]
25. ____ An inedible butterfly is a model for the edible *Dismorphia* [p.853]
26. ____ In the same area, drought-tolerant foxtail grasses have a shallow, fibrous root system, mallow plants have a deeper, taproot system, and smartweed has a taproot system that branches in topsoil and in soil below the roots of other species [p.849]
27. ____ Aggressively stinging yellowjackets are the likely model for nonstinging edible wasps [p.853]
28. ____ Two species of *Paramecium* requiring identical resources cannot coexist indefinitely [pp.848–849]
29. ____ Baboon on the run turns to give canine tooth display to a pursuing leopard [pp.850,852]
30. ____ Least bittern with coloration similar to surrounding withered reeds [p.852]

A. Predator–prey interactions
B. Camouflage
C. Warning coloration
D. Mimicry
E. Moment-of-truth defenses
F. Competitive exclusion
G. Resource partitioning
H. Predator responses to prey

47.6. PARASITE–HOST INTERACTIONS [pp.854–855]

47.7. FORCES CONTRIBUTING TO COMMUNITY STABILITY [pp.856–857]

47.8. FORCES CONTRIBUTING TO COMMUNITY INSTABILITY [pp.858–859]

47.9. FOCUS ON THE ENVIRONMENT: *Exotic and Endangered Species* [pp.860–861]

Selected Words: *ecto*parasites [p.854], *endo*parasites [p.854], microparasites [p.854], macroparasites [p.854], *holo*parasitic plants [p.854], *hemi*parasitic plants [p.854], *biological controls* [p.855], *natural* and *active* restoration of the climax community [p.857], *jump* dispersal [p.859], "killer bee" [p.859], kudzu [p.861]

Boldfaced, Page-Referenced Terms

[p.854] parasites ____________________

____

[p.855] social parasites ____________________

____

[p.855] parasitoids ____________________

____

[p.856] ecological succession ____________________

____

[p.856] pioneer species ____________________

____

[p.856] climax community ____________________

____

[p.856] primary succession ____________________

____

[p.856] secondary succession ____________________

____

[p.856] climax-pattern model ____________________

____

[p.858] keystone species ____________________

____

[p.859] geographic dispersal ____________________

____

[p.859] exotic species ____________________

____

[p.860] endangered species ____________________

____

Matching

Choose the most appropriate answer for each term.

1. ____ ectoparasites [p.854]
2. ____ holoparasitic plants [p.854]
3. ____ endoparasites [p.854]
4. ____ microparasites [p.854]
5. ____ hemiparasitic plants [p.854]
6. ____ macroparasites [p.854]
7. ____ social parasites [p.855]
8. ____ biological controls [p.855]
9. ____ parasitoids [p.855]

A. Parasites and parasitoids often touted as an alternative to chemical pesticides
B. Bacteria, viruses, protozoans, and sporozoans; microscopically small, reproduce rapidly
C. Live on a host's surface
D. Insect larvae that always kill what they eat; consume larvae or pupae of other insect species
E. Complete their life cycle by drawing on social behaviors of another species; the cowbird is a good example
F. Nonphotosynthetic; they withdraw nutrients and water from host plants
G. Large; includes many flatworms, roundworms (nematodes), and arthropods such as fleas and ticks
H. Live inside a host's body; some species live on or in one or more hosts for their entire life cycle
I. Retain the capacity for photosynthesis but still withdraw nutrients and water from host plants; mistletoe is an example

Fill-in-the-Blanks

When a community develops in a sequence from pioneers to a stable end array of species that remain in equilibrium over some region, this is known as ecological (10) ____________ [p.856]. (11) ____________ [p.856] species are opportunistic colonizers of vacant habitats that enjoy high dispersal rates and rapid growth. As time passes, the (12) ____________ [p.856] are replaced by species that are more competitive; these species are themselves then replaced, until the array of species stabilizes under the prevailing habitat conditions. This persistent array of species is the (13) ____________ [p.856] community. When pioneer species begin to colonize a barren habitat such as a volcanic island, this is known as (14) ____________ [p.856] succession. Once established, the pioneers improve conditions for other species and often set the stage for their own (15) ____________ [p.856].

When an area within a community is disturbed and then recovers to move again toward the climax state, it is known as (16) ____________ [p.856] succession. Some scientists hypothesize that the colonizers (17) ____________ [p.856] their own replacement, while other scientists subscribe to the idea that the sequence of (18) ____________ [p.856] depends on what species arrive first to compete against the species that could replace them. According to the (19) ____________-____________ [p.856] model, a community is adapted to a total pattern of environmental factors — topography, climate, soil, wind, species interactions, recurring disturbances, chance events, and so on — that vary in their influence over a region.

Small-scale and recurring changes contribute to the internal dynamics of the (20) ____________ [p.856] as a whole. For example, a tree falling in a tropical forest opens a gap in the forest canopy that allows more light to enter that gap. Here the growth of previously suppressed small trees and the germination of (21) ______________ [p.857] or shade-intolerant species is encouraged. Giant sequoia trees in climax communities of the California Sierra Nevada are best maintained by modest (22) ______________ [p.857] which eliminate trees and shrubs that compete with young sequoias but do not damage the older sequoias. Thus recurring small-scale changes are built into the overall workings of many communities. The example of secondary succession regrowth following Mount St. Helen's violent eruption in 1980 is classified as (23) ______________ [p.857] restoration of the climax community. Attempts to re-establish biodiversity in key areas adversely affected or lost as a result of agriculture and other human activities are known as (24) ______________ [p.857] restoration.

Choice

For questions 25–33, choose from the following terms: [p.856]

a. primary succession b. secondary succession

25. _____ Following a disturbance, a patch of habitat or a community moves once again toward the climax state
26. _____ Successional changes begin when a pioneer population colonizes a barren habitat
27. _____ A disturbed area within a community recovers and moves again toward the climax state
28. _____ Many plants in this succession arise from seeds or seedlings that are already present when the process begins
29. _____ Many types are mutualists with nitrogen-fixing bacteria and initially outcompete other plants in nitrogen-poor habitats
30. _____ Involves typically small plants, with brief life cycles, adapted to grow in exposed areas with intense sunlight, swings in temperature, and nutrient-deficient soil
31. _____ A successional pattern that occurs in abandoned fields, burned forests, and storm-battered intertidal zones
32. _____ May occur on a new volcanic island or on land exposed by the retreat of a glacier
33. _____ Once established, the pioneers improve conditions for other species and often set the stage for their own replacement

Fill-in-the-Blanks

A(n) (34) ______________ [p.858] species is a dominant species that dictates community structure. *Pisaster,* a sea star, is a(n) (35) ______________ [p.858] species that controls an abundance of mussels, limpets, chitons, and barnacles. If sea stars are removed from experimental plots, (36) ______________ [p.858], the main prey of sea stars, become the strongest competitors; they crowd out seven other species of invertebrates. When sea stars are removed, the community of (36) shrinks from 15 species to eight.

Matching

Choose the most appropriate answer for each term.

37. ____ community stability [p.858]
38. ____ endangered species [p.860]
39. ____ keystone species [p.858]
40. ____ geographic dispersal [p.859]
41. ____ jump dispersal [p.859]
42. ____ exotic species [p.860]

A. A rapid geographic dispersal mechanism, as when an insect might travel in a ship's hold from an island to the mainland
B. An outcome of forces that have achieved an uneasy balance
C. Resident of an established community that has moved from its home range and successfully taken up residence elsewhere; Nile perch, European rabbits, and kudzu are examples
D. Species that are now on rare- or threatened-species lists or have already become extinct
E. Residents of established communities move out from their home range and successfully take up residence elsewhere; the process may be slow or rapid, as in the case of expansion of home range or jump dispersal, or extremely slow, as in the case of continental drift
F. A dominant species that can dictate community structure; a sea star is an example

47.10. PATTERNS OF BIODIVERSITY [pp.862–863]

Boldfaced, Page-Referenced Terms

[p.863] distance effect ____________________

____

[p.863] area effect ____________________

____

Short Answer

1. List three factors responsible for creating the higher species-diversity values correlated with the distance of land and sea from the equator. [p.862]

   a. ____________________

   b. ____________________

   c. ____________________

Fill-in-the-Blanks

The number of coexisting species is highest in the (2) ______________ [p.862] and systematically declines toward the poles. In 1965, a volcanic eruption formed a new island southwest of (3) ______________ [p.862], which was named Surtsey. The new island served as a laboratory for studying (4) ______________ [p.862]. Within six months, bacteria, fungi, seeds, flies, and some seabirds became established on the island. After two years one vascular plant appeared, and two years after that a moss plant was observed. These organisms were all colonists from (5) ______________ [p.862].

Islands far from a source of potential colonists receive few colonizing species. The few that arrive are adapted for long-distance (6) ______________ [p.863]; this is called the (7) ______________ [p.863] effect. Larger islands tend to support more species than do small islands at equivalent distances from source areas; this is the (8) ______________ [p.863] effect. (9) ______________ [p.863] islands tend to have more and varied habitats, display more complex topography, and extend farther above sea level; thus, they favor species (10) ______________ [p.863]. In addition, being bigger (11) ______________ [p.863], they may intercept more colonists.

Most important, extinctions suppress (12) ______________ [p.863] on (13) ______________ [p.863] islands. There, the (14) ______________ [p.863] populations are far more vulnerable to storms, volcanic eruptions, diseases, and random shifts in birth and death rates. As for any island, the number of species reflects a balance between (15) ______________ [p.863] rates for new species and (16) ______________ [p.863] rates for established ones. Small islands that are distant from a source of colonists have low (17) ______________ [p.863] and high (18) ______________ [p.863] rates, so they support few species once the balance has been struck for their populations.

Problem

19. After considering the distance effect, the area effect, and species diversity patterns as related to the equator, answer the following question. There are two islands (B and C) of the same size and topography that are equidistant from the African coast (A), as shown in the following illustration. Which will have the higher species-diversity values? [pp.862–863] ______________________________

____

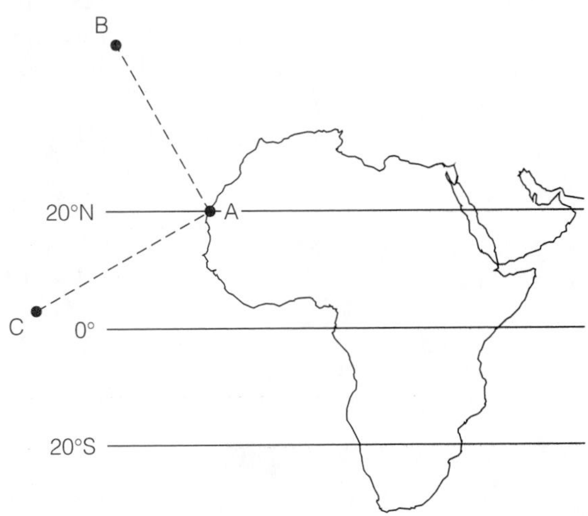

Self-Quiz

____ 1. All the populations of different species that occupy and are adapted to a given habitat are referred to as a(n) __________. [p.846]
   a. biosphere
   b. community
   c. ecosystem
   d. niche

____ 2. The range of all factors that influence whether a species can obtain resources essential for survival and reproduction is called the __________ of a species. [p.846]
   a. habitat
   b. niche
   c. carrying capacity
   d. ecosystem

____ 3. A one-way relationship in which one species benefits and the other is directly harmed is called __________. [p.846]
   a. commensalism
   b. competitive exclusion
   c. parasitism
   d. mutualism

____ 4. A lopsided interaction that directly benefits one species but does not harm or help the other much, if at all, is __________. [p.846]
   a. commensalism
   b. competitive exclusion
   c. predation
   d. mutualism

____ 5. An interaction in which both species benefit is best described as __________. [p.846]
   a. commensalism
   b. mutualism
   c. predation
   d. parasitism

____ 6. A striped skunk being pursued by a predator suddenly turns and releases its foul-smelling odor. This is an example of __________. [p.852]
   a. warning coloration
   b. mimicry
   c. camouflage
   d. moment-of-truth defense

____ 7. When an inexperienced predator attacks a yellow-banded wasp, the predator receives the pain of a stinger and will not attack again. This is an example of __________. [p.852]
   a. mimicry
   b. camouflage
   c. a prey defense
   d. warning coloration
   e. both c and d

____ 8. __________ is represented by foxtail grass, mallow plants, and smartweed because their root systems exploit different areas of the soil in a field. [p.849]
   a. Succession
   b. Resource partitioning
   c. A climax community
   d. A disturbance

____ 9. During the process of community succession, __________. [p.856]
   a. pioneer populations adapt to growing in habitats that cannot support most species
   b. pioneers set the stage for their own replacement
   c. species composition eventually is stable in the form of the climax community
   d. all of the above

____ 10. G. Gause used two species of *Paramecium* in a study that described __________. [p.848]
   a. interspecific competition and competitive exclusion
   b. resource partitioning
   c. the establishment of territories
   d. coevolved mutualism

_____ 11. The most striking patterns of species diversity on land and in the seas relate to the _______________. [p.862]
a. distance effect
b. area effect
c. immigration rate for new species
d. distance from the equator

_____ 12. The relationship between the yucca plant and the yucca moth that pollinates it is best described as _______________. [p.847]
a. camouflage
b. commensalism
c. competitive exclusion
d. obligatory mutualism

Chapter Objectives/Review Questions

1. The type of place where you normally find a maple is its _________. [p.846]
2. List five factors that shape the structure of a biological community. [p.846]
3. An organism's _________ is the sum of activities and relationships in which it engages to secure and use the resources necessary for its survival and reproduction. [p.846]
4. Contrast *fundamental niche* and *realized niche*. [p.846]
5. The relation between a bird nesting in a tree and the tree itself can be characterized as _________. [p.846]
6. In forms of _________, each of the participating species reaps benefits from the interaction. [p.846]
7. In _________ competition, disadvantages flow both ways between species. [p.846]
8. _________ and _________ are interactions that directly benefit one species and directly hurt the other. [p.846]
9. Commensalism, mutualism, and parasitism are all forms of _________, which means "living together." [p.846]
10. Define and cite an example of obligatory mutualism. [p.851]
11. Define the terms *intraspecific competition* and *interspecific competition*. [p.848]
12. Describe a study that demonstrates laboratory evidence supporting the competitive exclusion concept. [pp.848–849]
13. Cite one example of resource partitioning. [p.849]
14. A predator gets food from other living organisms, its _________. [p.850]
15. _________ take up residence in or on other living organisms — their hosts — and feed on specific host tissues for part of their life cycle. [p.850]
16. Many of the adaptations of predators (or parasites) and their victims arose through _________. [p.850]
17. Generally describe an idealized cycling of predator and prey abundances. [pp.850–851]
18. In relation to predator–prey cycles, explain what is meant by a *three-level interaction*. [p.851]
19. Completely define and give examples of the following prey defenses: warning coloration, mimicry, moment-of-truth defenses, and camouflage. [pp.852–853]
20. Describe two examples of how predators counter prey defenses with their own marvelous adaptations. [p.853]
21. What are the general body locations of ectoparasites and endoparasites? [p.854]
22. List the types of organisms classified as *microparasites* and those classified as *macroparasites*. [p.854]
23. Distinguish between holoparasitic plants and hemiparasitic plants. [p.854]
24. Discuss the advantages and disadvantages of using parasites as biological controls. [p.855]
25. Define the term *ecological succession*. [p.856]
26. A(n) _________ community is a stable, persistent array of species in equilibrium with one another and their habitat. [p.856]
27. Distinguish between primary and secondary succession. [p.856]
28. By the _________-_________ model, a community is adapted to a total pattern of environmental factors. [p.856]
29. Describe how fire disturbances positively affect a community of giant sequoias. [p.857]
30. Distinguish between natural and active restoration. [p.857]
31. Community _________ is an outcome of forces that have achieved an uneasy balance. [p.857]

32. Define what is meant by a *keystone species.* [p.858]
33. Give the reasons why Robert Paine was able to identify a sea star (*Pisaster*) as a keystone species. [p.859]
34. Explain how the introduction of exotic species can be disastrous. List five specific examples of species introductions into the United States that have had adverse results. [pp.860–861]
35. Describe the distance effect and the area effect. [p.863]
36. Estimate qualitatively the differences in species diversity and abundance of organisms likely to exist on two islands with the following characteristics: Island A has an area of 6,000 square miles and Island B has an area of 60 square miles; both islands lie at 10°N latitude and are equidistant from the same source area of colonizers. [pp.862–863]

Integrating and Applying Key Concepts

If you were Ruler of All People on Earth, how would you organize industry and human populations in an effort to solve our most pressing pollution problems?

Is there a *fundamental niche* that is occupied by humans? If you think so, describe the minimal abiotic and biotic conditions required by populations of humans in order to live and reproduce. (Note that thrive and be happy are not criteria.) If you do not think so, state why.

These minimal niche conditions can be viewed as categories of resources that must be protected by populations if they are to survive. Do you believe that the cold war between the United States and the Soviet Union primarily involved protection of minimal niche conditions, or do you believe that the cold war was based on other, more (or less) important factors? (a) If the former, how do you think *minimal* niche conditions might have been guaranteed for all humans willing and able to accept certain responsibilities as their contribution toward enabling this guarantee to be met? (b) If the latter, identify what you think those factors are, and explain why you consider them more (or less) important than minimal niche conditions.

48

ECOSYSTEMS

Interactive Exercises

Crêpes for Breakfast, Pancake Ice for Dessert [pp.866–867]

48.1. THE NATURE OF ECOSYSTEMS [pp.868–869]

48.2. THE NATURE OF FOOD WEBS [pp.870–871]

48.3. FOCUS ON SCIENCE: *Biological Magnification in Food Webs* [p.872]

48.4. STUDYING ENERGY FLOW THROUGH ECOSYSTEMS [p.873]

48.5. FOCUS ON SCIENCE: *Energy Flow at Silver Springs* [p.874]

Selected Words: conservation biology [p.867], *herbivores* [p.868], *carnivores* [p.868], *parasites* [p.868], *omnivores* [p.868], *scavengers* [p.868], *energy inputs* [p.868], *nutrient inputs* [p.868], *energy outputs* [p.868], *nutrient outputs* [p.868], *troph* [p.869], *cross-connecting* [p.869], *gross* primary productivity [p.873], *net* amount [p.873]

Boldfaced, Page-Referenced Terms

[p.868] primary producers ________________

[p.868] consumers ________________

[p.868] decomposers ________________

[p.868] detritivores ________________

[p.868] ecosystem ________________

[p.869] trophic levels ________________

[p.869] food chain ________________

[p.869] food webs ________________

[p.871] grazing food webs ________________

[p.871] detrital food webs ________________

[p.872] ecosystem modeling ________________

[p.872] biological magnification ________________

[p.873] primary productivity ________________

[p.873] net ecosystem production ________________

[p.873] biomass pyramid ________________

[p.873] energy pyramid ________________

Matching

Choose the most appropriate answer for each term.

1. ____ primary producers [p.868]
2. ____ consumers [p.868]
3. ____ herbivores [p.868]
4. ____ carnivores and parasites [p.868]
5. ____ decomposers [p.868]
6. ____ detritivores [p.868]
7. ____ omnivores [p.868]
8. ____ scavengers [p.868]
9. ____ ecosystem [p.868]
10. ____ energy inputs [p.868]
11. ____ nutrient inputs [p.868]
12. ____ energy outputs [p.868]
13. ____ nutrient outputs [pp.868–869]
14. ____ trophic levels [p.869]
15. ____ food chain [p.869]
16. ____ food webs [p.869]

A. Enters the ecosystem from the sun
B. Consumers that dine on animals, plants, fungi, even protistans and bacteria
C. Feed on the tissues of other organisms
D. Hierarchy of feeding relationships
E. A system of cross-connecting food chains
F. Feed on living animal tissues
G. Cannot be recycled; over time most is lost to the environment
H. The autotrophs
I. Animals that ingest nonliving plant and animal tissues
J. Refers to a straight-line series of steps by which energy stored in autotroph tissues passes on through higher trophic levels
K. Consumers that eat only plants
L. Generally cycled, but some slips away
M. Fed by nonliving products and remains of producers and consumers
N. An array of organisms and their physical environment, interacting through a one-way flow of energy and a cycling of materials
O. Heterotrophs that ingest decomposing bits of organic matter, such as leaf litter
P. An example would be a creek that delivers dissolved materials into a lake

Choice

For questions 17–31, choose from the following trophic levels in a tallgrass prairie: [pp.869–870]

a. primary producers b. first-level consumers c. second-level consumers d. third-level consumers e. fourth-level consumers

17. ____ Garter snake, crow
18. ____ Larvae, earthworm
19. ____ Marsh hawk, mites
20. ____ Grasses, composites
21. ____ Saprobic fungi, bacteria
22. ____ Badger, coyote
23. ____ Prairie vole, pocket gopher
24. ____ Ticks, parasitic flies
25. ____ Gopher, ground squirrel
26. ____ Nitrifying bacteria
27. ____ Grasshoppers, moths, butterflies
28. ____ Weasel, badger, coyote
29. ____ The only category lacking heterotrophs
30. ____ Meadow frog, spiders
31. ____ Green plants

Fill-in-the-Blanks

When researchers compared the chains of different food webs, a pattern emerged: In most cases, (32) ______________ [p.870] initially captured by producers passed through no more than four or five trophic levels. It simply makes no difference how much energy is available in the (33) ______________ [p.870]. Remember, energy transfers never are (34) ______________ [p.870] percent efficient; a bit of energy is lost at each step. In time, the energy required to capture an organism at a higher (35) ______________ [p.870] level would be more than the amount of energy obtainable from it. Even rich ecosystems that are able to support many species in complex food webs are not characterized by (36) ______________ [p.870] food chains.

Field studies and computer simulations of the food webs of marine, freshwater, and terrestrial ecosystems reveal more (37) ______________ [p.871]. For example, the chains in food webs tend to be (38) ______________ [p.871] when environmental conditions vary. By contrast, the chains are longer in (39) ______________ [p.871] environments, such as zones of the deep ocean. The most complex webs have the greatest number of (40) ______________ [p.871] species, but their chains are (41) ______________ [p.871]. Such webs are found in (42) ______________ [p.871]. The simplest food webs have more top (43) ______________ [p.871].

Energy from a primary source flows in one direction through two categories of food webs. In (44) ______________ [p.871] food webs, the energy flows from photoautotrophs to herbivores and then through carnivores. By contrast, in (45) ______________ [p.871] food webs, energy flows primarily from photoautotrophs through detritivores and decomposers. In nearly all ecosystems, grazing and detrital food webs (46) ______________ - ______________ [p.871]. The amount of energy moving through food webs differs from one ecosystem to the next and often varies with the (47) ______________ [p.871]. In most cases, however, the bulk of the net primary production passes through the (48) ______________ [p.871] food webs. Remains and wastes of (49) ______________ [p.871] and (50) ______________ [p.871] form the basis of detrital food webs.

(51) ______________ ______________ [p.872] is a method of identifying and combining crucial bits of information about an ecosystem through computer programs and models in order to predict the outcome of the next disturbance. (52) ______________ [p.872], a relatively stable hydrocarbon, is a synthetic organic pesticide. Because it is insoluble in water, winds carry DDT in vapor form, transporting fine particles of it. DDT is also highly soluble in (53) ______________ [p.872], so it can accumulate in the tissues of organisms. Thus, DDT may exhibit (54) ______________ ______________ [p.872], the situation in which a nondegradable or slowly degradable substance becomes more and more (55) ______________ [p.872] in the tissues of organisms at the higher trophic levels of a food (56) ______________ [p.872]. Most of the DDT from all organisms that a(n) (57) ______________ [p.872] feeds on during its lifetime ends up in its own tissues. DDT and modified forms of it disrupt (58) ______________ [p.872] activities and are often toxic to many aquatic and terrestrial animals.

The rate at which an ecosystem's primary producers capture and store a given amount of energy in a specified time interval is the primary (59) ______________ [p.873]. (60) ______________ [p.873] primary productivity is the total rate of photosynthesis for the ecosystem during the specified interval. The (61) ______________ [p.873] amount is the rate of energy storage in plant tissues in excess of the rate of aerobic respiration by the plants. Other factors such as seasonal patterns and distribution through a given habitat also influence the amount of (62) ______________ [p.873] primary production.

Ecologists often represent the trophic structure of an ecosystem in the form of an ecological (63) ______________ [p.873]. In such models, the primary (64) ______________ [p.873] form a broad base for successive tiers of consumers above them. Some pyramids are based on (65) ______________ [p.873], or the weight of all the members at each tier or trophic level. Sometimes a pyramid of (66) ______________ [p.873] may be upside down, with the (67) [check one] ☐ smallest ☐ largest [p.873] tier on the bottom. Another useful way of depicting the diminishing flow of energy through ecosystems is by means of a(n) (68) ______________ [p.873] pyramid. (69) ______________ [p.873] energy enters the pyramid's base, then diminishes through successive levels to its tip (the top carnivores). These pyramids have a (70) [check one] ☐ small ☐ large [p.873] energy base at the bottom and are always right-side up. The ecological study of energy flow at Silver Springs, Florida, found that about (71) [check one] ☐ 1 ☐ 5 [p.874] percent of all incoming solar energy was captured by producers prior to transfer to the next trophic level. It was also reported that about (72) [check one] ☐ 6 ☐ 8 [p.874] percent to (73) [check one] ☐ 10 ☐ 16 [p.874] percent of the energy entering one trophic level becomes available for organisms at the next level. In general, the study showed that the efficiency of the energy transfers is so (74) [check one] ☐ high ☐ low [p.874] that ecosystems have usually no more than (75) [check one] ☐ four ☐ six [p.874] consumer trophic levels.

48.6. BIOGEOCHEMICAL CYCLES — AN OVERVIEW [p.875]

48.7. HYDROLOGIC CYCLE [pp.876–877]

Selected Words: *atmospheric* cycles [p.875], *sedimentary* cycles [p.875]

Boldfaced, Page-Referenced Terms

[p.875] biogeochemical cycle ____

____

[p.876] hydrologic cycle ____

____

[p.876] watershed ____

____

Short Answer

1. In what form(s) are elements used as nutrients usually available to producers? [p.875] ______________

2. How is the ecosystem's reserve of nutrients maintained? [p.875] ______________

3. How does the amount of a nutrient being cycled through most major ecosystems compare with the amount entering or leaving in a given year? [p.875] ______________

4. What are the common environmental input sources for an ecosystem's nutrient reserves? [p.875] ______________

5. What are the output sources of nutrient loss for land ecosystems? [p.875] ______________

Complete the Table

6. Complete the following table summarizing the functions of the three types of biogeochemical cycles. [p.875]

Biogeochemical Cycle	*General Function(s)*
a. Hydrologic cycle	
b. Atmospheric cycles	
c. Sedimentary cycles	

Matching

Choose the most appropriate answer for questions 7–10; questions 11–14 may have more than one answer.

7. _____ solar energy [p.876]
8. _____ Earth's main water reservoirs [p.876]
9. _____ watershed [p.876]
10. _____ water and plants taking up water [p.877]
11. _____ forms of precipitation falling to land [p.876]
12. _____ deforestation [p.877]
13. _____ have important roles in the global hydrologic cycle [p.876]
14. _____ forms of atmospheric water [p.876]

A. Mostly rain and snow
B. Important in moving nutrients in biochemical cycles
C. Water vapor, clouds, and ice crystals
D. Where the precipitation of a specified region becomes funneled into a single stream or river
E. May have long-term disruptive effects on nutrient availability for an entire ecosystem
F. Slowly drives water from the ocean into the atmosphere, to land, and back to the ocean — the main reservoir
G. Ocean currents and wind patterns
H. The oceans

Fill-in-the-Blanks

A(n) (15) ______________ [p.876] is any region in which precipitation becomes funneled into a single stream or river. Most of the water that enters a watershed seeps into the (16) ______________ [p.877] or becomes surface runoff that enters (17) ______________ [p.877]. Plants withdraw water together with its dissolved minerals from the soil, then lose it by (18) ______________ [p.877]. Watershed studies have revealed the influence of (19) ______________ [p.877] cover in the movements of (20) ______________ [p.877] through the ecosystem phase of biogeochemical cycles.

Measurements of watershed inputs and outputs have many practical applications. In studies of young, undisturbed forests in the Hubbard Brook watersheds, each hectare lost only 8 kilograms or so of (21) ______________ [p.877]. Rainfall and the weathering of rocks provided replacements of this element. Tree roots were also "mining" the soil, so (22) ______________ [p.877] was being stored in a growing (23) ______________ [p.877] of tree tissues. In experimental watersheds in the Hubbard Brook Valley, (24) ______________ [p.877] caused a shift in nutrient outputs. Calcium and other nutrients (25) ______________ [p.877] so slowly that deforestation may disrupt nutrient availability for entire (26) ______________ [p.877]. This is especially the case for forests that cannot (27) ______________ [p.877] themselves over the short term; (28) ______________ [p.877] forests are like this.

48.8. CARBON CYCLE [pp.878–879]

48.9. FOCUS ON SCIENCE: *From Greenhouse Gases to a Warmer Planet?* [pp.880–881]

48.10. NITROGEN CYCLE [pp.882–883]

48.11. SEDIMENTARY CYCLES [p.884–885]

Selected Words: chlorofluorocarbons [p.880], "greenhouse gases" [p.880], root nodules [p.882]

Boldfaced, Page-Referenced Terms

[p.878] carbon cycle ____

____

[p.880] greenhouse effect ____

____

[p.880] global warming ____

____

[p.882] nitrogen cycle ____

____

[p.882] nitrogen fixation ____

____

[p.882] decomposition ____

____

[p.882] ammonification ____

____

[p.882] nitrification ____

____

[p.883] denitrification ____

____

[p.883] ion exchange ____

____

[p.884] phosphorus cycle ____

____

[p.885] eutrophication ____

____

Matching

Choose the most appropriate answer for each term.

1. ____ greenhouse gases [p.880]
2. ____ carbon dioxide fixation [p.878]
3. ____ carbon cycle [pp.878–879]
4. ____ ways carbon enters the atmosphere [p.878]
5. ____ greenhouse effect [p.880]
6. ____ carbon dioxide (CO_2) [p.878]
7. ____ oceans [p.878]
8. ____ global warming [p.880]

A. Form of most of the atmospheric carbon
B. Aerobic respiration, fossil fuel burning, and volcanic eruptions
C. CO_2, CFCs, CH_4, and N_2O
D. Photosynthesizers incorporate carbon atoms into organic compounds
E. Holds most of the carbon in dissolved form
F. Carbon reservoirs → atmosphere and oceans → through organisms → carbon reservoirs
G. An effect that may be caused by greenhouse gases contributing to long-term higher temperatures at Earth's surface
H. Warming of Earth's lower atmosphere due to accumulation of certain gases

Short Answer

9. Explain why the greenhouse effect is both necessary and harmful. ______________________________

____

____

____

Matching

Choose the most appropriate answer for each term.

10. ____ nitrogen cycle description [p.882]
11. ____ nitrogen fixation [p.882]
12. ____ decomposition and ammonification [p.882]
13. ____ nitrification [p.882]
14. ____ denitrification [p.883]

A. Ammonia or ammonium in soil is stripped of electrons, with nitrite (NO_2^-) the result; other bacteria convert nitrite to nitrate (NO_3^-)
B. Bacteria convert nitrate or nitrite to N_2 and a bit of nitrous oxide (N_2O)
C. Bacteria and fungi break down nitrogen-containing wastes and plant and animal remains; released amino acids and proteins are used for growth, with the excess given up as ammonia or ammonium ions that plants can use
D. Occurs in the atmosphere (largest reservoir); only certain bacteria, volcanic action, and lightning can convert N_2 into forms that can enter food webs
E. A few kinds of bacteria convert N_2 to ammonia (NH_3), which dissolves quickly in water to form ammonium (NH_4^-)

Short Answer

15. List reasons why an insufficient soil nitrogen supply is a problem for land plants. [p.883] ____________

____

____

____

Fill-in-the-Blanks

The Earth's crust is the largest (16) ______________ [p.884] for phosphorus, just as it is for other minerals. In rock formations on land, phosphorus is typically in the form of (17) ______________ [p.884]. By natural processes of weathering and soil erosion, phosphates enter rivers and streams that transport them to (18) ______________ [p.884] sediments. Phosphorus slowly accumulates, mainly on submerged (19) ______________ [p.884] of continents. Millions of years go by. Where movements of (20) ______________ [p.884] plates uplift part of the seafloor, phosphates become exposed on drained land surfaces. In time, weathering releases the phosphates from the exposed rocks, and the cycle's (21) ______________ [p.884] phase begins again.

The (22) ______________ [p.884] phase of the cycle is more rapid than the long-term geochemical phase. All (23) ______________ [p.884] require phosphorus for synthesizing phospholipids, NADPH, ATP, nucleic acids, and other compounds. Plants take up dissolved, (24) ______________ [p.884] forms of phosphate very rapidly. Bacterial and fungal decomposers in the (25) ______________ [p.884] release phosphates, then plants take up dissolved forms of the mineral. In this way, plants help (26) ______________ [p.884] phosphorus rapidly through the ecosystem. Of all minerals, (27) ______________ [p.884] is the most prevalent limiting factor in natural (28) ______________ [p.884] around the world. Soils hold little of it, and (29) ______________ [p.884] tie up most of it in aquatic ecosystems.

In developed countries, years of heavy use of phosphorus in agricultural applications have raised the level of phosphorus in (30) ______________ [p.885]. Without stringent monitoring, phosphorus becomes concentrated in eroded (31) ______________ [p.885] and in runoff from agricultural fields. Phosphorus is also present in outflows from sewage treatment plants and from industries. It is in (32) ______________ [p.885] from cleared land — even from lawns.

Dissolved phosphorus that enters streams, lakes, and estuaries can promote dense (33) ______________ [p.885] blooms. (34) ______________ [p.885] of the remains of all these algae depletes the water of (35) ______________ [p.885], which kills off the fish and other organisms. (36) ______________ [p.885] is the name for nutrient enrichment of an ecosystem that is naturally low in nutrients. This is a natural process, but phosphorus inputs (37) ______________ [p.885] it.

Self-Quiz

____ 1. An array of organisms and their physical environment, interacting through a one-way flow of energy and a cycling of materials, is a(n) ______________ . [p.868]
a. population
b. community
c. ecosystem
d. biosphere

____ 2. ______________ ingest decomposing particles of organic matter. [p.868]
a. Herbivores
b. Parasites
c. Detritivores
d. Carnivores

____ 3. The members of feeding relationships are structured in a hierarchy, the steps of which are called ______________ . [p.869]
a. organism level
b. energy source level
c. eating level
d. trophic levels

____ 4. In grazing food webs, energy flows from ______________ . [p.871]
a. photoautotrophs through detritivores and decomposers
b. primary consumers through detritivores and decomposers
c. photoautotrophs to herbivores, then through carnivores
d. primary consumers to herbivores, then through carnivores

____ 5. Which of the following is a primary consumer? [p.869]
a. cow
b. dog
c. hawk
d. all of the above

____ 6. In a natural community, the primary consumers are ______________ . [p.868]
a. herbivores
b. carnivores
c. scavengers
d. decomposers

____ 7. A straight-line series of steps of who eats whom in an ecosystem is sometimes called a(n) ____________. [p.869]
a. trophic level
b. food chain
c. ecological pyramid
d. food web

____ 8. Of the 1,700,000 kilocalories of solar energy that entered an aquatic ecosystem in Silver Springs, Florida, investigators determined that about ____________ percent of incoming solar energy was trapped by photosynthetic autotrophs. [p.874]
a. 1
b. 10
c. 25
d. 74

____ 9. A biogeochemical cycle that deals with phosphorus and other nutrients that do not have gaseous forms is the ____________ type. [p.875]
a. sedimentary
b. hydrologic
c. nutrient
d. atmospheric

____ 10. ____________ is a process in which nitrogenous waste products or organic remains of organisms are decomposed by soil bacteria and fungi that use the amino acids being released for their own metabolism and release the excess as ammonia or ammonium in decay products, which are taken up by plants. [p.882]
a. Nitrification
b. Ammonification
c. Denitrification
d. Nitrogen fixation

____ 11. In the carbon cycle, carbon enters the atmosphere through ____________. [p.878]
a. carbon dioxide fixation
b. respiration, burning, and volcanic eruptions
c. oceans and accumulation of plant biomass
d. release of greenhouse gases

____ 12. ____________ refers to an increase in concentration of a nondegradable (or slowly degradable) substance in organisms as it is passed along food chains. [p.872]
a. Ecosystem modeling
b. Nutrient input
c. Biogeochemical cycle
d. Biological magnification

Chapter Objectives/Review Questions

1. List the principal trophic levels in an ecosystem of your choice; state the source of energy for each trophic level and give one or two examples of organisms associated with it. [pp.868–869]
2. Distinguish among herbivores, carnivores, parasites, omnivores, and scavengers. [p.868]
3. A(n) ________ is an array of organisms and their physical environment, all interacting through a flow of energy and a cycling of materials. [p.868]
4. Explain why nutrients can be completely recycled but energy cannot. [pp.868–869]
5. In terms of an ecosystem, define *energy inputs, nutrient inputs, energy outputs*, and *nutrient outputs*. [p.868]
6. Members of an ecosystem fit somewhere in a hierarchy of energy transfers (feeding relationships) called ________ levels. [p.869]
7. Distinguish between food chains and food webs. [p.869]
8. Compare grazing food webs with detrital food webs. Give an example of each. [p.871]
9. Through ________ modeling, crucial bits of information about different ecosystem components are identified and used to build computer models for predicting outcomes of ecosystem disturbances. [p.872]
10. Describe how DDT damages ecosystems; discuss biological magnification. [p.872]
11. Explain how materials and energy enter, pass through, and exit an ecosystem. [p.873]

12. Ecological pyramids that are based on ________ are determined by the weight of all the members of each trophic level; ________ pyramids reflect the energy losses at each transfer to a different trophic level. [p.873]
13. Distinguish among primary productivity, gross primary productivity, and the net amount. [p.873]
14. Discuss the use of biomass pyramids and energy pyramids. [p.873]
15. In the study of energy flow at Silver Springs, the producers trapped ________ percent of the incoming solar energy, and only a little more than a(n) ________ of the amount became fixed in new plant biomass. The producers used more than ________ percent of the fixed energy for their own metabolism. [p.874]
16. In a(n) ________ cycle, ions or molecules of a nutrient are transferred from the environment into organisms, then back to the environment, part of which functions as a vast reservoir for them. [p.875]
17. Name and define the three categories of biogeochemical cycles. [p.875]
18. Discuss water movements through the hydrologic cycle. [pp.876–877]
19. Explain what studies in the Hubbard Brook watershed have taught us about the movement of substances (water, for example) through a forest ecosystem. [pp.876–877]
20. In a vital atmospheric cycle, carbon moves through the atmosphere and food ________ on its way to and from the ocean, sediments, and rocks. [p.878]
21. Certain gases cause heat to build up in the lower atmosphere, a warming action known as the ________ effect. [p.880]
22. As many researchers suspect, greenhouse gases may be contributing to long-term higher temperatures at Earth's surface, an effect called global ________. [p.880]
23. The four greenhouse gases are carbon dioxide, CFCs, methane, and ________ oxide. [p.880]
24. A major element found in all proteins and nucleic acids moves in an atmospheric cycle called the ________ cycle. [p.882]
25. Specify the chemical events that occur during nitrogen fixation, decomposition and ammonification, and nitrification. [p.882]
26. Through the process of ________, certain bacteria convert nitrate or nitrite to N_2 and N_2O. [p.883]
27. List the ways in which humans impact the nitrogen cycle. [p.883]
28. Describe the geochemical and ecosystem phases of the phosphorus cycle. [p.884]
29. Define the term *eutrophication* and discuss the causes of this process. [pp.884–885]

Integrating and Applying Key Concepts

In 1971, *Diet for a Small Planet* was published. Its author, Frances Moore Lappé, felt that people in the United States of America wasted protein and ate too much meat. She said, "We have created a national consumption pattern in which the majority, who can pay, overconsume the most inefficient livestock products [cattle] well beyond their biological needs (even to the point of jeopardizing their health), while the minority, who cannot pay, are inadequately fed, even to the point of malnutrition." Cases of marasmus (a nutritional disease caused by prolonged lack of food calories) and kwashiorkor (caused by severe, long-term protein deficiency) have been found in Nashville, Tennessee, and on an Indian reservation in Arizona, respectively. Lappé's partial solution to the problem was to encourage people to get as much of their protein as possible directly from plants and to supplement that with less meat from the more efficient converters of grain to protein (chickens, turkeys, and hogs) and with seafood and dairy products. Most of us realize that feeding the hungry people of the world is not just a matter of distributing the abundance that exists; it is also a matter of political, economic, and cultural factors. Yet it is still worthwhile to consider applying Lappé's idea to our everyday living. Devise two full days of breakfasts, lunches, and dinners that would enable you to exploit the lowest acceptable trophic levels to sustain yourself healthfully.

49

THE BIOSPHERE

Interactive Exercises

Does a Cactus Grow in Brooklyn? [pp.888–889]

49.1. AIR CIRCULATION PATTERNS AND REGIONAL CLIMATES [pp.890–891]

49.2. THE OCEANS, LANDFORMS, AND REGIONAL CLIMATES [pp.892–893]

Selected Words: hydrosphere [p.889], lithosphere [p.889], "ozone layer" [p.890], *currents* [p.892], *leeward* [p.893], *windward* [p.893], minimonsoons [p.893]

Boldfaced, Page-Referenced Terms

[p.888] biogeography ____________________

____

[p.889] biosphere ____________________

____

[p.889] atmosphere ____________________

____

[p.889] climate ____________________

____

[p.890] temperature zones ____________________

____

[p.892] ocean ____________________

____

[p.893] rain shadow ____________________

____

[p.893] monsoons ____________________

____

Matching

Choose the most appropriate answer for each term.

1. ____ biogeography [p.888]
2. ____ biosphere [p.889]
3. ____ hydrosphere [p.889]
4. ____ lithosphere [p.889]
5. ____ atmosphere [p.889]
6. ____ climate [p.889]
7. ____ ozone layer [p.890]
8. ____ temperature zones [p.890]
9. ____ oceans [p.892]
10. ____ rain shadow [p.893]
11. ____ leeward [p.893]
12. ____ windward [p.893]
13. ____ monsoons [p.893]
14. ____ minimonsoons [p.893]

A. One continuous body of water that covers 71 percent of Earth's surface
B. Made up of gases and airborne particles that envelop Earth
C. Of the world, defined by differences in solar heating at different latitudes and the modified air circulation patterns
D. The sum total of all places in which organisms live
E. Refers to the direction not facing a wind
F. An atmospheric region between 17 and 25 kilometers above sea level
G. Recurring sea breezes along coastlines; warmed air above land rises and cooler marine air moves in
H. The waters of Earth, including the ocean, polar ice caps, and other forms of liquid and frozen water
I. Refers to the direction from which the wind is blowing
J. Average weather conditions, such as temperature, humidity, wind speed, cloud cover, and rainfall, over time
K. The outer, rocky layer of Earth
L. An arid or semiarid region of sparse rainfall on the leeward side of high mountains
M. The study of the distribution of organisms, past and present, and of diverse processes that underlie the distribution patterns
N. Patterns of air circulation that affect conditions on the continents lying poleward of warm oceans; intensely heated land draws in moisture-laden air that forms above ocean water

Fill-in-the-Blanks

The numbered items in the illustrations on the following page represent missing information. Fill in the corresponding answer blanks to complete the following narrative.

The sun's rays are more concentrated in (15) ______________ [p.890] regions than at the poles. The global pattern of air circulation starts as (16) ______________ [p.890] equatorial air (17) ______________ [p.890] and spreads northward and southward, giving up a large amount of (18) ______________ [p.890] as precipitation (warm air can hold more moisture than cold air), which supports luxuriant growth of tropical forests. The cooled, drier air then (19) ______________ [p.891] at latitudes of about 30° N and S and becomes warmer and drier in areas where deserts form. Air even farther from the equator picks up some (20) ______________ [p.891] and (21) ______________ [p.891] to higher altitudes, cools, and then gives up (22) ______________ [p.891] at latitudes of about 60° N and S to create another moist belt. The cooled, dry air then (23) ______________ [p.891] at the polar regions, where the low temperatures and almost nonexistent precipitation give rise to the cold, dry polar deserts.

Earth's rotation then modifies the air circulation by deflecting it into worldwide belts of prevailing (24) ______________ [p.891] and (25) ______________ [p.891] winds.

The world's temperature zones, beginning at the equator and moving toward the poles, are the (26) ______________ [p.891], the (27) ______________ [p.891] temperate, the (28) ______________ [p.891] temperate, and the (29) ______________ [p.891]. Finally, the amount of the (30) ______________ [p.891] radiation reaching the surface varies annually, owing to Earth's (31) ______________ [p.891] around the sun. This leads to seasonal changes in daylength, prevailing wind directions, and temperature. These factors influence the locations of different ecosystems.

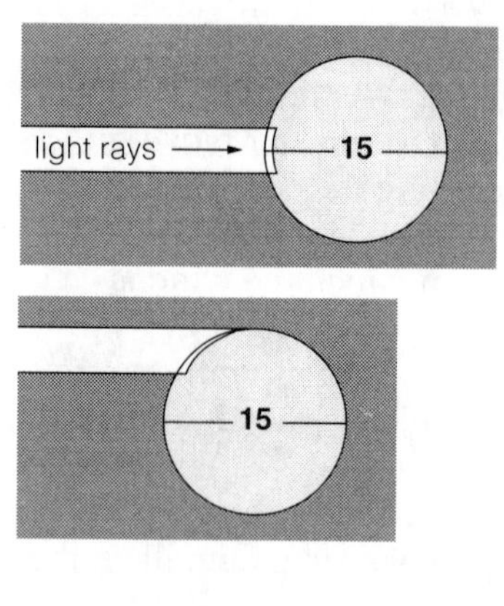

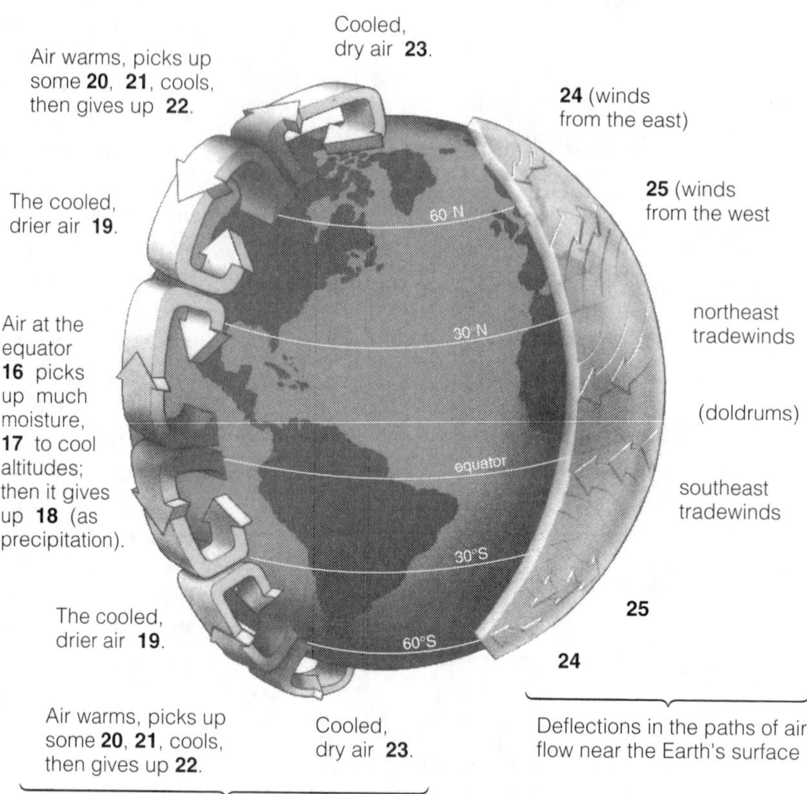

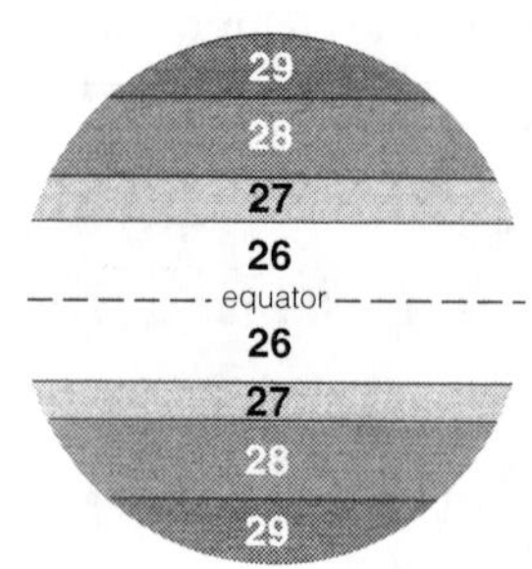

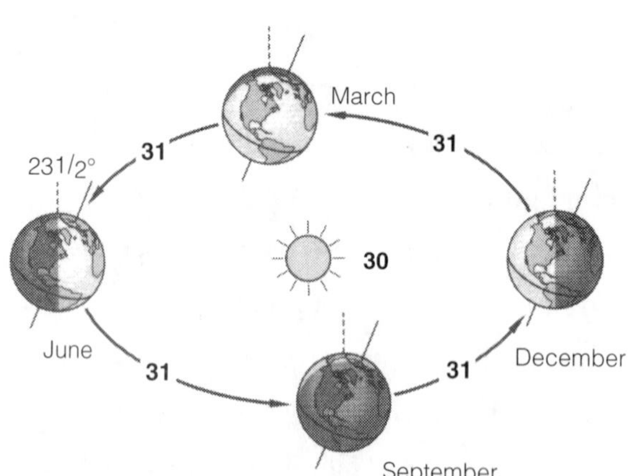

49.3. REALMS OF BIODIVERSITY [pp.894–895]

49.4. SOILS OF MAJOR BIOMES [p.896]

49.5. DESERTS [p.897]

49.6. DRY SHRUBLANDS, DRY WOODLANDS, AND GRASSLANDS [pp.898–899]

Selected Words: soil *profile* [p.896], *shortgrass* prairie [p.898], *tallgrass* prairie [p.898], *monsoon* grasslands [p.899]

Boldfaced, Page-Referenced Terms

[p.895] biogeographic realms ______________________________

____

[p.895] biome ______________________________

____

[p.895] ecoregions ______________________________

____

[p.896] soils ______________________________

____

[p.897] deserts ______________________________

____

[p.897] desertification ______________________________

____

[p.898] dry shrublands ______________________________

____

[p.898] dry woodlands ______________________________

____

[p.898] grasslands ______________________________

____

[p.898] savannas ______________________________

____

Matching

Choose the most appropriate answer for each term.

1. ____ biogeographic realms [p.895]
2. ____ biome [p.895]
3. ____ soils [p.896]
4. ____ deserts [p.897]
5. ____ desertification [p.897]
6. ____ dry shrublands [p.898]
7. ____ soil profile [p.896]
8. ____ dry woodlands [p.898]
9. ____ grasslands [p.898]
10. ____ ecoregions [p.895]
11. ____ savannas [p.898]
12. ____ monsoon grasslands [p.899]

A. Mixtures of mineral particles and variable amounts of decomposing organic material (humus)
B. Areas receiving less than 25 to 60 centimeters of rain per year; local names include *fynbos* and *chaparral*
C. The conversion of grasslands and other productive biomes to dry wastelands
D. Sweep across much of the interior of continents, in the zones between deserts and temperate forests; warm temperatures prevail in summer, winters are extremely cold
E. Areas that dominate when annual rainfall is about 40 to 100 centimeters; dominant trees can be tall but do not form a dense, continuous canopy
F. Subdivision of biogeographic realms; a large region of land characterized by the climax vegetation of the ecosystems within its boundaries
G. Formed in land regions where the potential for evaporation exceeds sparse rainfall
H. Six vast land areas on the Earth, each with distinguishing plants and animals
I. Broad belts of grasslands with a smattering of shrubs or trees; rainfall averages 90 to 150 centimeters a year, with prolonged seasonal droughts common
J. The layered structure of soils
K. Form in southern Asia where heavy rains alternate with a dry season; dense stands of tall, coarse grasses form, then die back and often burn in the dry season
L. Large areas representative of globally important biomes and water provinces that are vulnerable to extinction

Matching

Choose the most appropriate answer for each term. [p.896]

13. ____ tropical rain forest soil
14. ____ deciduous forest soil
15. ____ desert soil
16. ____ grassland soil
17. ____ coniferous soil

A. A horizon: alkaline, deep, rich in humus
B. O horizon: scattered litter; A horizon: rich in organic matter above humus layer unmixed with minerals
C. O horizon: sparse litter; A–E horizons: continually leached
D. O horizon: pebbles, little organic matter; A horizon: shallow, poor soil
E. O horizon: well-defined, compacted mat of organic deposits resulting mainly from activity of fungal decomposers

Choice

For questions 18–27, choose from the following:

a. deserts b. dry shrublands c. dry woodlands d. grasslands e. savannas

18. ____ Monsoon type that forms dense stands of tall, coarse plants in parts of southern Asia where heavy rains alternate with a dry season [p.899]

19. ____ Biome where the potential for evaporation greatly exceeds rainfall [p.897]

20. ____ Steinbeck's *Grapes of Wrath* speaks eloquently of the disruption of this biome [p.898]

21. ____ Local names for this biome include *fynbos* and *chaparral;* dominant plants often have hardened, tough, evergreen leaves [p.898]

22. ____ A biome in which dominant trees can be tall but do not form a dense canopy; includes eucalyptus woodlands of southwestern Australia and oak woodlands of California and Oregon biome [p.898]

23. ____ Home to deep-rooted evergreen shrubs, fleshy-stemmed, shallow-rooted cacti, saguaro, short prickly pear, and ocotillo biome [p.897]

24. ____ Broad belts of grasslands with a smattering of shrubs or trees; prolonged seasonal droughts are common biome [p.898]

25. ____ The dominant animals are grazing and burrowing species; grazing and periodic fires maintain the fringes of this biome [p.898]

26. ____ Within this biome, fast-growing grasses dominate where rainfall is low, but acacia and other shrubs grow where there is slightly more moisture [p.898]

27. ____ In summer, lightning-sparked, wind-driven firestorms can sweep through these biomes and swiftly burn shrubs with highly flammable leaves to the ground [p.898]

49.7. TROPICAL RAIN FORESTS AND OTHER BROADLEAF FORESTS [pp.900–901]

49.8. CONIFEROUS FORESTS [p.902]

49.9. ARCTIC AND ALPINE TUNDRA [p.903]

Selected Words: *tropical* deciduous forests [p.901], *monsoon* forests [p.901], *temperate* deciduous forests [p.901], *taiga* [p.902], *tuntura* [p.903], *arctic* tundra [p.903], *alpine* tundra [p.903]

Boldfaced, Page-Referenced Terms

[p.900] evergreen broadleaf forests ______________________________

____

[p.900] tropical rain forest ______________________________

____

[pp.900–901] deciduous broadleaf forests ______________________________

____

[p.902] coniferous forests ______________________________

____

[p.902] boreal forests ____

____

[p.902] southern pine forests ____

____

[p.903] tundra ____

____

[p.903] permafrost ____

____

Choice

For questions 1–16, choose from the following biomes:

a. deciduous broadleaf forests b. coniferous forests
c. evergreen broadleaf forests d. tundra

1. ____ Nearly continuous sunlight in summer; short plants grow and flower profusely, with rapidly ripening seeds [p.903]
2. ____ Sweep across tropical zones of Africa, the East Indies and Malay Archipelago, Southeast Asia, South America, and Central America biome [p.900]
3. ____ Highly productive forest; decomposition and mineral cycling are rapid in the hot, humid climate biome; tropical rain forests [p.900]
4. ____ Within this temperate biome, complex forests of ash, beech, chestnut, elm, and deciduous oaks once stretched across northeastern North America [p.901]
5. ____ Boreal forests or taiga; conifers are the primary producers [p.902]
6. ____ Spruce and balsam fir dominate the northern part [p.902]
7. ____ Biome of the temperate zone; cold winter temperatures; many trees drop all their leaves in winter [p.901]
8. ____ Great treeless plain between the polar ice cap and belts of boreal forests in Europe, Asia, and North America [p.903]
9. ____ Soils are highly weathered, humus-poor, and not good nutrient reservoirs [pp.900,903]
10. ____ Forests that dominate the coastal plains of the south Atlantic and Gulf states; dominant plants are adapted to the dry, sandy, nutrient-poor soil and to natural fires or controlled burns [p.902]
11. ____ Annual rainfall can exceed 200 centimeters and is never less than 130 centimeters [p.900]
12. ____ Includes tropical deciduous forests, monsoon forests, and temperate deciduous forests [pp.900–901]
13. ____ Cone-bearing trees are the primary producers; most have needle-shaped leaves with a thick cuticle and recessed stomata — adaptations that help the trees conserve water through dry times [p.902]
14. ____ Pines, scrub oak, and wiregrass grow in New Jersey; palmettos grow below loblolly and other pines in the Deep South [p.902]
15. ____ Just beneath the surface is a perpetually frozen layer, the permafrost [p.903]
16. ____ One type is known as alpine; it prevails at high elevations throughout the world, although there is no permafrost beneath the soil [p.903]

49.10. FRESHWATER PROVINCES [pp.904–905]

49.11. THE OCEAN PROVINCES [pp.906–907]

Selected Words: littoral zone [p.904], limnetic zone [p.904], profundal zone [p.904], *phyto*plankton [p.904], *zoo*plankton [p.904], *thermocline* [p.904], *oligotrophic* lakes [p.905], *eutrophic* lakes [p.905], *riffles* [p.905], *pools* [p.905], *runs* [p.905], *benthic* province [p.906], *pelagic* province [p.906]

Boldfaced, Page-Referenced Terms

[p.904] lake ______________________________

____

[p.904] spring overturn ______________________________

____

[p.904] fall overturn ______________________________

____

[p.905] eutrophication ______________________________

____

[p.905] streams ______________________________

____

[p.906] ultraplankton ______________________________

____

[p.906] marine snow ______________________________

____

[p.906] hydrothermal vents ______________________________

____

Fill-in-the-Blanks

A(n) (1) ______________ [p.904] is a body of standing fresh water with three zones. The shallow, usually well-lit (2) ______________ [p.904] zone extends around the shore to the depth at which rooted aquatic plants stop growing. The diversity of organisms is greatest here. The (3) ______________ [p.904] zone is the open, sunlit water past the littoral and extends to a depth where photosynthesis is insignificant. Aquatic communities of (4) ______________ [p.904] abound here. The (5) ______________ [p.904] zone includes all open water below the depth at which wavelengths of light suitable for photosynthesis can penetrate. Bacterial decomposers in bottom sediments of this zone enrich the water with nutrients.

Water is densest at 4°C; at this temperature, it sinks to the bottom of its basin, displacing the nutrient-rich bottom water upward and giving rise to spring and fall (6) ______________ [p.904]. In spring, ice melts, daylength increases, and the surface waters of a lake slowly warm to (7) ______________ °C [p.904]. Surface winds cause a(n) (8) ______________ [p.904] overturn, in which strong vertical movements of water carry dissolved oxygen from a lake's surface layer to its depths, while nutrients released by decomposition are brought from the bottom sediments to the surface. The surface layer warms above 4°C by midsummer and becomes less dense, and the lake develops a middle layer, the (9) ______________ [p.904], that prevents vertical mixing. When autumn comes, the upper layer (10) ______________ [p.904], becoming denser, then sinks, and the thermocline vanishes. This is termed the (11) ______________ [p.904] overturn. Water then mixes vertically, allowing dissolved oxygen to move (12) [check one] ☐ up ☐ down [p.904] and nutrients to move (13) [check one] ☐ up ☐ down [p.904]. Primary productivity of a lake corresponds with the seasons. Following a spring overturn, longer daylengths and cycled nutrients support (14) [check one] ☐ lower ☐ higher [p.904] rates of photosynthesis. By late summer, nutrient shortages are limiting photosynthesis. After the fall overturn, nutrient cycling drives a (15) [check one] ☐ short ☐ long [p.905] burst of primary activity. Not until spring will (16) ______________ [p.905] productivity increase again.

Lakes are trophic in nature. (17) ______________ [p.905] lakes are deep, nutrient-poor, and low in primary productivity. Lakes that are (18) ______________ [p.905] are often shallow, nutrient-enriched, and high in primary productivity. Human activities can determine the trophic condition of lakes. The term (19) ______________ [p.905] refers to nutrient enrichment of a lake resulting in reduced water transparency and a community rich in phytoplankton. (20) ______________ [p.905] are flowing-water ecosystems that begin as freshwater springs or seeps. Three habitat types are found between headwaters and the river's end: riffles, pools, and (21) ______________ [p.905].

Labeling and Matching

Label each numbered item in the accompanying illustration. Complete the exercise by matching and entering the letter of the proper description in the parentheses after each label. [p.906]

22. _____ province ( )
23. _____ province ( )
24. _____ zone ( )
25. _____ zone ( )

A. The entire volume of ocean water
B. All the water above the continental shelves
C. Includes all sediments and rocks of the ocean bottom; begins with continental shelves and extends to deep-sea trenches
D. Water of the ocean basin

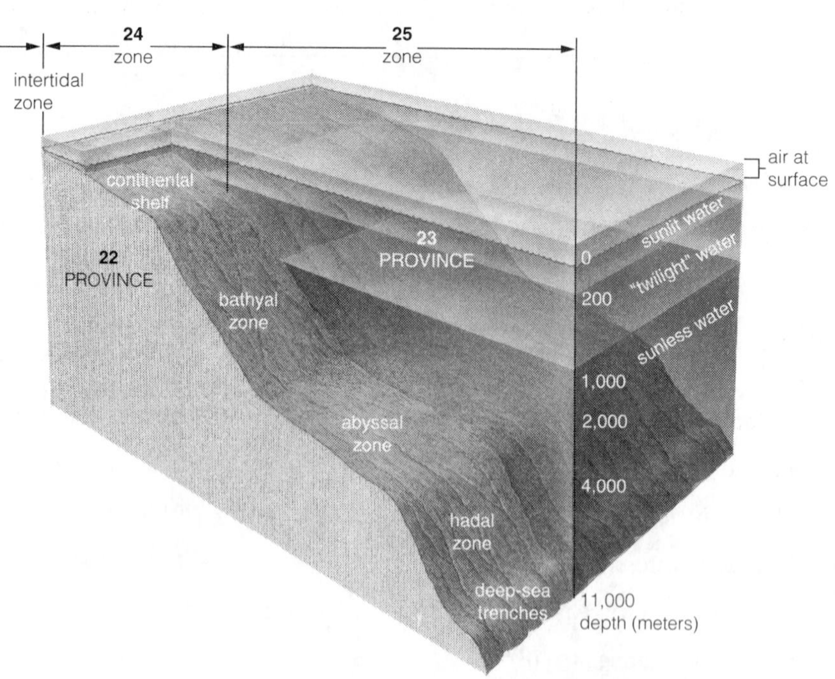

Choice

For questions 26–45, choose from the following:

a. stream ecosystems b. lake ecosystems c. ocean

26. ____ Riffles, pools, and runs [p.905]
27. ____ Hydrothermal vent communities of chemoautotrophic bacteria in the abyssal zone; possible sites of life's origin [pp.906–907]
28. ____ Ultraplankton contribute 70 percent of the primary productivity [p.906]
29. ____ Can be oligotrophic or eutrophic [p.905]
30. ____ Submerged mountains, valleys, and plains [p.906]
31. ____ Begin as freshwater springs or seeps [p.905]
32. ____ Sewage discharges on adjacent lands can contribute to eutrophication of this water [p.905]
33. ____ Average flow volume and temperature depend on rainfall, snowmelt, geography, altitude, and even shade cast by plants [p.905]
34. ____ Spring and fall overturns [pp.904–905]
35. ____ Has benthic and pelagic provinces [p.906]
36. ____ Especially in forests, these waters import most of the organic matter that supports food webs [p.905]
37. ____ Created by geologic processes such as retreating glaciers [p.904]
38. ____ They grow and merge as they flow downslope and then often combine [p.905]
39. ____ Primary productivity there varies seasonally, just as it does on land [p.906]
40. ____ The final successional stage is a filled-in basin [p.905]
41. ____ Vast "pastures" of phytoplankton and zooplankton become the basis of detrital food webs [p.906]
42. ____ Since cities formed, these waters have been sewers for industrial and municipal wastes [p.905]
43. ____ Has a thermocline by midsummer [p.904]
44. ____ Littoral, limnetic, and profundal zones [p.904]
45. ____ Bathyal, abyssal, and hadal zones [p.906]

49.12. WETLANDS AND THE INTERTIDAL ZONE [pp.908–909]

49.13. EL NIÑO AND THE BIOSPHERE [pp.910–911]

49.14. CONNECTIONS: *Rita in the Time of Cholera* [pp.912–913]

Selected Words: *rocky, sandy,* and *muddy* shores [p.909], "downwelling" [p.910], *El Niño Southern Oscillation* (ENSO) [p.910], *Vibrio cholerae* [p.912]

Boldfaced, Page-Referenced Terms

[p.908] mangrove wetland ____________________

____

[p.908] estuary ____________________

____

[p.909] intertidal zone ____

____

[p.910] upwelling ____

____

[p.910] El Niño ____

____

[p.911] La Niña ____

____

Choice

For questions 1–17, choose from the following:

a. estuary b. intertidal zone c. coastal upwelling
d. ENSO e. mangrove wetlands

1. ____ The fogbanks that form along the California coast are one outcome [p.910]
2. ____ Waves batter its resident organisms [p.909]
3. ____ Massive dislocations in global rainfall patterns characterize this event, which corresponds to changes in sea surface temperatures and air circulation patterns [p.910]
4. ____ Organisms living there must constantly contend with the tides [p.909]
5. ____ Primary producers are phytoplankton, salt-tolerant plants that withstand submergence at high tide, and algae living in mud and on plants [p.908]
6. ____ In general, an upward movement of cold, deep, often nutrient-rich ocean waters occurring in equatorial currents as well as along the coasts of continents in both hemispheres [p.910]
7. ____ A partly enclosed coastal region where seawater swirls and slowly mixes with nutrient-rich fresh water from rivers, streams, and runoff from the surrounding land [p.908]
8. ____ Salt marshes are common [p.908]
9. ____ El Niño [p.910]
10. ____ Wind friction causes surface waters to begin moving, and under the force of Earth's rotation, the slow-moving water is deflected west, away from a coast [p.910]
11. ____ Commercial fishing industries of Peru and Chile depend on it [p.910]
12. ____ A recurring seesaw in atmospheric pressure in the western equatorial Pacific [p.910]
13. ____ Forests in sheltered regions along tropical coasts [p.908]
14. ____ Sandy and muddy shores [p.909]
15. ____ In the Southern Hemisphere, commercial fishing industries of Peru depend on a wind-induced form of it [p.910]
16. ____ Ecosystems found along rocky and sandy coastlines [p.909]
17. ____ Plants have shallow, spreading roots or branching prop roots that extend from the trunk; many have pneumatophores [p.908]

Self-Quiz

_____ 1. The distribution of different types of ecosystems is influenced by _________. [pp.891,893]
   a. global air circulation patterns
   b. variation in the amount of solar radiation reaching Earth through the year
   c. surface ocean currents
   d. all of the above

_____ 2. In a(n) _____________, nutrient-rich fresh water draining from the land mixes with seawater carried in on tides. [p.910]
   a. pelagic province
   b. rift zone
   c. upwelling
   d. estuary

_____ 3. A biome with broad belts of grasslands and scattered trees adapted to prolonged dry spells is known as a _____________. [p.898]
   a. warm desert
   b. savanna
   c. tundra
   d. taiga

_____ 4. The _____________ biome is located at latitudes of about 30° N and S, has limited vegetation, and undergoes rapid surface cooling at night. [p.897]
   a. shrublands
   b. savanna
   c. taiga
   d. desert

_____ 5. In tropical rain forests, _____________. [p.900]
   a. productivity is high
   b. litter does not accumulate
   c. soils are weathered and humus poor and have poor nutrient reservoirs
   d. decomposition and mineral cycling are extremely rapid
   e. all of the above

_____ 6. In a lake, the open, sunlit water with its suspended phytoplankton is referred to as its _____________ zone. [p.904]
   a. epileptic
   b. limnetic
   c. littoral
   d. profundal

_____ 7. The lake's upper layer cools, the thermocline vanishes, lake water mixes vertically, and once again dissolved oxygen moves down and nutrients move up. This describes the _____________. [p.904]
   a. spring overturn
   b. summer overturn
   c. fall overturn
   d. winter overturn

_____ 8. The _____________ is a permanently frozen, water-impermeable layer just beneath the surface of the _____________ biome. [p.903]
   a. permafrost; alpine tundra
   b. hydrosphere; alpine tundra
   c. permafrost; arctic tundra
   d. taiga; arctic tundra

_____ 9. _____________ are air circulation patterns that influence the continents north or south of warm oceans; low pressure causes moisture-laden air above the neighboring ocean to move inland, resulting in heavy rains. [p.893]
   a. Geothermal ecosystems
   b. Upwelling
   c. Taigas
   d. Monsoons

_____ 10. All the water above the continental shelves is in the _____________. [p.906]
   a. neritic zone of the benthic province
   b. oceanic zone of the pelagic province
   c. neritic zone of the pelagic province
   d. oceanic zone of the benthic province

_____ 11. Complex forests of ash, beech, birch, chestnut, elm, and oak are found in the _____________. [p.901]

a. tropical deciduous forest
b. monsoon forest
c. temperate deciduous forest
d. evergreen broadleaf forest

_____ 12. Chemoautotrophic bacteria are the starting point for _____________. [p.907]

a. hydrothermal vent communities
b. desert communities
c. lake communities
d. coniferous forest communities

Chapter Objectives/Review Questions

1. _________ is the study of the distribution of organisms, past and present, together with the diverse processes that underlie the distribution patterns. [p.888]
2. The _________ is the sum total of all the places in which organisms live. [p.889]
3. _________ refers to average weather conditions, such as temperature, humidity, wind speed, cloud cover, and rainfall. [p.889]
4. State the reason that most forms of life depend on the ozone layer. [p.890]
5. _________ energy drives Earth's weather systems. [p.890]
6. Explain the causes of global air circulation patterns. [pp.890–891]
7. Describe how the tilt of Earth's axis affects annual variation in the amount of incoming solar radiation. [p.891]
8. Air _________ patterns, ocean _________, and landforms, in combination with global air circulation patterns, influence regional climates and help distribute nutrients in marine ecosystems. [p.893]
9. Mountains, valleys, and other aspects of topography influence _________ regional climates. [p.893]
10. Describe the cause of the rain shadow effect. [p.893]
11. Broadly, there are six distinct land realms, the _________ realms that were named first by W. Sclater and then by Alfred Wallace. [p.895]
12. Realms are further divided into _________. [p.895]
13. _________ are mixtures of mineral particles and variable amounts of decomposing organic material. [p.896]
14. List the major biomes and briefly characterize them in terms of climate, topography, and organisms. [pp.897–903]
15. The wholesale conversion of grasslands and other productive biomes to desertlike wastelands is known as _________. [p.897]
16. A(n) _________ is a standing body of fresh water with littoral, limnetic, and profundal zones. [p.904]
17. Define the terms *plankton, phytoplankton,* and *zooplankton*. [p.904]
18. Describe the spring and fall overturn in a lake in terms of causal conditions and physical outcomes. [pp.904–905]
19. _________ lakes are often deep, poor in nutrients, and low in primary productivity; _________ lakes are often shallow, rich in nutrients, and high in primary productivity. [p.905]
20. _________ refers to nutrient enrichment of a lake or some other body of water. [p.905]
21. Describe a stream ecosystem. [p.905]
22. Fully describe the benthic and pelagic provinces of the ocean. [p.906]
23. Within the pelagic province, all the water above the continental shelves is the _________ zone; the _________ zone is the water of the ocean basin. [p.906]
24. As much as 70 percent of the ocean's primary productivity may be the contribution of _________. [p.906]
25. Describe the unusual hydrothermal vent ecosystems. [pp.906–907]
26. At tropical latitudes, a different kind of nutrient-rich, saltwater ecosystem forms in tidal flats near the sea; it is called a(n) _________ wetland. [p.908]
27. Descriptively distinguish between estuaries and intertidal zones. [pp.908–909]
28. Explain the significance of ocean upwelling. [p.910]

29. Describe conditions of ENSO occurrence and how this phenomenon interrelates ocean surface temperatures, the atmosphere, and the land. [pp.902–911]
30. Explain how and why cholera outbreaks correlate with El Niño episodes in Bangladesh. [pp.912–913]

Integrating and Applying Key Concepts

One species, *Homo sapiens,* uses about 40 percent of all of Earth's productivity, and its representatives have invaded every biome, either by living there or by dumping waste products there. Many of Earth's residents are being denied the minimal resources they need to survive, while human populations continue to increase exponentially. Can you suggest a better way of keeping Earth's biomes healthy while satisfying at least the minimal needs of all Earth's residents (not just humans)? If so, outline the requirements of such a system and devise a way in which it could be established.

50

PERSPECTIVE ON HUMANS AND THE BIOSPHERE

Interactive Exercises

An Indifference of Mythic Proportions [pp.916–917]

50.1. AIR POLLUTION—PRIME EXAMPLES [pp.918–919]

50.2. OZONE THINNING—GLOBAL LEGACY OF AIR POLLUTION [p.920]

Selected Words: "ozone hole" [p.920]

Boldfaced, Page-Referenced Terms

[p.918] pollutants ______________________________

____

[p.918] thermal inversion ______________________________

____

[p.918] industrial smog ______________________________

____

[p.918] photochemical smog ____

____

[p.918] peroxyacyl nitrates (PANs) ____

____

[p.918] dry acid deposition ____

____

[p.918] acid rain ____

____

[p.920] ozone thinning ____

____

[p.920] chlorofluorocarbons (CFCs) ____

____

Fill-in-the-Blanks

(1) ______________ [p.918] are substances with which ecosystems have had no prior evolutionary experience. Adaptive mechanisms are not in place to deal with them. Air pollutants are premier examples; they include (2) ______________ ______________ [p.918], oxides of nitrogen and sulfur, and (3) ______________ [p.918]. The United States alone releases over (4) ______________ [p.918] metric tons of air pollutants each day. Whether these remain concentrated at the source or are dispersed depends on local climate and (5) ______________ [p.918].

When weather conditions trap a layer of relatively dense, cool air beneath a layer of warmer air, the situation is known as a(n) (6) ______________ ______________ [p.918]; this has been a key factor in some of the worst air pollution disasters. Where climates are cold and wet, (7) ______________ ______________ [p.918] develops as a gray haze over industrialized cities that burn coal and other fossil fuels for manufacturing, heating, and generating electric power. In warm climates, (8) ______________ ______________ [p.918] develops as a brown haze over large cities located in natural basins. The key culprit is nitric oxide. After release from vehicles, it reacts with oxygen in the air to form (9) ______________ ______________ [p.918]. When exposed to sunlight it reacts with hydrocarbons, and (10) ______________ [p.918] oxidants result. Most hydrocarbons come from spilled or partially burned (11) ______________ [p.918]. The main oxidants are ozone and (12) ______________ ______________ [p.918]. Even traces of these can sting eyes, irritate lungs, and damage crops.

Oxides of (13) ______________ [p.918] and (14) ______________ [p.918] are among the worst pollutants. Coal-burning power plants, metal smelters, and factories emit most (15) ______________ [p.918] dioxides. Motor vehicles, gas- and oil-burning power plants, and nitrogen-rich fertilizers produce (16) ______________ [p.918] oxides. During dry weather, fine particles of oxides may be briefly airborne and then fall to Earth as dry (17) ______________ [p.918] deposition. When the oxides dissolve in atmospheric water, they form (18) ______________ ______________. [p.918]. The deposited acids eat away at marble buildings, metals, rubber, plastics, and even nylon stockings. They also have the potential to disrupt the physiology of organisms, the chemistry of (19) ______________ [p.919], and biodiversity.

Choice

For questions 20–41, choose from the following aspects of atmospheric pollution:

a. thermal inversion [p.918] b. industrial smog [p.918] c. photochemical smog [p.918]
d. acid deposition [pp.918–919] e. chlorofluorocarbons [p.920] f. ozone layer [p.920]

20. _____ Develops as a brown, smelly haze over large cities
21. _____ Includes the "dry" and "wet" types
22. _____ Contributes to ozone reduction more than any other factor
23. _____ Where winters are cold and wet, this develops as a gray haze over industrialized cities that burn coal and other fossil fuels
24. _____ Weather conditions trap a layer of cool, dense air under a layer of warm air
25. _____ The cause of London's 1952 air pollution disaster, in which 4,000 people died
26. _____ Each year, from September through mid-October, it thins down at higher altitudes
27. _____ Methyl bromide, a fungicide, will account for about 15 percent of its thinning if production does not stop
28. _____ Intensifies a phenomenon called smog
29. _____ Today most of this forms in cities of China, India, and other developing countries, as well as in coal-dependent countries of eastern Europe
30. _____ Contains airborne pollutants, including dust, smoke, soot, ashes, asbestos, oil, bits of lead and other heavy metals, and sulfur oxides
31. _____ Depending on soils and vegetation cover, some regions are more sensitive than others to this
32. _____ Have been key factors in some of the worst local air pollution disasters
33. _____ Chemically attack marble buildings, metals, mortar, rubber, plastic, and even nylon stockings
34. _____ Its reduction allows more ultraviolet radiation to reach Earth's surface
35. _____ Reaches harmful levels where the surrounding land forms a natural basin, as it does around Los Angeles and Mexico City
36. _____ Tall smokestacks were added to power plants and smelters in an unsuccessful attempt to solve this problem
37. _____ As it thins, it lets more ultraviolet radiation reach Earth
38. _____ A dramatic rise in skin cancers, eye cataracts, immune system weakening, and harm to photosynthesizers is related to its reduction
39. _____ Found in refrigerators and air conditioners (as the coolants), solvents, and plastic foams
40. _____ The key culprit is nitric oxide, produced mainly by vehicles
41. _____ Oxides of sulfur and nitrogen dissolve in water to form weak solutions of sulfuric acid and nitric acid; winds may disperse them over great distances and they may fall to Earth with rain or snow

50.3. WHERE TO PUT SOLID WASTES? WHERE TO PRODUCE FOOD? [p.921]

50.4. DEFORESTATION—AN ASSAULT ON FINITE RESOURCES [pp.922–923]

50.5. FOCUS ON BIOETHICS: *You and the Tropical Rain Forest* [p.924]

Selected Words: *subsistence* agriculture [p.921], *animal-assisted* agriculture [p.921], *mechanized* agriculture [p.921]

Boldfaced, Page-Referenced Terms

[p.921] green revolution ____

____

[p.922] deforestation ____

____

[p.922] shifting cultivation ____

____

Matching

Choose the most appropriate answer for each term.

1. ____ throwaway mentality [p.921]
2. ____ green revolution [p.921]
3. ____ shifting cultivation [p.922]
4. ____ animal-assisted agriculture [p.921]
5. ____ deforestation [p.921]
6. ____ recycling [p.921]
7. ____ watersheds of forested regions [p.922]
8. ____ subsistence agriculture [p.921]
9. ____ new genetic resources [p.924]
10. ____ mechanized agriculture [p.921]

A. Agriculture in developing countries runs on energy inputs from sunlight and human labor
B. An affordable, technologically feasible alternative to "throwaway technology"
C. Act like giant sponges that absorb, hold, and gradually release water
D. Potential benefits to be obtained by genetic engineering and tissue culture methods in the rain forests
E. Runs on energy inputs from oxen and other draft animals
F. An attitude prevailing in the United States and other developed countries that greatly adds to solid waste accumulation
G. Requires massive inputs of fertilizers, pesticides, fossil fuel energy, and ample irrigation to sustain high-yield crops
H. Research directed toward improving crop plants for higher yields and exporting modern agricultural practices and equipment to developing countries
I. Trees are cut and burned, then ashes tilled into the soil; crops are grown for one to several seasons on quickly leached soils that become infertile
J. Removal of all trees from large land tracts; leads to loss of fragile soils and disrupts watersheds; greatest today in Brazil, Indonesia, Colombia, and Mexico

50.6. WHO TRADES GRASSLANDS FOR DESERTS? [p.925]

50.7. A GLOBAL WATER CRISIS [pp.926–927]

Selected Words: *primary, secondary,* and *tertiary* wastewater treatment [p.927]

Boldfaced, Page-Referenced Terms

[p.925] desertification ____________________

____

[p.925] desalinization ____________________

____

[p.925] salinization ____________________

____

[p.925] water table ____________________

____

[p.925] wastewater treatment ____________________

____

Dichotomous Choice

Circle one of two possible answers given between parentheses in each statement.

1. Conversion of large tracts of grasslands, or rain-fed or irrigated croplands to a more barren state is known as (subsistence agriculture/desertification). [p.925]
2. Presently, (too many cattle in the wrong places/overgrazing on marginal lands) is the main cause of large-scale desertification. [p.925]
3. In Africa, (domestic cattle/native wild herbivores) trample grasses and compact the soil surfaces as they wander about looking for water. [p.925]
4. A 1978 study by biologist David Holpcraft demonstrated that African range conditions improved in land areas where (domestic cattle/native wild herbivores) were ranched. [p.925]
5. Without irrigation and conservation practices, grasslands that were converted for agriculture often end up as (deserts/forested watersheds). [p.925]

Fill-in-the-Blanks

Earth has a tremendous supply of water, but most is too (6) ____________ [p.926] for human consumption or for agriculture. The removal of salt from seawater is called (7) ____________ [p.926]. For most countries this process is impractical, due to the costly fuel (8) ____________ [p.926] necessary to drive it. Large-scale (9) ____________ [p.926] accounts for nearly two-thirds of the human population's use of fresh water. Irrigation of otherwise useless soil can result in salt buildup, or (10) ____________ [p.926], caused by evaporation in areas of poor soil drainage. Land that drains poorly also becomes waterlogged, thus raising the (11) ____________ ____________ [p.926]. When this is too close to the surface, soil becomes saturated with (12) ____________ [p.926] water, which can damage plant roots.

Another large problem is the fact that water tables are subsiding. For example, overdrafts have depleted half of the Ogallala aquifer, which supplies irrigation water for (13) ____________ [p.926] percent of the croplands in the United States. Inputs of sewage, animal wastes, toxic chemicals, agricultural runoff, sediments, pesticides, and plant nutrients are all sources of water (14) ______________ [p.926], which amplifies the problem of water scarcity. There are three levels of (15) ______________ [p.927] treatment: primary, secondary, and (16) ______________ [p.927] treatment. Most wastewater is not treated adequately. If the current rates of population growth and water depletion hold, the amount of fresh water available for each person on the planet will be (17) ______________ [p.927] percent less than it was in 1976. Water, not (18) ______________ [p.927], may become the most important fluid of the twenty-first century. Meanwhile national, regional, and global (19) ______________ [p.927] for water usage and water rights have yet to be developed.

Complete the Table

20. Complete the following table, which summarizes three levels of treatment methods for restoring the water quality of polluted wastewater. [p.927]

Treatment Method	*Description*
a.	Chlorination
b.	Screens and settling tanks remove sludge
c.	Microbial populations break down organic matter

50.8. A QUESTION OF ENERGY INPUTS [pp.928–929]

50.9. ALTERNATIVE ENERGY SOURCES [p.930]

50.10. BIOLOGICAL PRINCIPLES AND THE HUMAN IMPERATIVE [p.939]

Selected Words: *total* energy [p.928], *net* energy [p.928], "photovoltaic cells" [p.930]

Boldfaced, Page-Referenced Terms

[p.928] fossil fuels ____

____

[p.928] meltdown ____

____

[p.930] solar–hydrogen energy ____

____

[p.930] wind farms ____

____

[p.930] fusion power ____

____

Choice

For questions 1–10, choose from the following answers:

a. nuclear energy [p.928] b. fossil fuels [p.928] c. fusion power [p.930] d. wind energy [p.930] e. solar–hydrogen energy [p.930]

1. ____ The waste material from this process would be water vapor
2. ____ The same process as occurs in the sun's core
3. ____ Concerns over meltdowns have slowed development of this resource
4. ____ The carbon-containing remains of plants
5. ____ The waste products may persist in the environment for 250,000 years
6. ____ The technology to use this may not be available for 50 years
7. ____ Has a potential for development in areas of North and South Dakota
8. ____ Uses photovoltaic cells to split water
9. ____ Uses seawater and sunlight as the input materials
10. ____ Major reserves of these will be depleted within the next century

---

Self-Quiz

____ 1. Which of the following processes is not generally considered a component of secondary wastewater treatment? [p.927]
   a. Screens and settling tanks remove sludge.
   b. Microbial populations are used to break down organic matter.
   c. All nitrogen, phosphorus, and toxic substances are removed.
   d. Chlorine is often used to kill pathogens in the water.

____ 2. When fossil-fuel burning gives off dust, smoke, soot, ashes, asbestos, oil, bits of lead, other heavy metals, and sulfur oxides, it forms ________. [p.918]
   a. photochemical smog
   b. industrial smog
   c. a thermal inversion
   d. both a and c

____ 3. ________ result(s) when nitrogen dioxide and hydrocarbons react in the presence of sunlight. [p.918]
   a. Photochemical smog
   b. Industrial smog
   c. A thermal inversion
   d. Both a and c

____ 4. When weather conditions trap a layer of cool, dense air under a layer of warm air, ________ occurs. [p.918]
   a. photochemical smog
   b. a thermal inversion
   c. industrial smog
   d. acid deposition

____ 5. A major concern about the use of nuclear energy is ________. [pp.928–929]
   a. excessive temperature
   b. meltdown
   c. waste disposal
   d. all of the above

____ 6. Nitrogen oxides dissolve in atmospheric water to form a weak solution of sulfuric acid and nitric acid; this describes ________. [p.918]
   a. photochemical smog
   b. industrial smog
   c. ozone and PANs
   d. acid rain

_____ 7. Which of the following statements is *false*? [p.926]
   a. Ozone reduction allows more ultraviolet radiation to reach the Earth's surface.
   b. CFCs enter the atmosphere and resist breakdown.
   c. Salinization of soils aids plant growth and increases yields.
   d. CFCs already in the air will be there for over a century.

_____ 8. "Adequately reduces pollution but is largely experimental and expensive" describes _______________ wastewater treatment. [p.927]
   a. quaternary
   b. secondary
   c. tertiary
   d. primary

_____ 9. The statement "Photovoltaic cells exposed to sunlight produce an electric current that splits water molecules into oxygen and hydrogen gas" refers to _______________. [p.930]
   a. fusion power
   b. wind energy
   c. water power
   d. solar–hydrogen energy

_____ 10. Energy inputs from sunlight and human labor are the basis of _______________. [p.921]
   a. animal-assisted agriculture
   b. subsistence agriculture
   c. the green revolution
   d. mechanized agriculture

Chapter Objectives/Review Questions

1. Define *pollutant.* [p.918]
2. Distinguish between industrial and photochemical smog. [p.918]
3. During a _________ _________, weather conditions trap a layer of relatively cool, dense air under a layer of warmer air; trapped pollutants may reach dangerous levels. [p.918]
4. Describe the process by which acid rain is formed and what it does to an ecosystem. [pp.918–919]
5. Discuss the significance of the effects of the ozone layer's thinning to life on Earth. [p.920]
6. List the key sources of air pollutants as related to ozone layer thinning. [p.920]
7. Under the banner of the _________ _________, research has been directed toward improving the genetics of crop plants for higher yields and exporting modern agricultural practices and equipment to the developing countries. [p.921]
8. _________ agriculture runs on energy inputs from sunlight and human labor; _________-_________ agriculture runs on energy inputs from oxen and other draft animals; _________ agriculture requires massive inputs of fertilizers, pesticides, and ample irrigation to sustain high-yield crops. [p.921]
9. Explain the repercussions of deforestation that are evident in soils, water quality, and genetic diversity in general. [pp.922–923]
10. _________ _________ involves cutting and burning trees, tilling ashes with soil, planting crops from one to several seasons, and then abandoning the clear plots. [pp.922–923]
11. _________ is the name for the conversion of large tracts of natural grasslands to a more desertlike condition. [p.925]
12. Define <u>primary, secondary,</u> and <u>tertiary wastewater treatment,</u> and list some of the methods used in each. [p.926]
13. Explain the meaning of "the coming water wars." [p.927]
14. Explain why desalinization is not practical on a large scale. [p.926]
15. Explain how the practice of heavy irrigation influences the water table and the process of salinization. [p.926]
16. _________ energy refers to the amount left over after subtracting the energy that is used to locate, extract, transport, store, and deliver energy to consumers. [p.928]
17. List the sources of fossil fuels and tell how long the known reserves will last. [p.928]
18. Describe the dangers accompanying a meltdown. [pp.928–929]

19. Briefly characterize solar–hydrogen energy, wind energy, and fusion power as alternative sources of energy. [p.930]
20. List five ways in which you could become personally involved in ensuring that institutions serve the public interest in a long-term, ecologically sound way. [p.931, together with the whole chapter]

Integrating and Applying Key Concepts

1. If you were Ruler of All People on Earth, how would you encourage people to depopulate the cities and adopt a way of life by which they could supply their own resources from the land and dispose of their own waste products safely on their own land?
2. Explain why some biologists believe that the endangered species list now includes all species.
3. Despite the advantages of solar–hydrogen power, most industries have been slow to adopt the process. Speculate on why this may be so, and propose some steps that could be taken to encourage its use.

ANSWERS

Chapter 1 Concepts and Methods in Biology

Why Biology? [pp.2–3]

1.1. DNA, ENERGY, AND LIFE [pp.4–5]
1. cell; 2. DNA; 3. proteins; 4. amino acids; 5. enzymes; 6. RNA; 7. DNA; 8. RNA; 9. protein; 10. inheritance; 11. reproduction; 12. development; 13. energy; 14. *Metabolism*; 15. photosynthesis; 16. ATP; 17. enzymes; 18. aerobic respiration; 19. responses; 20. receptors; 21. stimuli; 22. stimulus; 23. hormones; 24. internal; 25. receptors; 26. homeostasis.

1.2. ENERGY AND LIFE'S ORGANIZATION [pp.6–7]
1. G; 2. E; 3. H; 4. J; 5. F; 6. C; 7. L; 8. B; 9. M; 10. I; 11. K; 12. D; 13. A; 14. M; 15. G; 16. A; 17. J; 18. L; 19. E; 20. D; 21. B; 22. C; 23. H; 24. I; 25. F; 26. K; 27. producers; 28. into; 29. consumers; 30. decomposers; 31. energy; 32. cycling.

1.3. IF SO MUCH UNITY, WHY SO MANY SPECIES? [pp.8–9]
1. species; 2. genus; 3. species; 4. genus; 5. species; 6. domains; 7. kingdoms; 8. fungi; 9. *prokaryotic*; 10. *eukaryotic*; 11. a. Protistans; b. Plants; c. Fungi; d. Archaebacteria; e. Eubacteria; f. Animals.

1.4. AN EVOLUTIONARY VIEW OF DIVERSITY [pp.10–11]
1. a; 2. b; 3. b; 4. b; 5. a; 6. b; 7. b; 8. a; 9. a; 10. b.

1.5. THE NATURE OF BIOLOGICAL INQUIRY [pp.12–13]
1.6. FOCUS ON SCIENCE: ***The Power of Experimental Tests*** [pp.14–15]
1.7. THE LIMITS OF SCIENCE [p.15]
1. G; 2. A; 3. D; 4. C; 5. E; 6. B; 7. F; 8. O; 9. O; 10. C; 11. O; 12. C; 13. a. inductive logic; b. control group; c. hypotheses; d. scientific theory; e. scientific experiments; f. variables; g. prediction; h. deductive logic; 14. experiment; 15. control; 16. hypothesis; 17. prediction; 18. belief; 19. inductive logic; 20. deductive logic; 21. variable; 22. test predictions; 23. experimental design; 24. sampling error; 25. subjective; 26. supernatural; 27. conviction.

Self-Quiz
1. d; 2. a; 3. b; 4. c; 5. c; 6. d; 7. b; 8. d; 9. a; 10. c; 11. d.

Chapter 2 Chemical Foundations for Cells

How Much Are You Worth? [pp.20–21]

2.1. REGARDING THE ATOMS [p.22]
2.2. FOCUS ON SCIENCE: ***Using Radioisotopes to Track Chemicals and Save Lives*** [p.23]
1. L; 2. C; 3. H; 4. D; 5. M; 6. B; 7. F; 8. J; 9. K; 10. E; 11. G; 12. A; 13. I.

2.3. WHAT HAPPENS WHEN ATOM BONDS WITH ATOM? [pp.24–25]
1. C; 2. E; 3. I; 4. H; 5. A; 6. D; 7. F; 8. B; 9. G; 10. a. sodium, 11, 12, 11; b. 20, 40; c. 12, 6, 6; d. hydrogen, 1; e. 8, 8; f. 7, 14, 7, 7; g. chlorine, 18, 17; 11. equation; 12. formula; 13. reactants; 14. products; 15. six; 16. compound; 17. mixture;

18.

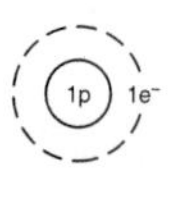

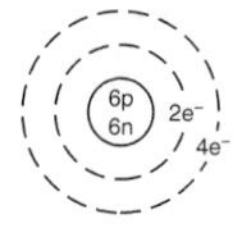

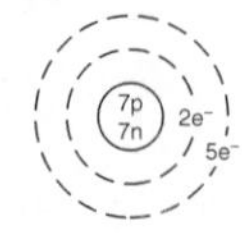

8p 8n $2e^-$ $6e^-$

O

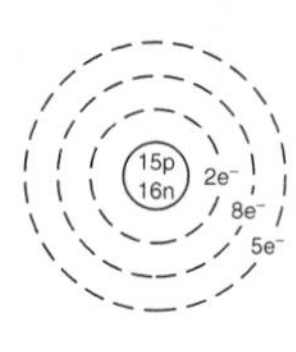

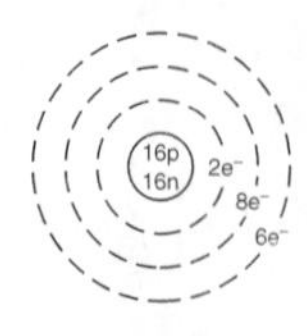

2.4. IMPORTANT BONDS IN BIOLOGICAL MOLECULES [pp.26–27]

1.

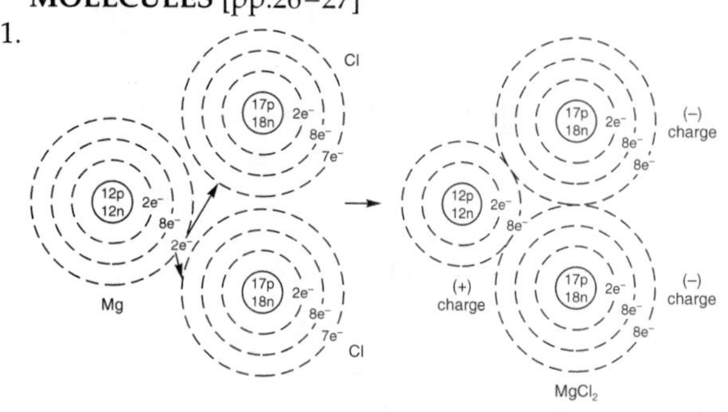

2.

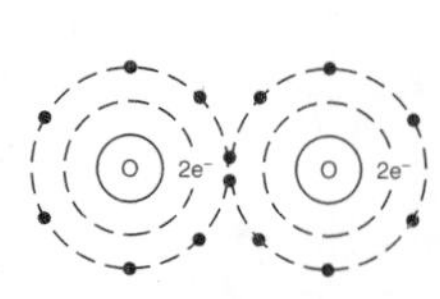

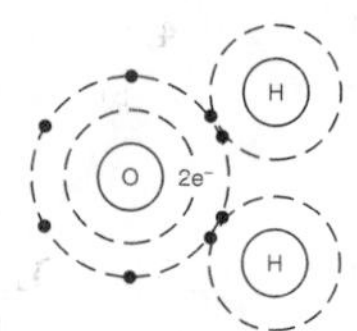

3. In a nonpolar covalent bond, there is no difference in charge between the two ends of the bond. An example is hydrogen gas (H_2). In a polar bond, there is an unequal attraction of electrons by the atoms, so that one end of the atom is negative in relation to the other. An example is water (H_2O); 4. Hydrogen bonds hold the two strands of the DNA molecule together and stabilize its structure.

2.5. PROPERTIES OF WATER [pp.28–29]

1. polarity; 2. hydrophilic; 3. hydrophobic; 4. *temperature*; 5. hydrogen; 6. stabilize; 7. Evaporation; 8. ice; 9. cohesion; 10. solvent; 11. solute; 12. dissolved; 13. hydration.

2.6. ACIDS, BASES, AND BUFFERS [pp.30–31]

1. M; 2. J; 3. E; 4. I; 5. N; 6. G; 7. C; 8. H; 9. A; 10. B; 11. K; 12. D; 13. F; 14. L; 15. a. 9, Base; b. 3, Acid; c. 6, Acid; d. 7.3–7.5, Base; e. 1.0–3.0, Acid; f. 7, Neutral.

Self-Quiz

1. c; 2. a; 3. d; 4. d; 5. c; 6. c; 7. a; 8. c; 9. a; 10. e; 11. e; 12. c.

Chapter 3 Carbon Compounds in Cells

Carbon, Carbon, in the Sky—Are You Swinging Low and High? [pp.34–35]

3.1. THE MOLECULES OF LIFE—FROM STRUCTURE TO FUNCTION [pp.36–37]

3.2. OVERVIEW OF FUNCTIONAL GROUPS [p.38]

1. global warming; 2. photosynthesis; 3. declines; 4. autumn; 5. fossil fuels; 6. cells; 7. ecosystems; 8. cells; 9. organic compound; 10. hydrocarbon; 11. functional groups; 12. four; 13. backbones; 14. ball and stick, space-filling, ribbon; 15. complexity; 16. A; 17. C; 18. D; 19. B; 20. E; 21. F; 22. methyl; 23. hydroxyl; 24. carbonyl (ketone); 25. amino; 26. phosphate; 27. carboxyl; 28. carbonyl (aldehyde).

3.3. HOW DO CELLS BUILD ORGANIC COMPOUNDS? [p.39]

1. A; 2. G; 3. C; 4. B; 5. H; 6. F; 7. D; 8. I; 9. E;

10.

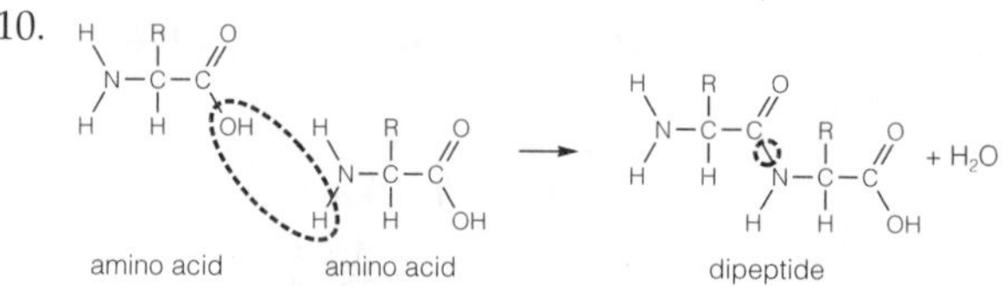

3.4. CARBOHYDRATES—THE MOST ABUNDANT MOLECULES OF LIFE [pp.40–41]

1.

enzyme (synthesis) (hydrolysis) enzyme + H_2O

glucose (a monosaccharide) glucose (a monosaccharide) maltose (a disaccharide) water

2. b; 3. b; 4. a; 5. c; 6. a; 7. c; 8. b; 9. b; 10. a. Glucose; b. starch; c. sucrose; d. glycogen; e. chitin; f. cellulose; g. ribose.

3.5. GREASY, OILY—MUST BE LIPIDS [pp.42–43]

1. a. unsaturated; b. saturated; c. unsaturated;

oleic acid **a.** stearic acid **b.** linolenic acid **c.**

2. glycerol three fatty acids yields triglyceride (a complete fat molecule) + $3H_2O$

3. a; 4. c; 5. d; 6. a; 7. a; 8. d; 9. b; 10. d; 11. c; 12. d; 13. a; 14. a; 15. a; 16. c; 17. a; 18. b; 19. d.

3.6. A STRING OF AMINO ACIDS: PROTEIN PRIMARY STRUCTURE [pp.44–45]

3.7. HOW DOES A PROTEIN'S FINAL STRUCTURE EMERGE? [pp.46–47]

3.8 WHY IS PROTEIN STRUCTURE SO IMPORTANT? [pp.48–49]

1. a. R group; b. amino group; c. carboxyl group; 2. a. carboxyl group; b. amino group; c. R group;

3.

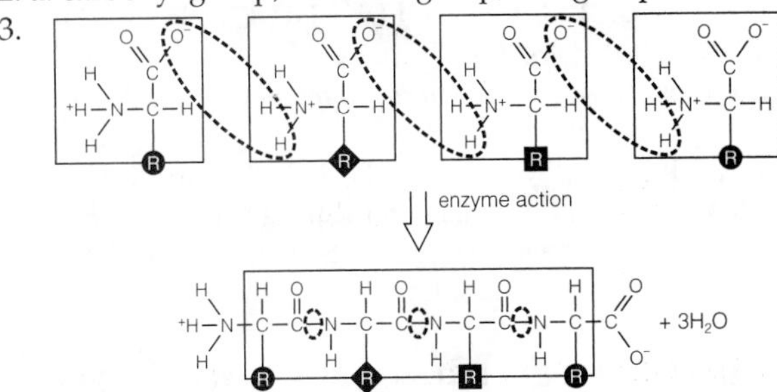

4. K; 5. N; 6. F; 7. B; 8. J; 9. L; 10. A; 11. D; 12. I; 13. G; 14. E; 15. C; 16. H; 17. M; 18. O; 19. quaternary; 20. globin; 21. heme; 22. beta; 23. valine; 24. sickle-cell anemia; 25. attracted; 26. rupture; 27. circulatory.

3.9. NUCLEOTIDES AND NUCLEIC ACIDS [pp.50–51]

1. a. phosphate group; b. five-carbon sugar; c. nitrogenous base; 2. three;

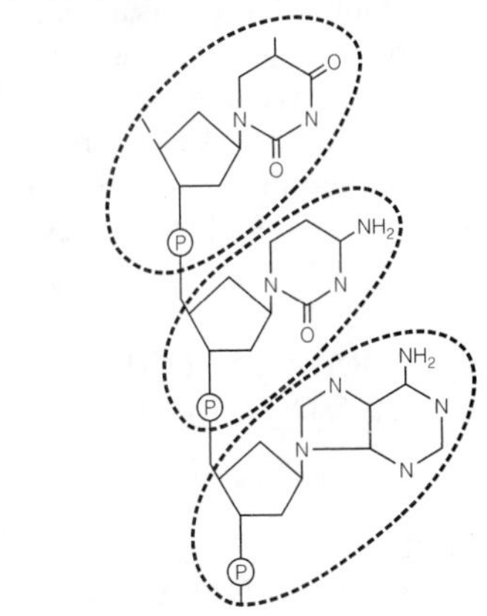

3. B; 4. C; 5. E; 6. A; 7. D.

Self-Quiz

1. c; 2. a; 3. d; 4. b; 5. d; 6. a; 7. b; 8. c; 9. a; 10. d; 11. a; 12. d; 13. e; 14. d; 15. c; 16. c; 17. b; 18. e; 19. c; 20. b; 21. b.

Chapter 4 Cell Structure and Function

Animalcules and Cells Fill'd With Juices [pp.54–55]

4.1. BASIC ASPECTS OF CELL STRUCTURE AND FUNCTION [pp.56–57]

4.2. FOCUS ON SCIENCE: ***Microscopes—Gateways to the Cell*** [pp.58–59]

1. C; 2. H; 3. J; 4. F; 5. I; 6. K; 7. D; 8. B; 9. A; 10. G; 11. E; 12. If a cell expands in diameter during growth, then its volume will increase faster than its surface area; 13. E; 14. F; 15. C; 16. B; 17. A; 18. G; 19. D.

4.3. DEFINING FEATURES OF EUKARYOTIC CELLS [pp.58–61]

1. a. Golgi body; b. Nucleus; c. Mitochondria; d. Cytoskeleton; e. Vesicles; f. Ribosome; g. Endoplasmic reticulum; 2. Plant cells contain chloroplasts, a cell wall, and a central vacuole; these are absent in animal cells.

4.4. THE NUCLEUS [pp.62–63]

1. First, it physically separates the DNA from the cytoplasm. Second, the membrane serves as a boundary that helps to control the movement of substances; 2. a. Nucleolus; b. Nuclear envelope; c. Chromatin; d. Nucleoplasm; e. Chromosome; 3. ribosomes; 4. cytoplasm; 5. endomembrane; 6. endoplasmic; 7. Golgi; 8. DNA's; 9. proteins; 10. Lipids; 11. enzymes; 12. Vesicles.

4.5. THE ENDOMEMBRANE SYSTEM [pp.64–65]

1. H; 2. A; 3. B; 4. C; 5. C; 6. D; 7. F; 8. K; 9. J; 10. E; 11. D; 12. G; 13. I.

4.6. MITOCHONDRIA [p.66]

4.7. SPECIALIZED PLANT ORGANELLES [p.67]

1. b; 2. a; 3. c; 4. a; 5. e; 6. a; 7. d; 8. b; 9. e; 10. d; 11. b; 12. b; 13. a; 14. c; 15. a; 16. b; 17. d; 18. a; 19. b.

4.8. SUMMARY OF TYPICAL FEATURES OF EUKARYOTIC CELLS [pp. 68–69]

1. a. Modifies new polypeptide chains, synthesizes lipids; b. Protects and structurally supports the cell; c. Produces sugars by photosynthesis; d. Digests and recycles materials; e. Controls substances and signals into and out of cell; f. Modifies, sorts, and ships new proteins and lipids; g. Organizes DNA molecules; h. Forms ATP by aerobic respiration.

4.9. EVEN YOUR CELLS HAVE A SKELETON [pp.70–71]

4.10. HOW DO CELLS MOVE? [pp.72–73]

4.11. CELL SURFACE SPECIALIZATIONS [pp.74–75]

1. Protein; 2. movement; 3. tubulin; 4. plus; 5. MTOCs; 6. taxol; 7. actin; 8. cytoplasmic streaming; 9. *cortex*; 10. Intermediate filaments; 11. motor; 12. ATP; 13. flagella; 14. cilia; 15. centriole; 16. centrioles; 17. microtubules; 18. dynein; 19. pseudopods; 20. microfilaments; 21. F; 22. E; 23. B; 24. A; 25. C; 26. D; 27. One cell signaling another to change its activities.

4.12. PROKARYOTIC CELLS [pp.76–77]

1. nucleus; 2. eukaryotic; 3. micrometers; 4. wall; 5. permeable; 6. polysaccharides; 7. Pathogenic; 8. mediate; 9. proteins; 10. ribosomes; 11. DNA; 12. *nucleoid*; 13. *chromosome*; 14. *flagella*; 15. pili; 16. protein; 17. archaebacteria; 18. eubacteria; 19. prokaryotic.

Self-Quiz

1. Golgi body (J); 2. vesicle (G); 3. microfilaments (L); 4. mitochondrion (O); 5. chloroplast (M); 6. microtubules (H); 7. central vacuole (C); 8. rough endoplasmic reticulum (P); 9. ribosomes (R); 10. smooth endoplasmic reticulum (D); 11. DNA + nucleoplasm (Q); 12. nucleolus (K); 13. nuclear envelope (A); 14. nucleus (I); 15. plasma membrane (N); 16. cell wall (B); 17. microfilaments (L); 18. microtubules (H); 19. plasma membrane (N); 20. mitochondrion (O); 21. nuclear envelope (A); 22. nucleolus (K); 23. DNA + nucleoplasm (Q); 24. nucleus (I); 25. vesicle (G); 26. lysosome (E); 27. rough endoplasmic reticulum (P); 28. ribosomes (R); 29. smooth endoplasmic reticulum (D); 30. vesicle (G); 31. Golgi body (J); 32. centrioles (F); 33. d; 34. d; 35. d; 36. b; 37. a; 38. c; 39. c; 40. a; 41. b; 42. c; 43. b; 44. d; 45. c; 46. d; 47. b,c,d,e; 48. a,b,c,d; 49. b,c,d,e; 50. b,c,d,e; 51. c,d; 52. a; 53. a,b,d; 54. b,d; 55. b,c,d,e; 56. b,c,d,e; 57. b,c,d,e; 58. b,c,d,e; 59. a,b,c,d,e; 60. a,b,c,d,e.

Chapter 5 A Closer Look at Cell Membranes

One Bad Transporter and Cystic Fibrosis [pp.80–81]

5.1. MEMBRANE STRUCTURE AND FUNCTION [pp.82–83]
5.2. A GALLERY OF MEMBRANE PROTEINS [pp.84–85]
5.3. FOCUS ON SCIENCE: *Do Membrane Proteins Stay Put?* [p.86]
1. D; 2. F; 3. H; 4. L; 5. A; 6. E; 7. M; 8. I; 9. G; 10. K; 11. B; 12. J; 13. C; 14. A mutated CFTR protein; 15. Microbial populations that are anchored to a lining by a polysaccharide coating.

5.4. THINK DIFFUSION [pp.86–87]
5.5. TYPES OF CROSSING MECHANISMS [p.88]
5.6. HOW DO THE TRANSPORTERS WORK? [pp.88–89]
1. permeability; 2. gradient; 3. concentration gradient; 4. Diffusion; 5. steep; 6. equilibrium; 7. higher; 8. size; 9. electric; 10. pressure; 11. passive; 12. facilitated; 13. active; 14. against; 15. energy; 16. endocytosis; 17. exocytosis; 18. b; 19. a; 20. b; 21. c; 22. c; 23. b; 24. c; 25. b; 26. c; 27. a; 28. a; 29. a; 30. a. osmosis; b. diffusion; c. active transport; d. facilitated diffusion; e. active transport; f. diffusion; g. active transport; h. facilitated diffusion; i. endocytosis; j. exocytosis.

5.7. WHICH WAY WILL WATER MOVE? [pp.90–91]
1. C; 2. F; 3. A; 4. I; 5. E; 6. B; 7. J; 8. D; 9. H; 10. G; 11. T; 12. T; 13. tonicity; 14. hypotonic; 15. T; 16. isotonic; 17. less; 18. bulk transport; 19. T.

5.8. MEMBRANE TRAFFIC TO AND FROM THE CELL SURFACE [pp.92–93]
1. D; 2. E; 3. C; 4. A; 5. B.

Self-Quiz
1. d; 2. a; 3. d; 4. c; 5. b; 6. d; 7. c; 8. d; 9. b; 10. b; 11. c; 12. a.

Chapter 6 Ground Rules of Metabolism

Growing Old with Molecular Mayhem [pp.96–97]

6.1. ENERGY AND THE UNDERLYING ORGANIZATION OF LIFE [pp.98–99]
1. energy; 2. work; 3. metabolism; 4. ATP; 5. kinetic energy; 6. heat; 7. I; 8. I; 9. II; 10. I; 11. II; 12. I; 13. I; 14. II; 15. b; 16. e; 17. d; 18. a; 19. c; 20. T; 21. T; 22. T; 23. T; 24. increasing; 25. T; 26. free radical; 27. alter; 28. superoxide dismutase; 29. hydrogen peroxide; 30. catalase; 31. antioxidants; 32. age spots.

6.2. ENERGY INPUTS, OUTPUTS, AND CELLULAR WORK [pp.100–101]
6.3. CELLS JUGGLE SUBSTANCES AS WELL AS ENERGY [pp.102–103]
6.4. ELECTRON TRANSFER CHAINS IN THE MAIN METABOLIC PATHWAYS [p.104]
1. exergonic; 2. endergonic; 3. exergonic; 4. exergonic; 5. exergonic; 6. ATP; 7. phosphate; 8. adenine; 9. ribose; 10. phosphates; 11. covalent; 12. phosphate; 13. energy; 14. energy; 15. ATP; 16. phosphorylation; 17. triphosphate; 18. ribose; 19. adeneine; 20. adenosine triphosphate; 21. electron; 22. oxidized; 23. reduced; 24. energy; 25. most; 26. energy; 27. work; 28. ATP; 29. E; 30. B; 31. C; 32. A; 33. D; 34. products; 35. reactants; 36. reversible; 37. equilibrium; 38. rate; 39. C; 40. B; 41. D; 42. F; 43. E; 44. A; 45. B; 46. A; 47. A; 48. B; 49. F; 50. G (A); 51. I; 52. D; 53. J; 54. A; 55. E; 56. C; 57. B (A); 58. H; 59. intermediate; 60. enzyme; 61. electron transfer; 62. coenzymes; 63. linear; 64. cyclic; 65. branched.

6.5. ENZYMES HELP WITH ENERGY HILLS [p.105]
6.6. HOW DO ENZYMES LOWER ENERGY HILLS? [pp.106–107]
6.7. ENZYMES DON'T WORK IN A VACUUM [pp. 108–109]
6.8. FOCUS ON HEALTH: *Beer, Enzymes, and Your Liver* [p.109]
6.9. CONNECTIONS: *Light Up the Night — and the Lab — with Enzymes* [pp.110–111]
1. enzyme; 2. Enzymes; 3. equilibrium; 4. substrate; 5. active site; 6. induced-fit; 7. activation energy; 8. collision; 9. water; 10. cofactors; 11. rate; 12. activation energy without enzyme; 13. activation energy with enzyme; 14. catalyzed; 15. uncatalyzed; 16. energy released by the reaction; 17. 40°C; 18. 60°C; 19. C; 20. B; 21. A; 22. Temperature (pH, saltiness); 23. pH (saltiness); 24. metabolism; 25. bonds; 26. 7; 27. tryptophan; 28. amino acids; 29. falls; 30. feedback inhibition; 31. enzyme; 32. allosteric; 33. allosteric binding;

34. hormones; 35. liver; 36. detoxifying; 37. damage; 38. alcoholic hepatitis; 39. alcoholic cirrhosis; 40. luciferase; 41. bioluminescence; 42. metabolism; 43. living.

Self-Quiz

1. d; 2. c; 3. e; 4. d; 5. a; 6. c; 7. d; 8. c; 9. d; 10. a.

Chapter 7 How Cells Acquire Energy

Sunlight and Survival [pp.114–115]

7.1. PHOTOSYNTHESIS — AN OVERVIEW [pp.116–117]

1. Autotrophs; 2. carbon dioxide; 3. Photosynthetic; 4. Englemann; 5. violet; 6. Heterotrophs; 7. animals; 8. aerobic respiration; 9. $12H_2O + 6CO_2 \rightarrow 6O_2 + C_6H_{12}O_6 + 6H_2O$; 10. a. Twelve; b. carbon dioxide; c. oxygen; d. glucose; e. six; 11. light-dependent (light-independent); 12. light-independent (light-dependent); 13. Carbon dioxide; 14. water; 15. glucose; 16. thylakoid; 17. grana; 18. hydrogen; 19. stroma; 20. sucrose; 21. starch; 22. cellulose; 23. b; 24. b; 25. a; 26. a; 27. b; 28. a; 29. a; 30. b; 31. a; 32. O_2; 33. ATP; 34. NADPH; 35. CO_2; 36. chloroplasts; 37. thylakoid membrane system; 38. stroma.

7.2. SUNLIGHT AS AN ENERGY SOURCE [pp.118–119]
7.3. THE RAINBOW CATCHERS [pp.120–121]

1. thylakoid; 2. photon; 3. pigments; 4. chlorophylls; 5. red (blue); 6. blue (red); 7. Carotenoids; 8. photosystem; 9. G; 10. F; 11. D; 12. E; 13. A; 14. B; 15. H; 16. I; 17. C.

7.4. THE LIGHT-DEPENDENT REACTIONS [pp.122–123]
7.5. CASE STUDY: *A Controlled Release of Energy* [p.124]

1. a. A pigment cluster dominated by P700; b. Electrons representing energy are ejected from P700 to an electron acceptor but move over the electron transport system, where some of the energy is used to produce ATP; c. A special chlorophyll molecule that absorbs wavelengths of 700 nanometers and then ejects electrons; d. A molecule that accepts electrons ejected from chlorophyll P700 and then passes electrons down the electron transport system; e. Electrons flow through this system, which is composed of a series of molecules bound in the thylakoid membrane that drive photophosphorylation; f. ADP undergoes photophosphorylation in cyclic photophosphorylation to become ATP; 2. electron acceptor molecule; 3. electron transport system; 4. photosystem II; 5. photosystem I; 6. photolysis; 7. NADPH; 8. ATP; 9. electrons; 10. thylakoid; 11. gradients; 12. water (H_2O); 13. ATP synthase; 14. ATP; 15. chemiosmotic; 16. NADPH; 17. electrons; 18–53. The following numbers should have a checkmark (✓): 19, 21, 25, 27, 29, 30, 31, 32, 33, 35, 38, 39, 40, 42, 47, and 48. All others should be blank.

7.6. THE LIGHT-INDEPENDENT REACTIONS [p.125]
7.7. FIXING CARBON — SO NEAR, YET SO FAR [pp.126–127]
7.8. CONNECTIONS: *Autotrophs, Humans, and the Biosphere* [p.128]

1. carbon dioxide (D); 2. carbon fixation (E); 3. phosphoglycerate (F); 4. adenosine triphosphate (H); 5. NADPH (G); 6. phosphoglyceraldehyde (A); 7. phosphorylated glucose (B); 8. Calvin–Benson cycle (I); 9. ribulose bisphosphate (C); 10. ATP (NADPH); 11. NADPH (ATP); 12. carbon dioxide; 13. ribulose bisphosphate; 14. PGA; 15. fixation; 16. PGA; 17. PGAL; 18. six; 19. RuBP; 20. carbon dioxide; 21. PGALs; 22. phosphorylated glucose; 23. fixation; 24. light-dependent; 25. ATP (NADPH); 26. NADPH (ATP); 27. Phosphorylated glucose; 28. photorespiration; 29. C3; 30. food (sugar); 31. oxygen; 32. oxaloacetate; 33. succulent; 34. night; 35. carbon dioxide; 36. oceans; 37. microscope; 38. photo-; 39. global warming; 40. chemo-; 41. organic; 42–85. The following numbers should have a checkmark (✓): 42, 47, 52, 54, 55, 56, 57, 59, 61, 65, 66, 71, 72, 73, 82, and 83. All others should be blank.

Self-Quiz

1. a; 2. b; 3. c; 4. a; 5. c; 6. d; 7. a; 8. b; 9. c; 10. d.

Chapter 8 How Cells Release Stored Energy

The Killers Are Coming! The Killers Are Coming! [pp.132–133]

8.1. HOW DO CELLS MAKE ATP? [pp.134–135]
1. Adenosine triphosphate (ATP); 2. Oxygen withdraws electrons from the electron transport system and joins with H^+ to form water (acts as terminal electron acceptor); 3. Glycolysis followed by fermentation or anaerobic electron transport. Some organisms (including humans) use fermentation pathways when oxygen supplies are low; many microbes rely exclusively on anaerobic pathways; 4. ATP; 5. photosynthesis; 6. aerobic (or cell) respiration; 7. glycolysis; 8. pyruvate; 9. Krebs; 10. water; 11. ATP; 12. electrons; 13. electron transfer; 14. phosphorylation; 15. ATP; 16. Oxygen; 17. anaerobic; 18. Fermentation; 19. electron transfer; 20. $C_6H_{12}O_6 + 6O_2 \rightarrow 6CO_2 + 6H_2O$; 21. One molecule of glucose plus six molecules of oxygen (in the presence of appropriate enzymes) yield six molecules of carbon dioxide plus six molecules of water; 22. G; 23. I; 24. H; 25. D; 26. F; 27. J; 28. A; 29. B; 30. C; 31. E.

8.2. GLYCOLYSIS: FIRST STAGE OF ENERGY-RELEASING PATHWAYS [pp.136–137]
1. Autotrophic; 2. Glucose; 3. pyruvate; 4. ATP; 5. NADH; 6. glucose; 7. ATP; 8. PGAL; 9. phosphate; 10. hydrogen; 11. ATP; 12. water; 13. phosphate; 14. ATP; 15. substrate-level; 16. glucose; 17. pyruvate; 18. three; 19. D; 20. F; 21. B; 22. H (G); 23. G (H); 24. A; 25. E; 26. C.

8.3. SECOND STAGE OF THE AEROBIC PATHWAY [pp.138–139]
8.4. THIRD STAGE OF THE AEROBIC PATHWAY [pp.140–141]
1. acetyl-CoA; 2. Krebs; 3. electron transfer; 4. 36 (38); 5. ATP; 6. carbon dioxide; 7. electrons; 8. NAD^+ (FAD); 9. FAD (NAD^+); 10. inner compartment; 11. inner membrane; 12. outer compartment; 13. outer membrane; 14. cytoplasm; 15. ATP; 16. oxygen (O_2); 17. $FADH_2$; 18. NADH; 19. electron transfer chain; 20. three; 21. two; 22. electron transfer; 23. chemiosmotic; 24. synthases; 25. ATP; 26. oxygen; 27. mitochondria; 28. 36 (38); 29. 38 (36); 30. c; 31. a, b; 32. c; 33. a; 34. b; 35. a; 36. b; 37. b; 38. a, b, c; 39. b; 40. c; 41. a, b; 42. b; 43. c; 44. a; 45. c; 46. c; 47. a; 48. b; 49. a; 50. c.

8.5. ANAEROBIC ROUTES OF ATP FORMATION [pp.142–143]
1. oxygen (O_2); 2. fermentation (anaerobic); 3. lactate; 4. ethanol; 5. carbon dioxide; 6. Anaerobic; 7. electron; 8. Glycolysis; 9. pyruvate; 10. NADH; 11. pyruvate; 12. lactate; 13. acetaldehyde; 14. carbon dioxide; 15. ethanol; 16. glycolysis; 17. NAD^+; 18. bacteria; 19. ATP; 20. sulfate; 21–72. With a checkmark (✓): 21, 23, 28, 29, 31, 32, 34, 38, 43, 46, 49, 51, 54, 56, 58, 60, 65; all others lack a checkmark.

8.6. ALTERNATIVE ENERGY SOURCES IN THE HUMAN BODY [pp.144–145]
8.7. CONNECTIONS: ***Perspective on the Molecular Unity of Life*** [p.146]
1. Figure 8.12 in text shows how any complex carbohydrate or fat can be broken down, and at least part of those molecules can be fed into the glycolytic pathway; 2. Energy needs are usually met first; excess is usually sent to long-term storage as fat; 3. T; 4. Page 97 (last paragraph) in text tells us that energy flows through time in one direction — from organized to less organized forms; thus energy cannot be completely recycled; 5. Page 132 in text tells us that Earth's first organisms were anaerobic fermenters; 6. fatty acids; 7. glycerol; 8. glycolysis; 9. amino acids; 10. Krebs cycle; 11. acetyl-CoA formation; 12. pyruvate; 13–102. With a checkmark (✓): 13, 15, 16, 17, 19, 22, 23, 25, 29, 31, 32, 39, 46, 47, 48, 49, 53, 54, 65, 66, 70, 71, 72, 80, 84, 87, 95, 102; all others lack a checkmark; 103. e; 104. c; 105. d; 106. a; 107. c; 108. h; 109. e; 110. c; 111. b; 112. h; 113. i; 114. e; 115. c; 116. h; 117. g.

Self-Quiz
1. c; 2. c; 3. d; 4. b; 5. c; 6. d; 7. a; 8. d; 9. c; 10. d; 11. C; 12. A, B, D; 13. B, C, D; 14. C, E; 15. B, C; 16. A, D; 17. A; 18. B, D; 19. E; 20. A, E.

Chapter 9 Cell Division and Mitosis

From Cell to Silver Salmon [pp.150–151]

9.1. DIVIDING CELLS: THE BRIDGE BETWEEN GENERATIONS [pp.152–153]
9.2. THE CELL CYCLE [pp.154–155]
1. I; 2. L; 3. C; 4. E; 5. B; 6. A; 7. K; 8. J; 9. H; 10. D; 11. G; 12. F; 13. G-1 phase; 14. S phase; 15. G-2 phase; 16. prophase; 17. metaphase; 18. anaphase; 19. telophase; 20. cytokinesis; 21. interphase; 22. mitosis; 23. daughter cells; 24. 15; 25. 22; 26. 14; 27. 13; 28. 21; 29. 20; 30. 21; 31. 22.

9.3. MITOSIS [pp.156–157]
1. interphase daughter cells (F); 2. anaphase (A); 3. late prophase (G); 4. metaphase (D); 5. cell at interphase (E); 6. early prophase (C); 7. transition to metaphase (B); 8. telophase (H); 9. (5) interphase; 10. (6) early prophase; 11. (3) late prophase; 12. (7) transition to metaphase; 13. (4) metaphase; 14. (2) anaphase; 15. (8) telophase; 16. (1) interphase daughter cells

9.4. DIVISION OF THE CYTOPLASM [pp.158–159]
9.5. FOCUS ON SCIENCE: *Henrietta's Immortal Cells* [p.160]
1. a; 2. b; 3. a; 4. a; 5. b; 6. a; 7. b; 8. b; 9. a; 10. a; 11. A self-perpetuating line of human cells used in research.

Self-Quiz
1. a; 2. c; 3. a; 4. d; 5. c; 6. c; 7. e; 8. b; 9. c; 10. d; 11. d.

Chapter 10 Meiosis

Octopus Sex and Other Stories [pp.162–163]

10.1. COMPARING SEXUAL WITH ASEXUAL REPRODUCTION [p.164]
10.2. HOW MEIOSIS HALVES THE CHROMOSOME NUMBER [pp.164–165]
1. b; 2. a; 3. a; 4. b; 5. a; 6. b; 7. a; 8. b; 9. a; 10. b; 11. Meiosis; 12. gamete; 13. Diploid; 14. Diploid; 15. Meiosis; 16. sister chromatids; 17. sister chromatids; 18. one; 19. four; 20. two; 21. chromosome; 22. haploid; 23. meiosis II; 24. 23; 25. 3; 26. 5; 27. 2; 28. 4; 29. 1.

10.3. A VISUAL TOUR OF THE STAGES OF MEIOSIS [pp.166–167]
10.4. A CLOSER LOOK AT KEY EVENTS OF MEIOSIS I [pp.168–169]
1. anaphase II (H); 2. metaphase II (F); 3. metaphase I (A); 4. prophase II (B); 5. telophase II (C); 6. telophase I (G); 7. prophase I (E); 8. anaphase I (D); 9. F; 10. I; 11. G; 12. D; 13. B; 14. J; 15. A; 16. H; 17. C; 18. E.

10.5. FROM GAMETES TO OFFSPRING [pp.170–171]
1. b; 2. c; 3. a; 4. c; 5. b; 6. a; 7. b; 8. a; 9. b; 10. a; 11. b; 12. c; 13. b; 14. a; 15. 2 ($2n$); 16. 5 (n); 17. 4 (n); 18. 1 ($2n$); 19. 3 (n); 20. A; 21. B; 22. E; 23. D; 24. C; 25. crossing-over and recombination during prophase I; independent assortment of chromosomes during metaphase I; random fertilization of gametes.

10.6. MEIOSIS AND MITOSIS COMPARED [pp.172–173]
1. a. mitosis; b. mitosis; c. meiosis; d. meiosis; e. meiosis; f. meiosis; g. meiosis; h. mitosis; i. meiosis; j. mitosis; k. meiosis; 2. C; 3. F; 4. D; 5. A; 6. B; 7. E; 8. 4; 9. 8; 10. 4; 11. 8; 12. 2.

Self-Quiz
1. a; 2. a; 3. d; 4. a; 5. c; 6. b; 7. a; 8. a; 9. b; 10. d.

Chapter 11 Observable Patterns of Inheritance

A Smorgasbord of Ears and Other Traits [pp.176–177]

11.1. MENDEL'S INSIGHT INTO PATTERNS OF INHERITANCE [pp.178–179]
1. I; 2. B; 3. D; 4. J; 5. H; 6. M; 7. O; 8. E; 9. F; 10. K; 11. A; 12. G; 13. N; 14. C; 15. L; 16. P.

11.2. MENDEL'S THEORY OF SEGREGATION [pp.180–181]
11.3. INDEPENDENT ASSORTMENT [pp.182–183]
1. C; 2. B; 3. G; 4. A; 5. D; 6. F; 7. E; 8. a. 1/2 *Tt*, 1/2 *tt*, b. 1/2 tall, 1/2 dwarf; 9. a. _ tall, _ dwarf; 1/2 *Tt*, 1/2 *tt*, b. All tall; _ *TT*, _ *Tt*, c. All dwarf; All *tt*, d. _ tall, _ dwarf;

_TT, 2/4 Tt, _ tt, e. _ tall, _ dwarf; 1/2 Tt, 1/2 tt, f. All tall; All Tt, g. All tall; All TT, h. All tall; _ TT, _ Tt, 10. a. 9/16; b. 3/16; c. 3/16; d. 1/16 (note Punnett square below):

	BR	Br	bR	br
BR	BBRR	BBRr	BbRR	BbRr
Br	BBRr	BBrr	BbRr	Bbrr
bR	BbRR	BbRr	bbRR	bbRr
br	BbRr	Bbrr	bbRr	bbrr

11. Albino = *aa*, normal pigmentation = *AA* or *Aa*. The woman of normal pigmentation with an albino mother is genotype *Aa*; the woman received her recessive gene (*a*) from her mother and her dominant gene (*A*) from her father. It is likely that half of the couple's children will be albinos (*aa*) and half will have normal pigmentation but be heterozygous (*Aa*); 12. a. *F*1: black trotter; *F*2: nine black trotters, three black pacers, three chestnut trotters, one chestnut pacer; b. black pacer ; c. *BbTt*; d. *bbtt*, chestnut pacers and *BBTT*, black trotters.

11.4. DOMINANCE RELATIONS [p.184]
11.5. MULTIPLE EFFECTS OF SINGLE GENES [p.185]
11.6. INTERACTIONS BETWEEN GENE PAIRS [pp.186–187]

1. a. multiple allele system; b. incomplete dominance; c. codominance; d. pleiotropy; e. epistasis; 2. a. All pink; All *RR'*, b. All white; All *R'R'*, c. _ red, _ pink; _*RR'*, _ *RR*, d. All red; All *RR*; 3. The man must have sickle-cell trait with the genotype Hb^AHb^S, and the woman he married would have a normal genotype, Hb^AHb^A. The couple could be told that the probability is 1/2 that any child would have sickle-cell trait and 1/2 that any child would have the normal genotype; 4. Both the man and the woman have the genotype Hb^AHb^S. The probability of children from this marriage is: 1/4 normal, Hb^AHb^A; 1/2 sickle-cell trait, Hb^AHb^S; 1/4 sickle-cell anemia, Hb^SHb^S; 5. genotypes: 1/4 I^AI^A, 1/4 I^AI^B, 1/4 I^Ai, 1/4 I^Bi, phenotypes: 1/2 A, 1/4 AB, 1/4 B; 6. genotypes: 1/4 I^AI^B; _ I^Bi; 1/4 I^Ai; 1/4 *ii*, phenotypes: 1/4 AB, 1/4 B, 1/4 A, 1/4 O; 7. genotypes: all I^Ai; phenotypes: all A; 8. genotypes: all *ii*; phenotypes: all O; 9. genotypes: 1/4 I^AI^A, 1/2 I^AI^B, 1/4 I^BI^B, phenotypes: 1/4 A, 1/2 AB, 1/4 B; 10. 1/4 color, 3/4 white; 11. 3/4 color, 1/4 white; 12. 1/4 color, 3/4 white; 13. 3/8 black, 1/2 yellow, 1/8 brown; 14. The genotype of the male parent is *RrPp* and the genotype of the female parent is *rrpp*. The offspring are 1/4 walnut comb, *RrPp*; 1/4 rose comb, *Rrpp*; 1/4 pea comb, *rrPp*; and 1/4 single comb, *rrpp*; 15. The genotype of the walnut-combed male is *RRpp* and the genotype of the single-combed female is *rrpp*. All offspring are rose comb with the genotype *Rrpp*.

11.7. HOW CAN WE EXPLAIN LESS PREDICTABLE VARIATIONS? [pp.188–189]
11.8. ENVIRONMENTAL EFFECTS ON PHENOTYPE [pp.190–191]

1. b; 2. b; 3. a; 4. b; 5. a.

Self-Quiz

1. d; 2. b; 3. a; 4. c; 5. d; 6. b; 7. d; 8. a; 9. a; 10. c; 11. d.

Chapter 12 Human Genetics

The Philadelphia Story [pp.194–195]

12.1. CHROMOSOMES AND INHERITANCE [p.196]
12.2. FOCUS ON SCIENCE: ***Karyotyping Made Easy*** [p.197]

1. Philadelphia; 2. Leukemias; 3. karyotype; 4. spectral; 5. genes; 6. homologous; 7. Alleles; 8. wild; 9. crossing over; 10. recombination; 11. sex; 12. autosomes; 13. karyotype; 14. in vitro; 15. colchicine; 16. metaphase; 17. centrifugation; 18. centromeres.

12.3. SEX DETERMINATION IN HUMANS [pp.198–199]
12.4. WHAT MENDEL DIDN'T KNOW: CROSSOVERS AND RECOMBINATION [pp.200–201]

1. E; 2. C; 3. A; 4. D; 5. B; 6. Two blocks of the Punnett square should be *XX* and two blocks should be *XY*, 7. sons; 8. mothers; 9. daughters; 10. a. F_1 flies all have red eyes: 1/2 heterozygous red females: 1/2 red-eyed males; b. F_2 Phenotypes: females all have red eyes; males 1/2 red eyes, 1/2 white eyes; Genotypes: females are X^+X^+; X^+X^W; males are X^+Y, X^WY; 11. d.; 12. Crossing over would be expected to occur twice as often between genes *A* and *B* as it would between genes *C* and *D*; 13. The son has genotype *Ab/Ab* because a crossover must have occurred during meiosis in his mother, resulting in the recombination *Ab*.

12.5. HUMAN GENETIC ANALYSIS [pp.202–203]

1. A; 2. E; 3. B; 4. F; 5. C; 6. G; 7. D; 8. A genetic abnormality is nothing more than an uncommon version of a trait that society may judge as either abnormal or merely interesting; a genetic disorder is an inherited condition that sooner or later causes mild to severe medical problems; a syndrome is a recognized set of symptoms that characterize a given disorder; a genetic disease is illness caused by a person's genes; 9. a. Autosomal recessive; b. Autosomal dominant;

c. X-linked recessive; d. X-linked recessive; e. Autosomal dominant; f. Changes in chromosome number; g. Autosomal dominant; h. Changes in chromosome structure; i. Changes in chromosome number; j. X-linked recessive; k. X-linked recessive; l. Autosomal dominant; m. Changes in chromosome number; n. X-linked recessive; o. Autosomal recessive.

12.6. EXAMPLES OF INHERITANCE PATTERNS [pp.204–205]
12.7. FOCUS ON HEALTH: *Progeria — Too Young to Be Old* [p.206]
1. b; 2. a and c; 3. b; 4. c; 5. a; 6. a and b; 7. b; 8. b; 9. a; 10. c; 11. c; 12. b; 13. b; 14. a; 15. b; 16. a, b and c; 17. a and b; 18. The woman's mother is either homozygous normal, *GG*, or heterozygous normal, *Gg*; the woman's father is homozygous recessive, *gg*, the woman is heterozygous normal, *Gg*. The man with galactosemia, *gg*, has two heterozygous normal parents, *Gg*. The two normal children are heterozygous normal, *Gg*; the child with galactosemia is *gg*; 19. Assuming the father is heterozygous with Huntington disorder and the mother normal, the chances are 1/2 that the son will develop the disease; 20. If only male offspring are considered, the probability is 1/2 that a son of the couple will be color-blind; 21. The probability is that 1/2 of the sons will have hemophilia; the probability is 0 that a daughter will express hemophilia; the probability is that 1/2 of the daughters will be carriers; 22. If the woman marries a normal male, the chance that her son will be color-blind is 1/2. If she marries a color-blind male, the chance that her son will be color-blind is also 1/2.

12.8. CHANGES IN CHROMOSOME STRUCTURE [pp.206–207]
12.9. CHANGES IN CHROMOSOME NUMBER [pp.208–209]
12.10. CASE STUDIES: CHANGES IN THE NUMBER OF SEX CHROMOSOMES [pp. 210–211]
1. duplication (B); 2. inversion (C); 3. deletion (A); 4. translocation (D); 5. a. With aneuploidy, individuals have one extra or one missing chromosome; a major cause of human reproductive failure; b. With polyploidy, individuals have three or more of each type of chromosome; common in flowering plants and some animals but lethal in humans; c. Nondisjunction is caused by a failure of one or more pairs of chromosomes to separate in mitosis or meiosis; some or all forthcoming cells will have too many or too few chromosomes; 6. All gametes will be abnormal; 7. One-half of the gametes will be abnormal; 8. About half of all flowering plant species are polyploids, but this condition is lethal for humans; 9. A tetraploid ($4n$) cell has four of each type of chromosome; a trisomic ($2n + 1$) individual will have three of one type of chromosome and two of every other type; a monosomic ($2n - 1$) individual will have only one of one type of chromosome but two of every other type; 10. c; 11. b; 12. c; 13. d; 14. a; 15. b; 16. d; 17. c; 18. a; 19. d.

12.11. FOCUS ON BIOETHICS: *Prospects In Human Genetics* [pp.210–211]
1. a. Prenatal diagnosis; b. Abortion; c. Genetic counseling; d. Preimplantation diagnosis; e. Phenotypic treatments; f. Genetic screening.

Self-Quiz
1. d; 2. d; 3. b; 4. c; 5. b; 6. a; 7. c; 8. b; 9. b; 10. c.

Chapter 13 DNA Structure and Function

Cardboard Atoms and Bent-Wire Bonds [pp.216–217]

13.1. DISCOVERY OF DNA FUNCTION [pp.218–219]
1. a. identified "nuclein" from nuclei of pus cells and fish sperm; discovered DNA; b. discovered the transforming principle in *Streptococcus pneumoniae*, live, harmless R cells were mixed with dead S cells; R cells became S cells; c. reported that the transforming substance in Griffith's bacteria experiments was probably DNA, the substance of heredity; d. worked with radioactive sulfur (protein) and phosphorus (DNA) labels; T4 bacteriophage and *E. coli* demonstrated that labeled phosphorus was in bacteriophage DNA and contained hereditary instructions for new bacteriophages; 2. virus; 3. bacterial; 4. viruses (bacteriophages); 5. proteins; 6. ^{32}P; 7. bacteriophage (viral); 8. ^{35}S; 9. ^{32}P; 10. genetic material; 11. DNA; 12. proteins.

13.2. DNA STRUCTURE [pp.220–221]
13.3. FOCUS ON BIOETHICS: *Rosalind's Story* [p.222]
1. A five-carbon sugar called *deoxyribose*, a phosphate group, and one of the four nitrogen-containing bases; 2. guanine (pu); 3. cytosine (py); 4. adenine (pu); 5. thymine (py); 6. deoxyribose (B); 7. phosphate group (G); 8. purine or guanine (C); 9. pyrimidine or thymine (A); 10. purine or adenine (E); 11. pyrimidine or cytosine (D); 12. nucleotide (F); 13. T; 14. T; 15. sugar; 16. T; 17. T; 18. pairing; 19. constant; 20. sequence; 21. different; 22. Living organisms have so many diverse body structures and behave in different ways because the many different habitats of Earth have selected those genotypes most able to survive in those habitats. The remaining genotypes have perished. The directions that code for the building of those body structures and that enable the specific successful behaviors reside in DNA

or, in a few cases, RNA. All living organisms follow the same rules for base-pairing between the two nucleotide strands in DNA: adenine always pairs with thymine in undamaged DNA, and cytosine always pairs with guanine. All living organisms must extract energy from food molecules, and the reactions of glycolysis occur in virtually all of Earth's species. This means that similar enzyme sequences enable similar metabolic pathways to occur. Although virtually all living organisms on Earth use the same code and the same enzymes during replication, transcription, and translation, the particular array of proteins being formed differs from individual to individual, even of the same species, according to the sequence of nitrogenous bases that make up the individual's chromosome(s). Therein lies the key to the enormous diversity of life on Earth: no two individuals have the exact same array of proteins in their phenotypes; 23. If you observe the sugar–phosphate sides of the DNA "ladder," you see that one strand runs from 5′to 3′ and the other runs from 3′ to 5′. The numerals 3′ and 5′ are used to identify specific carbon atoms in each deoxyribose molecule.

13.4. DNA REPLICATION AND REPAIR [pp.222–223]
13.5. FOCUS ON SCIENCE: *Cloning Mammals — A Question of Reprogramming DNA* [p.224]
1. T–A T–A G–C G–C A–T A–T C–G C–G C–G C–G C–G C–G 2. adenine bonds to thymine (during replication) or uracil (during transcription); 3. semiconservative; 4. T; 5. polymerases; 6. T; 7. The nucleus is removed from an egg cell; a cell from the animal to be cloned is inserted into this enucleated egg; electric shock or some other trigger is used to cause fusion of the egg and the added cell; the embryo is cultured until it is ready for implantation in a host animal.

Self-Quiz
1. d; 2. d; 3. a; 4. d; 5. b; 6. d; 7. b; 8. a; 9. d; 10. d.

Chapter 14 From DNA to Proteins

Beyond Byssus [pp.226–227]

14.1. HOW IS DNA TRANSCRIBED INTO RNA? [pp.228–229]
1. sequence; 2. gene; 3. transcription (translation); 4. translation (transcription); 5. transcription; 6. translation; 7. protein; 8. folded; 9. structural (functional); 10. functional (structural); 11. a. rRNA; RNA molecule that associates with certain proteins to form the ribosome, the "workbench" on which polypeptide chains are assembled; b. mRNA; RNA molecule that moves to the cytoplasm, complexes with the tRNA and ribosome, and carries the code for the amino acid sequence of the protein; c. tRNA; RNA molecule that moves into the cytoplasm, picks up a specific amino acid, and moves it to the ribosome, where tRNA pairs with a specific mRNA code word for that amino acid; 12. RNA molecules are single-stranded, whereas DNA has two strands; uracil substitutes in RNA molecules for thymine in DNA molecules; ribose sugar is found in RNA, while DNA has deoxyribose sugar; 13. Both DNA replication and transcription follow base-pairing rules; nucleotides are added to a growing RNA strand one at a time, as in DNA replication; 14. Only one region of a DNA strand serves as a template for transcription; transcription requires different enzymes (three types of RNA polymerase); the results of transcription are single-stranded RNA molecules, but replication results in DNA, a double-stranded molecule; 15. C; 16. B; 17. E; 18. A; 19. D; 20. A-U-G-U-U-C-U-A-U-U-G-U-A-A-U-A-A-A-G-G-A-U-G-G-C-A-G-U-A-G; 21. DNA (E); 22. introns (B); 23. cap (F); 24. exons (A); 25. tail (D); 26. [mature] mRNA (C).

14.2. DECIPHERING THE mRNA TRANSCRIPTS [pp.230–231]
14.3. HOW IS mRNA TRANSLATED? [pp.232–233]
1. F; 2. B; 3. G; 4. H; 5. C; 6. A; 7. E; 8. D; 9. a. initiation; b. [chain] elongation; c. [chain] termination; 10. mRNA transcript: AUG UUC UAU UGU AAU AAA GGA UGG CAG UAG; 11. tRNA anticodons: UAC AAG AUA ACA UUA UUU CCU ACC GUC AUC; 12. amino acids: met phe tyr cys asn lys gly try gln; 13. amino acids; 14. three; 15. one; 16. mRNA; 17. codon; 18. mRNA; 19. assembly (synthesis); 20. Transfer; 21. amino acid; 22. protein (polypeptide); 23. codon; 24. anticodon; 25. initiation; 26. elongation; 27. termination; 28. release; 29. enzyme; 30. polysome; 31. endoplasmic reticulum.

14.4. DO MUTATIONS AFFECT PROTEIN SYNTHESIS? [pp.234–235]
14.5. SUMMARY [p.236]
1. mutagens; 2. substitution; 3. amino acid; 4. one; 5. frameshift; 6. transposable; 7. mutation; 8. mutagens; 9. free; 10. ultraviolet; 11. thymine (cytosine); 12. cytosine (thymine); 13. alkylating; 14. mutation; 15. carcinogens; 16. protein; 17. evolutionary; 18. DNA (H); 19. transcription (J); 20. intron (E); 21. exon (A); 22. mRNA (B); 23. rRNAs (F); 24. tRNAs (C); 25. ribosomal subunits (G); 26. amino acids (D); 27. anticodon (K); 28. translation (I); 29. polypeptide (L).

Self-Quiz
1. c; 2. b; 3. b; 4. c; 5. a; 6. b; 7. a; 8. a; 9. d; 10. d.

Chapter 15 Controls over Genes

When DNA Can't Be Fixed [pp.238–239]

15.1. TYPES OF CONTROL MECHANISMS [p.240]
15.2. BACTERIAL CONTROL OF TRANSCRIPTION [pp.240–241]

1. a. Blocks transcription by preventing RNA polymerases from binding to DNA; this is negative transcription control; b. Activator proteins; c. Specific base sequences on DNA that serve as binding sites for RNA polymerases; d. Operators; 2. In any organism, gene controls operate in response to chemical changes within the cell or its surroundings; 3. transcription; 4. regulatory gene; 5. promoter; 6. negative control; 7. low; 8. RNA polymerase (mRNA transcription); 9. blocks; 10. repressor; 11. operator; 12. needed (required); 13. regulatory gene (I); 14. operator (B); 15. lactose enzyme genes (F); 16. promoter (H); 17. lactose operon (A); 18. repressor protein (G); 19. repressor–operator complex (E); 20. lactose (C); 21. mRNA (J); 22. RNA polymerase (D).

15.3. GENE CONTROLS IN EUKARYOTIC CELLS [pp.242–243]
15.4. TYPES OF CONTROL MECHANISMS [pp.244–245]

1. All cells in the body descend from the same zygote; as cells divide to form the body, they become specialized in composition, structure, and function — they differentiate through selective gene expression; 2. pretranscriptional control (D); 3. transcriptional control (E); 4. transcript processing control (B); 5. translational control (A); 6. post-translational control (C); 7. DNA (genes); 8. cell differentiation; 9. selective; 10. controls; 11. regulatory; 12. activators; 13. Transcription; 14. Barr; 15. mosaic; 16. anhidrotic ectodermal dysplasia; 17. selective; 18. The genes are very similar, with only minor conservative changes in all eukaryotes; 19. T.

15.5. EXAMPLES OF SIGNALING MECHANISMS [pp.246–247]
15.6. FOCUS ON SCIENCE: *Lost Controls and Cancer* [pp.248–249]

1. signals (molecules); 2. hormones; 3. receptors; 4. activator protein; 5. enhancer; 6. ecdysone; 7. amplification; 8. polytene; 9. prolactin; 10. receptors; 11. phytochrome; 12. cancer; 13. tumor; 14. metastasis; 15. T; 16. increase, high, start; 17. T; 18. T; 19. bring about; 20. T; 21. abnormal.

Self-Quiz

1. d; 2. b; 3. d; 4. b; 5. b; 6. c; 7. c; 8. b; 9. b; 10. c.

Chapter 16 Recombinant DNA and Genetic Engineering

Mom, Dad, and Clogged Arteries [pp.252–253]

16.1. A TOOLKIT FOR MAKING RECOMBINANT DNA [pp.254–255]

1. The bacterial chromosome, a circular DNA molecule, contains all the genes necessary for normal growth and development. Plasmids — small, circular molecules of "extra" DNA — carry only a few genes and are self-replicating; 2. small circles of DNA in bacteria; 3. T; 4. mutations; 5. recombinant DNA; 6. species; 7. amplify; 8. protein; 9. research; 10. Genetic engineering; 11. T; 12. a and d; 13. same; 14. T; 15. some of; 16. B; 17. F; 18. E; 19. C; 20. A; 21. D; 22. protein; 23. engineered (changed, altered); 24. introns; 25. transcriptase; 26. mRNA; 27. cDNA; 28. Enzyme; 29. DNA; 30. cDNA; 31. transcript.

16.2. PCR — A FASTER WAY TO AMPLIFY DNA [p.256]
16.3. FOCUS ON BIOETHICS: *DNA Fingerprints* [p.257]
16.4. HOW IS DNA SEQUENCED? [p.258]
16.5. FROM HAYSTACKS TO NEEDLES — ISOLATING GENES OF INTEREST [p.259]

1. a. A process that determines the order of nucleotides in a cloned or amplified DNA fragment; b. Short, synthesized sequences of nucleotides that base-pair with any complementary sequences in DNA; recognized as START tags by DNA polymerases; c. A collection of DNA fragments produced by restriction enzymes and incorporated into plasmids; d. An electrical field forces molecules (generally DNA or proteins) to move through a viscous medium and separate from each other according to their different physical and chemical properties; e. A method for amplifying DNA fragments in a test tube (see Fig. 16.5 in the text); f. Several procedures allow researchers to determine the nucleotide sequence of a DNA fragment (see Fig. 16.6 in the text); g. A very short length of DNA labeled with a radioisotope so that it is distinguishable from other DNA molecules in a sample; h. A unique array of RFLPs; 2. B; 3. D; 4. E; 5. C; 6. A; 7. F.

16.6. USING THE GENETIC SCRIPTS [p.260]
16.7. DESIGNER PLANTS [pp.260–261]
16.8. GENE TRANSFERS IN ANIMALS [pp.262–263]
16.9. SAFETY ISSUES [p.263]
16.10. CONNECTIONS: *Bioethics in a Brave New World* [pp.264–265]

1. a. *E. coli,* which produce human insulin, hemoglobin, interferon, and blood-clotting factors; b. We now have bacterial factories that produce various substances—even plastic—as well as oils and textile fibers; c. Herbicide-resistant cotton plants. Tobacco plants that produce hemoglobin; d. Bacteria that break down crude oil into less toxic compounds. Bacteria that absorb excess phosphates or heavy metals; 2. The more closely related two species are, the greater the extent of nucleic acid hybridization and the more similar their metabolic pathways; 3. If we learn about the genes, we may understand better the nature of a pathogen's attack strategy and be able to have advance warning about its likely plan of attack; 4. In 1970, a new strain of the fungus that causes Southern corn leaf blight destroyed most of the corn crop in the United States. Because corn plants were genetically similar, most were killed. If different kinds had been planted, more of the crop would more likely have survived; 5. Geneticists splice the genes into the Ti plasmid from *Agrobacterium tumefaciens,* which infects many species of flowering plants. Sometimes electric shocks or chemicals can deliver modified genes into plant cells. Some researchers blast microscopic particles coated with DNA into the plant cells. 6. Human serum albumin helps control blood pressure. It would be much easier to obtain large quantities of the protein from abundant supplies of milk instead of having to separate it from large quantities of donated human blood; 7. Researchers are working to sequence the estimated 3 billion nucleotides present in human chromosomes; 8. The plasmid of *Agrobacterium* can be used as a vector to introduce desired genes into cultured plant cells; *Agrobacterium* was used to deliver a firefly gene into cultured tobacco plant cells; 9. Certain cotton plants have been genetically engineered for resistance to worm attacks; 10. If human collagen can be produced, it may be used to correct various skin, cartilage, and bone disorders; 11. In separate experiments, the rat and human somatotropin genes became integrated into the mouse DNA. The mice grew much larger than their normal littermates; 12. Bacteria that have specific proteins on their cell surfaces facilitate ice crystal formation on whatever substrate the bacteria are located; bacteria without the ability to synthesize such proteins ("ice-minus") have been genetically engineered and were sprayed on strawberry plants. Nothing bad happened; 13. In 1996, researchers produced a genetic duplicate of an adult ewe by fusing a reprogrammed mammary gland cell with an egg from which the nucleus had been removed. When the fused cell was implanted into a surrogate mother, Dolly was born. Her existence means that genetically engineered clones of domestic animals may supply reliable quantities of specific proteins for research and for medical uses; 14. Comparative genomics; 15. ice-minus; 16. body cells; 17. gene therapy; 18. eugenic engineering.

Self-Quiz

1. a; 2. b; 3. c; 4. b; 5. a; 6. d; 7. d; 8. a; 9. d; 10. b.

Chapter 17 Microevolution

Designer Dogs [pp.270–271]

17.1. EARLY BELIEFS, CONFOUNDING DISCOVERIES [pp.272–273]

1. G; 2. H; 3. J; 4. I; 5. D; 6. E; 7. F; 8. C; 9. B; 10. A; 11. K.

17.2. A FLURRY OF NEW THEORIES [pp.274–275]
17.3. DARWIN'S THEORY TAKES FORM [pp.276–277]

1. b; 2. a; 3. b; 4. a; 5. a; 6. c; 7. b; 8. a; 9. c; 10. b; 11. a; 12. b; 13. b; 14. c; 15. a. John Henslow; b. HMS *Beagle;* c. Charles Lyell; d. Thomas Malthus; e. Galápagos Islands; f. Alfred Wallace; g. Cambridge University; h. artificial selection; i. natural selection; 16. F (D); 17. D (F); 18. B; 19. E; 20. A; 21. C; 22. G.

17.4. INDIVIDUALS DON'T EVOLVE—POPULATIONS DO [pp.278–279]
17.5. FOCUS ON SCIENCE: *When Is a Population Not Evolving?* [pp.280–281]

1. individuals; 2. populations; 3. species; 4. morphological; 5. Physiological; 6. behavioral; 7. sexually; 8. vary; 9. qualitatively; 10. polymorphisms; 11. quantitatively; 12. gene pool; 13. alleles; 14. frequency; 15. equilibrium; 16. mutation; 17. mutation rate; 18. lethal; 19. neutral; 20. rare; 21. allele; 22. D; 23. C; 24. E; 25. B; 26. A; 27. No genes are undergoing mutation; a very, very large population; the population is isolated from other populations of the species; all members of the population survive and reproduce; mating is random; 28. a. 0.64 *BB,* 0.16 *Bb,* 0.16 *Bb,* and 0.04 *bb;* b. genotypes: 0.64 *BB,* 0.32 *Bb,* and 0.04 *bb;* phenotypes: 96% black, 4% gray; c. (see following table)

Parents	*B sperm*	*b sperm*
0.64 *BB*	0.64	0
0.32 *Bb*	0.16	0.16
0.04 *bb*	0	0.04
Totals =	0.80	0.20

29. a. $2pq = 2 \times (0.9) \times (0.1) = 2 \times (0.09) = 0.18 = 18\%$, which is the percentage of heterozygotes; b. $p^2 = 0.81$, $p = 0.9$ = the frequency of the dominant allele; c. $p + q = 1, q = 1.00 - 0.9 = 0.1$ = the frequency of the recessive allele; 30. a. homozygous dominant = $p^2 \times 200 = (0.8)^2 \times 200 = 0.64 \times 200 = 128$ individuals; b. $q = (1.00 - p) = 0.20$; homozygous recessive = $q^2 \times 200 = (0.2)^2 \times 200 = (0.04) \times (200) = 8$ individuals; c. heterozygotes = $2pq \times 200 = 2 \times 0.8 \times 0.2 \times 200 = 0.32 \times 200 = 64$ individuals. Check: $128 + 8 + 64 = 200$; 31. If $p = 0.70$, since $p + q = 1$, $0.70 + q = 1$; then $q = 0.30$, or 30 percent; 32. If $p = 0.60$, since $p + q = 1$, $0.60 + q = 1$; then $q = 0.40$; thus, $2pq = 0.48$, or 48 percent; 33. Hardy-Weinberg; 34. *p*; 35. *q*; 36. allele frequencies; 37. mutation; 38. large; 39. mating; 40. reproduce; 41. remain stable.

17.6. NATURAL SELECTION REVISITED [p.281]
17.7. DIRECTIONAL CHANGE IN THE RANGE OF VARIATION [pp.282–283]
17.8. SELECTION AGAINST OR IN FAVOR OF EXTREME PHENOTYPES [pp.284–285]

1. a. directional selection, allele frequencies shift in a consistent direction; b. disruptive selection, intermediate forms of a trait are selected against; c. stabilizing selection, intermediate forms of a trait are favored; 2. a; 3. b; 4. b; 5. c; 6. b; 7. a; 8. b; 9. c; 10. Both are examples of directional selection that favors individuals who have resistance to a toxin that is used to control the population; 11. It allows scientists to study the effects of selection on a marked subset of the population.

17.9. MAINTAINING VARIABILITY IN A POPULATION [pp.286–287]
17.10. GENE FLOW [p.287]
17.11. GENETIC DRIFT [pp.288–289]

1. sexually; 2. phenotype; 3. dimorphisms; 4. Sexual; 5. selection; 6. balancing; 7. polymorphism; 8. sickle-cell anemia; 9. Hb^A; 10. homozygous; 11. malaria; 12. a; 13. b; 14. a; 15. b; 16. a; 17. a; 18. a; 19. a; 20. b; 21. a; 22. b; 23. A bottleneck is a severe reduction in population brought on by intense pressure. The founder effect is a limited form of bottleneck in which a few individuals of a population leave to form a new population. Unless the bottleneck phenomenon is severe, with only a very few individuals surviving, the founder effect will tend to have the smaller amount of genetic diversity; 24. genetic drift; 25. small; 26. Gene; 27. bottleneck; 28. allele frequencies; 29. inbreeding; 30. homozygous.

Self-Quiz

1. c; 2. d; 3. b; 4. c; 5. c; 6. d; 7. c; 8. b; 9. a; 10. d; 11. e; 12. a; 13. b; 14. a; 15. a.

Chapter 18 Speciation

The Case of the Road-Killed Snails [pp.292–293]

18.1. ON THE ROAD TO SPECIATION [pp.294–295]

1. C; 2. B; 3. A; 4. D; 5. E; 6. C (pre); 7. A (pre); 8. H (post); 9. G (pre); 10. B (post); 11. E (pre); 12. D (pre); 13. F (post); 14. f; 15. c; 16. a; 17. e; 18. d; 19. b; 20. a. one; b. two; c. B; d. D; e. B and C.

18.2. SPECIATION IN GEOGRAPHICALLY ISOLATED POPULATIONS [pp.296–297]
18.3. MODELS FOR OTHER SPECIATION ROUTES [pp.298–299]

1. b; 2. c; 3. a; 4. b; 5. c; 6. a; 7. a; 8. c; 9. c; 10. c; 11. a. allopatric; b. earthworms, yes; hawks, no; plants that disperse their seeds using wind, no; plants that disperse their seeds with the assistance of small mammals, yes.

18.4. PATTERNS OF SPECIATION [pp.300–301]

1. b; 2. c; 3. c; 4. a; 5. c; 6. J; 7. D; 8. G; 9. I; 10. E; 11. A; 12. F; 13. B; 14. C; 15. H.

Self-Quiz

1. b; 2. b; 3. d; 4. a; 5. e; 6. c; 7. b; 8. a; 9. c; 10. b; 11. d; 12. e; 13. a; 14. d; 15. c.

Chapter 19 The Macroevolutionary Puzzle

Measuring Time [pp.304–305]

19.1. FOSSILS — EVIDENCE OF ANCIENT LIFE [pp.306–307]
19.2. FOCUS ON SCIENCE: ***Dating Pieces of the Macroevolutionary Puzzle*** [pp.308–309]
1. G; 2. B; 3. D; 4. A; 5. C; 6. F; 7. E; 8. geologic time scale; 9. fossil sequences; 10. macroevolution; 11. radioactive decay; 12. unstable; 13. decays; 14. half-life; 15. 5,370; 16. a. Mesozoic, Triassic, 3; b. Paleozoic, Devonian, 2; c. Mesozoic, Cretaceous, 4; d. Cenozoic, Quaternary, 5; e. Proterozoic, N/A, 1; 17. 0.5 grams; 16,110 years.

19.3. EVIDENCE FROM BIOGEOGRAPHY [pp.310–311]
19.4. EVIDENCE FROM COMPARATIVE MORPHOLOGY [pp.312–313]
19.5. EVIDENCE FROM PATTERNS OF DEVELOPMENT [pp.314–315]
19.6. EVIDENCE FROM COMPARATIVE BIOCHEMISTRY [pp.316–317]
1. C; 2. D; 3. E; 4. F; 5. A; 6. B; 7. d; 8. b; 9. a; 10. c; 11. d; 12. d; 13. b; 14. a; 15. d; 16. b; 17. comparative morphology; 18. homologous structures; 19. genetically; 20. gene flow; 21. morphological; 22. macroevolution; 23. independently; 24. environmental; 25. selection; 26. morphological convergence; 27. analogous; 28. c; 29. e; 30. a; 31. d; 32. a; 33. b; 34. a; 35. a.

19.7. HOW DO WE INTERPRET THE EVIDENCE? [pp.318–319]
19.8. FOCUS ON SCIENCE: ***Constructing a Cladogram*** [pp.320–321]
19.9. ON INTERPRETING AND MISINTERPRETING THE PAST [pp.322–323]
1. D; 2. E; 3. F; 4. B; 5. H; 6. A; 7. G; 8. C; 9. I; 10. kingdom, phylum, class, order, family, genus, species; 11. Classical taxonomy uses degrees of morphological divergence to construct evolutionary trees, while cladistic taxonomy is more concerned with the branch points within the evolutionary tree; 12. The three-domain system represents the primordial branching of organisms into three major groups: eukaryotes, bacteria, and archaeans. The six-kingdom system gives greater diversity to the eukaryotes, breaking the group into protistans, fungi, plants, and animals; 13. B; 14. E; 15. A; 16. D; 17. C; 18. morphological, physiological, biochemical, or genetic differences (answers may vary); 19. 8; 20. 4; 21. 5; 22. they have diverged from each other more recently; 23. A; 24. E and F are a monophyletic group from derived trait 9 (answers may vary); 25. a; 26. a; 27. c; 28. a; 29. b.

Self-Quiz
1. d; 2. a; 3. b; 4. d; 5. c; 6. d; 7. a; 8. b; 9. a; 10. b; 11. c; 12. c; 13. d.

Chapter 20 The Origin and Evolution of Life

In the Beginning . . . [pp.326–327]

20.1. CONDITIONS ON THE EARLY EARTH [pp.328–329]
20.2. EMERGENCE OF THE FIRST LIVING CELLS [pp.330–331]
1. f; 2. e; 3. a (b); 4. c; 5. d; 6. b; 7. f (c); 8. c; 9. b; 10. a; 11. d; 12. f; 13. e; 14. c; 15. a; 16. b; 17. e; 18. c; 19. a; 20. b.

20.3. ORIGIN OF PROKARYOTIC AND EUKARYOTIC CELLS [pp.332–333]
20.4. WHERE DID ORGANELLES COME FROM? [pp.334–335]
1. b; 2. b; 3. a; 4. b; 5. a; 6. a; 7. b; 8. a; 9. a; 10. b; 11. a; 12. b; 13. a; 14. a; 15. b; 16. organelles; 17. gene; 18. plasma; 19. channels; 20. endoplasmic reticulum; 21. nuclear; 22. genes; 23. replication; 24. transcription; 25. prokaryotic; 26. mitochondria; 27. chloroplasts; 28. endosymbiosis; 29. inside; 30. eukaryotic; 31. oxygen; 32. Electron; 33. oxygen; 34. bacteria; 35. ATP; 36. host; 37. incapable; 38. mitochondria; 39. ATP; 40. DNA; 41. bacterial; 42. mitochondrial; 43. chloroplasts; 44. host; 45. bacteria; 46. DNA; 47. protistans; 48. prokaryotes; 49. eubacteria; 50. eukaryotes; 51. archaebacteria; 52. mitochondria; 53. chloroplasts; 54. eukaryotes; 55. methanogens; 56. animals; 57. fungi; 58. plants; 59. eubacteria.

20.5. LIFE IN THE PALEOZOIC ERA [pp.336–337]
20.6. LIFE IN THE MESOZOIC ERA [pp.338–339]
20.7. FOCUS ON SCIENCE: ***Horrendous End to Dominance*** [p.340]
20.8. LIFE IN THE CENOZOIC ERA [pp.340–341]
1. G; 2. H; 3. E; 4. K; 5. D; 6. L; 7. F; 8. M; 9. A; 10. N; 11. C; 12. J; 13. B; 14. I; 15. G,H,E; 16. E,K; 17. K,D,L,F,M; 18. F,M,A; 19. M,A,N,C,J,B; 20. a; 21. a; 22. b; 23. b; 24. a; 25. c; 26. a; 27. c; 28. a; 29. b; 30. b; 31. c.

Self-Quiz
1. b; 2. a; 3. b; 4. b; 5. e; 6. d; 7. a; 8. d; 9. c; 10. b; 11. c.

Chapter 21 Prokaryotes and Viruses

The Unseen Multitudes [pp.346–347]

21.1. CHARACTERISTICS OF PROKARYOTIC CELLS [pp.348–349]
21.2. PROKARYOTIC GROWTH AND REPRODUCTION [pp.350–351]
1. a; 2. c; 3. b; 4. d; 5. prokaryotic; 6. plasmids; 7. wall; 8. capsule (glycocalyx); 9. micrometers; 10. prokaryotic fission; 11. cocci; 12. bacilli; 13. spirilla; 14. positive.

21.3. PROKARYOTIC CLASSIFICATION [p.351]
21.4. MAJOR PROKARYOTIC GROUPS [p.352]
21.5. ARCHAEBACTERIA [pp.352–353]
21.6. EUBACTERIA — THE TRUE BACTERIA [pp.354–355]
1. archaebacteria; 2. fatty acids; 3. eubacteria; 4. cyanobacteria; 5. nitrogen-fixation; 6. sulfur; 7. nitrogen; 8. *Rhizobium*; 9. endospores; 10. temperature; 11. *Clostridium botulinum* (*C. tetani*); 12. *Clostridium tetani* (*C. botulinum*); 13. Lyme disease; 14. Rocky Mountain spotted fever; 15. *Borrelia* (*Rickettsia*); 16. membrane receptors; 17. sunlight; 18. decomposers; 19. Actinomycetes; 20. K; 21. d, L; 22. c, C; 23. c, J; 24. a, G; 25. c, E; 26. a, A; 27. b, I; 28. c, H; 29. e, B; 30. c, K; 31. c, N; 32. c, D; 33. c, M; 34. a, F.

21.7. THE VIRUSES [pp.356–357]
21.8. VIRAL MULTIPLICATION CYCLES [pp.358–359]
21.9. CONNECTIONS: *Evolution and Infectious Diseases* [pp.360–361]
1. a. Nonliving, infectious agents, smaller than the smallest cells; require living cells to act as hosts for their replication; not acted on by antibiotics; b. The core can be DNA or RNA; the capsid can be protein and/or lipid; c. Bacteriophage viruses may use the lytic pathway, in which the virus quickly subdues the host cells and replicates itself, releasing descendants as the cell undergoes lysis; or they may use a temperate pathway, in which viral genes remain inactive inside the host cell during a period of latency, which may be a long time, before activation and lysis; 2. There can be multiple answers (see Table 22.3 in text); a. Possible answers include *Herpes simplex* (a herpesvirus), rhinovirus (a picornavirus), and HIV (a retrovirus); b. *Herpes simplex*: DNA virus. Initial infection is a lytic cycle that causes herpes (sores) on mucous membranes of mouth or genitals. Recurrent infections are lysogenic. Most cells are in nerves and skin. No immunity. No cure. Rhinovirus: RNA virus. Causes the common cold. Host cells are generally mucus-producing cells of respiratory tract. Ebola virus: RNA virus. Kills 70–90 percent of its victims. No vaccine or treatment. The disease starts with high fever and flulike aches. Within a few days nausea, vomiting, and diarrhea begin. Blood vessels are destroyed as virus damages circulatory tissues. Blood seeps into the tissues surrounding blood vessels and out through all of the body's openings. Patients die of circulatory shock. HIV: RNA virus. Host cells are specific white blood cells. Lysogenic cycle has a latency period that may last longer than a year before host tests positive for HIV. As white blood cells are destroyed, the host's immune system is progressively destroyed (AIDS). No cure exists; 3. virus; 4. nucleic acid; 5. protein coat (viral capsid); 6. Viruses; 7. *Herpesvirus* (or *Varicella*); 8. Retroviruses (HIV); 9. lysogenic; 10. latency; 11. viroids; 12. prions; 13. Nanometers; 14. micrometers; 15. 86,000; 16. Rhinoviruses, RNA; 17. Retroviruses, RNA; 18. Herpesviruses, DNA; 19. C; 20. H; 21. E; 22. D; 23. K; 24. I; 25. G; 26. L; 27. F; 28. B; 29. A; 30. J.

Self-Quiz
1. d; 2. a; 3. A, C; 4. A; 5. A; 6. B, E; 7. B, F; 8. A, D; 9. A, D; 10. E; 11. F; 12. D; 13. B; 14. C; 15. A.

Crossword Puzzle

1 B		2 P	H	O	T	O	A	U	T	O	T	R	O	P	H	S		3 F	I	4 S	S	I	O	N		
A		E																		P						5 G
C		6 P	A	T	H	O	G	E	N	S		7 C	O	N	J	U	G	A	T	I	O	N				R
I		T																		R						A
L		I					8 A	N	T	I	B	I	O	T	I	9 C		10 F	L	A	G	E	L	11 L	U	M
12 L	A	D														H				L				Y		
U		O		13 A			14 C	Y	A	N	O	B	A	C	T	E	R	I	A					T		15 C
S		G		R												M								I		H
	16 G	L	Y	C	O	C	A	L	Y	X				17 P	R	O	K	A	R	Y	O	T	I	C		E
		Y		H												H										M
18 T		C		A				19 P	H	O	T	O	H	E	T	E	R	O	T	R	O	P	H	S		O
H		A		E												T										A
E		N		20 B	A	C	T	E	R	I	O	P	H	A	G	E								21 L		U
R				A												R		22 P						Y		T
M		23 M	I	C	R	O	O	R	G	A	N	I	S	M		O		L		24 P	I	L	U	S		O
O				T												T		A						O		T
25 P	A	N	D	E	M	I	C		26 E	N	D	O	S	P	O	R	E	S						G		R
H				R												O		M						E		O
I		27 E	P	I	D	E	M	I	C			28 H	A	L	O	P	H	I	L	E				N		P
L				A												H		D						I		H
E						29 M	E	T	H	A	N	O	G	E	N	S					30 C	O	C	C	U	S

Chapter 22 Protistans

Confounding Critters at the Crossroads [pp.364–365]

22.1. CONNECTIONS: ***An Emerging Evolutionary Road Map*** [pp.366–367]

1. a. E; b. P; c. E; d. B; e. E; f. B; g. E; 2. single; 3. multicellular; 4. monophyletic; 5. Chlorophytes (green algae); 6. chlorophyll a; 7. chlorophyll b (carotenoids); 8. flagellated protozoans; 9. pellicle; 10. heterotrophs; 11. vitamin B_{12}.

22.2. ANCIENT LINEAGES OF FLAGELLATED PROTOZOANS [p.368]

22.3. AMOEBOID PROTOZOANS [p.369]
22.4. THE CILIATES [pp.370–371]
22.5. THE SPOROZOANS [p.372]
22.6. FOCUS ON HEALTH: ***Malaria and the Night-Feeding Mosquitoes*** [p.373]

1. trypanosomes; 2. *Trichomonas vaginalis*; 3. Giardiasis; 4. pseudopods; 5. Foraminiferans; 6. radiolarians; 7. heliozoans; 8. freshwater; 9. contractile vacuoles; 10. gullet; 11. enzyme-filled vesicles (food vacuoles); 12. *Plasmodium*; 13. mosquito; 14. gametocytes (gametes); 15. B, C; 16. B, C, I; 17. A, J; 18. F, H, K; 19. G, H; 20. D, L; 21. D, E, K; 22. A; 23. D; 24. C; 25. B; 26. E.

22.7. THE CELL FROM HELL AND OTHER DINOFLAGELLATES [p.374]
22.8. OOMYCOTES — ANCIENT STRAMENOPILES [p.375]
22.9. PHOTOSYNTHETIC STRAMENOPILES — CHRYSOPHYTES AND BROWN ALGAE [pp.376–377]
22.10. GREEN ALGAE AND THEIR CLOSEST RELATIVES [pp.378–379]
22.11. RED ALGAE [p.380]
22.12. SLIME MOLDS [p.381]

1. producers; 2. algal blooms; 3. red tides; 4. *Pfiesteria piscicida*; 5. Oomycotes; 6. water molds; 7. downy mildews; 8. Chrysophytes; 9. diatoms; 10. fucoxanthin (carotenoids); 11. silica (glass); 12. diatom shells (diatomaceous earth); 13. filtering; 14. brown algae; 15. ecosystems; 16. algins; 17. photosynthetic; 18. cellulose (pectins, polysaccharides); 19. starch; 20. cyanobacteria; 21. Agar; 22. zygote, F; 23. resistant zygote, B; 24. meiosis and germination, H; 25. asexual reproduction, C; 26. gamete production, E; 27. gametes meet, G; 28. cytoplasmic fusion, A; 29. fertilization, D; 30. a. chlorophyll a, c_1, c_2, fucoxanthin, and other carotenoids; b. diatomaceous earth for abrasives, filtering, etc.; c. *Synura*; d. chlorophyll a, c_1, c_2, fucoxanthin, and other carotenoids; e. algin, used as a thickener, emulsifier, and stabilizer of foods, cosmetics, medicines, paper, and floor polish; also are sources of mineral salts and fertilizer; f. *Postelsia* (sea palm), *Sargassum, Laminaria, Macrocystis*; g. chlorophylls a and b; h. chlorophytes form much of the phytoplankton base of many food webs that support humans; i. *Volvox, Ulva, Spirogyra, Chlamydomonas*; j. chlorophyll a, phycobilins; k. agar, used as a moisture-preserving agent and culture medium; carrageenan is a stabilizer of emulsions; l. *Bonnemaisonia, Eucheuma, Porphyra*; 31. (+): a, b, c, d, e, g; (−): f, h, i, j.

Self-Quiz

1. b; 2. c; 3. b; 4. b; 5. a; 6. a; 7. d; 8. c; 9. a; 10. b; 11. d; 12. A, E; 13. A, I; 14. A, B; 15. A, E, J; 16. A, D; 17. A, H; 18. A, F, G; 19. A, C; 20. B; 21. E; 22. D; 23. A; 24. C.

Chapter 23 Plants

Pioneers in a New World [pp.384–385]

23.1. TRENDS IN PLANT EVOLUTION [pp.386–387]

1. C; 2. D; 3. E; 4. A; 5. B; 6. a. lignin; b. spores; c. heterospory; d. pollen grains; e. seed; f. root systems; g. xylem and phloem; h. gametophytes; i. sporophytes; j. cuticle; k. stomata; l. shoot systems.

23.2. BRYOPHYTES [pp.388–389]

1. air; 2. T; 3. rhizoids; 4. T; 5. mosses; 6. sporophytes; 7. T; 8. Peat; 9. water; 10. nonvascular; 11. sporophyte (2*n*); 12. meiosis; 13. spores (*n*); 14. sperm-producing structure of gametophyte (*n*); 15. egg-producing structure of gametophyte (*n*); 16. fertilization; 17. zygote (2*n*).

23.3. EXISTING SEEDLESS VASCULAR PLANTS [pp.390–391]
23.4. FOCUS ON THE ENVIRONMENT: ***Ancient Carbon Treasures*** [p.392]

1. c; 2. b; 3. e; 4. a; 5. d; 6. d; 7. d; 8. c; 9. a; 10. d; 11. e; 12. d; 13. c; 14. d; 15. e; 16. b; 17. d; 18. d; 19. b; 20. e; 21. sporophyte (2*n*); 22. rhizome (2*n*); 23. sorus (2*n*); 24. spore (*n*); 25. gametophyte (*n*); 26. egg (*n*); 27. sperm (*n*); 28. Carboniferous; 29. lignin-reinforced; 30. lycophyte; 31. 20; 32. sea level; 33. peat; 34. pressure; 35. coal; 36. fossil fuels; 37. nonrenewable.

23.5. THE RISE OF THE SEED-BEARING PLANTS [p.393]

1. C; 2. F; 3. E; 4. A; 5. B; 6. G; 7. D.

23.6. GYMNOSPERMS — PLANTS WITH "NAKED" SEEDS [pp.394–395]
23.7. A CLOSER LOOK AT THE CONIFERS [pp.396–397]

1. b; 2. d; 3. a; 4. b; 5. c; 6. a; 7. e; 8. b; 9. e; 10. d; 11. d; 12. e; 13. sporophyte (2*n*); 14. female cone (2*n*); 15. male strobilus (2*n*); 16. microspores (*n*); 17. megaspores (*n*); 18. zygote (2*n*); 19. Large areas that have been planted with a single species of tree, usually pine; 20. The removal of all trees from large tracts of land by clear-cutting.

23.8. ANGIOSPERMS — THE FLOWERING, SEED-BEARING PLANTS [pp.398–399]
23.9. SEED PLANTS AND PEOPLE [pp.400–401]

1. A; 2. B; 3. C; 4. D; 5. F; 6. E; 7. seed coat; 8. embryo; 9. endosperm; 10. seed; 11. sporophyte; 12. pollen sac; 13. ovules; 14. pollen grains (microspores); 15. egg; 16. gametophyte (female); 17. gametophyte (male); 18. G; 19. F; 20. I; 21. B; 22. J; 23. E; 24. K; 25. A; 26. D; 27. H; 28. C.

Self-Quiz

1. a. gametophyte, no, no; b. sporophyte, yes, no; c. sporophyte, yes, no; d. sporophyte, yes, no; e. sporophyte, yes, yes; f. sporophyte, yes, yes; 2. b; 3. e; 4. a; 5. c; 6. d; 7. b; 8. a; 9. e; 10. d; 11. c; 12. d.

Chapter 24 Fungi

Ode to the Fungus among Us [pp.404–405]

24.1. CHARACTERISTICS OF FUNGI [p.406]
1. Symbiosis; 2. mutualism; 3. lichen; 4. mycorrhiza; 5. decomposers; 6. extracellular digestion and absorption; 7. E; 8. C; 9. G; 10. F; 11. B; 12. A; 13. D.

24.2. CONSIDER THE CLUB FUNGI [pp.406–407]
1. decomposers; 2. symbionts; 3. fungal rusts; 4. *Agaricus brunnescens*; 5. honey mushroom (*Armillaria ostoyae*); 6. basidia (n); 7. nuclear fusion; 8. basidium ($2n$); 9. meiosis; 10. spore (n); 11. cytoplasmic fusion; 12. cap (n); 13. stalk (n); 14. hyphal cells (n); 15. gill (n).

24.3. SPORES AND MORE SPORES [pp.408–409]
1. nuclear fusion; 2. zygospore ($2n$); 3. meiosis; 4. spores (n); 5. spore sac (n); 6. rhizoids (n); 7. asexual reproduction (mitosis); 8. zygospore (n); 9. E; 10. G; 11. H; 12. D; 13. A; 14. F; 15. B; 16. C.

24.4. THE SYMBIONTS REVISITED [pp.410–411]
24.5. FOCUS ON SCIENCE: ***A Look at the Unloved Few*** [p.412]
1. Symbiosis; 2. mutualism; 3. lichen; 4. mycobiont; 5. photobiont; 6. sac; 7. hypha; 8. cytoplasm; 9. mycobiont; 10. photobiont; 11. layers; 12. hostile; 13. fungus; 14. photobiont; 15. photobiont's; 16. shelter; 17. parasite; 18. mycorrhizae; 19. efficiently; 20. ectomycorrhizae; 21. temperate; 22. club; 23. endomycorrhizae; 24. penetrate; 25. zygomycetes; 26. absorptive; 27. soil; 28. D; 29. B; 30. E; 31. C; 32. A; 33. F.

Self-Quiz
1. scarlet hood (C); 2. *Pilobus* (B); 3. *A. ocreata* (C); 4. morel (A); 5. lichen (E); 6. lichen (E); 7. big laughing mushroom (C); 8. scarlet cup fungus (A); 9. b; 10. d; 11. d; 12. c; 13. a; 14. a; 15. c; 16. c; 17. d; 18. a; 19. a, c; 20. b.

Chapter 25 Animals: The Invertebrates

Madeleine's Limbs [pp.414–415]

25.1. OVERVIEW OF THE ANIMAL KINGDOM [pp.416–417]
25.2. PUZZLES ABOUT ORIGINS [p.418]
25.3. SPONGES — SUCCESS IN SIMPLICITY [pp.418–419]
1. O; 2. I; 3. H; 4. E; 5. P; 6. A; 7. F; 8. K; 9. C; 10. J; 11. D; 12. G; 13. N; 14. M; 15. L; 16. B; 17. a. Placozoa; b. Sponges; c. Cnidaria; d. Turbellarians, flukes, tapeworms; e. Nematoda; f. Rotifera; g. Snails, slugs, clams, squids, octopuses; h. Annelida; i. Crustaceans, spiders, insects; j. Echinodermata; 18. c; 19. a; 20. c; 21. d; 22. c; 23. b; 24. c; 25. a; 26. c; 27. b; 28. H; 29. K; 30. D; 31. J; 32. B; 33. I; 34. C; 35. E; 36. A; 37. G; 38. F.

25.4. CNIDARIANS — TISSUES EMERGE [pp.420–421]
1. Cnidaria; 2. nematocysts; 3. medusa; 4. polyp; 5. gastrodermis; 6. epidermis; 7. epithelium; 8. nerve; 9. contractile; 10. nerve; 11. mesoglea; 12. Jellyfishes; 13. hydrostatic; 14. mesoglea; 15. both (two); 16. gonads; 17. planulas; 18. Portuguese; 19. toxin; 20. Atlantic; 21. planula; 22. teams; 23. reef; 24. calcium; 25. skeletons; 26. nutrient; 27. dinoflagellate; 28. feeding polyp; 29. reproductive polyp; 30. female medusa; 31. planula.

25.5. ACOELOMATE ANIMALS — AND THE SIMPLEST ORGAN SYSTEMS [pp.422–423]
1. b, c; 2. c; 3. c; 4. a; 5. a; 6. b; 7. a; 8. c; 9. a; 10. c; 11. c; 12. c; 13. branching gut; 14. pharynx (protruding); 15. brain; 16. nerve cord; 17. ovary; 18. testis; 19. planarian (genus name = *Dugesia*); 20. no; 21. yes; 22. no coelom (acoelomate).

25.6. ROUNDWORMS [p.423]
25.7. FOCUS ON HEALTH: ***A Rogue's Gallery of Worms*** [pp.424–425]
25.8. ROTIFERS [p.425]
1. roundworm; 2. no; 3. a false coelom (pseudocoelom); 4. rotifer; 5. bilateral; 6. yes; 7. false coelom; 8. Rotifers have a crown of cilia at the head end that assists in swimming and in wafting food toward the mouth; its rhythmic motions reminded early microscopists of a turning wheel; 9. b; 10. a; 11. a; 12. a; 13. b; 14. a; 15. a; 16. b; 17. a; 18. a; 19. b; 20. b; 21. a; 22. b; 23. a; 24. b; 25. a; 26. a; 27. eggs; 28. larvae; 29. snail; 30. larvae; 31. human; 32. Larvae; 33. intermediate; 34. human; 35. small intestine; 36. proglottids; 37. organs; 38. proglottids; 39. feces; 40. larval; 41. intermediate.

25.9. A MAJOR DIVERGENCE [p.426]
25.10. A SAMPLING OF MOLLUSKS [pp.426–427]

25.11. EVOLUTIONARY EXPERIMENTS WITH MOLLUSCAN BODY PLANS [pp.428–429]
1. b; 2. b; 3. a; 4. b; 5. a; 6. b; 7. a; 8. a; 9. b; 10. a; 11. III; 12. I; 13. II; 14. mouth; 15. anus; 16. gill; 17. heart; 18. radula (mouth); 19. foot; 20. shell; 21. stomach; 22. mouth; 23. gill; 24. mantle; 25. muscle; 26. foot; 27. radula; 28. internal shell; 29. mantle; 30. reproductive organ; 31. gill; 32. ink sac; 33. tentacle; 34. mollusk; 35. shell; 36. mantle; 37. gills (ctenidia); 38. foot; 39. radula; 40. eyes (tentacles); 41. tentacles (eyes); 42. d; 43. b; 44. b; 45. a; 46. d; 47. b; 48. c; 49. d; 50. b; 51. d; 52. d; 53. b; 54. d; 55. c; 56. b; 57. d; 58. c; 59. d; 60. b; 61. d; 62. b; 63. d; 64. c; 65. b.

25.12. ANNELIDS — SEGMENTS GALORE [pp.430–431]
1. F; 2. K; 3. B; 4. D; 5. G; 6. M; 7. C; 8. J; 9. I; 10. A; 11. H; 12. E; 13. L; 14. The earthworm uses its hydrostatic skeleton to extend longitudinal muscles in the anterior end while segments in the posterior end have their setae fixed into the wall of the burrow. This pushes the anterior end of the worm forward. Next the worm fixes the anterior setae and releases the posterior setae while shortening its segments. This pulls the posterior end forward. The worm then fixes the posterior setae and repeats the motion. 15. coelom; 16. cuticle; 17. nerve cord; 18. seta; 19. nephridium; 20. body wall; 21. hearts; 22. vessels; 23. mouth; 24. brain; 25. nerve cord; 26. earthworm; 27. Annelida; 28. segmentation and a closed circulatory system; 29. yes; 30. bilateral; 31. yes.

25.13. ARTHROPODS — THE MOST SUCCESSFUL ORGANISMS ON EARTH [p.432]
1. e; 2. f; 3. a; 4. b; 5. f; 6. a; 7. f; 8. d; 9. b; 10. c; 11. e; 12. d; 13. a; 14. Arthropods, as a group, comprise the largest number of species, occupy the most habitats, and have very efficient defenses against predators and competitors, as well as the capacity to exploit the greatest amounts and kinds of food.

25.14. A LOOK AT SPIDERS AND THEIR KIN [p.433]
25.15. A LOOK AT THE CRUSTACEANS [pp.434–435]
25.16. HOW MANY LEGS? [p.435]
1. E; 2. G; 3. F; 4. D; 5. C; 6. B; 7. A; 8. an exoskeleton; 9. lobsters and crabs; 10. similar; 11. Lobsters and crabs; 12. Barnacles; 13. Copepods; 14. barnacles; 15. barnacles; 16. molts; 17. millipedes; 18. centipedes; 19. Millipedes; 20. Centipedes; 21. cephalothorax; 22. abdomen; 23. swimmerets; 24. legs; 25. cheliped; 26. antennae; 27. lobster; 28. crustaceans; 29. exoskeleton and jointed legs; 30. yes; 31. bilateral; 32. yes; 33. poison gland; 34. brain; 35. heart; 36. spinnerets; 37. book lung; 38. chelicerates.

25.17. A LOOK AT INSECT DIVERSITY [pp.436–437]
25.18. FOCUS ON HEALTH: ***Unwelcome Arthropods*** [pp.438–439]
1. J; 2. F; 3. E; 4. B; 5. I; 6. A; 7. D; 8. G; 9. H; 10. C; 11. d; 12. b; 13. a; 14. c; 15. a; 16. b; 17. c; 18. b; 19. d; 20. b.

25.19. THE PUZZLING ECHINODERMS [pp.440–441]
1. deuterostomes; 2. echinoderms; 3. calcium carbonate; 4. skeleton; 5. radial; 6. bilateral; 7. brain; 8. nervous; 9. arm; 10. tube; 11. water; 12. ampulla; 13. muscle; 14. whole; 15. digesting; 16. anus; 17. lower stomach; 18. upper stomach; 19. anus; 20. gonad; 21. coelom; 22. digestive gland; 23. eyespot; 24. tube feet; 25. starfish; 26. deuterostome; 27. water vascular; 28. a. brittle star; b. sea urchin; c. feather star (crinoid); d. sea cucumber; 29. Echinodermata; 30. A water-vascular system and a body wall with spines, spicules, or plates; 31. radial.

Self-Quiz
1. a; 2. c; 3. a; 4. d; 5. b; 6. c; 7. d; 8. c; 9. d; 10. a; 11. a; 12. b; 13. g, G; 14. b, DHJ; 15. f, BE; 16. h, F; 17. d, K; 18. i, M; 19. e, C; 20. a, I; 21. c, A.

Chapter 26 Animals: The Vertebrates

So You Think the Platypus Is a Stretch [pp.444–445]

26.1. THE CHORDATE HERITAGE [pp.446–447]
26.2. CONNECTIONS: TRENDS IN VERTEBRATE EVOLUTION [p.448]
26.3. EXISTING JAWLESS FISHES [p.449]
26.4. EXISTING JAWED FISHES [pp.450–451]
1. traits (characteristics); 2. nerve cord; 3. gill slits; 4. notochord; 5. invertebrate; 6. vertebrates; 7. lancelets; 8. filter-feeding; 9. gill slits; 10. Tunicates (sea squirts); 11. tadpoles; 12. notochord; 13. metamorphosis; 14. lancelets; 15. vertebral column; 16. fish; 17. jaws; 18. brain; 19. Lobed (Fleshy); 20. lungs; 21. circulatory; 22. lancelet; 23. pharynx with gill slits; 24. anus; 25. notochord; 26. dorsal tubular nerve cord; 27. early jawless fish (agnathan); 28. supporting structure; 29. gill slit; 30. placoderm; 31. jaws; 32. hagfishes; 33. lampreys; 34. hagfishes; 35. mucus; 36. lampreys; 37. sharks; 38. gill slits; 39. bony; 40. ray-finned; 41. lobe-finned; 42. lungfishes; 43. swim bladders; 44. Coelocanths;

45. legs; 46. C; 47. F; 48. J; 49. G; 50. K; 51. B; 52. L; 53. D; 54. E; 55. H; 56. I; 57. A; 58. all jawless; 59. filter feeders; 60. jaws; 61. cartilage; 62. bone; 63. ray-finned fishes; 64. lobe-finned fishes; 65. branch 5; 66. branch 3; 67. Silurian; 68. about 375 million years ago.

26.5. AMPHIBIANS [pp.452–453]
26.6. THE RISE OF AMNIOTES [pp.454–455]
26.7. A SAMPLING OF EXISTING REPTILES [pp.456–457]

1. lungs; 2. fins; 3. brains; 4. balance; 5. Circulatory; 6. oxygen; 7. ATP; 8. insects; 9. salamanders; 10. water; 11. reproduce; 12. respiratory surface; 13. Reptiles; 14. eggs; 15. scaly skin; 16. internal; 17. lungs; 18. Carboniferous; 19. turtles; 20. Carboniferous; 21. Triassic; 22. Cretaceous; 23. birds; 24. temperature; 25. synapsid; 26. Carboniferous; 27. G; 28. E; 29. B; 30. D; 31. C; 32. A(F); 33. F(A); 34. a. dry, scaly skin; b. four-chambered heart; c. feather development; d. hair development; e. loss of limbs.

26.8. BIRDS [pp.458–459]
26.9. THE RISE OF MAMMALS [pp.460–461]
26.10. A PORTFOLIO OF EXISTING MAMMALS [pp.462–463]

1. embryo (notochord); 2. albumin; 3. yolk sac; 4. reptiles; 5. 2.25; 6. ostrich; 7. feathers; 8. sternum (breastbone); 9. air cavities; 10. amniote; 11. parental; 12. scales; 13. metabolic; 14. oxygen (blood); 15. four; 16. hair; 17. mammary glands; 18. dentition; 19. placenta; 20. synapsids; 21. therians; 22. dinosaurs; 23. platypus; 24. marsupials (pouched); 25. placental; 26. Convergent evolution.

26.11. CONNECTIONS: TRENDS IN PRIMATE EVOLUTION [pp.464–465]
26.12. FROM PRIMATES TO HOMINIDS [pp.466–467]
26.13. EMERGENCE OF EARLY HUMANS [pp.468–469]
26.14. EMERGENCY OF MODERN HUMANS [p.470]
26.15. FOCUS ON SCIENCE: ***Out of Africa — Once, Twice, or . . .*** [p.470]

1. primates; 2. anthropoids; 3. monkeys; 4. daytime; 5. forward-facing; 6. smell; 7. prehensile; 8. opposable; 9. grip; 10. tools; 11. upright walking; 12. teeth; 13. brain; 14. behaviors; 15. culture; 16. language; 17. Africa; 18. australopiths; 19. *Homo*; 20. 1.7; 21. Asia; 22. Europe; 23. brain; 24. toolmaker; 25. 100,000; 26. *Homo sapiens*; 27. Neandertals; 28. 2,000,000; 29. African emergence; 30. *Homo erectus*; 31. *Homo sapiens*; 32. multiregional; 33. *Homo sapiens*; 34. anthropoids; 35. hominoids; 36. hominids; 37. gorilla; 38. chimpanzee; 39. *Australopithecus afarensis*; 40. *Homo erectus.*

Self-Quiz

1. d; 2. a; 3. c; 4. d; 5. c; 6. a; 7. b; 8. a; 9. a; 10. b; 11. d; 12. B; 13. H; 14. C; 15. D; 16. A; 17. E; 18. G; 19. F; 20. d, H; 21. b, B; 22. i, F; 23. g, D; 24. h, A; 25. c, E; 26. a, I; 27. e, G; 28. f, C; 29. early amphibian, amphibian; 30. Arctic fox, mammal; 31. soldier fish, bony fish; 32. Ostracoderm, jawless fish; 33. owl, bird; 34. turtle, reptile; 35. shark, cartilaginous fish; 36. coelacanth, bony fish (lobe-finned fish); 37. reef ray, cartilaginous; 38. tunicate, urochordate; 39. lancelet, cephalochordate.

Crossword Puzzle

1 P	R	I	M	2 A	T	E		3 H	O	4 M	I	N	O	5 I	D
				U						A				N	
6 H			7 A	S	T	E	R	8 S		M		9 B		C	
O				T				A		M		10 I	R	I	S
11 M	O	L	A	R				P		A		P		S	
I				A				I		L		E		O	
N		12 C	U	L	T	U	R	E			13 A	D	O	R	E
I				O				N				A			
14 D	R	Y	O	P	I	T	H	S		15 F		L		16 T	
				I					17 C	A	N	I	N	E	S
18 E	R	E	C	T	U	S				C		S		E	
				H						E		M		T	
	19 P	R	O	S	I	M	I	A	N	S				H	

Chapter 27 Biodiversity in Perspective

The Human Touch [pp.474–475]

27.1. ON MASS EXTINCTIONS AND SLOW RECOVERIES [pp.476–477]
27.2. THE NEWLY ENDANGERED SPECIES [pp.478–479]
27.3. FOCUS ON THE ENVIRONMENT: *Case Study: The Once and Future Reefs* [pp.480–481]
27.4. FOCUS ON SCIENCE: *Rachel's Warning* [p.482]
1. a. erosion, overfarming; b. used for firewood and canoes; c. depleted for food; d. lack of food, war, cannibalism; 2. meteorite impact or global cooling; 3. asteroid impact; 4. asteroid impact; 5. humans; 6. E; 7. A; 8. C; 9. B; 10. F; 11. D; 12. G; 13. Coral reefs; 14. marine; 15. dinoflagellates; 16. protection; 17. oxygen; 18. stressed; 19. bleaching; 20. dumping raw sewage; 21. oil spills; 22. dredging; 23. mining; 24. Rachel Carson; 25. pesticides; 26. environmental.

27.5. CONSERVATION BIOLOGY [pp.482–483]
27.6. RECONCILING BIODIVERSITY WITH HUMAN DEMANDS [pp.484–485]
1. 9; 2. developing; 3. need; 4. conservation biology; 5. survey; 6. methods; 7. Hot spots; 8. ecoregion; 9. 238; 10. pharmaceutical; 11. tropical rain forests; 12. sustainable; 13. strip logging; 14. corridor; 15. contours; 16. road; 17. saplings; 18. nutrients; 19. growth; 20. riparian; 21. flood; 22. runoffs; 23. endemic; 24. ungulates; 25. watering.

Self-Quiz
1. a; 2. b; 3. c; 4. c; 5. e; 6. a; 7. c; 8. b; 9. d.

Chapter 28 How Plants and Animals Work

On High-Flying Geese and Edelweiss [pp.488–489]

28.1. LEVELS OF STRUCTURAL ORGANIZATION [pp.490–491]
28.2. THE NATURE OF ADAPTATION [pp.492–493]
1. F; 2. A; 3. E; 4. B; 5. C; 6. D; 7. a. organs; b. tissues; c. organ systems; 8. so that conditions are correct for performing necessary metabolic functions; 9. Adaptation; 10. Short-term; 11. individual; 12. heritable; 13. surviving; 14. reproducing; 15. outcome.

28.3. MECHANISMS OF HOMEOSTASIS IN ANIMALS [pp.494–495]
28.4. DOES THE CONCEPT OF HOMEOSTASIS APPLY TO PLANTS? [pp.496–497]
1. D; 2. B; 3. A; 4. G; 5. C; 6. E; 7. H; 8. F; 9. b; 10. a; 11. a; 12. b; 13. a; 14. a; 15. b; 16. b; 17. a.

28.5. COMMUNICATION AMONG CELLS, TISSUES, AND ORGANS [pp.498–499]
28.6. RECURRING CHALLENGES TO SURVIVAL [pp.500–501]
1. talk; 2. prokaryotic; 3. eukaryotic; 4. activation; 5. transduction; 6. functional; 7. genes; 8. signaling; 9. environmental; 10. hormones; 11. organ-identity; 12. ABC; 13. gene transcription; 14. signal; 15. axons; 16. Chemical; 17. diffuse; 18. myelin; 19. multiple sclerosis; 20. D; 21. C; 22. A; 23. B.

Self-Quiz
1. d; 2. b; 3. c; 4. d; 5. e.

Chapter 29 Plant Tissues

Plants versus the Volcano [pp.504–505]

29.1. OVERVIEW OF THE PLANT BODY [pp.506–507]
1. ground tissue (C); 2. vascular tissues (E); 3. dermal tissues (D); 4. shoot system (A); 5. root system (B); 6. E; 7. D; 8. B; 9. F; 10. A; 11. C; 12. radial; 13. tangential; 14. transverse (cross); 15. shoot apical meristem (D); 16. shoot primary meristematic tissues (B); 17. root primary meristematic tissues (E); 18. root apical meristem (A); 19. lateral meristems (C); 20. a. Epidermis, primary; b. Ground tissues, primary; c. Vascular tissues, primary; d. Vascular tissues, secondary; e. Periderm, secondary.

29.2. TYPES OF PLANT TISSUES [pp.508–509]
1. sclerenchyma; 2. sclerenchyma; 3. collenchyma; 4. sclerenchyma; 5. parenchyma; 6. b; 7. c; 8. a; 9. a; 10. b; 11. b; 12. c; 13. a; 14. a; 15. c; 16. pits (xylem);

17. cytoplasm (xylem); 18. tracheids (xylem); 19. vessel (xylem); 20. vessel (xylem); 21. sieve (phloem); 22. companion (phloem); 23. sieve (phloem); 24. sieve (phloem); 25. a. Vessel members and tracheids; no; conduct water and dissolved minerals absorbed from soil, provide mechanical support; b. Sieve-tube members and companion cells; yes; transport sugar and other solutes; 26. a. Primary plant body; cutin in the cuticle layer over epidermal cells restricts water loss and resists microbial attack; openings (stomata) permit water vapor and gases to enter and leave the plant; b. Secondary plant body; replaces epidermis to cover roots and stems; 27. a. One; in threes or multiples thereof; usually parallel; one pore or furrow; distributed throughout ground stem tissue; b. Two, in fours or fives or multiples thereof; usually netlike; three pores or pores with furrows; positioned in a ring in the stem.

29.3. PRIMARY STRUCTURE OF SHOOTS [pp.510–511]

1. immature; 2. apical meristem; 3. bud; 4. shoot; 5. leaf; 6. procambium; 7. protoderm; 8. procambium; 9. ground meristem; 10. epidermis; 11. cortex; 12. pith; 13. primary xylem; 14. primary phloem; 15. epidermis; 16. ground tissue; 17. vascular bundle; 18. sclerenchyma cells; 19. air space; 20. xylem vessel; 21. sieve tube; 22. companion cell; 23. xylem; 24. epidermis; 25. cortex; 26. vascular bundle; 27. pith; 28. xylem vessels; 29. meristematic cells; 30. sieve tube; 31. companion cells.

29.4. A CLOSER LOOK AT LEAVES [pp.512–513]

1. blade; 2. petiole (leaf stalk); 3. axillary bud; 4. node; 5. stem; 6. sheath; 7. dicot; 8. monocot; 9. palisade mesophyll (D); 10. spongy mesophyll (B); 11. lower epidermis (A); 12. stoma (C); 13. leaf vein (E).

29.5. PRIMARY STRUCTURE OF ROOTS [pp.514–515]

1. root hair (H); 2. endodermis (G); 3. pericycle (B); 4. epidermis (E); 5. cortex (D); 6. apical meristem (F); 7. root cap (A); 8. endodermis (G); 9. pericycle (B); 10. primary phloem (C); 11. primary; 12. lateral; 13. taproot; 14. adventitious (lateral); 15. fibrous; 16. hairs; 17. vascular; 18. pericycle; 19. cortex; 20. pith; 21. air; 22. oxygen; 23. endodermis; 24. control; 25. pericycle; 26. lateral; 27. cortex.

29.6. ACCUMULATED SECONDARY GROWTH— THE WOODY PLANTS [pp.516–517]

29.7. A LOOK AT WOOD AND BARK [pp.518–519]

1. C; 2. K; 3. M; 4. F; 5. E; 6. H; 7. G; 8. B; 9. A; 10. J; 11. L; 12. D; 13. I; 14. N; 15. vessel (E); 16. early wood (F); 17. late wood (B); 18. bark (D); 19. one (C); 20 and 21. two and three (A); 22. vascular cambium (G).

Self-Quiz

1. b; 2. a; 3. b; 4. c; 5. a; 6. d; 7. d; 8. d; 9. c; 10. c; 11. b; 12. b.

Chapter 30 Plant Nutrition and Transport

Flies for Dinner [pp.522–523]

30.1. PLANT NUTRIENTS AND THEIR AVAILABILITY IN SOILS [pp.524–525]

1. F; 2. C; 3. I; 4. B; 5. H; 6. A; 7. J; 8. E; 9. G; 10. D; 11. hydrogen; 12. 13; 13. ionic; 14. calcium; 15. macronutrients; 16. micronutrients; 17. a. sulfur; Macronutrient; b. potassium; Macronutrient; c. phosphorus; Macronutrient; d. nitrogen; Macronutrient; e. calcium; Macronutrient; f. copper; Micronutrient; g. iron; Micronutrient; h. chlorine; Micronutrient; i. manganese; Micronutrient; j. magnesium; Macronutrient; k. zinc; Micronutrient; l. molybdenum; Micronutrient; m. boron; Micronutrient.

30.2. HOW DO ROOTS ABSORB WATER AND MINERAL IONS? [pp.526–527]

1. exodermis (B); 2. root hair (E); 3. epidermis (H); 4. vascular cylinder (J); 5. cortex (C); 6. endodermis (I); 7. cytoplasm (A); 8. water movement (G); 9. Casparian strip (D); 10. endodermal cell wall (F); 11. mutualism; 12. Nitrogen "fixed" by bacteria; 13. root nodules; 14. scarce minerals; 15. sugars and nitrogen-containing compounds.

30.3. HOW IS WATER TRANSPORTED THROUGH PLANTS? [pp.528–529]

1. water; 2. xylem; 3. tracheids; 4. vessel members; 5. transpiration; 6. hydrogen; 7. cohesion; 8. tension; 9. roots; 10. xylem.

30.4. HOW DO STEMS AND LEAVES CONSERVE WATER? [pp.530–531]

1. 90 percent; 2. T; 3. T; 4. stomata; 5. T; 6. T; 7. T; 8. decrease; 9. stoma; 10. T; 11. night; 12. day.

30.5. HOW ARE ORGANIC COMPOUNDS DISTRIBUTED THROUGH PLANTS? [pp.532–533]

1. photosynthesis; 2. starch; 3. proteins; 4. seeds; 5. fats; 6. Protein; 7. starches; 8. solutes; 9. sucrose; 10. sucrose; 11. D; 12. F; 13. B; 14. A; 15. C; 16. E.

Self-Quiz

1. c; 2. e; 3. c; 4. d; 5. b; 6. c; 7. b; 8. a; 9. b; 10. c.

Chapter 31 Plant Reproduction

A Coevolutionary Tale [pp.536–537]

31.1. REPRODUCTIVE STRUCTURES OF FLOWERING PLANTS [pp.538–539]

31.2. FOCUS ON THE ENVIRONMENT: *Pollen Sets Me Sneezing* [p.539]

1. sporophyte (C); 2. flower (B); 3. meiosis (D); 4. gametophyte (F); 5. gametophyte (E); 6. fertilization (A); 7. sepal; 8. petal; 9. stamen; 10. filament; 11. anther; 12. carpel; 13. stigma; 14. style; 15. ovary; 16. ovule; 17. receptacle; 18. I; 19. C; 20. G; 21. F; 22. B; 23. E; 24. H; 25. A; 26. D.

31.3. A NEW GENERATION BEGINS [pp.540–541]

1. anther; 2. pollen sac; 3. microspore mother; 4. meiosis; 5. microspores; 6. pollen tube; 7. sperm-producing; 8. pollen; 9. stigma; 10. male gametophyte; 11. ovule; 12. integuments; 13. meiosis; 14. megaspores; 15. megaspore; 16. megaspore; 17. mitosis; 18. eight; 19. embryo sac; 20. female gametophyte; 21. two; 22. endosperm; 23. egg; 24. double fertilization; 25. endosperm; 26. embryo; 27. seed; 28. coat; 29. Chemical and molecular cues guide a pollen tube's growth through tissues of the style and the ovary, toward the egg chamber and sexual destiny; 30. The embryo sac is the site of double fertilization; 31. a. Fusion of one egg nucleus (n) with one sperm nucleus (n); the plant embryo ($2n$); eventually develops into a new sporophyte plant; b. Fusion of one sperm nucleus (n) with the endosperm mother cell ($2n$); endosperm tissues ($3n$); nourishes the embryo within the seed.

31.4. FROM ZYGOTE TO SEEDS AND FRUITS [pp.542–543]

1. a. "Seed leaves" that develop from two lobes of meristematic tissue of the embryo; b. Seeds are mature ovules; c. Integuments of the ovule harden into the seed coat; d. A fruit is a mature ovary; 2. seed coat; 3. shoot tip (apical meristem); 4. cotyledons; 5. endosperm; 6. root tip; 7. mitotic; 8. sporophyte; 9. ovule; 10. fruit; 11. cotyledons; 12. two; 13. one; 14. endosperm; 15. germinates; 16. thin; 17. enzymes; 18. seedling; 19. ovule; 20. endosperm; 21. ovary; 22. coat; 23. seed; 24. ovule; 25. fruits; 26. simple; 27. aggregate; 28. multiple; 29. accessory; 30. endocarp; 31. mesocarp; 32. exocarp; 33. pericarp; 34. e; 35. b; 36. d; 37. c; 38. f; 39. b; 40. a; 41. c.

31.5. DISPERSAL OF FRUITS AND SEEDS [p.545]

31.6. FOCUS ON SCIENCE: *Why So Many Flowers and So Few Fruits?* [p.546]

1. c; 2. c; 3. b; 4. a; 5. a; 6. c; 7. b; 8. b; 9. b; 10. a; 11. Suppose the presumed excess flowers are formed strictly to produce pollen for export to other plants. Pollen grains are small and, in terms of energy, inexpensive to produce, compared to large, calorie-rich, seed-containing fruits. Thus, for a fairly small investment, a plant might reap a large reward in offspring that carry its genes.

31.7. ASEXUAL REPRODUCTION OF FLOWERING PLANTS [pp.546–547]

1. E; 2. G; 3. I; 4. F; 5. H; 6. D; 7. B; 8. C; 9. A; 10. D; 11. E; 12. A; 13. B; 14. C.

Self-Quiz

1. b; 2. c; 3. d; 4. c; 5. c; 6. d; 7. b; 8. c; 9. c; 10. d.

Chapter 32 Plant Growth and Development

Foolish Seedlings and Gorgeous Grapes [pp.550–551]

32.1. PATTERNS OF EARLY GROWTH AND DEVELOPMENT—AN OVERVIEW [pp.552–553]

1. embryo; 2. germination; 3. environmental; 4. imbibition; 5. ruptures; 6. aerobic; 7. meristematic; 8. root; 9. primary root; 10. germination; 11. heritable (genetic); 12. genes; 13. zygote; 14. genes; 15. cytoplasmic; 16. metabolic; 17. selective; 18. hormones; 19. Interactions; 20. cotyledons; 21. hypocotyls; 22. primary; 23. cotyledons; 24. foliage; 25. primary; 26. cotyledons; 27. coleoptile; 28. branch; 29. primary; 30. foliage; 31. stem; 32. adventitious; 33. branch; 34. primary; 35. prop; 36. foliage; 37. coleoptile.

32.2. WHAT THE MAJOR PLANT HORMONES DO [pp.554–555]

32.3. ADJUSTING THE DIRECTION AND RATES OF GROWTH [pp.556–557]

1. e; 2. c; 3. a; 4. d; 5. a; 6. e; 7. a; 8. b; 9. a; 10. c; 11. e; 12. c; 13. e; 14. d; 15. I; 16. F; 17. D; 18. H; 19. E; 20. C; 21. G; 22. B; 23. J; 24. A; 25. a; 26. c; 27. d; 28. b; 29. d; 30. b; 31. a; 32. a; 33. c; 34. b.

32.4. HOW DO PLANTS KNOW WHEN TO FLOWER? [pp.558–559]
1. D; 2. G; 3. H; 4. J; 5. B; 6. A; 7. I; 8. C; 9. F; 10. E; 11. Pr; 12. Pfr; 13. Pfr; 14. Pr; 15. response; 16. Long-day; 17. Short-day; 18. day-neutral; 19. night; 20. dawn; 21. dusk; 22. 10; 23. light.

32.5. LIFE CYCLES END, AND TURN AGAIN [pp.560–561]

32.6. GROWING CROPS AND A CHEMICAL ARMS RACE [p.562]
1. b; 2. e; 3. d; 4. a; 5. a; 6. d; 7. c; 8. a; 9. c; 10. d; 11. a; 12. e; 13. B; 14. A; 15. C; 16. D; 17. E.

Self-Quiz
1. b; 2. d; 3. a; 4. c; 5. a; 6. d; 7. a; 8. d; 9. a; 10. b; 11. c.

Chapter 33 Animal Tissues and Organ Systems

Meerkats, Humans, It's All the Same [pp.566–567]

33.1. EPITHELIAL TISSUE [pp.568–569]
1. C; 2. F; 3. B; 4. H; 5. G; 6. D; 7. A; 8. E; 9. Epithelial; 10. Simple epithelium; 11. Stratified epithelium; 12. Tight; 13. Adhering; 14. Gap; 15. Tight; 16. peptic; 17. Exocrine; 18. epithelial; 19. Endocrine; 20. hormones; 21. bloodstream.

33.2. CONNECTIVE TISSUE [pp.570–571]
33.3. MUSCLE TISSUE [p.572]
33.4. NERVOUS TISSUE [p.573]
33.5. FOCUS ON SCIENCE: *Frontiers in Tissue Research* [p.573]
1. b; 2. c; 3. a; 4. c; 5. b; 6. a; 7. c; 8. a; 9. c; 10. b; 11. a. adipose; b. cartilage; c. blood; d. bones; 12. smooth; 13. cardiac; 14. smooth; 15. skeletal; 16. Skeletal; 17. smooth; 18. striped; 19. Skeletal; 20. move internal organs; 21. cardiac; 22. Nervous; 23. neurons; 24. Neuroglia; 25. neuron; 26. neurons; 27. a designer organ; 28. connective, D, 8, 10; 29; epithelial, G, 1, 5, 11; 30. muscle, I, 7, (10); 31. muscle, J, 4, (10); 32. connective, E, 6, 10, (13); 33. connective, B, 9; 34. epithelial, H, 1, 5, 11; 35. connective, K, 1, 14; 36. nervous, L, 2; 37. muscle, C, 12; 38. epithelial, F, 1, 5, 11; 39. connective, A, 3, 13.

33.6. ORGAN SYSTEMS [pp.574–575]
1. cranial; 2. spinal; 3. thoracic; 4. abdominal; 5. pelvic; 6. a. Mesoderm; b. Endoderm; c. Ectoderm; 7. superior; 8. distal; 9. proximal; 10. posterior; 11. transverse; 12. inferior; 13. anterior; 14. frontal; 15. circulatory (D); 16. respiratory (E); 17. urinary (= excretory) (A); 18. skeletal (C); 19. endocrine (J); 20. reproductive (G); 21. digestive (I); 22. muscular (H); 23. nervous (B); 24. integumentary (F); 25. lymphatic (K).

Self-Quiz
1. d; 2. b; 3. c; 4. c; 5. d; 6. c; 7. c; 8. d; 9. a; 10. d; 11. a; 12. A; 13. G; 14. E; 15. J; 16. H; 17. I; 18. F; 19. B; 20. D; 21. C.

Chapter 34 Integration and Control: Nervous Systems

Why Crack the System? [pp.586–587]

34.1. NEURONS — THE COMMUNICATION SPECIALISTS [pp.576–577]
34.2. HOW ARE ACTION POTENTIALS, TRIGGERED AND PROPAGATED? [pp.578–579]
1. B; 2. F; 3. C; 4. E; 5. D; 6. A; 7. neuron; 8. dendrites; 9. trigger; 10. axon; 11. output; 12. B; 13. C; 14. E; 15. A; 16. D; 17. ion; 18. graded; 19. local; 20. intense; 21. trigger; 22. action potential; 23. threshold; 24. sodium; 25. positive; 26. all-or-nothing; 27. millisecond; 28. sodium; 29. potassium; 30. voltage; 31. sodium–potassium; 32. sodium; 33. open; 34. self-propagating; 35. magnitude; 36. insensitive.

34.3. CHEMICAL SYNAPSES [pp.584–585]
34.4. PATHS OF INFORMATION FLOW [pp.586–587]
1. D; 2. H; 3. E; 4. C; 5. A; 6. F; 7. B; 8. G; 9. graded; 10. EPSP; 11. hyperpolarizing; 12. synaptic integration; 13. synaptic; 14. summed; 15. suppressed; 16. synaptic clefts; 17. diffuse; 18. Enzymes; 19. Transport; 20. G; 21. A; 22. E; 23. C; 24. B; 25. F; 26. D; 27. sensory; 28. inter-; 29. motor; 30. receptor endings or dendrites; 31. peripheral axon; 32. cell body; 33. axon; 34. axon endings; 35. B; 36. C; 37. F; 38. A; 39. E; 40. D; 41. G; 42. E–D–C–A–B–F–G.

34.5. INVERTEBRATE NERVOUS SYSTEMS [pp.588–589]

1. nervous; 2. neurons; 3. body; 4. seas; 5. nervous; 6. radial; 7. nerve; 8. sensory; 9. reflex; 10. body; 11. bilateral; 12. midsagittal; 13. two; 14. ganglion; 15. sensory; 16. bilateral; 17. plexuses; 18. sensory; 19. selection; 20. Cephalization.

34.6. VERTEBRATE NERVOUS SYSTEMS — AN OVERVIEW [pp.590–591]

1. a; 2. c; 3. b; 4. c; 5. b; 6. B; 7. D; 8. G; 9. I; 10. J; 11. F; 12. C; 13. E; 14. A; 15. K; 16. H.

34.7. WHAT ARE THE MAJOR EXPRESSWAYS? [pp.592–593]

1. autonomic; 2. sympathetic; 3. parasympathetic; 4. midbrain; 5. medulla oblongata; 6. cervical; 7. thoracic; 8. lumbar; 9. sacral; 10. spinal cord; 11. ganglion; 12. nerve; 13. vertebra; 14. intervertebral disk; 15. meninges; 16. gray matter; 17. white matter; 18. c; 19. a; 20. b; 21. d; 22. a; 23. c; 24. d; 25. d; 26. b; 27. d; 28. b; 29. b; 30. a; 31. b; 32. d; 33. c; 34. d.

34.8. THE VERTEBRATE BRAIN [pp.594–595]

1. D; 2. B; 3. F; 4. A; 5. E; 6. H; 7. G; 8. C; 9. I; 10. a; 11. a; 12. c; 13. b; 14. b; 15. b; 16. a.

34.9. THE HUMAN CEREBRUM [pp.596–597]
34.10. FOCUS ON SCIENCE: *Sperry's Split-Brain Experiments* [p.598]

1. cerebrum; 2. hypothalamus; 3. thalamus; 4. pineal; 5. medulla oblongata; 6. pons; 7. cerebellum; 8. midbrain; 9. optic; 10. corpus callosum; 11. a. right cerebral hemisphere; b. cerebral cortex, frontal lobe; c. cerebral cortex; d. corpus callosum; e. cerebral cortex, parietal lobe; f. cerebral cortex, occipital lobe; g. left cerebral hemisphere; h. limbic system; i. cerebral cortex, temporal lobe; 12. Sperry demonstrated that signals across the corpus callosum coordinate the function of the two cerebral hemispheres, each of which responds to visual signals from the opposite side of the body.

34.11. HOW ARE MEMORIES TUCKED AWAY? [p.599]
34.12. REFLECTIONS ON THE NOT-QUITE-COMPLETE TEEN BRAIN [pp.600–601]
34.13. FOCUS ON HEALTH: *Drugging the Brain* [pp.602–603]

1. touch; 2. hearing; 3. vision; 4. hippocampus; 5. amygdala; 6. smell; 7. prefrontal cortex; 8. basal ganglia; 9. thalamus and hypothalamus; 10. motor cortex; 11. corpus striatum; 12. Memory; 13. Learning; 14. Short-term; 15. long-term; 16. sensory; 17. irrelevant; 18. skills; 19. long-term; 20. Skills; 21. input; 22. amygdala (hippocampus); 23. hippocampus (amygdala); 24. amygdala; 25. hippocampus; 26. fact; 27. basal; 28. long-term; 29. Skill; 30. sensory; 31. corpus striatum; 32. cerebellum; 33. Amnesia; 34. skills; 35. Parkinson's; 36. Alzheimer's; 37. hippocampus; 38. information; 39. a; 40. a; 41. a; 42. c; 43. b; 44. c; 45. d; 46. d; 47. a; 48. a.

Self-Quiz

1. a; 2. a; 3. b; 4. d; 5. a; 6. b; 7. c; 8. d; 9. e; 10. b; 11. e; 12. d; 13. c; 14. e.

Chapter 35 Sensory Reception

Different Strokes for Different Folks [pp.606–607]

35.1. OVERVIEW OF SENSORY PATHWAYS [pp.608–609]

1. sensory receptors; 2. nerve pathways; 3. brain regions; 4. sensation; 5. perception; 6. stimulus; 7. Chemoreceptors; 8. mechanoreceptors; 9. photoreceptors; 10. thermoreceptors; 11. action potentials; 12. brain; 13. stimulus intensity; 14. frequency; 15. number; 16. frequency; 17. sensory adaptation; 18. change; 19. Stretch; 20. length; 21. somatic sensations; 22. D; 23. C; 24. A; 25. B; 26. A; 27. E; 28. E; 29. B; 30. D; 31. B; 32. B; 33. A; 34. C; 35. B.

35.2. SOMATIC SENSATIONS [pp.610–611]
35.3. SENSES OF TASTE AND SMELL [p.612]

1. somatosensory cortex; 2. head; 3. ear; 4. cerebral hemisphere; 5. mouth; 6. hand; 7. touch; 8. cold; 9. skin; 10. skeletal; 11. mechanoreceptors; 12. free; 13. Pain; 14. receptors; 15. somatic pain; 16. referred pain; 17. Mechanoreceptors; 18. skin; 19. Chemo; 20. sensory; 21. chemo (olfactory); 22. nose; 23. 5; 24. olfactory; 25. Pheromones; 26. olfactory (chemo); 27. taste buds.

35.4. SENSE OF BALANCE [p.613]
35.5. SENSE OF HEARING [pp.614–615]

1. mechanoreceptors; 2. balance; 3. amplitude; 4. frequency; 5. higher; 6. middle; 7. cochlea; 8. organ of Corti; 9. middle ear bones (hammer, anvil, stirrup); 10. cochlea; 11. auditory nerve; 12. tympanic membrane/eardrum; 13. oval window; 14. basilar membrane; 15. tectorial membrane.

35.6. SENSE OF VISION [pp.616–617]
35.7. STRUCTURE AND FUNCTION OF VERTEBRATE EYES [pp.618–619]
35.8. CASE STUDY: FROM SIGNALING TO VISUAL PERCEPTION [pp.620–621]

35.9. FOCUS ON HEALTH: ***Disorders of the Human Eye*** [pp.622–623]
1. photons; 2. Photoreception; 3. Vision; 4. Ocelli; 5. Eyes; 6. cornea; 7. retina; 8. ommatidia; 9. focal point; 10. Visual accommodation; 11. rhodopsin; 12. Cone; 13. fovea; 14. vitreous body; 15. cornea; 16. iris; 17. lens; 18. aqueous humor; 19. ciliary muscles; 20. retina; 21. fovea; 22. optic nerve; 23. blind spot/optic disk; 24. sclera; 25. Astigmatism; 26. nearsightedness; 27. radial keratotomy; 28. Cataracts; 29. glaucoma.

Self-Quiz
1. d; 2. c; 3. d; 4. e; 5. d; 6. a; 7. c; 8. b; 9. b; 10. a.

Chapter 36 Integration: Endocrine Control

Hormone Jamboree [pp.626–627]

36.1. THE ENDOCRINE SYSTEM [pp.628–629]
1. E; 2. A; 3. B; 4. F; 5. C; 6. D; 7. a. 1, six releasing and inhibiting hormones; synthesizes ADH, oxytocin; b. 2, ACTH, TSH, FSH, LH, GSH; c. 2, stores and secretes two hypothalamic hormones, ADH and oxytocin; d. 3, sex hormones of opposite sex, cortisol, aldosterone; e. 3, epinephrine, norepinephrine; f. 4, estrogens, progesterone; g. 5, testosterone; h. 6, melatonin; i. 7, thyroxine and triiodothyronine; j. 8, parathyroid hormone (PTH); k. 9, thymosins; l. 10, insulin, glucagon, somatostatin.

36.2. SIGNALING MECHANISMS [pp.630–631]
1. a; 2. b; 3. a; 4. b; 5. b; 6. a; 7. b; 8. b; 9. b; 10. b.

36.3. THE HYPOTHALAMUS AND PITUITARY GLAND [pp.632–633]
1. A (H); 2. P (G); 3. A (A); 4. A (I); 5. A (D); 6. I or (B); 7. P (E); 8. A (C); 9. A (F); 10. hypothalamus; 11. posterior; 12. anterior; 13. intermediate; 14. releasers; 15. inhibitors.

36.4. FOCUS ON HEALTH: ***Abnormal Pituitary Output*** [p.634]
1. a. Gigantism; b. Pituitary dwarfism; c. Diabetes insipidus; d. Acromegaly.

36.5. SOURCES AND EFFECTS OF OTHER HORMONES [p.635]
1. a. D (c); b. I (e); c. F (h); d. A (g); e. H (k); f. E (d); g. C (a); h. K (j); i. B (i); j. J (f); k. G (b); l. F (l); m. A (m); n. B (o); o. C (n).

36.6. FEEDBACK CONTROL OF HORMONAL SECRETIONS [pp.636–637]
1. E; 2. F; 3. D; 4. H; 5. I; 6. J; 7. A; 8. G; 9. B; 10. C; 11. thyroid; 12. metabolic (metabolism); 13. iodine; 14. decrease; 15. TSH; 16. goiter; 17. Hypothyroidism; 18. overweight; 19. Hyperthyroidism; 20. gonads; 21. hormones; 22. testes; 23. ovaries; 24. testosterone; 25. secondary; 26. gametes.

36.7. DIRECT RESPONSES TO CHEMICAL CHANGES [pp.638–639]
1. parathyroid; 2. PTH; 3. calcium; 4. calcium; 5. reabsorption; 6. D_3; 7. intestinal; 8. rickets; 9. a. Glucagon, Causes glycogen (a storage polysaccharide) and amino acids to be converted to glucose in the liver (glucagon raises the glucose level); b. Insulin, Stimulates glucose uptake by liver, muscle, and adipose cells; promotes synthesis of proteins and fats, and inhibits protein conversion to glucose (lowers the glucose level); c. Somatostatin, Helps control digestion; can block secretion of insulin and glucagon; 10. rises; 11. excessive; 12. energy; 13. ketones; 14. insulin; 15. Glucagon; 16. type 1 diabetes; 17. type 1 diabetes; 18. type 2 diabetes; 19. Type 2 diabetes.

36.8. HORMONES AND THE EXTERNAL ENVIRONMENT [pp.640–641]
1. b; 2. a; 3. a; 4. a; 5. b; 6. a; 7. a; 8. a; 9. a; 10. b; 11. The thyroid gland. Preliminary evidence suggests that frog embryos raised in hot-spot water (with as many as 20 kinds of dissolved pesticides—and with high deformity rates) had few or no deformity symptoms when supplied with extra thyroid hormones.

Self-Quiz
1. a; 2. e; 3. d; 4. b; 5. e; 6. d; 7. b; 8. a; 9. c; 10. a; 11. N; 12. E; 13. Q; 14. K; 15. D; 16. F; 17. J; 18. A; 19. I; 20. B; 21. R; 22. G; 23. L; 24. C; 25. P; 26. O; 27. M; 28. H.

Chapter 37 Protection, Support, and Movement

Of Men, Women, and Polar Huskies [pp.644–645]

37.1. EVOLUTION OF VERTEBRATE SKIN [pp.646–647]
37.2. FOCUS ON HEALTH: *Sunlight and Skin* [p.648]
1. epidermis; 2. dermis; 3. connective; 4. keratinization; 5. glands; 6. amphibians; 7. keratinocytes; 8. melanocytes; 9. keratin; 10. melanin; 11. UV; 12. sensory neuron; 13. sweat gland; 14. smooth muscle; 15. blood vessels; 16. hair follicle; 17. oil gland; 18. hypodermis; 19. dermis; 20. epidermis; 21. hair; 22. B; 23. A; 24. C; 25. E; 26. F; 27. G; 28. D; 29. A; 30. D; 31. B; 32. E; 33. C; 34. F.

37.3. TYPES OF SKELETONS [p.649]
37.4. EVOLUTION OF THE VERTEBRATE SKELETON [pp.650–651]
37.5. A CLOSER LOOK AT BONES AND JOINTS [pp.652–653]
1. c; 2. a; 3. d; 4. b; 5. a; 6. c; 7. cartilaginous; 8. endoskeletons; 9. 206; 10. appendicular; 11. pelvic; 12. skull; 13. ribs; 14. vertebrae; 15. intervertebral disks; 16. herniated; 17. a. Pelvic girdle; b. Encloses and protects internal organs; c. Femur; d. Phalanges; e. Absorb movement-related stress; f. Enclose and protect the brain; g. Fibula; h. Clavicle; i. Radius and ulna; j. Enclose and protect the spinal cord; support the skull; k. Sternum; l. Bones of the hands; 18. Haversian system; 19. connective tissue; 20. blood vessel; 21. compact bone tissue; 22. spongy bone tissue; 23. F; 24. C; 25. E; 26. D; 27. H; 28. G; 29. A; 30. B; 31. a. Rheumatoid arthritis; b. Strain; c. Sprain; d. Osteoporosis; e. Osteoarthritis.

37.6. SKELETAL–MUSCULAR SYSTEMS [pp.654–655]
37.7. HOW DOES SKELETAL MUSCLE CONTRACT? [pp.656–657]
37.8. WHAT CONTROLS CONTRACTION? [pp.658–659]
37.9. ENERGY FOR CONTRACTION [p.659]
37.10. PROPERTIES OF WHOLE MUSCLES [p.660]
37.11. FOCUS ON HEALTH: *A Bad Case of Tetanic Contraction* [p.661]
1. skeletal; 2. muscle; 3. contract; 4. gravity; 5. tendon; 6. bone; 7. lever; 8. joint; 9. force; 10. promote; 11. opposition; 12. 600; 13. a. Raises the arm; b. Tibialis anterior; c. Extends and rotates the thigh outward; d. Draws thigh backward, bends knee; e. Pectoralis major; f. Sartorius; g. Lifts the shoulder blade, draws the head back; h. Bends the thigh at the hip; i. Rectus abdominis; j. Straightens the forearm at the elbow; k. Flexes the thigh at the hips; 14. F; 15. D; 16. H; 17. G; 18. E; 19. A; 20. C; 21. B; 22. 3; 23. 5; 24. 4; 25. 6; 26. 2; 27. 1; 28. electric; 29. excitable; 30. action potential; 31. tubular; 32. actin; 33. sarcomeres; 34. sarcoplasmic reticulum; 35. calcium; 36. outward; 37. myofibrils; 38. clear; 39. contraction; 40. tropomyosin filaments; 41. troponin; 42. glycolysis alone; 43. aerobic respiration; 44. dephosphorylation of creatine phosphate; 45. ATP; 46. oxygen; 47. glucose and glycogen from bloodstream; 48. E; 49. D; 50. F; 51. A; 52. B; 53. C; 54. H; 55. G; 56. Aerobic exercise increases the number of capillaries and the number of mitochondria. Strength training makes fast-acting muscle cells form more myofibrils and enzymes; 57. Isotonically contracting muscle shortens and moves loads, while isometrically contracting muscle develops tension but does not shorten. In lengthening contracting, a muscle lengthens when the load is greater than its tension.

Self-Quiz
1. e; 2. c; 3. c; 4. d; 5. b; 6. c; 7. a; 8. c; 9. d; 10. b; 11. e; 12. d; 13. a; 14. e; 15. b; 16. c; 17. b; 18. d; 19. a; 20. e; 21. c.

Chapter 38 Circulation

Heartworks [pp.664–665]

38.1. EVOLUTION OF CIRCULATORY SYSTEMS [pp.666–667]
1. nutrients (food); 2. wastes; 3. closed; 4. Blood; 5. heart; 6. interstitial fluid; 7. heart; 8. rapidly; 9. capillary; 10. solutes; 11. lymphatic; 12. digestive (respiratory); 13. respiratory (digestive); 14. respiratory; 15. urinary; 16. heart(s); 17. blood vessels; 18. hearts; 19. open circulatory system; blood is pumped into short tubes that open into spaces in the body's tissues, mingles with tissue fluids, then is reclaimed by open-ended tubes that lead back to the heart; 20. closed circulatory system; blood flow is confined within blood vessels that have continuously connected walls and is pumped by five pairs of "hearts."

38.2. CHARACTERISTICS OF BLOOD [pp.668–669]
38.3. FOCUS ON HEALTH: *Blood Disorders* [p.670]
38.4. BLOOD TRANSFUSION AND TYPING [pp.670–671]
1. a. Plasma proteins; b. Red blood cells; c. Neutrophils; d. Lymphocytes; e. Platelets; 2. connective; 3. pH; 4. iron; 5. cell count; 6. 50 to 60; 7. bone marrow; 8. Stem cells;

9. Neutrophils; 10. Platelets; 11. nucleus; 12. four; 13. nucleus; 14. nine; 15. AB; 16. Agglutination (clumping); 17. stem (F); 18. red blood (E); 19. platelets (C); 20. neutrophils (A); 21. B (D); 22. T (D); 23. monocytes; 24. macrophages (A); 25. A; 26. I; 27. D; 28. G; 29. F; 30. H; 31. B; 32. C; 33. E; 34. E.

38.5. HUMAN CARDIOVASCULAR SYSTEM [pp.672–673]
38.6. THE HEART IS A LONELY PUMPER [pp.674–675]
1. pulmonary; 2. systemic; 3. oxygen; 4. oxygen; 5. heart; 6. atria; 7. ventricles; 8. systole; 9. diastole; 10. ventricles; 11. atrial; 12. jugular; 13. superior vena cava; 14. pulmonary; 15. hepatic portal; 16. renal; 17. inferior vena cava; 18. iliac; 19. femoral; 20. femoral; 21. iliac; 22. abdominal; 23. renal; 24. brachial; 25. coronary; 26. pulmonary; 27. carotid; 28. aorta; 29. left pulmonary veins; 30. left semilunar valve; 31. left ventricle; 32. inferior vena cava; 33. right atrioventricular valve; 34. right pulmonary artery; 35. superior vena cava.

38.7. BLOOD PRESSURE IN THE CARDIOVASCULAR SYSTEM [pp.676–677]
38.8. FROM CAPILLARY BEDS BACK TO THE HEART [pp.678–679]
38.9. FOCUS ON HEALTH: ***Cardiovascular Disorders*** [pp.680–681]
1. aorta (arteries); 2. Arteries; 3. ventricles; 4. pressure; 5. resistance; 6. elastic; 7. little; 8. does not drop much; 9. arterioles; 10. nervous; 11. Arterioles; 12. pressure; 13. medulla oblongata; 14. beat more slowly; 15. contract less forcefully; 16. vasodilation; 17. capillary; 18. endothelial; 19. capillary bed (diffusion zone); 20. interstitial; 21. diffusion; 22. bulk flow (blood pressure); 23. Veins; 24. Venules; 25. Veins; 26. venules; 27. valves; 28. 50–60; 29. vein; 30. artery; 31. arteriole; 32. capillary; 33. smooth muscle, elastic fibers; 34. valve; 35. pressure; 36. blood pressure cannot remain constant because it passes through various kinds of vessels that have varied structures; 37. hypertension; 38. silent killer; 39. atherosclerosis; 40. cholesterol; 41. atherosclerotic plaque; 42. platelets; 43. embolus.

38.10. HEMOSTASIS [p.682]
38.11. LYMPHATIC SYSTEM [pp.682–683]
1. hemostasis; 2. platelet plug formation; 3. coagulation; 4. collagen; 5. insoluble; 6. d; 7. c; 8. e; 9. b; 10. a; 11. tonsils; 12. thymus; 13. thoracic; 14. spleen; 15. lymph node(s); 16. lymphocytes; 17. bone marrow; 18. Lymph; 19. fats; 20. small intestine.

Self-Quiz
1. d; 2. e; 3. c; 4. e; 5. e; 6. a; 7. b; 8. a; 9. a; 10. a; 11. I; 12. G; 13. M; 14. H; 15. B; 16. O; 17. F; 18. N; 19. E; 20. L; 21. A; 22. J; 23. D; 24. C; 25. K.

Chapter 39 Immunity

Russian Roulette, Immunological Style [pp.686–687]

39.1. THREE LINES OF DEFENSE [p.688]
39.2. COMPLEMENT PROTEINS [p.689]
39.3. INFLAMMATION [pp.690–691]
1. mucous; 2. Lysozyme; 3. Diarrhea; 4. bacterial; 5. phagocytic; 6. clotting; 7. Phagocytic (Macrophage); 8. complement system; 9. histamine; 10. capillaries; 11. neutrophils; 12. monocytes (macrophages); 13. chemotaxins; 14. endogenous pyrogen; 15. secrete histamine and prostaglandins that change permeability of blood vessels in damaged or irritated tissues; 16. attack parasitic worms by secreting corrosive enzymes; 17. the most abundant white blood cells; they quickly phagocytize bacteria and reduce them to molecules that can be used for other purposes; 18. slow, "big eaters"; engulf and digest foreign agents and clean up dead and damaged tissues.

39.4. OVERVIEW OF THE IMMUNE SYSTEM [pp.692–693]
39.5. HOW LYMPHOCYTES FORM AND DO BATTLE [pp.694–695]
39.6. ANTIBODY-MEDIATED RESPONSE [pp.696–697]
39.7. CELL-MEDIATED RESPONSE [pp.698–699]
1. physical barriers; 2. immune system; 3. MHC marker (surface antigens); 4. nonself; 5. lymphocytes; 6. B cell; 7. T cell; 8. thymus; 9. viruses; 10. tumor; 11. identity; 12. antigen; 13. antigen-presenting; 14. effector; 15. helper T; 16. cytotoxic T; 17. cell-mediated; 18. antibodies; 19. antibody-mediated; 20. memory; 21. a; 22. d; 23. e; 24. b; 25. c; 26. C; 27. E; 28. F; 29. B; 30. D; 31. H; 32. G; 33. A; 34. secondary immune response; 35. memory cells; 36. antigens; 37. IgA; 38. IgE; 39. IgG; 40. IgM. 41. natural killer; 42. perforins.

39.8. FOCUS ON SCIENCE: ***Cancer and Immunotherapy*** [p.699]
39.9. DEFENSES ENHANCED, MISDIRECTED, OR COMPROMISED [pp.700–702]

39.10. FOCUS ON HEALTH: *AIDS — The Immune System Compromised* [pp.702–703]
1. immunotherapy; 2. *cancer*, 3. monoclonal antibodies; 4. hybrid; 5. lymphocytes; 6. lymphokine (interleukin); 7. b, d, e, h, i, j; 8. b, d, e, g; 9. a, b, c, f, j; 10. g, i, j; 11. g, i, j; 12. b, d, e, g, i, j; 13. b, d, f, i; 14. g, i, j; 15. b, d, e, h, i, j; 16. b, d, e, g, I, j; 17. b, d, e, h, i; 18. Allergy; 19. anaphylactic shock; 20. Autoimmune disease; 21. multiple sclerosis; 22. Rheumatoid arthritis; 23. women; 24. human immunodeficiency virus; 25. male homosexuals; 26. retrovirus; 27. reverse transcriptase; 28. body fluids; 29. helper T (antigen-presenting); 30. AZT; 31. ddI; 32. protease inhibitors; 33. vaccines.

Self-Quiz
1. d; 2. b; 3. b; 4. a; 5. a; 6. e; 7. e; 8. e; 9. a; 10. d; 11. H; 12. D; 13. E; 14. J; 15. B; 16. G; 17. C; 18. I; 19. F; 20. A.

Chapter 40 Respiration

Of Lungs and Leatherbacks [pp.706–707]

40.1. THE NATURE OF RESPIRATION [p.708]
40.2. INVERTEBRATE RESPIRATION [p.709]
40.3. VERTEBRATE RESPIRATION [pp.712–713]
1. oxygen; 2. respiration; 3. internal; 4. carbon dioxide; 5. aerobic; 6. organic; 7. Respiratory; 8. homeostasis; 9. oxygen; 10. carbon dioxide; 11. carbon dioxide; 12. oxygen; 13. digestive system; 14. water and solutes; 15. urinary system; 16. D; 17. B; 18. A; 19. C; 20. H; 21. F; 22. E; 23. G; 24. b; 25. c; 26. a; 27. a; 28. c; 29. a; 30. c; 31. b; 32. a; 33. b.

40.4. HUMAN RESPIRATORY SYSTEM [pp.712–713]
40.5. CYCLIC REVERSALS IN AIR PRESSURE GRADIENTS [pp.714–715]
1. a. intercostal muscles (6); b. larynx (11); c. trachea (12); d. alveolar sac (3); e. epiglottis (10); f. oral cavity (4); g. pharynx (9); h. diaphragm (7); i. nasal cavity (8); j. pleural membrane (5); k. alveoli (1); l. bronchiole (2); m. lung (13); 2. inhalation; 3. exhalation; 4. active; 5. diaphragm; 6. increase; 7. atmospheric; 8. intrapulmonary; 9. downward; 10. outward; 11. lower; 12. into; 13. passive; 14. decreases; 15. greater; 16. out; 17. exercise; 18. contract; 19. increasing; 20. Tidal volume is the amount of air moving into or out of the lungs during each respiratory cycle, while vital capacity is the volume of air that can move out of the lungs after maximal inhalation.

40.6. GAS EXCHANGE AND TRANSPORT [pp.716–717]
40.7. FOCUS ON HEALTH: *When the Lungs Break Down* [pp.718–719]
40.8. HIGH CLIMBERS AND DEEP DIVERS [pp.720–721]
1. b; 2. a; 3. a; 4. b; 5. a; 6. b; 7. blood; 8. rhythm; 9. magnitude; 10. medulla oblongata; 11. pons; 12. chemoreceptors; 13. diaphragm; 14. G; 15. F; 16. D; 17. C; 18. A; 19. E; 20. B.

Self-Quiz
1. I; 2. F; 3. K; 4. L; 5. E; 6. H; 7. O; 8. G; 9. M; 10. B; 11. D; 12. A; 13. C; 14. N; 15. J; 16. a; 17. e; 18. a; 19. c; 20. a; 21. a; 22. a; 23. a; 24. d; 25. e.

Chapter 41 Digestion and Human Nutrition

Lose It — And It Finds Its Way Back [pp.724–725]

41.1. THE NATURE OF DIGESTIVE SYSTEMS [pp.726–727]
41.2. VISUAL OVERVIEW OF THE HUMAN DIGESTIVE SYSTEM [p.728]
41.3. INTO THE MOUTH, DOWN THE TUBE [p.729]
1. Nutrition; 2. carbohydrates; 3. particles; 4. molecules; 5. absorbed; 6. circulatory; 7. respiratory; 8. carbon dioxide (CO_2); 9. urinary; 10. incomplete; 11. circulatory; 12. complete; 13. opening; 14. Movements; 15. secretion; 16. ruminants; 17. crown; 18. stomach; 19. small intestine; 20. anus; 21. accessory; 22. pancreas; 23. salivary amylase; 24. epiglottis; 25. esophagus; 26. salivary glands; 27. liver; 28. gallbladder; 29. pancreas; 30. anus; 31. large intestine (colon); 32. small intestine; 33. stomach; 34. esophagus; 35. pharynx; 36. mouth (oral cavity).

41.4. DIGESTION IN THE STOMACH AND SMALL INTESTINE [pp.730–731]
41.5. ABSORPTION IN THE SMALL INTESTINE [pp.732–733]
41.6. DISPOSITION OF ABSORBED ORGANIC COMPOUNDS [p.734]

41.7. THE LARGE INTESTINE [p.735]

1. a. Mouth; b. Salivary glands; c. Stomach; d. Small intestine; e. Pancreas; f. Liver; g. Gallbladder; h. Large intestine; i. Rectum; 2. starches; 3. salivary amylase; 4. disaccharide; 5. stomach; 6. small intestine; 7. amylase; 8. small intestine; 9. disaccharidases; 10. stomach; 11. pepsins; 12. small intestine; 13. pancreas; 14. amino acids; 15. Lipase; 16. small intestine; 17. fatty acid; 18. Bile; 19. gallbladder; 20. lipase; 21. Pancreatic nucleases; 22. small intestine; 23. segmentation (peristalsis); 24. T; 25. T; 26. T; 27. small intestine; 28. constructing hormones, nucleotides, proteins, and enzymes; 29. monosaccharides, free fatty acids, and glycerol; 30. The three uses are (a) to construct components of cells and storage forms (such as glycogen) and specialized derivatives such as steroids and acetylcholine; (b) to convert to amino acids as needed; and (c) to serve as a source of energy.

41.8. HUMAN NUTRITIONAL REQUIREMENTS [pp.736–737]
41.9. VITAMINS AND MINERALS [pp.738–739]
41.10. FOCUS ON SCIENCE: ***Weighty Questions, Tantalizing Answers*** [pp.740–741]

1. a. 2,070; b. 2,900; c. 1,230; 2. 700; 3. Complex carbohydrates; 4. 40 to 60; 5. Phospholipids; 6. energy reserves; 7. 20 to 25; 8. essential fatty acids; 9. Proteins; 10. essential; 11. complete; 12. incomplete; 13. Vitamins; 14. Minerals; 15. Consulting Figure 41.16 (men's column, 6′ 0″) yields 178 pounds as his ideal weight. 195 − 178 = 17 lbs. overweight; 16. Multiply 178 times 10 (see text p.726) to obtain 1780 kilocalories (the daily number of calories that *maintains* weight in the correct size range). The excess 17 lbs. should be lost gradually by adopting an everyday exercise program that over many months would gradually eliminate the excess kilocalories that are stored mostly in the form of fat; The smallest range of serving sizes shown in Figure 41.14 will help keep the total caloric intake to about 1600 kcal; 17. a. 6 servings; b. bread, cereal, rice, pasta; 18. a. 2 servings; b. fruits; 19. a. 3 servings; b. vegetables; 20. a. 2 servings; b. milk, yogurt, or cheese; 21. a. 2 servings; b. legumes, nuts, poultry, fish, or meats; 22. a. Scarcely any; b. added fats and simple sugars; 23. food pyramid; 24. carbohydrates; 25. bread; 26. 6–11; 27. vegetable; 28. 3–5; 29. fruit; 30. 2–4; 31. apples; 32. berries; 33. meat; 34. proteins; 35. 2–3; 36. amino acids; 37. milk; 38. 2–3; 39. 0.

Self-Quiz

1. b; 2. b; 3. a; 4. e; 5. d; 6. c; 7. c; 8. c; 9. d; 10. c; 11. D; 12. C; 13. H; 14. J; 15. B; 16. A; 17. I; 18. F; 19. E; 20. G.

Chapter 42 The Internal Environment

Tale of the Desert Rat [pp.744–745]

42.1. URINARY SYSTEM OF MAMMALS [pp.746–747]

1. interstitial; 2. blood; 3. extracellular; 4. urinary; 5. volume; 6. homeostatic; 7. a, b; 8. a, c; 9. b, d; 10. b, d; 11. a, c; 12. a; 13. d; 14. d; 15. urine; 16. organic; 17. Urea; 18. liver; 19. carbon dioxide; 20. uric acid; 21. hemoglobin; 22. kidney; 23. ureter; 24. urinary bladder; 25. urethra; 26. kidney cortex; 27. kidney medulla; 28. ureter; 29. kidneys; 30. 1; 31. urine; 32. ureter; 33. urinary bladder; 34. urethra; 35. reflex; 36. Skeletal; 37. voluntary; 38. nephrons; 39. Bowman's capsule; 40. renal corpuscle; 41. proximal tubule; 42. loop of Henle; 43. collecting duct; 44. ureter; 45. renal corpuscle; 46. Bowman's capsule; 47. glomular capillaries; 48. proximal tubule; 49. loop of Henle; 50. distal tubule; 51. collecting duct.

42.2. URINE FORMATION [pp.748–749]
42.3. FOCUS ON HEALTH: ***When Kidneys Break Down*** [p.750]
42.4. THE ACID–BASE BALANCE [p.750]
42.5. ON FISH, FROGS, AND KANGAROO RATS [p.751]

1. c; 2. a; 3. b; 4. c; 5. a; 6. B; 7. E; 8. D; 9. C; 10. A; 11. a. uremic toxicity; b. kidney stones; c. renal failure; d. glomerulonephritis; e. hemodialysis; f. metabolic acidosis; g. peritoneal dialysis.

42.6. HOW ARE CORE TEMPERATURES MAINTAINED? [pp.752–753]
42.7. TEMPERATURE REGULATION IN MAMMALS [pp.754–755]

1. D; 2. B; 3. A; 4. C; 5. G; 6. F; 7. E; 8. a; 9. c; 10. c; 11. b; 12. a; 13. c; 14. c; 15. b; 16. b; 17. d.

Self-Quiz

1. c; 2. e; 3. d; 4. b; 5. d; 6. b; 7. d; 8. e; 9. b; 10. a; 11. d.

Chapter 43 Principles of Reproduction and Development

From Frog to Frog and Other Mysteries [pp.758–759]

43.1. THE BEGINNING: REPRODUCTIVE MODES [pp.760–761]
43.2. STAGES OF DEVELOPMENT—AN OVERVIEW [pp.762–763]
1. asexual; 2. environmental; 3. variation; 4. reproductive timing; 5. Viviparous; 6. ovoviviparous; 7. oviparous; 8. Gamete formation; 9. egg; 10. fertilization; 11. zygote; 12. Cleavage; 13. blastomeres; 14. Gastrulation; 15. nervous; 16. gut; 17. skeleton; 18. embryonic; 19. Development; 20. F, rich in lipids and proteins; 21. T; 22. T; 23. T; 24. T; 25. F; 26. B; 27. C; 28. A; 29. E; 30. D; 31. a. mesoderm; b. ectoderm; c. endoderm; d. mesoderm; e. ectoderm; f. mesoderm; g. endoderm; h. mesoderm; i. mesoderm.

43.3. EARLY MARCHING ORDERS [pp.764–765]
43.4. HOW DO SPECIALIZED TISSUES AND ORGANS FORM? [pp.766–767]
1. F, maternal control begins as soon as the zygote forms; while paternal controls are not seen until gastrulation; 2. F, all body cells have the same genes; 3. T; 4. T; 5. T; 6. sperm entry; 7. gray crescent; 8. body axis; 9. cleavage; 10. blastula; 11. maternal messages; 12. cytoplasmic localization; 13. gastrulation; 14. primary; 15. gastrula; 16. yolk; 17. three; 18. Cell differentiation; 19. genes; 20. activated; 21. neural tube; 22. Cell differentiation; 23. morphogenesis; 24. Morphogenesis; 25. active cell migration; 26. Adhesive; 27. microtubules; 28. microfilaments; 29. apoptosis; 30. ectodermal sheets.

43.5. PATTERN FORMATION [pp.768–769]
43.6. FOCUS ON SCIENCE: *To Know a Fly* [pp.770–771]
43.7. WHY DO ANIMALS AGE? [p.772]
43.8. CONNECTIONS: *Death in the Open* [p.773]
1. Pattern formation; 2. embryonic induction; 3. embryonic induction; 4. Morphogens; 5. master (homeotic); 6. disaster (problems); 7. *Drosophila* (fruit fly); 8. telomeres; 9. C; 10. D; 11. E; 12. B; 13. A; 14. internal; 15. T.

Self-Quiz
1. c; 2. b; 3. c; 4. a; 5. c; 6. e; 7. d; 8. b; 9. e; 10. a.

Chapter 44 Human Reproduction and Development

Sex and the Mammalian Heritage [pp.776–777]

44.1. REPRODUCTIVE SYSTEM OF HUMAN MALES [pp.778–779]
44.2. MALE REPRODUCTIVE FUNCTION [pp.780–781]
1. mitotic (mitosis); 2. seminiferous; 3. meiosis (spermatogenesis); 4. sperm; 5. epididymis; 6. vas deferens; 7. urethra; 8. Seminal vesicles; 9. Prostate gland; 10. Bulbourethral; 11. Leydig; 12. Testosterone; 13. testosterone; 14. anterior; 15. hypothalamus; 16. decrease; 17. LH; 18. Sertoli; 19. increase; 20. hypothalamus; 21. anterior pituitary; 22. Sertoli cells; 23. Leydig cells.

44.3. REPRODUCTIVE SYSTEM OF HUMAN FEMALES [pp.782–783]
44.4. FEMALE REPRODUCTIVE FUNCTION [pp.784–785]
44.5. VISUAL SUMMARY OF THE MENSTRUAL CYCLE [p.786]
1. ovary; 2. oviduct; 3. uterus; 4. cervix; 5. myometrium; 6. endometrium; 7. vagina; 8. labia majora; 9. labia minora; 10. clitoris; 11. urethra; 12. Meiosis; 13. 300,000; 14. follicle; 15. zona pellucida; 16. hypothalamus; 17. anterior pituitary; 18. estrogens; 19. ovulation; 20. LH; 21. menstruation; 22. endometrial; 23. corpus luteum; 24. progesterone; 25. blastocyst; 26. endometrium; 27. menstrual flow.

44.6. PREGNANCY HAPPENS [p.787]
44.7. FORMATION OF THE EARLY EMBRYO [pp.788–789]
44.8. EMERGENCE OF THE VERTEBRATE BODY PLAN [p.790]
44.9. WHY IS THE PLACENTA SO IMPORTANT? [p.791]
44.10. EMERGENCE OF DISTINCTLY HUMAN FEATURES [pp.792–793]
44.11. FOCUS ON HEALTH: *Mother as Protector, Potential Threat* [pp.794–795]
1. oviduct; 2. implantation; 3. placenta; 4. eighth; 5. fetus; 6. embryonic disk; 7. amniotic cavity; 8. four; 9. pharyngeal arches; 10. somites; 11. six; 12. forelimb; 13. D; 14. B; 15. A; 16. C; 17. human chorionic gonadotropin (HCG); 18. corpus luteum; 19. gastrulation; 20. primitive streak; 21. neural tube; 22. Somites; 23. pharyngeal arches; 24. placenta; 25. umbilical cord; 26. nutrients; 27. wastes; 28. arms; 29. head; 30. muscles; 31. 95; 32. folate (folic acid); 33. rubella; 34. alcohol; 35. nervous system; 36. streptomycin.

44.12. FROM BIRTH ONWARD [pp.796–797]
44.13. FOCUS ON BIOETHICS: *Control of Human Fertility* [p.798]
44.14. BIRTH CONTROL OPTIONS [pp.798–799]
44.15. FOCUS ON HEALTH: *Sexually Transmitted Diseases* [pp.800–801]
44.16. FOCUS ON BIOETHICS: *To Seek or End Pregnancy* [p.802]

1. 15,000; 2. 480,000; 3. more than 1,000,000; 4. abstinence; 5. Condoms; 6. diaphragm; 7. estrogens (progesterones); 8. progesterones (estrogens); 9. anterior pituitary; 10. tubal ligation; 11. A, F; 12. A, C; 13. D, E; 14. F; 15. F; 16. D; 17. C; 18. F; 19. C; 20. C, F; 21. C; 22. F; 23. D, F; 24. B; 25. A, B, C, D, E, F.

Self-Quiz

1. d; 2. e; 3. a; 4. b; 5. c; 6. e; 7. d; 8. a; 9. c; 10. a; 11. b; 12. d.

Chapter 45 Population Ecology

Tales of Nightmare Numbers [pp.806–807]

45.1. CHARACTERISTICS OF POPULATIONS [p.808]
45.2. FOCUS ON SCIENCE: *Elusive Heads to Count* [p.819]
45.3. POPULATION SIZE AND EXPONENTIAL GROWTH [pp.810–811]

1. K; 2. H; 3. D; 4. I; 5. B; 6. F; 7. A; 8. G; 9. J; 10. L; 11. C; 12. E; 13. 2.5 (3); 14. capture was not random, some individuals die during study period, no migration, animals learn to avoid trap; 15. immigration; 16. emigration; 17. migrations; 18. zero population; 19. per capita; 20. 0.5; 21. 0.1; 22. reproduction; 23. time; 24. 0.4; 25. J; 26. exponential; 27. doubling; 28. biotic potential; 29. a. it increases; b. it decreases; c. it must increase; 30. population growth rate; 31. it must decrease; 32. 100,000; 33. a. 100,000; b. 300,000.

45.4. LIMITS ON THE GROWTH OF POPULATIONS [pp.812–813]
45.5. LIFE HISTORY PATTERNS [pp.814–815]
45.6. NATURAL SELECTION AND THE GUPPIES OF TRINIDAD [pp.816–817]

1. limit; 2. Carrying capacity; 3. logistic; 4. carrying capacity; 5. increases; 6. decreases; 7. density-dependent; 8. density-independent; 9. dependent; 10. carrying capacity; 11. independent; 12. life history; 13. insurance; 14. cohort; 15. life; 16. survivorship; 17. Survivorship; 18. III; 19. I; 20. II; 21. I; 22. b; 23. a; 24. a; 25. a; 26. a; 27. b.

45.7. HUMAN POPULATION GROWTH [pp.818–819]
45.8. CONTROL THROUGH FAMILY PLANNING [pp.820–821]
45.9. POPULATION GROWTH AND ECONOMIC DEVELOPMENT [pp.822–823]
45.10. SOCIAL IMPACT OF NO GROWTH [p.823]

1. a. around 1962–1963; b. around 2025; c. answers may vary; 2. 6 billion; 3. T; 4. short; 5. T; 6. sidestepped; 7. cannot; 8. 9; 9. natural resources; 10. pollution; 11. Family planning; 12. birth; 13. fertility; 14. two; 15. female; 16. three; 17. 6.5; 18. baby-boomers; 19. one-third; 20. thirties; 21. China; 22. stabilize; 23. reproductive; 24. b; 25. a; 26. b; 27. a; 28. b; 29. a; 30. B; 31. D; 32. A; 33. C; 34. industrial; 35. decreasing; 36. smaller; 37. transitional; 38. transitional; 39. population; 40. immigration; 41. 16; 42. 4.7; 43. 21; 44. 25; 45. 50; 46. 25; 47. 1; 48. 3; 49. 3; 50. 12.9; 51. 258; 52. growth; 53. social; 54. older; 55. economic; 56. postponed; 57. cultural; 58. carrying capacity.

Self-Quiz

1. d; 2. a; 3. b; 4. a; 5. d; 6. d; 7. a; 8. b; 9. a; 10. a; 11. c; 12. c; 13. b; 14. d; 15. b; 16. a; 17. a.

Chapter 46 Social Interactions

Deck the Nest with Sprigs of Green Stuff [pp.826–827]

46.1. BEHAVIOR'S HERITABLE BASIS [pp.828–829]
46.2. LEARNED BEHAVIOR [p.830]

1. C; 2. F; 3. E; 4. B; 5. G; 6. A; 7. D; 8. H; 9. the banana slug; 10. ate; 11. inland; 12. were not; 13. genetic; 14. pineal gland; 15. increase; 16. estrogen; 17. Hormones; 18. a. Young cuckoos instinctively remove the natural-born offspring from a nest and then receive the undivided attention of the unsuspecting foster parents; b. Young toads must learn to leave bumblebees alone by first being stung by a bee; 19. B; 20. E; 21. C; 22. F; 23. A; 24. D.

46.3. THE ADAPTIVE VALUE OF BEHAVIOR [p.831]
1. F; 2. G; 3. A; 4. B; 5. E; 6. C; 7. D.

46.4. COMMUNICATION SIGNALS [pp.832–833]
1. h; 2. e; 3. i; 4. a; 5. d; 6. j; 7. b; 8. c; 9. f; 10. e; 11. d; 12. g.

46.5. MATES, PARENTS, AND INDIVIDUAL REPRODUCTIVE SUCCESS [pp.834–835]
1. a. Caspian terns; b. Hangingflies; c. Bison, lions, elephant seals, sheep, elk; d. Sage grouse.

46.6. COSTS AND BENEFITS OF LIVING IN SOCIAL GROUPS [pp.836–837]
46.7. THE EVOLUTION OF ALTRUISM [pp.838–839]
46.8. FOCUS ON SCIENCE: *Why Sacrifice Yourself?* [p.840]
1. A; 2. D; 3. C; 4. B; 5. subordinate; 6. self-sacrificing (altruistic); 7. termite; 8. genes; 9. inclusive fitness; 10. parenting; 11. indirect; 12. reproduce; 13. altruistic; 14. genes; 15. vertebrates; 16. cooperative; 17. queen; 18. perpetuate; 19. genotypes.

46.9. AN EVOLUTIONARY VIEW OF HUMAN SOCIAL BEHAVIOR [p.841]
1. human; 2. a trait valuable in gene transmission; 3. redirecting adaptive behaviors; 4. strangers; 5. can; 6. will; 7. nonrelated; 8. is; 9. are; 10. nonrelative.

Self-Quiz
1. c; 2. d; 3. b; 4. d; 5. a; 6. b; 7. b; 8. c; 9. a; 10. c; 11. b; 12. d; 13. c; 14. a.

Chapter 47 Community Interactions

No Pigeon Is an Island [pp.844–845]

47.1. WHICH FACTORS SHAPE COMMUNITY STRUCTURE? [p.846]
47.2. MUTUALISM [p.847]
1. habitat; 2. community; 3. niche; 4. fundamental; 5. realized; 6. commensalistic; 7. mutualism; 8. interspecific; 9. Predation (Parasitism); 10. parasitism (predation); 11. symbiosis; 12. a. It cannot complete its life cycle in any other plant, and its larvae eat only yucca seeds; b. The yucca moth is the plant's only pollinator.

47.3. COMPETITIVE INTERACTIONS [pp.848–849]
47.4. PREDATOR–PREY INTERACTIONS [pp.850–851]
47.5. CONNECTIONS: *An Evolutionary Arms Race* [pp.852–853]
1. Intraspecific; 2. Interspecific; 3. Interspecific; 4. competitive exclusion; 5. resource-partitioning; 6. pigeons; 7. size; 8. root; 9. Predators; 10. prey; 11. prey; 12. carrying capacity; 13. coevolution; 14. camouflage; 15. warning coloration; 16. *Mimicry*; 17. *Moment-of-truth*; 18. adaptations; 19. E; 20. C; 21. G; 22. A; 23. H; 24. B; 25. D; 26. G; 27. D; 28. F; 29. E; 30. B.

47.6. PARASITE–HOST INTERACTIONS [pp.854–855]
47.7. FORCES CONTRIBUTING TO COMMUNITY STABILITY [pp.856–857]
47.8. FORCES CONTRIBUTING TO COMMUNITY INSTABILITY [pp.858–859]
47.9. FOCUS ON THE ENVIRONMENT: *Exotic and Endangered Species* [pp.860–861]
1. C; 2. F; 3. H; 4. B; 5. I; 6. G; 7. E; 8. A; 9. D; 10. succession; 11. Pioneer; 12. pioneers; 13. climax; 14. primary; 15. replacement; 16. secondary; 17. facilitate; 18. succession; 19. climax-pattern; 20. community; 21. pioneers; 22. fires; 23. natural; 24. active; 25. b; 26. a; 27. b; 28. b; 29. a; 30. a; 31. b; 32. a; 33. a; 34. keystone; 35. dominant; 36. mussels; 37. B; 38. D; 39. F; 40. E; 41. A; 42. C.

47.10. PATTERNS OF BIODIVERSITY [pp.862–863]
1. a. Resource availability tends to be higher and more reliable. Tropical latitudes have more sunlight of greater intensity, rainfall amount is higher, and the growing season is longer. Vegetation grows all year long to support diverse herbivores, etc.; b. Species diversity might be self-reinforcing. When a greater number of plant species compete and coexist, a greater number of herbivore species evolve because no herbivore can overcome the chemical defenses of all kinds of plants. Then more predators and parasites evolve in response to the diversity of prey and hosts; c. The rates of speciation in the tropics have exceeded those of background extinction. At higher latitudes, biodiversity has been suppressed during times of mass extinction; 2. tropics; 3. Iceland; 4. biodiversity; 5. Iceland; 6. dispersal; 7. distance; 8. area; 9. Larger; 10. diversity; 11. targets; 12. biodiversity; 13. small; 14. small; 15. immigration; 16. extinction; 17. immigration; 18. extinction; 19. Island C.

Self-Quiz
1. b; 2. b; 3. c; 4. a; 5. b; 6. d; 7. e; 8. b; 9. d; 10. a; 11. d; 12. d.

Chapter 48 Ecosystems

Crêpes for Breakfast, Pancake Ice for Dessert [pp.866–867]

48.1. THE NATURE OF ECOSYSTEMS [pp.868–869]
48.2. THE NATURE OF FOOD WEBS [pp.870–871]
48.3. FOCUS ON SCIENCE: ***Biological Magnification in Food Webs*** [p.872]
48.4. STUDYING ENERGY FLOW THROUGH ECOSYSTEMS [p.873]
48.5. FOCUS ON SCIENCE: ***Energy Flow at Silver Springs*** [p.874]

1. H; 2. C; 3. K; 4. F; 5. M; 6. O; 7. B; 8. I; 9. N; 10. A; 11. P; 12. G; 13. L; 14. D; 15. J; 16. E; 17. c; 18. b; 19. e; 20. a; 21. b; 22. d; 23. c; 24. e; 25. c; 26. a; 27. b; 28. d; 29. a; 30. c; 31. a; 32. energy; 33. ecosystem; 34. 100; 35. trophic; 36. lengthy; 37. patterns; 38. shorter; 39. stable; 40. herbivorous; 41. shortest; 42. grasslands; 43. carnivores; 44. grazing; 45. detrital; 46. cross-connect; 47. seasons; 48. detrital; 49. photosynthesizers; 50. consumers; 51. Ecosystem modeling; 52. DDT; 53. fats; 54. biological magnification; 55. concentrated; 56. web; 57. consumer; 58. metabolic; 59. productivity; 60. Gross; 61. net; 62. net; 63. pyramid; 64. producers; 65. biomass; 66. biomass; 67. smallest; 68. energy; 69. Sunlight; 70. large; 71. 1; 72. 6; 73. 16; 74. low; 75. four.

48.6. BIOGEOCHEMICAL CYCLES—AN OVERVIEW [p.875]
48.7. HYDROLOGIC CYCLE [pp.876–877]

1. Usually as mineral ions such as ammonium (NH_4^+); 2. Inputs from the physical environment and the cycling activities of decomposers and detritivores; 3. The amount of a nutrient being cycled through the ecosystem is greater; 4. Common sources are rainfall or snowfall, metabolism (such as nitrogen fixation), and weathering of rocks; 5. Loss of mineral ions occurs by runoff; 6. a. Oxygen and hydrogen move in the form of water molecules; b. A large portion of the nutrients is in the form of atmospheric gases such as carbon dioxide and nitrogen (mainly CO_2); c. Nutrients are not in gaseou forms; nutrients move from land to the seafloor and (return to land through geologic uplifting of very long duration; phosphorus is an example; 7. F; 8. H; 9. D; 10. B; 11. A (C); 12. E; 13. G (H); 14. C; 15. watershed; 16. soil; 17. streams; 18. transpiration; 19. vegetation; 20. nutrients; 21. calcium; 22. calcium; 23. biomass; 24. deforestation; 25. cycle; 26. ecosystems; 27. regenerate; 28. coniferous.

48.8. CARBON CYCLE [pp.878–879]
48.9. FOCUS ON SCIENCE: ***From Greenhouse Gases to a Warmer Planet?*** [pp.880–881]
48.10. NITROGEN CYCLE [pp.882–883]
48.11. SEDIMENTARY CYCLES [pp.884–885]

1. C; 2. D; 3. F; 4. B; 5. G; 6. A; 7. E; 8. D; 9. The gas layer helps to keep Earth warm enough to support life; but if too much warming occurs, rising sea levels, droughts, mudslides, spreading of diseases, and so on could be the consequence. 10. D; 11. E; 12. C; 13. A; 14. B; 15. Soil nitrogen compounds are vulnerable to being leached and lost from the soil; some fixed nitrogen is lost to air by denitrification; nitrogen fixation comes at a high metabolic cost to plants that are symbionts of nitrogen-fixing bacteria; loss of nitrogen is enormous in agricultural regions, through the tissues of harvested plants, soil erosion, and leaching processes; 16. reservoir; 17. phosphates; 18. ocean; 19. shelves; 20. crustal; 21. geochemical; 22. ecosystem; 23. organisms; 24. ionized; 25. soil; 26. cycle; 27. phosphorus; 28. ecosystems; 29. sediments; 30. soils; 31. sediments; 32. runoff; 33. algal; 34. Decomposition; 35. oxygen; 36. Eutrophication; 37. accelerate.

Self-Quiz

1. c; 2. c; 3. d; 4. c; 5. a; 6. a; 7. b; 8. a; 9. a; 10. b; 11. b; 12. d.

Chapter 49 The Biosphere

Does a Cactus Grow in Brooklyn? [pp.888–889]

49.1. AIR CIRCULATION PATTERNS AND REGIONAL CLIMATES [pp.890–891]
49.2. THE OCEANS, LANDFORMS, AND REGIONAL CLIMATES [pp.892–893]

1. M; 2. D; 3. H; 4. K; 5. B; 6. J; 7. F; 8. C; 9. A; 10. L; 11. E; 12. I; 13. N; 14. G; 15. equatorial; 16. warm; 17. rises (ascends); 18. moisture; 19. descends; 20. moisture; 21. ascends; 22. moisture; 23. descends; 24. east (easterlies); 25. west (westerlies); 26. tropical; 27. warm; 28. cool; 29. cold; 30. solar (sun's); 31. rotation.

49.3. REALMS OF BIODIVERSITY [pp.894–895]
49.4. SOILS OF MAJOR BIOMES [p.896]
49.5. DESERTS [p.897]

DRY WOODLANDS, AND
)9]
3; 7. J; 8. E; 9. D; 10. L; 11. I;
. A; 17. E; 18. d; 19. a; 20. d;
. d; 26. e; 27. b.

AIN FORESTS AND OTHER
ORESTS [pp.900–901]
OUS FORESTS [p.902]
AND ALPINE TUNDRA [p.903]
; 4. a; 5. b; 6. b; 7. a; 8. d; 9. c, d; 10. b; 11. c;
; 14. b; 15. d; 16. d.

FRESHWATER PROVINCES [pp.904–905]
. THE OCEAN PROVINCES [pp.906–907]
ake; 2. littoral; 3. limnetic; 4. plankton; 5. profundal;
. overturns; 7. 4; 8. spring; 9. thermocline; 10. cools;
11. fall; 12. down; 13. up; 14. higher; 15. short;
16. primary; 17. Oligotrophic; 18. eutrophic; 19. eutrophication; 20. Streams; 21. runs; 22. benthic (C); 23. pelagic (A); 24. neritic (B); 25. oceanic (D); 26. a; 27. c; 28. c; 29. b; 30. c; 31. a; 32. b; 33. a; 34. b; 35. c; 36. a; 37. b; 38. a; 39. c; 40. b; 41. c; 42. a; 43. b; 44. b; 45. c.

49.12. WETLANDS AND THE INTERTIDAL ZONE [pp.908–909]
49.13. EL NIÑO AND THE BIOSPHERE [pp.910–911]
49.14. CONNECTIONS: *Rita in the Time of Cholera* [pp.912–913]
1. c; 2. b; 3. d; 4. b; 5. a; 6. c; 7. a; 8. a; 9. d; 10. c; 11. c; 12. d; 13. e; 14. b; 15. d; 16. b; 17. e.

Self-Quiz
1. d; 2. d; 3. b; 4. d; 5. e; 6. b; 7. c; 8. c; 9. d; 10. c; 11. c; 12. a.

Chapter 50 Perspective on Humans and the Biosphere

An Indifference of Mythic Proportions [pp.916–917]

50.1. AIR POLLUTION—PRIME EXAMPLES [pp.918–919]
50.2. OZONE THINNING—GLOBAL LEGACY OF AIR POLLUTION [p.920]
1. Pollutants; 2. carbon dioxide; 3. chlorofluorocarbons; 4. 700,000; 5. topography; 6. thermal inversion; 7. industrial smog; 8. photochemical smog; 9. nitrogen dioxide; 10. photochemical; 11. gasoline; 12. peroxyacyl nitrates (PANs); 13. sulfur; 14. nitrogen; 15. sulfur; 16. nitrogen; 17. acid; 18. acid rain; 19. ecosystems; 20. c; 21. d; 22. e; 23. b; 24. a; 25. b; 26. f; 27. f; 28. a; 29. b; 30. b; 31. d; 32. a; 33. d; 34. f; 35. c; 36. d; 37. f; 38. f; 39. e; 40. c; 41. d.

50.3. WHERE TO PUT SOLID WASTES? WHERE TO PRODUCE FOOD? [p.921]
50.4. DEFORESTATION—AN ASSAULT ON FINITE RESOURCES [pp.922–923]
50.5. FOCUS ON BIOETHICS: *You and the Tropical Rain Forest* [p.924]
1. F; 2. H (G); 3. I; 4. E; 5. J; 6. B; 7. C; 8. A; 9. D; 10. G.

50.6. WHO TRADES GRASSLANDS FOR DESERTS? [p.925]
50.7. A GLOBAL WATER CRISIS [pp.926–927]
1. desertification; 2. overgrazing on marginal lands; 3. domestic cattle; 4. native wild herbivores; 5. deserts; 6. salty; 7. desalinization; 8. energy; 9. agriculture; 10. salinization; 11. water table; 12. saline; 13. 20; 14. pollution; 15. wastewater; 16. tertiary; 17. 55–66; 18. oil; 19. policies; 20. a. tertiary; b. primary; c. secondary.

50.8. A QUESTION OF ENERGY INPUTS [pp.928–929]
50.9. ALTERNATIVE ENERGY SOURCES [p.930]
50.10. BIOLOGICAL PRINCIPLES AND THE HUMAN IMPERATIVE [p.939]
1. e; 2. c; 3. a; 4. b; 5. a; 6. c; 7. d; 8. e; 9. e; 10. b.

Self-Quiz
1. a; 2. b; 3. a; 4. b; 5. d; 6. d; 7. c; 8. c; 9. d; 10. a.